BROADBAND ACCESS

BROADBAND ACCESS

WIRELINE AND WIRELESS – ALTERNATIVES FOR INTERNET SERVICES

Steve Gorshe

PMC-Sierra, Inc., USA

Arvind R. Raghavan

Blue Clover Devices, USA

Thomas Starr

Stefano Galli

ASSIA Inc., USA

Library of Congress Cataloging-in-Publication Data applied for

ISBN: 9780470741801

Set in 9/11pt TimesLTStd-Roman by Thomson Digital, Noida, India

1   2014

Contents

About the Authors

Steve Gorshe

Steve Gorshe is a Distinguished Engineer in the CTO organization of PMC-Sierra, Inc., where his work since 2000 has included technology development and telecommunications standards. He received his BSEE from the University of Idaho (1980) and both his MSEE (1982) and PhD (2002) degrees from Oregon State University. Since 1983, he has worked in product development, applied research, and systems architecture of telecommunications access and transport systems. His standards activity includes over 300 contributions across six standards bodies, serving as technical editor for nine North American and international standards, and currently serving as Associate Rapporteur for the Q11 group of ITU-T Study Group 15.

Steve is a Fellow of the IEEE. His IEEE activities include Communications Magazine Editor-in-Chief (2010–2012), Associate Editor-in-Chief (2006–2009), and Broadband Access Series co-editor (1999–2006). He has also served as the IEEE Communications Society Director of Magazines and Chair of the Transmission, Access and Optical Systems Technical Committee.

Steve has 37 patents issued or pending, over 24 published papers, and is co-author of two textbooks and co-author of chapters in three other textbooks.

Arvind R. Raghavan

Arvind R. Raghavan heads research and development at Blue Clover Devices, where he is involved with the design and implementation of innovative products for the Internet of Things, with current emphasis on Bluetooth Low Energy technology. Before joining Blue Clover Devices, he was part of the Radio Technology and Strategy group at AT&T Labs, where his work focused on the impact of QoS on LTE, design and analysis of heterogeneous networks, and advanced MIMO techniques for standardization in 3GPP. Prior to joining AT&T Labs, he played a lead role in the Systems Engineering group at ArrayComm, LLC, where they developed specifications for their multi-antenna signal processing products, conducted performance analyses, and made contributions to the standardization of WiMAX systems. Arvind holds MS and PhD degrees in Electrical Engineering from Clemson University.

Thomas Starr

Thomas Starr is a Lead Member of Technical Staff at AT&T Laboratories in Hoffman Estates, Illinois. Thomas is responsible for the development and standardization of local access and home networking technologies for AT&T's network. These technologies include ADSL, HDSL, SHDSL, VDSL and G.hn. In 2009, Thomas received the prestigious AT&T Science and Technology Medal. He serves as Chairman of the Broadband Forum and has also served as a member of the Board of Directors since its inception as the ADSL Forum in 1994. Thomas has been a distinguished fellow of the Broadband Forum From 1988 to 2000, has served as Chairperson of ANSI accredited standards working group T1E1.4, which develops xDSL standards for the United States, received the Committee T1 Outstanding Leadership Award in 2001, and now serves at ATIS COAST-NAI Chairman. In the ITU-T SG15, Thomas serves as Chairman of Working Party 1, addressing fiber, DSL, and home networking standards, and participates in the ITU SG15 Q4 group on xDSL international standards.

Thomas is a co-author of the books *DSL Advances*, published by Prentice Hall in 2003, and *Understanding Digital Subscriber Line Technology*, published by Prentice Hall in 1999. Thomas is also the author of the Science Fiction novel *Virtual Vengeance*. Thomas previously worked for 12 years at AT&T Bell Laboratories on ISDN and local telephone switching systems, and twenty US patents in the field to telecommunications have been issued to him. Thomas holds a MS degree in Computer Science and a BS degree in Computer Engineering from the University of Illinois in Urbana, Illinois.

Stefano Galli

Stefano Galli received his MS and PhD degrees in Electrical Engineering from the University of Rome "La Sapienza" (Italy) in 1994 and 1998, respectively. He is currently the Director of Technology Strategy of ASSIA – the leading developer of automated management and diagnostics tools for broadband networks. Prior to this position, he held the role of Director of Energy Solutions R&D for Panasonic Corporation and Senior Scientist at Bellcore.

Dr. Galli is serving as Chief Information Officer of the IEEE Communications Society (ComSoc), director of Smart Grid activities for the IEEE ComSoc Technical Committee on Power Line Communications, member of the Energy and Policy Committee of IEEE-USA, and as Editor for the IEEE Transactions on Communications and the IEEE Communications Magazine. Dr. Galli is also serving as Rapporteur for the ITU-T Q15/15 "Communications for Smart Grid" standardization group. Past positions include serving as Co-Chair of the "Communications Technology" Task Force of IEEE 2030 (Smart Grid), Leader of the "Theoretical and Mathematical Models" Group of IEEE 1901 (Broadband over Power Lines standard), Coexistence sub-group Chair of the SGIP/NIST PAP 15, elected Member-at-Large of the IEEE Communications Society (ComSoc) Board of Governors, and a variety of other leadership positions in the IEEE. He has also served as Founder and first Chair of the IEEE ComSoc Technical Committee on Power Line Communications.

Dr. Galli is a Fellow of the IEEE, has received the 2013 IEEE Donald G. Fink Best Paper Award for his paper on Smart Grid and Power Line Communications, the 2011 IEEE ComSoc Donald W. McLellan Meritorious Service Award, the 2011 Outstanding Service Award from the IEEE ComSoc Technical Committee on Power Line Communications, and the 2010 IEEE ISPLC Best Paper Award. He holds several issued and pending patents, has published over 90 peer-reviewed papers, has co-authored three book chapters on power line communications, and has made numerous standards contributions to the IEEE, the ITU-T, the Broadband Forum, and the UK NICC.

Acknowledgments

Thanks are given to the experts who provided assistance for the chapters on DSL technology: George Ginis, Ken Kerpez, Vladimir Oksman, Craig Schelp, Massimo Sorbara, and Arlynn Wilson. Thanks also are given to Marilynn Starr for her support and assistance.

Steve would like to thank the following people for their generous help, excellent comments and reviews for portions of his chapters: Frank Effenberger, Alon Bernstein, Chris Look, Onn Haran, Jeff Mandin, Lior Khermosh, Bob Murray, Valy Ossman, and Jim Dahl. Steve also wants to thank PMC-Sierra for allowing some of his white paper material to be adapted for this book.

Arvind would like to acknowledge the significant contributions of his wife, Sanchita Shetty, for painstakingly generating all the figures in the wireless chapters, and her unwavering support throughout the writing of this book. He would also like to express his heartfelt gratitude to Paul Chiuchiolo, Rich Kobylinski, Milap Majmundar, and Tom Novlan, for reviewing the wireless section of the book and providing excellent feedback for improving the quality and accuracy of the manuscript. Finally, he would like to thank his family and all his wonderful friends in Austin for their love and encouragement.

List of Abbreviations and Acronyms

2G	Second Generation
3G	Third Generation
3GPP	Third Generation Partnership Project
10GE	10 Gigabit/s Ethernet specified in IEEE 802.3
10G EPON	10 Gbit/s Ethernet Passive Optical Network specified in IEEE 802.3
ABS	Almost Blank Subframes
AC	Alternating Current
AC	Access Category
ACK	Acknowledgement
ACM	Adaptive Coding and Modulation
ADC	Analog-to-Digital Converter
ADSL	Asymmetric Digital Subscriber Line specified in ITU-T G.992.1
ADSL2	Asymmetric Digital Subscriber Line 2 specified in ITU-T G.992.3
ADSL2plus	Asymmetric Digital Subscriber Line 2plus specified in ITU-T G.992.5
AES	Advanced Encryption Standard
AFE	Analog Front End
AICH	Acquisition Indicator Channel
AM	Acknowledged Mode
AMI	Advanced Metering Infrastructure
A-MPDU	Aggregate MAC Protocol Data Unit
AMPS	Advanced Mobile Phone System
AMR	Automatic Meter Reading
A-MSDU	Aggregate MAC Service Data Unit
ANSI	American National Standards Institute
AP	Access Point
APD	Avalanche Photo Diode
APS	Automatic Protection Switching
ARIB	Association of Radio Industries and Businesses
ARP	Allocation and Retention Priority
ARQ	Automatic Repeat Request, Retransmission
AS	Access Stratum
ASE	Amplified Spontaneous Emission
ASF	DOCSIS Aggregated Service Flow
A-TDMA	Advanced TDMA (used with DOCSIS)
ATIS	Alliance for Telecommunications Industry Solutions
ATM	Asynchronous Transfer Mode protocol

AWG   American Wire Gauge
AWG   Arrayed Waveguide Grating WDM filter/multiplexer

BB   Broad Band
BCCH   Broadcast Control Channel
BCH   Broadcast Channel
BE   Best Effort service
BEMS   Building Energy Management System
BER   Bit Error Rate (or Ratio)
BIP   Bit Interleaved Parity
BMC   Broadcast Multicast Control
BMSC   Broadcast Multicast Service Center
B-ONU   DPoE Bridge ONU
BPL   Broadband over Power Lines
B-PON   FSAN/ITU-T Broadband PON protocol specified in the ITU-T G.983 series
BRI-ISDN   Basic Rate Integrated Services Digital network
BSS   Basic Service Set
BTS   Base Transceiver Station (for a wireless network)

CA   Carrier Aggregation
CAPEX   Capital Expense
CAPWAP   Control and Provisioning of Wireless Access Points
CATV   Community Access Television
CBR   Constant Bit Rate
CBS   Committed Burst Size
CC   Component Carrier
CCA   Clear Channel Assessment
CCCH   Common Control Channel
CCK   Complementary Code Keying
CCO   Capacity and Coverage Optimization
CDD   Cyclic-Delay Diversity
CDMA   Code Division Multiple Access
CENELEC   European Committee for Electotechnical Standardization
CEPCA   Consumer Electronics Powerline Alliance
CES   Circuit Emulation Service
CFP   Contention Free Period
CIF   Carrier Indicator Field
CIR   Committed Information Rate
CM   Cable Modem
CMCI   DOCSIS Cable Modem CPE Interface
CMTS   DOCSIS Cable Modem Terminating System
CN   Core Network
CO   Telephone company Central Office
CoMP   Cooperative Multi-Point
CP   Contention Period
CP   Cyclic Prefix
CPC   Continuous Packet Connectivity
CPE   Customer Premises Equipment
CPICH   Common Pilot Channel

CPRI	Common Public Radio Interface
CQI	Channel Quality Information
CRC	Cyclic Redundancy Check
CRE	Cell Range Expansion
CRS	Cell-specific Reference Signal
CS	Circuit Switched
CS	Channel Sensing
CSA	Carrier Serving Area
CS/CB	Coordinated Scheduling/Coordinated Beamforming
CSG	Closed Subscriber Group
CSI-RS	Channel State Information Reference Signal
CSM	Collaborative Spatial Multiplexing
CSMA/CA	Carrier Sense Multiple Access with Collision Avoidance
CSO	Cell Selection Offset
CTCH	Common Traffic Channel
CTS	Clear-to-send
CTS	Common Technical Specification for G-PON
CV	Code Violation
C-VID	Customer VLAN Identifier (Ethernet)
CWDM	Coarse Wavelength Division Multiplexing
DAC	Digital-to-Analog Converter
DAS	Distributed Antenna System
dB	Decibel, ten times the common logarithm of the ration of two powers
DBA	Dynamic Bandwidth Assignment
DBC	Dynamic Bonding Change (in DOCSIS 3.0)
DBG	Downstream Bonding Group (in DOCSIS 3.0)
DBR	Dynamic Bandwidth Report
DC	Direct Current
DCCH	Dedicated Control Channel
DCF	Distributed Coordination Function
DCH	Dedicated Channel
DCS	Downstream Channel Set (in DOCSIS 3.0)
DELT	Dual Ended Line Test
DEMARC	Carrier owned Demarcation device between the carrier and the CPE
DER	Distributed Energy Resources
DFE	Decision Feedback Equalizer
DFT	Discrete-time Fourier Transform
DHCP	Dynamic Host Configuration Protocol
DIFS	Distributed Interframe Spacing
DL	Downlink
DLC	Digital Loop Carrier
DLL	Data Link Layer
DL-SCH	Downlink Shared Channel
DM-RS	Demodulation Reference Signal
DMT	Discrete Multi Tone modulation
DOCSIS	Data Over Cable Service Interface Specification
D-ONU	DPoE ONU
Downstream	Data flowing towards the customer

DPB	Dynamic Point Blanking
DPCCH	Dedicated Physical Control Channel
DPDCH	Dedicated Physical Data Channel
DPoE	DOCSIS Protocol over Ethernet protocol
DPS	Dynamic Point Selection
DPSK	Differential Phase Shift Keying
DQPSK	Differential Quadrature Phase Shift Keying
DR	Demand Response
DRX	Discontinuous Reception
DS	Direct Sequence
DS1	Digital Signal level 1 in the North American asynchronous telephone network hierarchy
DS-CDMA	Direct Sequence Code Division Multiple Access
DSCP	DiffServ Code Point
DSID	Downstream Service ID (in DOCSIS 3.0)
DSL	Digital Subscriber Line
DSLAM	DSL Access Multiplexer
DSM	Dynamic Spectrum Management (in DSL)
DSM	Demand Side Management (in Smart Grid)
DSP	Digital Signal Processing
DSSS	Direct Sequence Spread Spectrum
DTX	Discontinuous Transmission
DVB	Digital Video Broadcast
DVB-RCS	Digital Video Broadcast Return Channel via Satellite
DVB-S2	Digital Video Broadcasting - Satellite - Second generation
DWDM	Dense Wavelength Division Multiplexing
E-AGCH	Enhanced Absolute Grant Channel
EBS	Excess Burst Size
ECH	Echo Cancelled Hybrid
eCM	embedded Cable Modem
EDCA	Enhanced Distributed Channel Access
E-DCH	Enhanced Dedicated Channel
EDFA	Erbium Doped Fiber Amplifier
EDGE	Enhanced Data-rates for GSM Evolution
E-DPCCH	Enhanced Dedicated Physical Control Channel
E-DPDCH	Enhanced Dedicated Physical Data Channel
E-HICH	Enhanced HARQ Indicator Channel
eICIC	Enhanced Inter-Cell Interference Coordination
EIR	Excess Information Rate
eMBMS	Enhanced Multimedia Broadcast and Multicast Service
EMC	Electro-Magnetic Compatibility
EMS	Element Management System
EO	Electrical to Optical signal conversion
eOAM	Extended OAM messages used in DPoE
EOC	Embedded Operations Channel
EONT	Embedded ONT
eNodeB	Evolved Node-B
EPC	Evolved Packet Core

EPON	Ethernet Passive Optical Network (1 Gbit/s rate)
EPS	Evolved Packet System
E-RGCH	Enhanced Relative Grant Channel
eSAFE	embedded Service/Application Functional Entity
ESP	Ethernet Service Path
ESS	Extended Service Set
ETSI	European Telecommunications Standards Institute
E-UTRAN	Evolved UMTS Terrestrial Radio Access Network
EVC	Ethernet Virtual Circuit
EVSE	Electric Vehicle Supply Equipment
FACH	Forward Access Channel
FBI	Feedback Information
FCC	Federal Communications Commission
FCS	Frame Check Sequence
FDD	Frequency Division Duplexing
FDM	Frequency Division Multiplexing
FDMA	Frequency Division Multiple Access
F-DPCH	Fractional Dedicated Physical Channel
FEC	Forward Error Correction
FeICIC	Further Enhanced Inter-Cell Interference Coordination
FEXT	Far End crosstalk
FFT	Fast Fourier Transform
FH	Frequency Hopping
FH-CDMA	Frequency Hopping Code Division Multiple Access
FHSS	Frequency Hopping Spread Spectrum
FITL	Fiber in the Loop
FN	Fiber Node (in a HFC network)
FSAN	Full Service Access Network industry consortium
FSK	Frequency Shift Keying
FTTC	Fiber to the Curb
FTTCab	Fiber to the Cabinet
FTTCell	Fiber to the Cell site
FTTH	Fiber to the Home
FTTN	Fiber to the Node
FTTO	Fiber to the Office
FTTP	Fiber to the Premises
G.hn	ITU-T G.9960/9961 home networking standard
G.hs	ITU-T G.994.1 DSL handshake protocol
G.lite	ITU-T G.992.2 reduced complexity ADSL
G.lt	ITU-T G.996.2 standard for DSL line test functions
G.test	ITU-T G.996.1 standard for testing of DSL modems
GBR	Guaranteed Bit Rate
GE	Gigabit/s Ethernet
GEM	G-PON Encapsulation Method
GERAN	GSM Edge Radio Access Network
GFP	Generic Framing Procedure specified in ITU-T G.7041
GGSN	Gateway GPRS Support Node

GMSC	Gateway Mobile Switching Center
GP	Guard Period
G-PON	FSAN/ITU-T Gigabit-capable PON protocol specified in the ITU-T G.984 series
GPRS	GSM Packet Radio System
gPTP	generalized Precision Timing Protocol
GSM	Global System for Mobile communications
GTC	G-PON Transmission Convergence
HAN	Home Area Network
HARQ	Hybrid Automatic Repeat Request
HCF	Hybrid Coordination Function
HD-PLC	High Definition Power Line Communication
HDR	High Data Rate
HDSL	High bit rate Digital Subscriber Line
HDSL2	High bit rate Digital Subscriber Line, 2 wire version
HDSL4	High bit rate Digital Subscriber Line, 4 wire version
HE	Head End
HEC	Header Error Check
HEMS	Home Energy Management System
HetNet	Heterogeneous Network
HF	High Frequency
HFC	Hybrid Fiber-Coaxial cable network
HLR	Home Location Register
HSDPA	High Speed Downlink Packet Access
HS-DPCCH	High Speed Dedicated Physical Control Channel
HS-DSCH	High Speed Downlink Shared Channel
HSPA	High Speed Packet Access
HS-PDSCH	High Speed Physical Downlink Shared Channel
HS-SCCH	High Speed Shared Control Channel
HSS	Home Subscriber Server
HSUPA	High Speed Uplink Packet Access
HV	High Voltage
IAD	Integrated Access Device
ICIC	Inter-cell Interference Coordination
IEC	International Electrotechnical Commission
IED	Intelligent Electronic Devices
IEEE	Institute of Electrical and Electronic Engineers
IETF	Internet Engineering Task Force
IFS	Inter-Frame Spacing
IGMP	Internet Group Management Protocol
IMT	International Mobile Telecommunications
IP	Internet Protocol
IP-HSD	DOCSIS IP High-Speed Data service
IPP	Inter-PHY Protocol
IPTV	Television delivered over Internet Protocol
IPv6	Internet Protocol version 6
IR	Infra-Red
IR	Incremental Redundancy

IRC	Interference Rejection Combining
IS-54	A second generation cellular standard
IS-136	A second generation cellular standard, an improvement on IS-54ISI Intersymbol interference
ISI	Inter-symbol Interference
ISM	Industrial, Scientific, and Medical
ISO	International Organization for Standardization
ISP	Internet Service Provider
ISP	IEEE 1901 Inter System Protocol
ITU-T	International Telecommunication Union – Telecommunication Standardization Sector
JP	Joint Processing
JT	Joint Transmission
kft	kilofeet (length of wire)
L1	Layer-1
L2	Layer-2
L3	Layer-3
LAN	Local Area Network
LDPC	Low Density Parity Check
LDR	Low Data Rate
LED	Light Emitting Diode
LF	Low Frequency
LLID	Ethernet Logical Link Identifier
LOF	Loss Of Frame
LoS	Line of Sight
LOS	Loss Of Signal
LSB	Least Significant Bit
LTE	Long Term Evolution (mobile telephone standard)
LV	Low Voltage
MAC	Medium Access Control
MAN	Metro Area Network
MBMS-GW	Multimedia Broadcast Multicast Service Gateway
MBR	Maximum Bit Rate
MBSFN	Multicast Broadcast Single Frequency Network
MCCA	MCF Controlled Channel Access
MCE	Multicell/Multicast Coordination Entity
MCF	Mesh Coordination Function
M-CMTS	Modular CMTS
MCS	Modulation and Coding Scheme
MEF	Metro Ethernet Forum
MELT	Metallic line test
MF	Medium Frequency
MF-TDMA	Multi-Frequency Time Division Multiple Access
MIB	Management Information Base
MIMO	Multiple Input Multiple Out
MLB	Mobility Load Balancing

MLME   MAC Layer Management Entity
MME   Mobility Management Entity
MMSE   Minimum Mean Squared Error
MoCA   Multimedia over Coax Alliance
Modem   Modulator/Demodulator, a transceiver
MPCPDU   Multi-Point Control Protocol PDU
MPDU   MAC Protocol Data Unit
MPEG   Motion Picture Experts Group video compression standards
MRC   Maximal Ratio Combining
MRO   Mobility Robustness Optimization
MSB   Most Significant Bit
MSC   Mobile Switching Center
MSDU   MAC Service Data Unit
MSO   Multiple System Operator (cable network operator)
MTA   Multimedia Terminal Adapter
MTL   Multi-Conductor Transmission Line
MU-MIMO   Multi-user Multiple Input Multiple Output
MV   Medium Voltage

NACK   Negative Acknowledgement
NAS   Non-Access Stratum
NAV   Network Allocation Vector
NB   Narrow Band
NE   Network Element
NEXT   near end crosstalk
NG-PON   FSAN/ITU-T Next Generation PON protocol
NI   Network Interface
NID   Network Interface Device
NMS   Network Management System
Node-B   Base Station in a third generation cellular system
nrt-PS   Non-real-time Poling Service (DOCSIS)
NRZ   Non-Return to Zero line code
NSR   Non-Status Reporting
NTU   Network Termination Units

OAM   Operations, Administration and Maintenance
OAM&P   Operations, Administration, Maintenance and Provisioning
OBSAI   Open Base Station Architecture Initiative
ODN   Optical Distribution Network
OE   Optical to Electrical signal conversion
OEO   Optical to Electrical to Optical signal conversion (repeater)
OFDM   Orthogonal Frequency Division Multiplexing
OFDMA   Orthogonal Frequency Division Multiple Access
OLT   Optical Line Terminal
OLU   Optical Line Unit
OMCC   ONU Management and Control Channel
OMCI   ONU Management and Control Interface
ONT   Optical Network Terminal
ONU   Optical Network Unit

OTN	Optical Transport Network (ITU-T G.709)
OVSF	Orthogonal Variable Spreading Factor
PAM	Pulse Amplitude Modulation
PAP	Priority Action Plan
PBCH	Physical Broadcast Channel
PBR	Prioritized Bit Rate
PCB	Physical layer Control Block
PCC	Primary Component Carrier
PCCH	Paging Control Channel
PCCPCH	Primary Common Control Physical Channel
PCF	Point Coordination Function
PCFICH	Physical Control Format Indicator Channel
PCH	Paging Channel
PCI	Pre-coder Indicator
PCMM	Packet Cable Multi-Media protocol
PCRF	Policy and Charging Rules Function
PDCCH	Physical Downlink Control Channel
PDCP	Packet Data Convergence Protocol
PDFA	Praseodymium Doped Fiber Amplifier
PDN	Packet Data Network
PDN	Premises Distribution Network
PDSCH	Physical Downlink Shared Channel
PDU	Protocol Data Unit
PEIN	Prolonged Electrical Impulse Noise
PF	Proportionally Fair
P-GW	PDN Gateway
PHEV	Plug-in (Hybrid) Electric Vehicles
PHICH	Physical HARQ Indicator Channel
PHS	Payload Header Suppression
PHY	Physical Layer
PIFS	PCF Inter-Frame Spacing
PIN	Photo diode constructed with P-type, Intrinsic, and N-type semiconductor regions
PL	Power Line
PLC	Power Line Communications
PLCP	Physical Layer Convergence Procedure
PLI	Payload Length Indicator
PLO	Physical Layer Overhead
PLOAM	Physical Layer OAM
PMCH	Physical Multicast Channel
PMD	Physical Medium Dependent sublayer
PMI	Precoding Matrix Indicator
PMS-TC	Physical media specific transmission convergence sublayer
PON	Passive Optical Network
POTS	Plain Old Telephone Service
PRACH	Physical Random Access Channel
PRB	Physical Resource Block
PRIME	Powerline Related Intelligent Metering
PS	Packet Switched

PSB	Physical Layer Synchronization Block
PSD	Power Spectral Density
PSS	Primary Synchronization Signal
PSTN	Public Switched Telephone Network
PTI	Payload Type Indicator
PTP	Precision Timing Protocol
PUCCH	Physical Uplink Control Channel
PUSCH	Physical Uplink Shared Channel
QAM	Quadrature Amplitude Modulation
QCI	QoS Class Identifier
QoS	Quality of Service
RACH	Random Access Channel
RAN	Radio Access Network
RAT	Radio Access Technology
RB	Resource Block
RCS	Ripple Carrier Signaling
RDI	Remote Defect Indication
RE	Resource Element
REIN	Repetitive Electrical Impulse Noise
RF	Radio Frequency
RFI	Radio Frequency Interference
RFoG	Radio Frequency over Glass
RI	Rank Indicator
RIT	Radio Interface Technology
RLC	Radio Link Control
RMS-DB	Root Mean Square - Delay Spread
RNC	Radio Network Controller
RoF	Radio over Fiber
RoHC	Robust Header Compression
R-ONU	RFoG Optical Network Unit
RP	Repeater
RP	Reception Point
RRC	Radio Resource Control
RRH	Remote Radio Head
RS	Reed Solomon
RSOA	Reflective Semiconductor Optical Amplifier
RT	Remote Terminal
RTD	Round Trip Delay
rt-PS	Real-time Poling Service (DOCSIS)
RTS	Request-to-send
RTT	Round Trip Time
SA	System Architecture
SAE	Society of Automotive Engineers
SAI	Serving Area Interface
SCADA	Supervisory Control and Data Acquisition
SCB	Single Copy Broadcast Ethernet frame

SCC	Secondary Component Carrier
SCCPCH	Secondary Common Control Physical Channel
SC-FDMA	Single-Carrier Frequency Division Multiple Access
SCH	Synchronization Channel
SCTE	Society of Cable Telecommunications Engineers
SDF	Service Data Flow
SDO	Standard Development Organization
SDU	Service Data Unit
SELT	Single Ended Line Test
SES	Severely Error Seconds
SF	DOCSIS Service Flow
SFBC	Space Frequency Block Coding
SFD	Ethernet Start of Frame Delimiter
SGSN	Serving GPRS Support Node
S-GW	Serving Gateway
SHDSL	Symmetric High bit rate Digital Subscriber Line, ITU-T G.991.2
SHINE	Short High amplitude Impulse Noise Event
SID	Service Identifier
SIEPON	Standard for Service Interoperability in Ethernet Passive Optical Networks
SIFS	Short Inter-Frame Spacing
SIM	Subscriber Identity Module
SINR	Signal-to-Interference-and-Noise Ratio
SIR	Signal-to-Interference Ratio
SLA	Service Level Agreement
SLF	Super Low Frequency
SMB	Small or Medium sized Business
SNMP	Simple Network Management Protocol
SNR	Signal to Noise Ratio
SOA	Semiconductor Optical Amplifier
SON	Self-Optimizing Network
S-ONU	DPoE Standalone ONU
SPS	Semi-Persistent Scheduling
SR	Status Reporting
SR	Scheduling Request
SRS	Sounding Reference Signal
S-SCMA	Synchronous CDMA (used with DOCSIS)
SSID	Service Set Identifier
SSS	Secondary Synchronization Signal
STA	Station
STB	Set-Top Box
STBC	Space Time Block Coding
STM	Synchronous Transfer Mode
SU-MIMO	Single-user Multiple Input Multiple Output
S-VID	Service VLAN Identifier (Ethernet)
T1	Repeatered 1.544 Mbit/s transmission line using Alternate Mark Inversion coding
T1E1.4	United States DSL standards committee now called COAST-NAI
TC	Transmission Convergence
TCM	Time Compression Multiplexing

TCM	Trellis Code Modulation
T-CONT	G-PON Transmission Container
TCP	Transmission Control Protocol
TC-PAM	Trellis Coded Pulse Amplitude Modulation
TDD	Time Division Duplexing
TDFA	Thulium Doped Fiber Amplifier
TDM	Time Division Multiplexing
TDMA	Time Division Multiple Access
TD-SCDMA	Time Division Synchronous Code Division Multiple Access
TFCI	Transport Format Combination Indicator
TFT	Traffic Flow Template
TFTP	Trivial File Transfer Protocol
TG	Task Group
TIA	Transimpedance Amplifier
TL	Transmission Line
TLV	Type-Length-Value field
TM	Transparent Mode
TM	Transmission Mode
ToD	Time of Day
TOS	Type of Service
TP	Transmission Point
TPC	Transmit Power Control
TR-069	Broadband Forum standard for remote management of CPE
TPS-TC	Transport protocol specific transmission convergence sublayer
TTI	Transmission Time Interval
TWACS	Two-Way Automatic Communications System
TWDM	Concurrent time and wavelength division multiplexing

UCD	DOCSIS Upstream Channel Descriptor
UE	User Equipment
UGS	Unsolicited Grant Service (DOCSIS)
UGS-AD	Unsolicited Grant Service with Activity Detection (DOCSIS)
UL	Uplink
ULF	Ultra Low Frequency
UL-SCH	Uplink Shared Channel
UM	Unacknowledged Mode
UMTS	Universal Mobile Telecommunication System
UNB	Ultra Narrowband
UNI	User-Network Interface
U-NII	Unlicensed National Information Infrastructure
UPBO	Upstream Power Back Off
Upstream	Data flowing from the customer
UTRAN	UMTS Terrestrial Radio Access Network

VBR	Variable Bit Rate
vCM	virtual Cable Modem
VCSEL	Vertical-Cavity Surface-Emitting Laser
VDSL1	Very high bit rate Digital Subscriber Line 1, ITU-T G.993.1
VDSL2	Very high bit rate Digital Subscriber Line 2, ITU-T G.993.2

VID	VLAN Identifier
VLAN	Ethernet Virtual LAN
VLF	Very Low Frequency
VoIP	Voice over Internet Protocol
VoLTE	Voice over Long Term Evolution
VSAT	Very Small Aperture Terminal

WAN	Wide Area Network
WARC	World Administrative Radio Conference
WBF	Wavelength Blocking Filter
W-CDMA	Wideband Code Division Multiple Access
WDM	Wavelength Division Multiplexing
WDMA	Wavelength Division Multiple Access
WG	Working Group
WiMAX	A fourth generation cellular standard based on OFDM/OFDMA
WLAN	Wireless Local Area Network
WSD	White-Space Device

XGEM	XG-PON Encapsulation Method
XG-PON	FSAN/ITU-T 10 Gbit/s PON protocol specified in the ITU-T G.987 series
XGTC	XG-PON Transmission Convergence

1

Introduction to Broadband Access Networks and Technologies

1.1 Introduction

In the mid-1990s, there were many doubts about the future of broadband access. No one was sure if the mass market needed or wanted more than 100 kbit/s; what applications would drive that need; what broadband access would cost to deploy and operate; what customers were willing to pay; whether the technology could provide reliable service in the real world; or which access technology would "win." Government regulation in many countries made it unclear if investment in broadband would yield profits. It seemed that broadband access would be available only to wealthy businesses. Fortunately, there were some people who had a vision of a broadband world and who also had the faith to carry on despite the doubts.

We now live in a world where broadband access is the norm and households without it are the exception. No one asks today why the average household would need broadband access. The answer is obvious: we need internet access, with its ever-growing number of applications, and VOD (video on demand). With more than 600 million customers connected to broadband networks, no one asks if the technology works or whether it can meet the customer's willingness to pay.

Furthermore, a growing application of broadband access is the support of femtocells, and small cells in general. Resorting to small cells has today become the most promising trend pursued for increasing wireless spectral efficiency, and the key to its success is the availability of a high capacity wired line to the home. Also, a growing fraction of cellular data is today generated indoors. In addition, it has become clear that no single broadband access technology will win the entire market, and that the market shares of the different technologies will change over time.

Each access technology has its strengths and weaknesses. A common constraint is that we can have it fast, low cost, and everywhere – but not all at the same time. In many cases, the choice of broadband access technology is driven by the legacy network infrastructure of the network provider. In other cases, national regulatory considerations are a significant factor. As a result, each access technology has its areas of dominance in terms of geography, applications, and political domains.

The book is divided into three sections:

- The chapters in the first section of the book cover technologies and standard protocols for broadband access over fiber-based access networks.

Broadband Access: Wireline and Wireless – Alternatives for Internet Services, First Edition.
Steven Gorshe, Arvind Raghavan, Thomas Starr, and Stefano Galli.
© 2014 John Wiley & Sons, Ltd. Published 2014 by John Wiley & Sons, Ltd.

- The chapters in the second section cover technologies and standards associated with non-fiber, non-wireless broadband access.
- The chapters in the final section of the book address wireless broadband access technology and standards. Some of these technologies have been widely deployed, while others are anticipated to see deployment soon.

1.2 A Brief History of the Access Network

The traditional access network consisted of point-to-point wireline connections between telephone subscribers and an electronic multiplexing or switching system. The early access network used a dedicated pair of wires (referred to as a copper line or "loop") between the subscriber and the central office (CO) switch.[1] As the cost of multiplexing technology decreased, it became more economical in many cases to connect subscribers to a remotely located terminal. This remote terminal (RT) would multiplex calls from multiple subscribers onto a smaller number of wires for the connection to the CO. Network cost was reduced by having far fewer pairs of wires from the CO to the remote areas. As the technology evolved from analog frequency domain multiplexing (FDM) to digital time domain multiplexing (TDM), the RT systems became known as digital loop carrier (DLC) systems.

Data access to the telephone network began with the introduction of voiceband modems that could transmit the data as a modulated signal within the nominally 4 kHz voiceband pass-band frequency. The shorter lines (loops) allowed by DLCs made increasingly efficient modulation technologies practical. However, as explained in Appendix 1.A, the maximum data capacity of voiceband modems was limited to 33.6 kbit/s, or 56 kbit/s under special circumstances. Modems and their evolution are also discussed briefly in Appendix 1.A.

As a result, out-of-band technologies were introduced that transmitted signals over the copper line at frequencies outside the voiceband. Since these technologies sent digital information in the out-of-band signals, they became known collectively as digital subscriber line (DSL) technology. DSL is discussed further in Section 1.3 and Chapters 7–10.

Since the subscriber lines are implemented with twisted wire pairs, with multiple lines sharing the same cable without being shielded from each other, there are limits on the bandwidth that is achievable with DSL. For this reason, network providers became interested in alternatives to the subscriber line for providing broadband access. The three main contending technologies are coaxial cable, fiber optic cable, and wireless radio frequency connections. Each of these technologies is reviewed in later chapters of this book.

Coaxial cable networks were deployed by community access cable television (CATV) companies to provide broadcast video distribution. Due to the high bandwidth capabilities of coaxial cables, they had the potential for offering broadband services to their subscribers. In order to offer broadband data services, CATV companies evolved their networks to support upstream data transmission, and introduced fiber optic cables for higher performance in the feeder portion of their networks. As discussed below and in Chapter 11, coaxial networks have their own challenges as well as advantages.

Telephone network providers responded to the potential broadband advantages of the CATV companies by deploying additional fiber in their access networks. Telephone companies have deployed fiber directly to each subscriber's premises in some areas. Others are deploying fiber to terminals near enough to the subscribers' premises that broadband services can be provided by the latest very high-speed DSL technologies. The most attractive aspect to fiber is its virtually unlimited bandwidth capability. The primary drawback has been the relatively high cost of the network and its associated optical components.

Wireless access had not originally been a significant contending technology for residential broadband access. However, as wireless mobile networks have become widely deployed, and new technologies and

---

[1] Of course, some of the earliest access lines were "party lines" where several subscribers were connected in parallel to the same loop. Due the inherent lack of privacy and decreased cost of providing access lines, party-lines have become a historical footnote.

protocols have been developed, wireless broadband access has become increasingly important. It is especially attractive in regions that lacked a legacy wireline infrastructure capable of evolution to broadband services. Examples of such regions include developing nations and rural areas. It also offers the very significant advantage of allowing mobile, ubiquitous service rather than being restricted to service at the subscriber's premises.

Since a limited amount of spectrum is available for use in broadband services, the networks to support it have become increasingly complex. For spectrum efficiency, wireless networks use grids of antenna, where each subscriber only needs enough signal power to reach the nearest antenna. The region covered by each antenna is referred to as a cell. The result is that the same frequencies can be used by subscribers in non-adjacent cells, since their signals should not propagate far enough to interfere with each other. The signal formats have been optimized in the latest protocols to approach the Shannon limit for data bits transmitted per Hertz of transmission channel bandwidth. Capacity is further increased by re-use of the spectrum through smaller cells and smart antenna technologies. Both add cost, and radio signals are always more vulnerable to various types of interference than wireline technologies. Wireless technologies are discussed further in Section 1.6 below, and in detail in Chapters 14–17.

1.3 Digital Subscriber Lines (DSL)

1.3.1 DSL Technologies and Their Evolution

DSL operates over a copper line at frequencies outside the voiceband, sending digital data directly from the subscriber, and thus avoiding the need for an analog to digital conversion. Since the telephone lines were designed to provide good quality for voiceband signals, they are often not particularly well suited for higher rate data signals. Reflections become a significant problem in the electrical domain at rates beyond the voiceband. One of the worst sources of reflections in North American networks is bridge taps. When the feeder cables are installed from the CO into the loop area, they go to splice boxes where the wires going to the subscribers are connected. When service is disconnected to a subscriber (e.g., due to the homeowner moving), a second pair of wires may be connected to the feeder cable to serve a different subscriber without removing the other line. The result is a bridge tap, and it is possible to have bridge taps at more than one location along the connection to a subscriber. The unterminated end of the unused line(s) causes electrical reflections of the DSL signals, and these reflections can cause destructive interference for certain frequencies (any impedance mismatch along the copper connection to the CO to the subscriber can cause harmful reflections, but the bridge taps are especially bad).

The first widely deployed services using a digital subscriber line were the Digital Data Service (DDS) from AT&T. DDS used baseband signals over the line and offered data rates including 2400, 4800 and 9600 bits/s, and 56 kbit/s. The lower rate signals were sometimes converted to analog signals at the CO and then mapped into a voiceband channel, thus avoiding any noise or distortion from the subscriber line. DDS required the end-to-end service be synchronized to a common atomic clock. DDS circuits also usually required that the line be groomed to remove impairments such as bridge taps. While DDS circuits were very valuable for some customers (e.g., banks using them for connections to ATM machines), they were too expensive to deploy to residential subscribers or even to many business subscribers.

The first serious attempt to provide higher data rates to subscribers was the basic rate interface of the Integrated Services Digital Network (ISDN-BRI). ISDN-BRI used baseband signals[2] over the subscriber line to offer bidirectional data rates of 144 kbit/s. ISDN-BRI was designed to operate over most subscriber lines of up to 18 000 feet without having to remove impairments such as bridge taps from the lines. The 144 kbit/s signal was typically divided into two 64 kbit/s bearer (B) channels and a 16 kbit/s data (D) channel. The B channels could be used for voice or data, while the D channel carried the connection signaling information, with its leftover bandwidth available to carry subscriber data packets. It was also

---

[2] Specifically ISDN BRI used the 2B1Q line code, which mapped two input data bits to a quaternary symbol (i.e., a symbol that has four possible amplitude values).

possible to merge the two B channels or merge the Bs and D channel into a single 144 kbit/s channel. The cost of ISDN-BRI was relatively expensive, however, and there were no driving subscriber applications to generate high demand. ISDN also required that the connection signaling protocol be processed by the CO switch, which meant a major upgrade to the switches. By the time that Internet connectivity became a driving application, much higher rates were practical for DSL.[3] In effect, ISDN BRI provided too little bandwidth, too late, with too much network complexity.[4]

DSL modems that were dedicated to data services began to be widely deployed instead of ISDN-BRI. Initially, there were two broad categories of DSL. The first was a high speed DSL (HDSL) that provided bidirectional symmetric service at half the DS1 rate over a single pair,[5] or symmetric full DS1 rate over two pairs (half on each pair). Although it would seem that HDSL had no advantage over T1[6] service, which also uses two pairs, HDSL was capable of operating over much longer line lengths than T1, and it could do so without requiring repeaters. The total cost of HDSL was less than half of T1 lines, mainly due to eliminating most of the labor needed to install repeaters and remove bridged taps. It became common for carriers to use HDSL as the primary technology for providing DS1 connections to business customers. The current generation of HDSL is HDSL2, which allows bidirectional symmetric transmission of up to 2.048 Mbit/s payloads over a single wire pair.

The second category is the DSL lines optimized for residential subscriber access. The first generation was called ADSL (asymmetric DSL) due to its use of asymmetric data rates in the upstream and downstream directions. Since residential subscriber are typically downloading more information than they are providing to the network, they typically require much higher data rates from the network (down-stream) than they do for upstream. This asymmetry in the desired data rates per direction was exploited to achieve the higher downstream rates. The service rate for ADSL is affected by several factors, but line length is the primary one. Over the past 25 years, telephone companies have tried to limit the line lengths to 12 000 feet.[7] Rates of 768 kbit/s downstream with 384 kbit/s upstream are possible over most of these lines. The actual rate is often determined adaptively as the system uses feedback to determine the frequency response of the line. In addition to higher data rates, another advantage of these DSL systems over ISDN-BRI was that they left the voiceband frequencies available for voice signals. This allowed analog POTS (Plain Old Telephone Service) signals to "ride underneath" the DSL data in its native format, which kept the voice and data signals separate within the network and allowed subscribers to use their existing telephone sets without conversion to digital signals at the subscriber premises.

ADSL rates and signal formats have been standardized by the ADSL Forum (now the Broadband Forum), by T1E1 (now ATIS COAST-NAI) and by the ITU-T SG15. SG15 is the primary body developing the current generation of DSL standards. The latest generation of ADSL is specified in the ITU-T G.992.5 standard for ADSL2plus which enables up to 20 Mb/s, with 12 Mb/s possible at 3000 feet.

Video delivery will require rates of 10–50 Mbit/s, depending on the service. For these rates, very high-speed DSL (VDSL) is required. VDSL requires lines lengths limited to 5000 feet. ITU-T SG15 has developed the VDSL2 standard, whose specifications are provided in ITU-T G.993.2 which enables rates up to 100 Mb/s upstream and downstream with 25 Mb/s downstream possible at 3000 feet. The ITU-T is developing the G.fast standard which promises to achieve bit rates up to 1 Gb/s over short copper lines.

---

[3] While DDS and ISDN BRI are digital subscriber loop technologies, it is most common to use the term DSL to refer to their successors that operated at higher data rates.

[4] Note that another early application of digital subscriber loops was to provide a second voice line over the same loop. This application is known as "pair gain" since it provides multiple voice channels over the same pair. Some carriers used ISDN technology to provide the simple pair gain service.

[5] When discussing subscriber loops, the term "pair" means a twisted pair of wires used for differential signal transmission.

[6] The term "T1" is commonly misused as being equivalent to a DS1. Strictly speaking, DS1 refers to the 1.544 Mbit/s signal and frame format, while T1 refers to a specific AT&T carrier system that transmits DS1 signals over 4-wire repeatered copper pairs.

[7] In the Bell System, the 12 000 ft. range was known as the Carrier Serving Area (CSA). Independent companies like GTE, who served more rural areas, specified their loop limits at 18 000 feet. Beyond 18 000 feet, inductive load coils need to be added to the loops to compensate their frequency response. DSL technology typically cannot operate through these load coils.

Both ADSL2plus and VDSL2 support transmission of packet transport mode (PTM), asynchronous transport mode, and synchronous transport mode (STM). ITU-T G.997.1 specifies management parameters for ADSL2plus and VDSL2.

1.3.2   DSL System Technologies

The first generation of DSL equipment connected DSL modems at the subscriber premises to DSL access multiplexers (DSLAMs) located in the central office. DSLAMs were next deployed in remote locations that were often co-located with DLC RTs. If the DLC RT was served by a SONET fiber connection, the DSLAM traffic would be multiplexed onto the same SONET signal as the DLC voice traffic. One of the challenges of co-locating the DSLAM and RT is that the DSLAMs require much more power per line than DLC equipment. This leads to heat dissipation issues when they shared the same cabinet, which can restrict the number of DSL lines that can be served. The DSLAM is not connected to the RTs backup batteries, however, since there is no requirement to maintain DSL service during a power outage.[8]

DSL was developed at a time when Asynchronous Transfer Mode (ATM) appeared to the preferred multiplexing technology for next generation networks. ATM provided adaptation techniques to carry a wide variety of packet-oriented data and constant bit rate (CBR) traffic such as voice signals. Hence, ATM was a natural choice for the encapsulation technology over the DSL line and for the multiplexing technology within the DSLAM. ATM allowed some statistical multiplexing for more efficient bandwidth utilization on the trunk from the remote DSLAM to the CO, or within the network.

There are two drawbacks to ATM, however. The first is that it adds at least five bytes of overhead to each 53-byte cell, causing a roughly 10% bandwidth overhead penalty. The bandwidth penalty is sometimes referred to as the ATM "cell tax." The other drawback to ATM is that it typically uses a rather complex signaling protocol that is overkill for purposes such as carrying connections to the Internet. Since most of the data going over DSL systems uses the Internet Protocol (IP) for Layer 3, it makes sense to use lower layer protocols that are more efficient with IP packets. Consequently, the emerging generation of DSLAMs is IP-based and uses Ethernet for the Layer 2 protocol instead ATM. These are commonly referred to as IP-DSLAMs.

1.4   Hybrid Fiber-Coaxial Cable (HFC)

While the telephone companies have focused on DSL, the CATV companies have deployed a network that is optimized for broadband broadcast traffic. As the demand for internet connectivity increased and the regulations allowed competition for providing telephone service, CATV companies have upgraded their networks to allow upstream data transmission from their subscribers.[9]

As illustrated in Figure 1.1, the CATV network uses a shared coaxial copper cable medium to connect to its subscribers. The coaxial segments are connected to remote equipment that provide the conversions to/from a fiber connection with a head-end office. These networks are called hybrid fiber coax (HFC). The bandwidth of the shared coaxial cable is divided into frequency bands, with one or more frequency bands being allocated for upstream transmission. Individual subscribers compete for the shared upstream bandwidth through a medium access control (MAC) protocol.

The most popular protocol for providing voice and data access is the Data Over Cable Service Interface Specification (DOCSIS[TM]) protocol developed by CableLabs, a laboratory that is jointly funded by multiple cable network providers. The downstream data is modulated into the RF channel slots that would otherwise have been used for carrying video signals. The upstream data is similarly transmitted using RF

---

[8] The assumption is that if the power is out at the DSLAM location, the power is also out at the subscriber premises, and hence there is no subscriber equipment operational to use the DSL line.

[9] An early challenge for most CATV systems was that their signal repeaters only worked in the downstream direction, and required upgrades to support upstream traffic. The deployment of fiber reduced the number of repeaters requiring upgrade, since the fiber systems were designed to support bidirectional traffic.

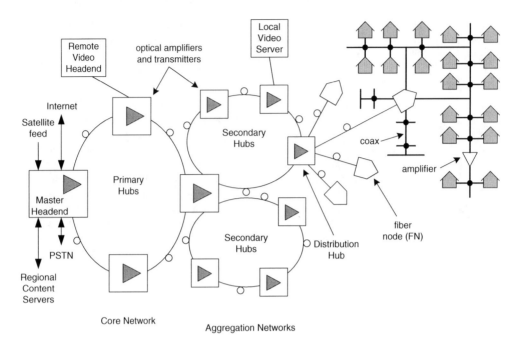

Figure 1.1 CATV network illustration

modulation into dedicated upstream frequency slots. The DOCSIS protocol assigns the frequency bands that are used by the cable modems at each customer's premises, and uses a shared-medium MAC protocol to determine the time slots in which cable modems can transmit their upstream data. The DOCSIS protocol also addresses the security issues associated with having a transmission shared among multiple subscribers where each can see the others' signals. The DOCSIS protocol is described in detail in Chapter 11.

Having been optimized for delivering video, the CATV networks are much better suited for delivering broadband video, including high-definition TV (HDTV) than are the telephone networks. A coaxial segment has typically been shared among several hundred subscribers. This degree of sharing inherently limits a CATV network's ability to provide high per-subscriber upstream bandwidth, and it also limits the number of video-on-demand (VOD) channels that can be provided. Reducing the coax sharing obviously increases their flexibility but also increases the cost of the CATV network. This is the primary tradeoff faced by the CATV providers in offering broadband access. To win in the market place, the cost and performance of the HFC networks must compete with the telephone company DSL networks and fiber to the home/curb networks (FTTH/C). One advantage HFC has over FTTH systems is that the copper coaxial cable allows a CATV company to provide power to the home telephone in the same manner that the telephone company does today.

1.5 Power Line Communications (PLC)

The use of power lines as a communication medium has been around for at least 100 years. This technology is generally referred to as Power Line Communications (PLC) and has sometimes enjoyed some degree of success over the years depending on the application it was used for.

The attractive feature of PLC is the high penetration of electrical infrastructure in the world, which in many areas is much higher than any other telecommunication infrastructure. As access to the internet today is becoming as indispensable as access to electrical power, and since devices that access the internet are normally plugged into an electrical outlet, the unification of these two networks always appeared to be a compelling option, despite the various technical challenges. As virtually every line-powered device can become the target of value-added services, PLC may be considered as the technological enabler of a variety of future applications that would probably not be available otherwise.

Among the various applications, today's interest for PLC spans several important applications: broadband internet access; indoor wired LAN for residential and business premises; in-vehicle data communications; smart grid applications (advanced metering and control, peak shaving, mains monitoring); and also municipal applications, such as traffic lights and lighting control and security.

In particular, smart grid applications have been and continue to be today a successful and promising area for PLC. Similarly, the interest in using PLC for home networking is increasing rapidly and, despite today's low penetration, many believe that home networking will be one of the most important areas of success for this technology. On the other hand, the great interest in the late 1990s for using PLC for providing broadband access to households has encountered many disappointments over the last two decades. Higher than anticipated costs in deploying PLC, growing EMC (Electro-Magnetic Compatibility) issues for the interference caused to radio services in the HF bands, its smaller capacity compared to DSL and cable, and the availability of other (and often cheaper) means to provide broadband access to consumers have made the initial enthusiasm in PLC for broadband access greatly diminish if not vanish.

There are very few PLC deployments in the world for broadband access and its use in industrialized countries, where the availability of other broadband access technology is abundant and cheaper and has made PLC a marginal technology. Perhaps the area where broadband access via PLC may still have some possibility of success is in third world countries, where access to the internet is essential to economic growth but there is no or very little telecom infrastructure. Similarly, rural areas in industrialized countries where it is very uneconomical to provide broadband services at competitive prices could also benefit from the deployment of PLC as most of these areas lack traditional telecom infrastructure but nevertheless have access to power.

Despite its failure to become a successful technology for broadband access, PLC will be addressed in Chapter 13. Because of its widespread use as a Smart Grid technology, the use of PLC in the power grid will also be addressed and its unique benefits for this application will be highlighted.

1.6 Fiber in the Loop (FITL)

Telephone company revenue from plain old telephone service (POTS) is declining as the result of losing some of their POTS customers to mobile phones and CATV companies. In order to increase their future revenue potential, the telephone companies believe that they need to be able to offer the best "triple-play" services, consisting of telephone, data (especially internet access), and video. At one time, telephone companies considered the idea of deploying the same type of HFC networks used by CATV providers. One drawback to this approach is that the coax networks typically have inferior reliability for voice service. The other main drawback is that they would only be "me-too" for video and data, thus providing no advantage over the CATV companies.

The model now preferred by telephone companies is based on their traditional approach of either avoiding fully shared media or limiting the amount of sharing. For non-shared subscriber medium, triple-play services typically will be provided through a fiber to the node (FTTN) architecture. With FTTN, as shown in Figure 1.2, a high-speed fiber connection[10] exists between the CO and a remote node that is close enough to the subscriber to allow individual VDSL connectivity to each subscriber served by that node. Variations on FTTN are Fiber to the Curb/Cabinet/Building (FTTC/FTTCab/FTTB). FTTN is very attractive when many subscribers are close enough together to be reached by VDSL (e.g., dense housing

---

[10] The high-speed fiber connection can either be a SONET/SDH link or a Gbit Ethernet link.

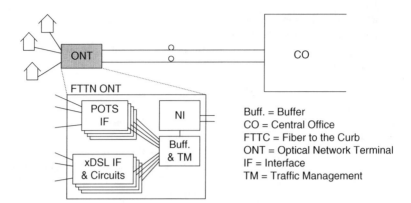

Figure 1.2 FTTN network illustration

neighborhoods or multi-tenant buildings). When subscribers are spaced further apart or require very high upstream bandwidth, fiber to the premises/home (FTTP/FTTH) becomes more attractive.

The passive optic network (PON) is the most attractive technology for FTTH/FTTP. PON systems share the fiber medium among a limited number of subscribers. Due to the directional nature of fiber optic transmission, only the downstream signals are visible to all subscribers on that PON. This simplifies the encryption processes required to ensure privacy relative to those required for shared coaxial cable or wireless networks.

Due to the relatively high cost of optical components (especially lasers and optical receivers), it is not cost effective to give each subscriber a separate fiber connection to the CO. The best way to reduce the number of optical components, as well as reducing the amount of fiber, is to have multiple subscribers share the same passive fiber network for their connection to the optical line terminal (OLT)[11] in the CO. The PON is illustrated in Figure 1.3. The terminal at the subscriber premises is typically called an optical network unit (ONU) or optical network terminal (ONT). Different generations of PON technology allow different numbers of ONUs to be connected to an individual PON, but 16 and 32 are typical numbers, with some systems connecting up to 64 and future systems being capable of higher numbers. Since passive optical splitters are used to divide (and merge) the optical signal among the ONTs, the number of ONTs connected to a fiber is often called the split ratio (e.g., 32-to-1).

PON systems typically transmit both upstream and downstream data over the same fiber. In some cases, only directional couplers are used to separate the upstream and downstream traffic, but higher speed systems typically use different wavelengths in each direction. The most common is course wave division multiplexing (CWDM), in which 1490 nm is used for the downstream direction and 1310 nm for the upstream. This wavelength assignment has the advantage of putting the less expensive 1310 nm lasers at the ONTs.

In the downstream direction, the OLT broadcasts the data for all ONUs. This downstream signal is comprised of the downstream data for all the ONTs and synchronization information for the upstream transmissions. The ONTs extract their downstream data based on either time slots or cell/packet address information.

In the upstream direction, the ONUs need a medium access control (MAC) protocol to share the PON. The most common MAC protocol is time domain multiple access (TDMA), which is similar to the protocols used by broadcast television satellites. With TDMA, the nodes are granted time slots in which to transmit their upstream data. In basic PON systems, each ONT is preassigned a fixed portion of the upstream bandwidth, and transmits its data at the appropriate time. In order to achieve greater efficiency,

---

[11] Another popular name for the OLT was host digital terminal (HDT). The OLT can either be located in the CO or at a remote (RT) site.

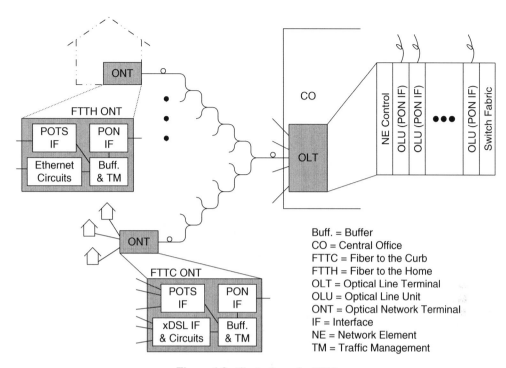

Figure 1.3 Illustration of a PON

PON systems now typically allow dynamic bandwidth allocation (DBA) among the ONTs. With DBA, each ONT uses part of its upstream transmission to inform the OLT of its bandwidth requirements.

For example, this information could be based its input queue fill level, including the levels for data in different classes of service. The OLT evaluates the requests from the ONTs, and assigns the bandwidth for the next upstream transmission frame. This bandwidth is typically communicated as a transmission start time and either a stop time or transmission duration time within the upstream frame. These bandwidth assignments are sent in the downstream transmission frame. The information used by the OLT in determining the appropriate bandwidth allocations can include the service level agreements (SLAs) associated with the ONT data flows. In some systems, the ONT is responsible for determining how to accommodate the relative priorities of its transmit data within the granted upstream transmission slot. The most popular TDMA PON protocols are described in detail in Chapters 3 and 4.

One alternative to TDMA is wavelength division multiple access (WDMA) in which each ONU has its own upstream and downstream wavelength for communication with the OLT. In other words, the separate wavelengths allow each ONU to have a point-to-point connection to the OLT over the shared PON fiber. The main drawback to WDM is that each ONU needs a unique wavelength, which would be very hard to administer if subscribers are allowed to buy their own ONUs. Tunable lasers would alleviate this problem, but they are currently too expensive. Other frequency selective technologies are being researched and developed for use at ONUs, but to date they have not been cost effective relative to TDMA technologies.

Another alternative is code division multiple access (CDMA). CDMA uses a spread spectrum approach where the subscriber bit stream modulates a code sequence, essentially in the same manner as is used for mobile phones. CDMA is very attractive since it can be implemented with entirely passive components at the transmitter and receiver. A further advantage of CDMA is that each subscriber can use a different native client interface. CDMA circuits, however, typically require optical amplifiers and precision receiver discriminator circuits to achieve the required signal to noise ratio. Other optical-domain medium access methods are also possible, but WDMA appears to be the most likely long-term approach. These optical domain technologies are described in Chapter 5.

There are also technology combinations that use a PON infrastructure in combination with a different technology, such as carrying the radio-frequency modulated CATV signal over a PON. These hybrid PON protocols and technologies are covered in Chapter 6.

1.7 Wireless Broadband Access

The mobile computing paradigm has seen phenomenal growth in the first decade of this century. Most services traditionally accessed on desktop PCs or dedicated networked hardware are being augmented with, or completely supplanted by, mobile access on tablets and smartphones. Mobile devices are clearly being powered by wireless technologies. However, before delving into the different wireless technology options, one must first establish a clear understanding of the role played by both wired and wireless technologies in delivering broadband access to the untethered end user.

Let us take a simplistic view of wireless access technologies initially, and divide wireless technologies into "long range" and "short range". Long-range wireless links (such as those used by cellular technologies) can serve users over a widely distributed geographical area, and can therefore be seen as a true alternative to the wired access options introduced in the previous sections. On the other hand, short-range wireless links only cover a small area such as a home or an office. Short-range wireless technologies therefore need to be augmented by wired backhaul access technologies in order to provide a complete solution for broadband access to the end user.

It is important to understand in this configuration that the "speed" of the broadband connection is actually determined by the smaller of the access rates of the wireless portion and the backhaul portion. To give a concrete example, a WiFi installation at a cafe may use the latest and fastest version of the standard, providing several hundred megabits of throughput. However, the cafe may use a DSL backhaul connection providing only a few tens of megabits of throughput, due to cost or availability considerations. In this example, the end user experience will be limited by the backhaul speed. The converse is also possible, where the wireless access speeds can limit the overall user experience, as we shall see.

Another possible distinction between wireless access technologies can be based on whether they provide "fixed" access or "mobile" access. In the early 2000s, several vendors developed systems based on a DOCSIS-like protocol for fixed access, where equipment would be installed at customers' premises and provide the long-range backhaul for Ethernet-based LANs. Early on, these systems were proprietary, but the need for a common standard soon became apparent. This led to the development of the IEEE 802.16 standard to provide long-range, fixed wireless access under the title "Wireless Metropolitan Area Networks" (Wireless-MAN).

However, fixed wireless systems, whether they were proprietary, or based on 802.16, were mainly restricted to smaller deployments in low-density population centers where the cost of installing wired access was seen as expensive for the corresponding revenue potential. Moreover, it also became apparent that a single technology, developed for both fixed and mobile access, would result in a more robust ecosystem with more applications and adoption potential. As a result, the 802.16 standard evolved to provide mobile access, but due to the deployment of competing cellular technologies, 802.16-based systems have not been able to gain any significant market share.

The two types of technologies that are most popular for providing broadband access today are the IEEE 802.11-based Wireless LAN (WLAN) standard, popularly known as WiFi, and the third and fourth generations of cellular technology. WLANs use unlicensed spectrum with restrictions on the transmit power and are, therefore, mostly used as a short-range technology. In addition, WiFi is expected to coexist with other systems in an unregulated environment, so it has been designed to be able to coexist and be robust in the presence of interference. Furthermore, the technology is designed to use a simple architecture that is easy to configure and install. The basic topology consists of an access point providing the broadband connectivity to several associated wireless clients that represent end users. It is this combination of the use of unlicensed spectrum, ease of installation and interference-robustness that led to the rapid adoption of this technology. It is ubiquitous today as the predominant access technology in low-mobility environments such as homes, offices, campuses, and other public spaces.

In contrast, cellular technologies took a very different evolution path to becoming an alternative for broadband access. While WiFi was designed from the start to provide access to data networking services, cellular technologies were initially designed with the sole goal of providing mobile voice service. The requirement to support seamless mobility for voice via handovers resulted in a more complex and expensive system. Furthermore, they were deployed in licensed spectrum to ensure that interference can be managed in a regulated fashion, thereby ensuring high reliability for voice services. As such, these systems are owned and operated by service providers, not by individuals or small enterprises. With transmit power not severely restricted, as in the case of unlicensed spectrum, cellular systems can cover much larger areas.

The basic topology of a cellular system is based on the concept of cells and frequency reuse. Early systems were not designed to be robust to interference, and therefore needed to use frequency separation and cells to manage interference, as shown in Figure 1.4. The figure shows a frequency reuse factor of three, because a separate carrier is needed in each cell in the repeating cluster of three cells in order to maintain a minimum interference separation. Other reuse factors are possible, with varying degrees of interference separation. Early cellular technologies, in the first and second generations required frequency reuse and supported mainly voice services. However, as the need for mobile data services grew with the growth in internet traffic, combination of scarce licensed spectrum resources, and the greater capacity needed for data services led to the design of reuse-one systems where all cells could use the same frequency, and the interference mitigation was carried out by a more robust physical layer designed to operate at lower signal-to-noise ratios.

In the wireless section of this book, comprising of Chapters 14–17, we take an in-depth look at the three most widely deployed technologies for mobile broadband access. Before delving into the details of the technologies, we first try to establish, in Chapter 14, the fundamental concepts that apply to all wireless systems. In addition, the various basic building blocks that are part of the air-interface of any broadband access technology are explained. Next, in Chapter 15, we discuss WiFi based on the IEEE 802.11 standard. Lastly, in Chapters 16 and 17, we discuss third and fourth generation cellular technologies. In Chapter 16, we focus on the technology based on Wideband Code Division Multiple Access (W-CDMA), and briefly discuss how it contrasts with the other third generation system based on CDMA-2000. In

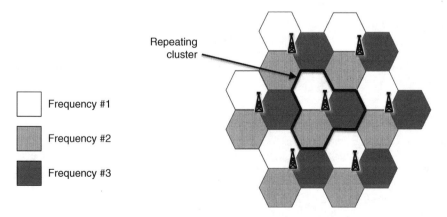

Figure 1.4 Cellular system with frequency-reuse factor = 3

Chapter 17, we discuss the fourth generation systems based on LTE and LTE-Advanced that are expected to be widely deployed, with a brief mention of WiMAX which is based on the IEEE 802.16.

All of these technologies are developed and implemented on the basis of specifications set forth by standards development organizations. As such the evolution of these technologies can be tracked by examining each new release of the specification in sequence. In discussing these technologies, we will first discuss the baseline features, network and protocol architecture of the technology. Next we will see how the technology evolved by taking a detailed look at each significant release of the standard and pointing out the key features and capabilities that were introduced in that release. Each chapter concludes with a summary that condenses all the material covered in the chapter into a few paragraphs to provide a quick review of the most noteworthy elements of the technology.

1.8   Direct Point-to-Point Connections

While direct point-to-point connections are not cost effective for residential subscribers, they will continue to be used for large corporate subscribers. Copper wireline connections can be DS1, E1, DS3, or Ethernet. Fiber connections include SONET/SDH, 1G, 10G, or 40G Ethernet, dark fiber, or a WDM wavelength. Wireless point-to-point connections are typically microwave radio links. The primary advantages of these direct connects are guaranteed bandwidth and security (since there is no shared medium).

While direct fiber connections are often not available to enterprise subscribers, DS1/E1 connection availability is ubiquitous. In North America, the regulatory environment can also create a price advantage for services providers to lease DS1/DS3 connections rather than fiber connections through the local exchange carrier networks. With the addition of virtual concatenation support for DS1/E1/DS3/E3 signals, copper connections through the traditional telecommunications infrastructure have become much more flexible. GFP then provides the transparent mapping for packet data services (see PMC-Sierra white paper PMC2041096). Previously, providing copper connections between the DS1 and DS3 rates required fractional DS3 or some relatively inflexible or inefficient method of combining DS1 s. These methods included inverse multiplexing with ATM (IMA), packet-specific techniques such as the IETF Multi-Link Point-to-Point Protocol (ML-PPP), or proprietary solutions.

1GE and 10GE fiber connections are becoming increasingly important as the UNI to enterprise subscribers. The telecommunications network provider may, in turn, use WDM for increased utilization, or map this data into its SONET or OTN infrastructure where TDM multiplexing allows even greater fiber utilization.

Appendix 1.A: Voiceband Modems

Voiceband modems began by using dual-tone frequency-shift key modulation for rates of 300 bit/s. As technology advanced, it became practical to us phase-shift key modulation and combinations of the amplitude modulation and phase modulation such as quadrature amplitude modulation (QAM) for greater efficiency. The capacity of any information channel is determined by the Shannon channel capacity theorem:

$$C = B \log_2(1 + S/N)$$

The capacity limits on the data rates for voiceband modems are primarily determined by the analog-to-digital conversion that takes place when the modem signal from the subscriber reaches the telephone network equipment (DLC or central office switch). Specifically, the 8 kHz sampling rate, and the quantization noise introduced when converting a voiceband signal to a 64 kbit/s digital signal determine the channel bandwidth (B) and the noise (N) terms of the Shannon capacity equation. The modem signal power (S) is limited by both the dynamic range of the analog-to-digital conversion, regulation, and the need to avoid crosstalk into other subscriber loops in the cable. The resulting capacity limit (C) for a voiceband modem is approximately 34 kbit/s, considering data transmission over voiceband channel with additive white Gaussian noise and assuming a nominal bandwidth of about 3.5 kHz and a signal-to-noise ratio of about 30 dB. Using efficient modulation techniques and error correction technologies, such as trellis coding, allowed standard voiceband modems to approach this limit with 33.6 kbit/s.

However, the value of 33.6 kbit/s was still far from the theoretically possible DS0 data rate of 64 kbit/s that could have been achieved with the same bandwidth but higher signal-to-noise ratio. The 64 kbit/s maximum value depended on the use of a 8ksample/s sampling rate and of 8 bits/sample in the analog-to-digital conversion.

In some circumstances, modems are indeed capable of approaching the theoretical maximum of 64 kbit/s if certain conditions of low quantization noise are met, for example, when a subscriber is connected via an analog line to a switched digital network and thus only one analog-to-digital conversion takes place. In some cases, the source of the data sent to a subscriber has a digital connection to the network (e.g., a DS1/T1 link) rather than a modem connection. Examples of such data sources include internet service providers. The digital-to-analog converter connecting to the subscriber loop creates a downstream signal that has none of the quantization noise that would have been created by an analog-to-digital conversion. If the other noise sources affecting that subscriber loop are small enough, then the channel capacity of the loop can be approached. In these circumstances, and sill using a sampling rate of 8ksample/s but encoding with data only seven bits of the 8-bit word in the analog-to-digital conversion,[12] then modems can achieve 56 kbit/s downstream rates. Since the upstream signal from the subscriber must go through the telephone network equipment's analog-to-digital conversion, the upstream signal rate of these modems is still limited to the standard 28.8 or 36.6 kbit/s rates.

---

[12] To improve the probability of error, only 128 PCM values are used.

2

Introduction to Fiber Optic Broadband Access Networks and Technologies

2.1 Introduction

Telephone companies and community access television (CATV) providers (also called "cable" providers) are competing to offer subscribers the triple play services of voice, video, and high-speed data access. Historically, both telephone and CATV networks have relied on copper cables to connect through the last mile to their subscribers, but a coaxial cable of the CATV companies has superior bandwidth capabilities relative to the twisted pair wiring from telephone companies. However, the coaxial cable must be shared by many subscribers in order to be economical. Clearly, the most flexible and future-proof medium is fiber optic, with its virtually unlimited bandwidth availability. For telephone network providers, fiber connections are attractive as the key to leapfrogging the capabilities of CATV providers. In response, CATV providers are also beginning to deploy all-fiber networks for enterprise customers and are considering it for residential customers.

Because providing a direct optical connection between the telephone company central office (CO)[1] and each subscriber is cost prohibitive in terms of cost, most optical access systems share a passive optical network (PON) among multiple subscribers. PON standards and technology are the focus of this section of the book. The section begins with a brief history of fiber optics in access systems, including early PON systems. This chapter also includes discussions of general PON topics and technologies that are largely independent of the specific PON protocol. These topics include an introduction to PONs, technology challenges, system powering issues, and protection for survivability. The remaining chapters of this section cover the different major families of PON protocols.

There have been two standards bodies developing protocols. Chapter 3 covers the IEEE PON protocols, which are the IEEE 802.3ah Ethernet PON (EPON) and 802.3av 10Gigibit/s EPON (10G EPON) standards. Chapter 4 covers the protocols developed by the Full Service Access Network (FSAN) consortium in conjunction with the International Telecommunications Union – Telecommunications

---

[1] Telephone network terminology is typically used in this chapter, since the telephone companies have driven much of the PON standards development and have deployed the majority of the PON networks. The equivalent cable provider terminology is used when appropriate.

Broadband Access: Wireline and Wireless – Alternatives for Internet Services, First Edition.
Steven Gorshe, Arvind Raghavan, Thomas Starr and Stefano Galli.
© 2014 John Wiley & Sons, Ltd. Published 2014 by John Wiley & Sons, Ltd.

Standards Sector (ITU-T). The FSAN consortium defines the requirements for these standards, which are then fully developed and published by the ITU-T. The FSAN/ITU-T standards include the G.983 series Broadband PON (B-PON), G.984 series Gigabit PON (G-PON), and the emerging G.987 XG-PON standards. FSAN and ITU-T are also beginning work on next generation (NG-PON) protocols that go beyond XG-PON.

Both the current IEEE and FSAN/ITU-T protocols are primarily based on time division multiple access (TDMA) for their medium access control (MAC) layer. Chapter 5 covers protocols that primarily used optical domain MAC techniques. These protocols include wave division multiple access (WDMA) and optical code division multiple access (CDMA) techniques, and some frequency-domain multiplexing techniques.

2.2   A Brief History of Fiber in the Loop (FITL)

FITL began with connecting remote terminals of a digital loop carrier (DLC) to the central office (CO) with a fiber instead of T1 lines[2]. The first serious interest in fiber to the home (FTTH) began in the late 1980s, as the telephone companies gained experience with Integrated Service Digital Network (ISDN) wideband services to subscribers. Rapid advances in the technology of optical transmitters, receivers and fibers made FTTH appear to be potentially just over the horizon. However, cost and powering issues with FTTH led to various fiber to the curb or cabinet (FTTC, FTTCab) systems as an alternative. The following discussion considers both the FTTH and FTTC/Cab technologies.

The first generation of FTTH systems attempted to replace the copper line (loop) directly with fiber. An optical network terminal (ONT)[3] was installed at (or near) the subscriber's premises. The telephone company side of the fiber was terminated on a line card in an optical line terminal (OLT) or a traditional DLC. The topology with such OLTs is called an active star, since there is an active fiber transceiver in the telephone company CO for each fiber radiating out to the subscriber. When DLCs were used, the topology was called an active double star since fibers from the CO connected to multiple DLC remote terminals (RTs) which, in turn, had active fiber transceivers for the connections to the subscribers. Most large equipment manufacturers built prototype or field trial versions of this type of system, for example, [1]. Typical bandwidth over the fiber to the subscriber on these systems ranged between the ISDN Basic Rate (160 kbit/s) to a DS1 or E1 signal.[4]

Passive optical networks (PONs) were also explored as a way to reduce the cost per subscriber by reducing the number of optical transceivers and fibers. As illustrated in Figure 2.1, a PON system uses a single optical transceiver at the OLT to serve multiple subscribers over a fiber tree constructed with passive optical signal splitters. The first serious trial PON system was developed and deployed by British Telecom [2]. First generation PON systems were also developed by both major equipment vendors and startup companies. NTT wrote its own standard for such PON systems for deployment in Japan. The evolution of PON systems is discussed in the next section.

Another approach to reducing the cost of FITL systems was to serve multiple subscribers from the same ONT. These systems were commonly called fiber to the curb (FTTC) systems, with 4–12 subscribers typically served from the same ONT. FTTC provided three major cost benefits. First, it reduced the

---

[2] Since the telephone connection to a subscriber is a pair of wires, the two-wire connection is often referred to as a "loop" or "subscriber loop."

[3] The EPON standard uses the term Optical Network Unit (ONU) exclusively to refer to the optical terminal closest to the subscriber. The GPON standards regard the ONU to be the more general term, with ONT referring to an ONU that serves a single user. From a PON protocol standpoint, there is no difference between ONUs serving single and multiple users. Consequently, the GPON standards use the terms OLU and ONT interchangeably.

[4] In the US regulatory environment of that time, the incumbent carriers were restricted regarding what data or video services they could offer. Consequently, the state public utility commissions required carriers to justify the cost of any new access technology on the basis of its cost for basic telephone service.

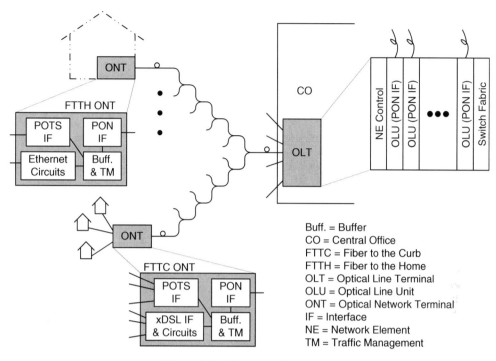

Figure 2.1 PON Network Example

number of optical components relative to FTTH (FTTC can be deployed as either an active star/double star or with a PON. Active stars were initially more common since serving multiple subscribers per ONT would quickly exhaust the capacity of early PON systems). The second cost advantage of FTTC is that it preserved the copper line connections from the "curb" over the last 1000 feet or so to the home. Installing fibers on this final subscriber drop is very expensive and opportunities for sharing costs on this portion are minimal. Since the short line length allows a high-speed DSL connection, most of the same services could potentially be delivered. The third advantage of FTTC was that it simplified the means for the network provider to continue to provide the power for the subscriber's phone (see Section 2.4.2 and the Appendix for further discussion of subscriber power issues).

Variations on the FTTC theme include fiber to the cabinet (FTTCab), where the cabinet serves more subscribers than a typical curb unit, and fiber to the premises (FTTP), where the premises are a multi-tenant building. FTTN (Fiber to the Node) has become a popular term for FTTC/P/Cab systems in the US. FTTC, while VDSL is the current preferred plan for some European and US carriers.

The cost-effectiveness of FTTC/FTTCab/FTTP systems depends on numerous factors, including:

- the relative cost of the number of ONTs served per OLT optical transceiver;
- the fiber and its installation cost;
- the cost of the DSL transceivers at the ONT and subscriber premise;
- the overall cost of powering the ONT;
- the real estate cost of placing the ONT.

FTTC/FTTCab/FTTP systems are clearly less flexible for high bandwidth services than FTTH systems, since there is much more network equipment impact whenever the subscriber wants a different service rate.

The first deployments of commercial PON systems targeted business customers. This initial application had a relatively small market, since it was uncommon to have a cluster of business customers wanting access bandwidth greater than DS1/E1 who were all reachable by the same PON. The circumstance that greatly accelerated PON deployment, beginning in the early 2000s, was the demand for high-speed internet access by residential customers. The cost of FTTH PON systems was still cost-prohibitive, so telephone companies relied on DSL technology to reach residential subscribers.

DSL rates were a good fit for the residential subscriber applications of the late 1990s and early 2000s. However, in order to make DSL ubiquitously available, the telephone companies needed to remove the T1 signals from the copper bundles due to spectral compatibility issues with DSL. As a result, PON became the most attractive option for serving business customers, freeing the copper cables for residential DSL service. In recent years, the cost of PON technology has decreased and the bandwidth requirements of residential customer applications have increased to the point where PON has finally become attractive for FTTH applications.

2.3   Introduction to PON Systems

2.3.1   PON System Overview

As discussed above and illustrated in Figure 2.1, a PON system consists of a passive fiber tree/bus network that connects multiple ONTs to a single OLT optical transceiver. For FTTH, the ONT is at the customer's premises, either mounted on an outside wall by the network interface or inside the house. The FTTH ONT optionally provides the POTS ("plain old telephone service") interface to the subscriber and an Ethernet interface for data services. For non-FTTH ONTs (e.g., FTTC), the ONT provides the final copper drop to the subscriber. This interface will typically use a DSL technology for the data service, with analog POTS sharing the copper drop in its native frequency range. With FTTC, the line lengths are typically less than 1000 feet, making VDSL very practical for video delivery. For G-PON, a mechanism exists to map the POTS channel directly into the G-PON frame. Alternatively, VoIP can be used to eliminate the need to carry the POTS signal separately from the data signal. VoIP is the method to provide the voice service with EPON systems.

As shown in Figure 2.1, the OLT consists of a number of PON interface units, a switch fabric for the data services (and potentially a simple switch fabric or multiplexer for the voice channels), and a NE controller. The ONTs are ultimately also managed by the OLT NE controller, which is responsible for all ONT provisioning and OAM&P reporting. The OLT and ONTs together form the PON system, enabling it to function logically as a single NE. In some ways, the fiber interconnection can be thought of as an extended backplane.

The OLT transmits the data for all ONTs in the downstream direction. This downstream signal comprises the downstream subscriber data for all the ONTs, the overhead for OAM&P, and the synchronization information for the upstream transmissions. The ONTs extract their downstream data based on either time slots, cell/packet frame address information, or wavelength.

In the upstream direction, the ONTs need a medium access control (MAC) protocol to share the PON. The most common MAC protocol is time domain multiple access (TDMA), which is similar to the protocols used by broadcast television satellites. With TDMA, the ONTs are each granted a time slot in which to transmit their upstream data. In basic PON systems, each ONT is pre-assigned a fixed portion of the upstream bandwidth and transmits its data at the appropriate time. More typically, in order to improve the PON bandwidth efficiency, the OLT dynamically grants the ONTs time slots in which to transmit their upstream data, and it communicates these bandwidth grants within the downstream signal.

As noted in Section 2.4.1, a guardband time is required between the upstream burst transmissions of the ONTs so that their transmissions do not overlap at the Optical Line Unit (OLU) receiver in the OLT. The ONT signals propagate through the fiber at the speed of light divided by the index of refraction of the fiber (approximately 2×10^8 m/s). With 20 km of fiber, which is a common maximum PON length, the round trip delay for just the fiber would be 200 μsec, which corresponds to 200 kbits for 1 Gbit/s and 2 Mbits for

10 Gbit/s line rates. Thus, to minimize the length of the guardband time, the relative fiber length from each ONT to the OLU should be taken into account. Beginning with second generation PON systems, most have a ranging protocol to measure this delay so that the ONT burst times can be assigned to allow a minimum guardband time between bursts arriving at the OLT.

As noted, PON systems now commonly allow dynamic bandwidth allocation (DBA) among the ONTs in order to achieve greater upstream bandwidth efficiency. With DBA, each ONT communicates its bandwidth requirements to the OLT. For example, this information could be based on its input queue fill level, including the levels for data in different classes of service. The OLT evaluates the requests from the ONTs and assigns the bandwidth for the next upstream transmission frame. The information used by the OLT in determining the appropriate bandwidth allocations can include the service level agreements (SLAs) pertaining to the data flows associated with the ONTs. These bandwidth assignments are sent in the overhead of the downstream transmission frame, and they are typically communicated as a transmission start and stop time within the upstream frame. In some systems, the ONT is responsible for determining how to accommodate the relative priorities of its transmit data within the granted upstream transmission slot. Allowing the ONT to decide how to fill its upstream bandwidth allocation allows it to minimize latency for higher priority traffic that has arrived at the ONT since it made its bandwidth request. It also distributes some of the processing load between the OLT and the ONTs.

In contrast to TDMA PON systems, wavelength domain multiple access (WDMA) systems use separate wavelengths to create virtual point-to-point connections between the OLT and each ONU. Optical filters within the PON are used to connect the appropriate wavelength(s) to each ONU. Each of these connections operates at the desired rate without the need for dynamic bandwidth assignment. While the bandwidth allocation is simpler with WDMA, the additional WDM optical components have made these systems more expensive than TDMA systems. This point will be discussed further in Chapter 5.

PON systems typically transmit both upstream and downstream data over the same fiber. In some cases, the same wavelength is used for both directions with only directional couplers to separate the upstream and downstream traffic. Higher speed systems typically use different wavelengths in each direction. The most common implementation is coarse wave division multiplexing (CWDM), in which the 1550 nm region is used for the downstream direction and the 1310 nm region for the upstream. This wavelength assignment has the advantage of putting the less expensive 1310 nm lasers at the ONTs.

Note that some PON systems use 1490 nm for the downstream PON signal, with analog video optionally overlaid and transmitted downstream at 1550 nm. Using WDM for video overlay provides a simple upgrade to existing deployments and increases the downstream capacity. Using a separate wavelength for analog video transmission also avoids such problems as lack of digital content and regulations involving digital content.

2.3.2   PON Protocol Evolution

As noted above, the first generation of PONs was based on TDM signals such as DS1/E1 signals. The downstream frame was a TDM frame, where each ONT's data was placed into time slots reserved for that ONT. With any TDMA protocol, the data transmitted upstream must be broken up into blocks that can be transmitted in bursts. These first generation PONs collected the data from their upstream TDM time slots and transmitted them at a higher rate during their assigned upstream burst time slot. For voice signals, this corresponded to a number of voice samples. For packet data, it was simply the number of bytes of the packet that would have been transmitted during that frame in a corresponding point-to-point TDM signal.

The second generation of PONs was based on Asynchronous Transfer Mode (ATM), which provided a convenient protocol for chopping the upstream data into blocks for the upstream transmission bursts. ATM supplied the mechanism for carrying TDM traffic, fragmenting large packets, and assisting QoS support. Also at this time, ATM was regarded as the likely basis for next generation networks and was already being used for broadband access in DSL systems. The upstream burst time slot allocated by the OLT to the ONT was simply the number of ATM cells it was allowed to send in that burst. NTT specified such an ATM-based PON (APON) system for use in their network. APON was also chosen by the Full-Service Access Network (FSAN) consortium, which publishes its standards through the ITU-T. The ITU-T G.983 series covers APON systems that are commonly referred to as Broadband PON (B-PON) systems. B-PON is described in Chapter 4.

With IP packets comprising more of the subscriber data, and Ethernet providing the typical connection to the CPE, it made sense to avoid ATM adaptation and to use packet technology (e.g., IP) for the packet routing. Consequently, the third generation of PON systems are based on, or optimized for, carrying Ethernet frames. The two primary high-speed third generation PON standards are Ethernet PON (EPON) from the IEEE (802.3ah) and Gigabit PON (G-PON) from the ITU-T (G.984 series). With EPON (described in detail in Chapter 3) the upstream transmission is a burst of one or more Ethernet frames. G-PON, which was developed after EPON, is much more flexible. As described in Chapter 4, a G-PON upstream burst can contain whole Ethernet frames, fragmented Ethernet frames, or TDM traffic bursts. Third generation systems thus avoid the protocol complexities and the bandwidth overhead associated with ATM adaptation.

The emerging generation of PON protocols uses a combination of higher speed TDMA and wave division multiplexing (WDM). With the 10 Gbit/s IEEE 10G EPON, the WDM is primarily used to allow EPON and 10G EPON ONUs to share the same physical PON fiber infrastructure. The 10G EPON OLT communicates with both types of ONUs. As discussed in Chapter 4, FSAN/ITU took the same approach with XG-PON in developing its next generation of PON protocols.

Future generations of PON protocols are expected to make more use of WDM to increase the PON capacity in terms both of the number of subscribers per PON and of the bandwidth per subscriber. As discussed in Chapter 5, the current optical technology does not make pure WDM PONs cost-competitive with TDMA PONs, and this situation is expected to persist for many years to come. Until WDM PON components become more cost-effective, high-capacity PONs will continue to use a mix of TDMA and WDM.

Some carriers envision PON technology as the means not only to provide new broadband services, but also as a means to reduce the cost of their metro and access networks. BT was the first carrier to announce such a vision as part of the CN21^{TM} plan. Central offices were originally built within less than 4–6 km of most subscribers[5] and DLC technology extended the reach, reducing the need for new COs. PON offers the potential to serve subscribers from fewer COs located much further away from the subscribers. To be economically feasible, the PON reach would need up to about 60 km, and each PON would need to connect to around 1000 subscribers. One carrier estimates that such a PON-based network would allow them to reduce the number of COs by a factor of ten. The financial gain from selling the COs that are no longer needed, combined with the greatly reduced network operating expenses, could easily justify the cost of building the new PON network.

Achieving the longer reaches, especially with higher split ratios, will likely require either optical amplifiers or repeaters on the PON. The ITU-T has defined the framework for this type of reach extension for G-PON systems [3]. An experimental "Super-PON" research system is described in Chapter 5 as one potential architecture for achieving the 1000 ONU capacity over 100 km of fiber.

Achieving 1000 subscribers per PON will require a mix of WDM and higher-speed TDMA. One of the advantages to WDM is that WDM optical splitters introduce much less loss than other passive optical splitters. A symmetric passive optical power splitter sends half the optical energy from all wavelengths down each of the two legs of the splitter, hence introducing about 3 dB of optical loss. Since this power split occurs for every 1 : 2 splitter, the ten stages of splitting (2^{10}) needed to serve 1000 subscribers would have 30 dB of optical loss. In contrast, a WDM splitter separates (preferentially directs) the energy of the different wavelengths into different legs of the WDM splitter with very little loss for each wavelength.

---

[5] Depending on the carrier, in North America this distance was typically limited to either 12,000 ft or 18,000 ft.

2.4   FITL Technology Considerations

2.4.1   Optical Components

FITL systems have always faced challenges related to the cost of the optical components that provided the desired capabilities.[6] Single mode fiber has been chosen for its higher bandwidth capabilities. The current generation of FTTx systems uses a combination of lasers in the 1260–1360 nm range for upstream and in the 1490–1590 nm range for downstream.

Fused optical power splitters may be produced relatively inexpensively. They are constructed by fusing fibers together so that their core areas are close to each other for some distance, with the fused area becoming an optical signal mixing region. As an optical signal passes through the mixing region, a portion of its light couples into the other fibers, going in the same direction. The fraction of the light that couples into each of the other fibers is determined by the construction of the splitter.

A common implementation for a 2×2 splitter is to have half the optical power coupled into the other fiber so that same optical power appears on each of the two output ports. $N \times N$ splitters may also be constructed, with the power on each output fiber now being $1/N$ of the input power $(= 10\log(N)$. In the case of a tree structured PON network, one of the ports on the OLT side of the splitter is not used (recall that the splitter is still a 2×2 device, so one still loses half the power to the 'unused' port). The result is that while the device is constructed as a 2×2, it is normally used as a 1×2 splitter, in which half of the downstream signal is transmitted over each output branch toward the ONUs. In the upstream direction through the same splitter, half of the upstream optical signal is coupled into the fiber towards the OLT. Consequently, the optical loss through the 1×2 splitter is at least 3 dB in each direction. In practice, power splitters introduce a fraction of a dB additional loss. Note that effectively the same type of splitters can be constructed with planar-integrated optics using close-proximity wave guides instead of fused fibers for the mixing region. The loss vs. number of ports behavior remains unchanged, however.

Wavelength filters are a primary component for WDM PON, but they are also used with TDMA PON systems to separate different signals (e.g., the downstream data signal and broadcast video signal). A wavelength filter routes input signals to different output ports based on their wavelength. Since there is no power splitting for any given wavelength, the filter introduces a smaller amount of insertion loss. The wavelength routing can be accomplished either by refraction (e.g., with a prism) or by using interference between multiple beams of light. An example of the latter is the Arrayed Waveguide Grating discussed in Chapter 5.

With TDMA-based PON systems, the various issues surrounding switching the upstream transmissions between the different ONUs are a challenge. One set of such issues involves the guard time that is required between transmission bursts of different ONUs. The factors that determine the guard time are often independent of the upstream data, so these factors have an increasing impact as the upstream data rate increases, since the burst transmission rate must increase more rapidly in order to achieve the desired

---

[6] First-generation FITL systems identified a number of technology challenges. Although multi-mode fiber was less expensive for the fiber drop to the subscriber, single-mode fiber was preferred, due to its superior bandwidth capabilities. Single-mode fiber, however, requires laser transmitters, since the core diameter is too small to couple adequate optical energy from incoherent light sources such as LED transmitters. There was initially some thought of using the inexpensive lasers used in CD players and CD-ROM drives, but the problem with these lasers is that their wavelengths (750 nm and 810 nm are typical) still propagate as multi-mode in glass single-mode fibers. The least expensive lasers that allowed single-mode transmission used 1310 nm, and these lasers cost several hundred dollars in the 1990 time frame. Fiber transceivers are still not cost-effective for direct point-to-point connections to each subscriber as of this writing. The development of fused fiber splitters dramatically reduced their cost, making PON more attractive, but any type of passive splitter divides the optical energy between the branches of the splitter. As the number of splitters between the OLT and the ONT increases, the power decreases quickly. For this reason, split ratios of ONTs per OLT transceiver are limited to between 16 to 1 and 64 to 1. Higher split ratios would require optical amplifiers. Since the cost of the optical amplifier can be shared among multiple subscribers, this can still be cost-effective. The other drawback of higher split ratios is that the bandwidth is shared among more subscribers, limiting the bandwidth that any one user can send. For this reason, split ratios higher than 64 to 1 will probably be used primarily on higher rate PON systems (e.g., future 10 Gbit/s PON systems).

overall data rate. One such factor had been the speed at which a light source can be turned on and turned off to the point where it has negligible output power, so that the collection of ONUs on the PON do not add too much interference to the signal from the ONU authorized to transmit. This is typically no longer a significant issue, since the current manufacturing techniques have reduced this time to a few nanoseconds.

Another factor in the guard times is the time required for the OLT to adjust its receiver to the difference in received power levels between ONUs that are at significantly different differences from the OLT. This adaptation to the different bursts includes the time required for OLT to achieve clock and data recovery synchronization. The OLT upstream receivers become even more challenging when different ONUs on the PON have different upstream data rates. For example, as discussed in Chapter 3, 10G EPON allows a mix of ONUs with some transmitting upstream at 1 Gbit/s and others at 10 Gbit/s. This requires the OLT burst mode receiver not only to adjust to the received power level, but also to adjust its received signal equalization and clock recovery circuits for the data rate of each burst.

The technology choices for dealing with mixed rate burst mode receivers can impact the effective optical link budget. For example, splitting the different rate signals in the optical domain introduces the 3 dB penalty of the optical splitter, while some circuits for handling the different signal rates in the electrical domain result in reduced receiver sensitivity.

As will be discussed further in Chapter 5, WDMA introduces additional challenges. From an operations perspective, it is not practical to build PONs with each ONU having a different fixed wavelength laser. Carriers would need to keep the optical modules for each ONU wavelength in their inventory and track all the different ONUs on each PON to ensure that none used the same wavelength. Building a cost-effective "colorless" ONU while maintaining adequate PON reach is a significant challenge.

Optical amplifiers can be used to overcome the optical loss associated with greater reach, higher split ratios, or other impairments. Reach extension techniques, including optical amplifiers, are discussed in Section 2.4.4.

2.4.2   Powering the Loop

The single biggest obstacle to FITL systems in many countries, however, has been not the cost of the optical components, but the challenge of providing reliable power to the subscriber telephone. As noted in the FTTC description, telephone companies have traditionally supplied power to the subscribers' telephones through a –48 Vdc power feed. The network-provided power virtually guarantees that telephone service remains available during the loss of utility service power. This high service availability is often referred to as 'lifeline POTS' since subscribers can count on the service being available for emergencies. A telephone company typically provides batteries to back up their equipment for about eight hours of typical usage in the event of a power outage from the power utility company.[7] The CO equipment uses batteries in the CO, and DLC RTs – which are powered by a local connection to the utility company – have batteries co-located at the RT. In the event that a major disaster (e.g., a flood or a hurricane) disrupts the power utility for more than eight hours, generators are used at the CO to provide the power needed for charging their batteries. The RT batteries can be charged by portable generators on trucks that are driven among the RT sites.

FTTH and FTTC complicate the power situation for several reasons. First, there are many more locations with active components (e.g., potentially 100–2000 more ONTs than DLC RTs to reach that many subscribers). Clearly, it is not possible to send trucks with portable generators to each ONT during prolonged power outages. More significantly, batteries have life spans of 5–10 years and thus must be replaced regularly, and replacing batteries is very costly and manpower-intensive. If the ONT is located within the subscriber premises, the telephone company must obtain permission to enter the home to install the replacement.

---

[7] While this lifeline POTS capability is mandated in the US by the FCC, other countries (e.g., Japan) do not require it.

A second factor is the number of connections to the power utility company. DLC RTs each have their own metered power utility connection. FTTC can also use per-ONT metered power utility connections, but the cost per subscriber increases. One alternative for FTTC is to use a separate 'power pedestal' that powers multiple ONTs and contains their back-up batteries. For FTTH, it is cost-prohibitive to provide per-ONT metered power. Power pedestals are one option, although more pedestals are required. The other option is to have the subscribers provide their own power, with the FTTH service cost effectively discounted to cover the subscribers' expense. More details of the power requirements and restrictions are provided in Appendix 2.A.

Supplying and backing up the power has been attacked on a number of fronts. Better batteries (e.g., longer life span, higher efficiency over the temperature range, higher capacity per unit size and cost) would mitigate many of the problems. Battery research (also motivated by the automotive industry) has brought some improvements, but has not solved the problems for FTTH/C. Solar power has been explored for the FTTH/C and power pedestals, but it is geographically limited to regions that have adequate reliable sunshine. More power-efficient electronic components have been developed, but the power consumption of the electronics was always a relatively small percentage of the overall power.

So, while power might seem like a mundane topic in the glamorous world of lasers, fiber, and high-speed data access, it is a substantial problem with no easy solution. A more recent phenomenon may have turned the tables, however. With the rapidly increasing use of cellular/mobile telephones, subscribers have become accustomed to providing their own power and maintaining their own batteries for their phone service. This situation may lead customers who desire broadband services over FTTH to be very willing to take responsibility for the power and battery back-up of their ONTs. Also, it is highly likely that an FTTH subscriber is already a mobile phone subscriber and can use that phone for lifeline service instead of the landline phone connection. This phenomenon, combined with the growing desire for broadband services, has opened the door for residential PON deployment.

2.4.3 System Power Savings

Another important extension to existing and future PON protocols is power savings. In one study, BT estimated that without some type of power savings capability, the additional power required for moving from current DSL to an ONU at each home would be equal to the output of a typical British power utility plant. An important driver for ONU power savings during normal operation is the European Union's Code of Conduct on Energy Consumption of Broadband Equipment (BBCoC). While BBCoC compliance is voluntary, the European carriers plan to work toward it. Another critical driver for ONU power consumption is the North American requirement, discussed above, for the ONU to continue to provide emergency POTS service for up to eight hours after the loss of utility power.

While power savings during normal operation and during battery backup operation are related, the criteria for battery operation are somewhat different. Battery backup operation implies that the subscriber has lost utility power and hence will typically be able only to use telephone service and not require video or data services.[8] This situation allows the ONU to activate only those circuits associated with supporting an active telephone connection. The more the power requirements during battery operation can be reduced, the smaller the battery the ONU requires.

While improvements in technology can reduce the power consumption of the ONU components, the expected increase in bandwidth and services will tend to increase the overall power requirements for the ONU. Achieving more substantial power savings will require shutting down those portions of the ONU that are not currently active. The ITU-T has been studying the different approaches to ONU power savings and has created a supplement document on the subject [4]. It groups the ONU power savings techniques into the three categories of ONU power shedding, dozing, and sleeping.

---

[8] With the increased use of laptop computers, subscribers may begin to request premium battery-backed services that allow a period of continued data service during power outages. In these instances, the router or gateway device would also require battery backup.

ONU power shedding: The most universal approach to ONU power savings is to use system-on-chip design techniques that idle portions of the chip such as a microcontroller when they are not in use. This type of ONU power savings is referred to as "power shedding." The subscriber interfaces and their associated circuits can also be shut down as long as there is a sure way to determine whether the subscriber is actively using them. For the telephone service, this is straightforward if the ONT provides the interface to the subscriber's telephones. For data services, however, including VoIP, it can be difficult to determine whether the interface is truly inactive or only experiencing a pause in the subscriber's use of a service.

ONU dozing: The ONU upstream interface circuits consume a significant portion of the ONU's power. Consequently, additional power can be saved if the ONU ceases its upstream transmissions when the ONU is not proving active services to the subscriber. This type of power savings approach requires coordination with the OLT so that it will not decide that the ONU has been removed from the PON. "ONU dozing" refers to the technique of shutting down the ONU's upstream interface during inactive periods, but always keeping the downstream PON interface awake. The characteristics of ONU dozing can be summarized as:

- The ONU continues to listen to the downstream PON signal.
- The OLT continues to send small upstream bandwidth allocations to the dozing ONU so that it can quickly signal a request for upstream bandwidth when the subscriber initiates a service. The ONU ignores these bandwidth grants if it has no information to send.
- The ONU wakes up and sends upstream information when either a subscriber service requiring upstream bandwidth becomes active or the OLT requests a response from the ONU. The requested responses are typically OAM&P-related reports.

An advantage to dozing is that it allows an ONU to respond almost immediately when a subscriber service becomes active. The ONU uses ongoing OLT upstream bandwidth grants to request appropriate upstream bandwidth quickly for the newly activated service. Another advantage is that it can work concurrently with services such as video that are only active in the downstream direction. A disadvantage to the dozing is that some downstream and upstream bandwidth is wasted due to the ongoing unused allocations to the dozing ONUs.

ONU sleeping: The power savings methods that save the most power are those that put all non-essential ONU functions into a sleep or standby state for periods of time when it is not carrying active traffic. The essential functions include timers and monitoring for the activation of subscriber services. Sleeping requires additional coordination between the ONU and OLT. For example, if another subscriber initiates a service connection to that ONU (e.g., dials that subscriber's telephone), then the OLT must buffer the downstream information until the ONU wakes up. Also, the OLT must schedule its OAM&P message exchanges with the ONUs so that they are awake to receive the messages. The characteristics of ONU sleeping can be characterized as:

- The OLT sends a periodic broadcast message to initiate the sleep cycle for all inactive ONUs on that PON.
- ONUs wake up after a pre-determined time when a local ONU timer expires. All ONUs have the same sleep time period, which results in all ONUs on the PON using the sleeping function go to sleep and to wake up at effectively the same time.
- The OLT buffers any new downstream service or OAM&P information for a sleeping ONU until it enters the awake portion of the sleep cycle.
- When the ONU detects a newly active service, it wakes up to buffer the information and sends the appropriate upstream information after its wake-up timer expires.
- When it is time for the sleeping ONUs to wake up, the OLT sends them upstream bandwidth grants so that they can make any necessary bandwidth requests for services that have become active since the ONU last entered its sleep state.

- When the ONUs are awake, the OLT sends them any information that it had buffered during their sleep period.

Synchronizing the sleep times of all the ONUs allows the OLT to maintain a single sleep cycle state table rather than per-ONU state tables.

The main advantage to ONU sleeping is that it provides the most power savings. Also, there is no downstream and upstream bandwidth lost due to the type of periodic bandwidth grants that the OLT must make to a dozing ONU. The main disadvantage to ONU sleeping is that there can be some lag in activating a new service with a sleeping ONU. It also requires some additional buffering at the OLT and ONU and additional OLT state tracking.

ONU sleeping can be further divided into "deep sleep" and "fast sleep" modes. The fast sleep mode operates in the manner described above. The OLT creates periodic sleep cycles, which are periods of activity followed by sleep periods. The periodic initiation of a sleeping period allows inactive ONUs to enter a sleep state soon after determining that they are inactive. The OLT can continue to send upstream allocations to sleeping ONUs in order either to recover from a potential mismatch between the ONU and OLT sleep state views or, potentially, to allow an ONU to awake early due to it detecting new subscriber activity.

A deep sleep period can be initiated by the ONU, for example in response to the customer powering down the ONU. It is critical that the ONU inform the OLT so that it does not generate alarms due to the subsequent loss of upstream information from that ONU. The original deep sleep specification had the ONU inform the OLT of its entry into deep sleep by sending a "dying gasp" message. It is also possible to use a handshake to communicate entry into the deep sleep mode. The OLT will periodically send upstream allocations to a deep-sleeping ONU, so that it will be able to communicate its exit from deep sleep.

2.4.4   PON Reach Extension

There are two general approaches to PON reach extension. One approach is to remain in the optical domain, using optical amplifiers, while the other is to use some type of repeater that operates in the electrical domain. There are tradeoffs to each method. Both methods are described in ITU-T Recommendation G.984.6 [3].

Since both the amplifier and repeater are active electronic elements that are subject to failure and have a view of the network status that is not visible to the OLT, it is important for the OLT to have a management communications link with them. G.984.6 refers to this function as an EONT (Embedded Optical Network Termination for management of the extender). The EONT function essentially behaves as another ONT on the PON except that its only bandwidth requirements are for the management communications.

2.4.4.1   Optical Domain Reach Extension

The typical target distance for PON reach extension has been 60 km, which corresponds to the logical delay limit supported by the GPON protocol for a purely passive optical distribution network (ODN) (see Chapter 4). Increasing the fiber from the typical 20 km limit of the current generation of PON protocols to 60 km adds $\approx$14.5 dB additional loss at 1310 nm and $\approx$12 dB at 1550 nm.

The primary advantage of optical domain techniques such as optical amplifiers is that they are virtually agnostic to the bit rates and formats of the signals they carry. This flexibility allows the potential to upgrade PON systems to carry future protocols with no changes to the ODN. However, there are also challenges in using optical amplifiers in PON systems. One general challenge is that optical amplifiers inherently generate noise in the form of amplified spontaneous emission (ASE). If the signal-to-noise-ratio (SNR) of the optical signal becomes too low prior to amplification, the ASE can significantly reduce the signal's SNR. It has been demonstrated that 10 Gbit/s transmission over 100 km with 1024-way splits is achievable using a cascade of optical amplifiers along the fiber [5].

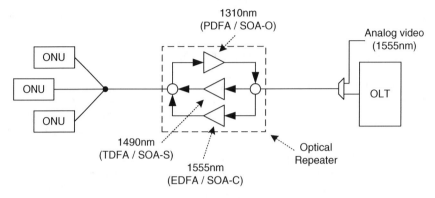

a) Discrete amplifiers per wavelength band

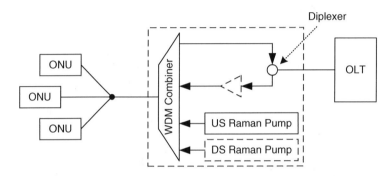

b) Raman amplifier for PON upstream and downstream

Figure 2.2 Illustration of reach extension with optical amplifiers

Since the TDMA PON systems typically use different wavelength bands for downstream, upstream, and broadcast video signals, as illustrated in Figure 2.2a, multiple different types of amplifiers would be required for each PON. The different wavelength bands and corresponding optical amplifier technologies are illustrated in Figure 2.3. The EPON and G-PON protocols would require separate amplifier types for the upstream signal (1310 nm, O-band) and downstream signal (1490 nm, S-band), and an additional amplifier type if broadcast video signals are being transmitted over the same PON (1555 nm, C-band). The EDFA (erbium doped fiber amplifier) amplifiers commonly used in telecommunications applications only cover the 1555 nm region. Special amplifiers are required for the O and S-bands.

The PDFA (praseodymium doped fiber amplifier) and TDFA (thulium doped fiber amplifier) technologies are similar to EDFA. One advantage to EDFA-type amplifiers is that they produce low noise and high gain. All doped fiber amplifiers operate by coupling an optical pump signal into a doped section of fiber, through which the data signal also passes. The pump signal wavelength excites the doped fiber, raising its energy level. Light of the data signal's wavelength causes the doped fiber to release its energy, which results in amplification of the data signal. The reliability of TDFA and PDFA has not been proven.

Another challenge for fiber doped amplifiers in the upstream direction is the "bursty" nature of the upstream signal, with gaps between the bursts. Also, the burst power level depends on the distance to the

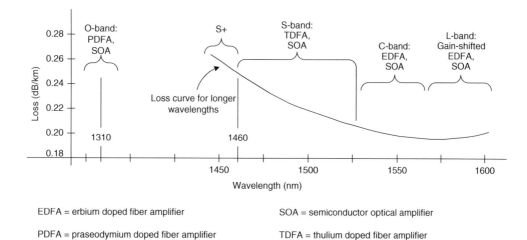

EDFA = erbium doped fiber amplifier   SOA = semiconductor optical amplifier

PDFA = praseodymium doped fiber amplifier   TDFA = thulium doped fiber amplifier

Figure 2.3 Illustration of wavelength bands and the associated optical amplifier technology

ONU and the wavelength varies slightly between ONUs. Input signal power variations occurring in timeframes of 10^{-4} to 10^{-3} seconds can cause the EDFA gain to react such that the burst can be distorted. Also, if a burst arrives at the EDFA after a relatively long period with no upstream activity, its gain would be too low and could take up to 10^{-3} seconds to reach the correct gain. Techniques exist to compensate for this, but they add cost and complexity.

An alternative optical amplifier technology is the semi-conductor optical amplifier (SOA), which uses a semiconductor for the gain region rather than a fiber section, and is electrically pumped rather than optically pumped. These are constructed similarly to Fabry-Perot lasers, except that they have non-reflective cavity endfaces that prevent them from operating as a laser.

The advantages to SOA amplifiers include being much smaller, using less power, and being able to be manufactured relatively easily for different wavelength ranges. SOAs are capable of operating in the 0.85–1.6 nm range and they also have faster gain dynamics than doped fiber amplifiers, making it easier for them to handle bursty upstream traffic.

The disadvantages of SOA are that they are typically 2–3 dB noisier than doped fiber amplifiers and have more limited gain (currently <13 dBm). The gain reacts rapidly to changes in the data signal power or the electrical pump power, and the gain changes cause phase changes that can distort the signal. This nonlinearity is the greatest performance challenge in this application. Like TDFA and PDFA, SOA have unproven field reliability. However, published data from vendors suggest that SOAs should have appropriate reliability. It is reasonable to expect that if there is sufficient demand, optical amplifiers at the typical PON wavelength bands can reach a deployable state. Another drawback to SOAs, though, remains their high cost.

An alternative optical amplifier arrangement makes use of Raman amplifiers [3]. This arrangement is illustrated in Figure 2.2b. The Raman pump output at the appropriate wavelength is coupled into the ODN to give reverse-pumped Raman gain for the upstream signal. For the downstream direction, either a separate optical amplifier can be used, or a Raman pump at a different wavelength can be used to give forward-pumped distributed Raman gain for the downstream signal. The downstream signals and Raman pump wavelengths are combined with a WDM combiner, which also separates the upstream signals. ASE noise can also be removed by the WDM combiner when it is used as an optical bandpass filter. The generic parameters for Raman amplifiers can be found in [6]. Some typical specific parameters for PON systems use a Raman pump laser wavelength of 1240(±0.5)nm, an upstream passband of 1300–1320 nm, and a downstream passband of 1480–1500 nm (Amendment 2 of G.984.6).

Amplifiers need to be located on the shared portion of the fiber, since placing amplifiers at the ONUs would be cost prohibitive. In order to keep the PON truly passive between the CO and subscriber, the amplifiers would need to be co-located with the OLT in the CO. However, having the amplifier at the OLT results in more noise power at the receiver than if it is located at the splitter. In the upstream direction, locating the amplifier at the OLT means it can only function essentially as a pre-amplifier, where it would not be as effective as if it were a transmit signal amplifier. The downstream amplification is limited by the physics of the fiber. When the optical signal power becomes too high, non-linear transmission effects occur in the fiber. There are also safety concerns about craftspeople working with systems with that intensity of light, since the Class M1 safety limit is +21.3 dBm.

Another potential reach extension technology is using WDM filters instead of power splitters, in order to reduce the optical loss due to the splitting. For example, a 1 : 32 split with power splitting introduces 18.4 dB of loss vs. 3.5 dB loss through a 1 : 32 wavelength multiplexer. This 15 dB difference corresponds to an additional ≈40 km reach. While this technology may be attractive for future new ODN deployment, it has two drawbacks: the first is that it is not compatible with current ODNs that are built with power splitters; while the second is that WDM PON requires colorless ONUs, which is currently a technical challenge. See Chapters 4 and 5 for additional discussion of WDM.

2.4.4.2 OEO (Optical-to-Electrical-to-Optical) PON Repeater Reach Extension

Converting the optical signals back into the electrical domain for regeneration avoids many of the cost and technology issues of all-optical reach extension. The primary drawback, however, is the lack of upgrade flexibility since the repeater must be implemented to operate at the bit rates of the PON. OEO repeaters also do not help with the analog video overlay signals, which would better suited for optical-domain amplification. Basic OEO repeater-based reach extension is illustrated in Figure 2.4. Note that the OLT provides the reference clock for the ONUs and the repeater.

A further complication of OEO repeater reach extension is handling the burst-mode nature of the upstream signal in TDMA protocols. For the basic repeater of Figure 2.4, the added complexity is primarily the need for a burst mode receiver similar to that at the OLT. It needs to be able to recover the clock and data for each ONU upstream burst, including adapting to the different signal levels that can exist for each due to their different distances from the repeater. A mechanism is required to report these signal levels to the OLT, so that the OLT can set the ONU transmit signal levels such that they are roughly the same when they arrive at the repeater. This received signal level reporting could be handled through the EONT function.

Since some of the upstream burst preamble may be effectively lost as the repeater attempts to achieve clock and data recovery for the burst, it may be necessary to use a longer preamble with repeater-type reach extension so that an adequate preamble remains when the signal arrives at the OLT.

The upstream repeater may also insert some bit pattern between the bursts rather than transmitting nothing (i.e., a steady string of 0 s). This pattern is typically under the control of the OLT, and it is chosen

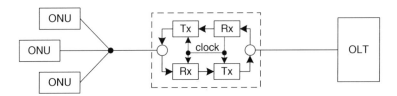

Figure 2.4 Basic OEO repeater PON reach extension illustration

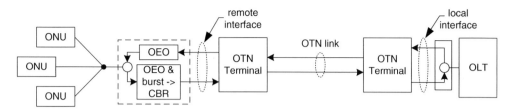

Figure 2.5 Illustration of an OTN link used for reach extension

to contain a balance of 0 s and 1 s and not resemble the preamble that begins a burst. The OLT can determine the arrival of a burst by looking for the preamble. Also, since the OLT schedules the upstream transmission explicitly, it knows when the burst is supposed to arrive.

Another variation on the repeater-based reach extension is to implement the repeater in a distributed manner by using a transmission link. The primary example of this is the use of an OTN transport link, as specified in G.984.6 for GPON and illustrated in Figure 2.5. Although G.984.6 only covers G-PON, the same approach could be adapted for Ethernet PON systems. The important differences between a basic repeater and a transmission link such as OTN include:

- The upstream signal must be converted to a continuous bit rate (CBR) signal by the remote burst mode receiver in order for it to be mapped into the transport channel protocol.
- Since transport channels are typically defined to have symmetrical bandwidth, the upstream burst to CBR conversion process may need to encode the signal to adapt it to the same effective bandwidth as the downstream signal.
- The transport transmission channel is typically a two-fiber interface rather than a single fiber.
- The transport signal can potentially be carried over portions of a carrier's metro transport network, sharing it with other signals in a consistent manner.
- The transmission link has its own OAM&P. In the case of OTN, this includes the capability to manage different wavelengths.

The coding conversion of the upstream CBR signal to match the downstream signal need not be strictly necessary if the burst-to-CBR converter is integrated into the transport node and if the upstream rate is an integer divisor of the downstream rate. Under these conditions, the upstream signal can effectively be over-sampled to create the upstream signal within the transport system. This condition would hold for G-PON, where the upstream rate of 1.24 416 Gbit/s is exactly half the 2.48 832 Gbit/s downstream rate. However, when a separate burst-to-CBR converter is used, the upstream interface labeled as "remote interface" in Figure 2.5 must have sufficient transition density for the OTN terminal input to lock onto the signal. Consequently, the Manchester code was chosen for the upstream CBR signal to provide a reliable input signal to the OTN terminal and to conveniently exactly double the bit rate of the upstream signal to match the downstream signal.[9]

The use of OTN could be particularly advantageous for carriers who plan to use PON to reduce the number of COs.[10] The OLTs can be pulled further into the network to a smaller number of COs, with their

---

[9] The Manchester line code, used with several other interfaces, including some 10BASE and 100BASE Ethernet, uses 01 to encode a data 0 and 10 to encode a data 1.

[10] BT, in their CN21™ plan, was the first carrier to announce a plan to use PON for a combination of higher bandwidth per subscriber and reducing the number of manned COs in maintains. Other carriers have also expressed interest in this approach.

metro OTN networks carrying the PON signals to a repeater location closer to the subscribers. In addition to allowing use of their metro networks, this approach has two distinct advantages. The first is that OTN provides a TDM mechanism for combining multiple PON signals onto a single wavelength. For example, four G-PON signals can be multiplexed onto a single 10 Gbit/s OTN link (ODU2). The second is that OTN is well suited for the future increasing use of WDM in PON networks since it was designed for efficient management of WDM networks.

2.4.4.3   Reach Extension through a Remote OLT

Although it is not reach extension is the strict sense of extending the ODN, the same effect can be achieved by moving (or keeping) the OLT closer to the subscriber. Using a remote OLT allows the ODN to be kept short enough to support the desired number of ONUs. While this may, at first, seem to be contrary to the desire for a passive outside plant, all of the reach extension technologies discussed above rely on some type of active electronics. The remote OLT is not substantially different than using a repeater in this sense, although the OLT circuitry is somewhat more complex. For example, the OLT could replace the burst-to-CBR converter in Figure 2.5, with the signal carried over the OTN being the OLT's network uplink.

Another potential variation on the remote OLT approach is to use cascaded PONs. For example, a 10 Gbit/s PON with its OLT in a CO could serve multiple remote OLTs that are attached to its 10 Gbit/s OLUs. These remote OLTs then connect to the subscriber ONUs, using 1 or 2.5 Gbit/s PONs.

2.5   Introduction to PON Network Protection

Telephone network providers and cable television providers both employ degrees of redundancy in their networks in order to protect their services against failures. Network redundancy strategies are chosen as a function of the cost of providing the redundancy relative to the amount of traffic, the number of subscribers, and the types of services affected by the potential failures. PON technology enables delivering an increasing number of services over the access portion of the network, with increasing bandwidth. PON is also increasingly used for connections to enterprise customers and wireless base stations. Consequently, the deployment of PONs is expected to increase the need and desire for resiliency in the access network. This section examines a reasonably comprehensive range of redundancy and protection options for different portions of the PON network. Some of the options are not expected to see substantial deployment, but they are presented here for completeness.

There are several different parts of the PON system to consider for protection. These parts and the extent of their impact are summarized in Table 2.1.

Table 2.1 PON system components and the impact of a fault on the component

PON System component	Sub-unit	Extent of the fault's impact

PON facility	Fiber and passive components
	Active components (e.g.,
	  extenders such as amplifiers, if used)
		The feeder fiber between the OLT and
		  splitters affects all subscribers on that
		  PON. The drop fiber between the final
		  splitter and an ONU affects just that ONU.

OLT	Network connection (uplink)	All subscribers served by the OLT

	Optical Line Units (OLUs) that
	  connect to the PON
		All subscribers on the PON(s) connected to
		  that OLU

	Common units/functions	Potentially all subscribers served by the OLT
	Entire OLT/CO	All subscribers served by the OLT
ONU	Optical interface	The subscriber(s) connected to the ONU
	Other ONU functions	Potentially the subscriber(s) connected to the ONU

After a brief background on protection, the remainder of this section is organized to discuss each of these aspects of PON protection outlined in Table 2.1. There are multiple options for protecting the different components. This section begins with the options that are the most complex and provide the highest reliability. The discussions move progressively to the least complex options that provide the lowest network reliability.

2.5.1 Background on Network Protection

The level of reliability in the telephone network has traditionally been a function of the number of subscribers affected by a single fault, the level of service to which the customer subscribes, and the expected failure rates of the network and equipment. For example, for economic reasons, no protection was typically provided against a fault that affected 48 or fewer residential telephone customers. Since enterprise customers typically pay a premium price for high reliability service, equipment and, in some cases, route redundancy is provided on their connections to protect against single faults.

Within the telephone network, constraints are placed on the time to detect a network fault and to restore the service through the redundant equipment and/or facilities. The maximum fault detection time is typically 10 ms and the subsequent maximum restoration time is typically 50 ms. These times, which are driven by legacy network equipment, insure that no telephone connections will be dropped due to the fault. The telephone network providers expect packet-based networks to offer similar high-speed protection in order to maintain traditional service quality when voice is carried as VoIP.

The ITU-T Study Group 15 (SG15) defined facility and optical interface protection architectures in Rec. G.983.5 [7]. The ITU-T documented further PON protection considerations in the informative supplement G.Sup51 [8]. The discussions in this section will refer to the G.983.5 and G.Sup51 architectures whenever appropriate. Note that G.Sup51 covers failure rate and availability calculations for the protection architectures that it describes. The IEEE SIEPON standard described in Chapter 3 specifies which of these protection modes should be used with EPON. Additional information on PON protection can be found in [9,10].

2.5.2 PON Facility Protection

The PON facility includes the fiber and any passive optical components such as splitters, filters and connectors. In the event that active optical amplifiers or repeaters are required to achieve the desired fiber split ratios and/or distance, the amplifiers are also considered as part of the PON facility. A fault on the feeder portion of the PON facility (between the OLT and the first splitter) affects all the subscribers connected to that PON, while faults on the final drop portion (between the last splitter and the subscriber) affect only that subscriber.

Note that most small and medium businesses (SMB) do not currently have redundancy on their physical access links to the carrier network. A typical SMB user network interface (UNI) is a single fiber or copper connection. Redundancy is typically used only for the portion of the access network shared by several subscribers, such as using rings to protect the interconnection between access network nodes and the CO. Furthermore, labor costs typically dominate the ONU installation. Fortunately, some of these labor costs could be shared for a redundant PON, and hence they may not be doubled to enable protection. This type of study would need to be performed by the carrier based on their own practices and cost structures for pulling fiber cables, terminating fibers, etc. The cost of protection installation must also take into account the potential additional revenue that could be derived by offering protection as a premium service.

2.5.2.1 Option 1 – Connect Each ONU to Two PONs, with Each of the Two PONs Connected to a Separate OLT

This option, illustrated in Figure 2.6, provides the highest reliability since it protects against failures of the entire optical path including the OLT, PON facility, and the ONU optical interface. The OLTs in this option may even be located in separate COs in order to protect against a failure affecting one of the COs.

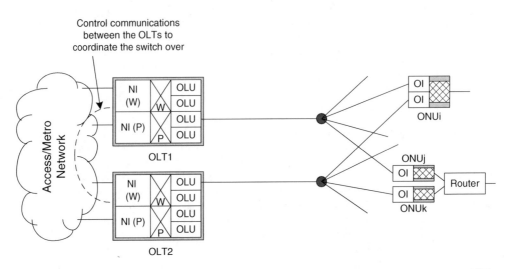

Figure 2.6 Illustration of protecting an ONU and the optical connection with two separate PONs and OLTs

This option can enable protection at either Layer 1 or Layer 2. Layer 1 protection would have only one PON active at a time (i.e., the working PON). Layer 2 protection would have both PONs active with the protection switching performed at Layer 2 outside the PON, and the potential need for communication between the two OLTs.

Two ONU protection examples for this option are illustrated in Figure 2.6. ONU protection is discussed in Section 2.5.4.

No OLU protection is required in this option, since it inherently locates the redundant OLU in a different OLT. This approach is more expensive than using OLU protection within the same OLT if it is used for all ONUs, but it can be a good option for an ONU carrying critical data. This option essentially enhances the Type C protection architecture of G.983.5 [7] and G.Sup51 [8] by adding dual OLT parenting. Fault recovery is complex, since it affects both the physical layer and the Layer 2 routing from the OLTs (or subscriber router) into the metro network.

2.5.2.2 Option 2 – Connect Each ONU to Two PONs that are Both Connected to the Same OLT

This option is similar to the first, except that both PONs are connected to the same OLT (see Figure 2.7). Assuming that the OLT contains protection for its common function and network interface units, and that the two PONs have diverse fiber routing, this option provides virtually the same reliability as the first option. The only fault it will not cover is the total failure of the OLT (e.g., due to a catastrophe that disables the CO where the OLT is located). Here, the ONU is registered only on one PON at a time. G.983.5 refers to this protection option as Type C protection architecture.

The recovery is much simpler with this option than with Option 1. Here, the protection only affects the PON side of the OLT and has no Layer 2 effect on the network side of the OLT.

2.5.2.3 Option 3 – A Single PON with Redundant OLT Equipment and Feeder Fiber

A compromise PON protection architecture is illustrated in Figure 2.8. This architecture protects the OLUs and the feeder portion of the PON between the OLUs and the distribution splitter, and is referred to in G.Sup51 as Type B protection architecture. The OLT equipment redundancy can be either through redundant OLUs in the same OLT, or through separate OLTs. G.Sup51 refers to the configuration with

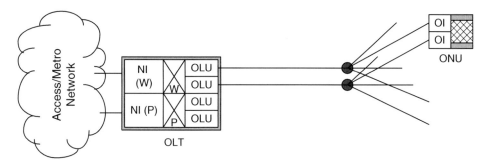

Figure 2.7 Illustration of protecting an ONU with separate PONs and a common OLT

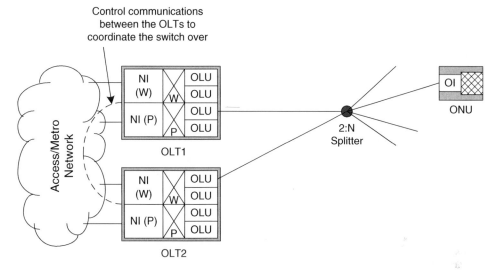

Figure 2.8 Illustration of protecting the OLTs and feeder portion of the PON (Type B protection architecture)

separate OLTs as "dual-parented" Type B protection. A primary advantage of this protection architecture is that it is less expensive than fully redundant PONs and yet, with the exception of the splitter, it protects all of the elements that are shared by multiple ONUs.[11] This architecture also exploits the fact that the optical loss through a 2 : N splitter is essentially the same as that through a 1 : N splitter.

Each ONU protected by dual-parenting is registered on both OLTs. This allows the protection OLT quickly to replicate the logical state of the working OLT.

The proponents of this architecture have typically preferred to dual-home the feeder legs of the PON to separate OLTs in separate locations, as illustrated in Figure 2.8.

The most difficult aspect of the dual-parented version of this protection option is the communication that is required between the working and protection OLTs. Since the OLTs only see the upstream signals, the protection OLT cannot determine directly the reason for a lack of upstream traffic. For example, there can be periods with no upstream traffic when the working OLT opens a discovery window for new ONUs

---

[11] If the carrier locates the splitters close to the subscribers, as is typical in many carrier networks, this architecture protects most of the PON fiber.

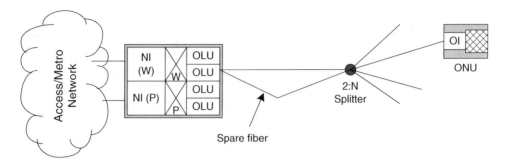

Figure 2.9 Illustration of protecting the feeder portion of the PON (Type A architecture)

to announce themselves. If the protection OLT begins to transmit while the working OLT is still transmitting, it would create interference and potentially overwhelm or cause harm to the ONU receivers. In the example of Figure 2.8, the fault status communication connection between the OLTs goes through the access/metro Ethernet network. Another alternative is for the Network Management System (NMS) to control the switchover. The NMS would use the alarm reports from the OLTs as the protection trigger and would implement the switch by commands to the OLTs. Switch times are a potential issue for any approach using separate OLTs, especially if they are controlled by the NMS.

Note that it is also possible to implement the Type B architecture by connecting the redundant feeder fibers of multiple PON ODNs to a centralized optical switch that can connect one of these ODNs to an OLT that provides shared protection for this group of ODNs. This is effectively the same configuration as the $1:N$ OLU protection illustrated in Figure 2.13, except that here it protects the entire OLT. While this $1:N$ configuration reduces the amount of OLT protection equipment, the optical switch offsets some of the cost savings and brings significant additional control complexity.

Note that it is still possible to support ONU redundancy with this protection architecture by connecting two of the PON's drop fibers to the same subscriber.

Note also that there are additional considerations when the Type B protection is applied to protocols that use dynamic wavelength or frequency assignment. Examples of such PON protocols include the NG-PON2 protocol described in Chapter 4 and OFDMA protocols described in Chapter 5. The ONU parameters are constrained by the need for them to remain constant during the switch.

2.5.2.4   Option 4 – Protecting Just the Feeder Portion of the PON

This architecture, illustrated in Figure 2.9, is similar to the previously described architecture, except that the only the feeder fiber is protected. The OLT equipment is not protected, so therefore no additional OLT port is required. This architecture is useful in applications where the feeder fiber is vulnerable to failures. For example, it could be used if the feeder fiber is deployed in an aerial configuration, where it is potentially vulnerable to falling branches or trees during storms.

With this configuration, referred to as a Type A protection architecture in G.Sup51, the protection can either be performed by an optical switch on the line card, or through a manual intervention.

2.5.2.5   Option 5 – Use a Single Unprotected PON

While this approach provides no PON redundancy, the PON is highly reliable and should not typically need protection.[12] A PON will typically only fail due to a fiber cut. The other redundant PON options are

---

[12] The fiber and passive splitters have very low failure rates. Using optical amplifiers to increase the reach or split ratio of the PON could potentially increase the failure rate enough to make redundant PONs more desirable.

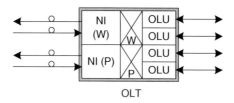

Figure 2.10   Illustration OLT network interface protection architecture

only more reliable if the two PONs are each routed in physically separate cables so that a cable cut cannot affect both PONs.

A variation on this option is to have redundant optical interfaces at the OLT that are connected to the single PON through a passive splitter. G.983.5 refers to this option as a variation on the Type B protection architecture discussed above. This option is covered in the section on OLT OLU protection.

2.5.3   OLT Function Protection

OLT protection includes the OLUs that connect to the PON, the network interface, and the common functions such as systems clocks, management and switch fabrics. Each of these has multiple protection options.

2.5.3.1   Network Interface (Uplink)

The OLT network interface (NI) connects it to the metro network or WAN. Each of the protection options uses a pair of NI units, each connected to a separate bidirectional facility. EPON systems use Ethernet links for NI. For G-PON systems, the NI can use either SONET/SDH or Ethernet for the physical layer. If SONET/SDH interfaces are used, any of the SONET/SDH protection mechanisms are available to protect the uplink. However, Ethernet interfaces are becoming more common for G-PON.[13]

The Ethernet interface is typically GE, multiple GEs, or 10GE. Since this interface carries all the OLT data, it is assume that it will be redundant, with redundant interface units, each connected to separate fiber pairs. See Figure 2.10.

For Ethernet interfaces, the following options are possible:

- 1 : 1 unit protection (bidirectional). Only one of the NI units is active at a time. If a failure is detected on this unit or the facility to which it is connected, all traffic will be transferred to the other unit. The failure is detected as the inability to communicate with the node at the other end of the link. The spanning tree protocol can be used to select the active interface.
- Ethernet Link Aggregation (LAG). With this option, both the working and protection NI units and their associated physical links carry traffic under control of Ethernet LAG. If one of the interface units or physical links fails, the network interface continues to function with half the bandwidth over the remaining link and NI unit.

    Note that Multiple Spanning tree (as a static method to control bandwidth between links) usually replaces LAG on fast links.
- IEEE 801.17 Resilient Packet Ring (RPR). RPR is protocol for a ring topology network that provides fair access to the ring's bandwidth. See [11] or [12] for a full tutorial on RPR. As the name implies,

---

[13] When SONET/SDH is used for Layer 1, the payload of the SONET/SDH signal is an Ethernet packet stream. Therefore, the Ethernet uplink protection methods discussed here are still applicable if they are used instead of the SONET/SDH layer protection mechanisms.

RPR was designed to use the inherent route diversity of the ring topology in order to provide protection for the traffic on the ring. Although RPR is independent of the Layer 2 protocol being carried, it is optimized for Ethernet transport.

One of the primary advantages of RPR is that its fairness mechanisms can be used to guarantee the QoS for the traffic on the ring. This is especially useful if the ring is used to backhaul the data from multiple OLTs. In this application, the bridging capability of RPR allows logical connections for direct data exchange between OLTs rather than performing the routing at a centralized location further into the network. The main disadvantage to RPR is that it has not seen wide deployment to date.

- ITU-T G.8031 [13] Ethernet Linear Protection or G.8032 [14] Ethernet Ring Protection. G.8031 specifies a mechanism for fast protection of point-to-point VLAN-based Ethernet network links that is modeled on traditional telecommunications network protection. G.8032 expands the protection mechanism to include ring topologies. One of the motivations for G.8031 and G.8032 is to protect VoIP connections fast enough to minimize interworking issues with traditional voice equipment in the network. The principle behind this mechanism is that all nodes periodically transmit a continuity check messages (CCM) to their neighbor node(s) and use a pre-determined protection path to route around a failure. The CCM allows a fast detection of a failure, and the pre-determined protection path allows immediate re-routing of the data without the need for running a spanning tree protocol. G.8031 supports both $1+1$ and $1:1$ protection architectures with both unidirectional and bidirectional switching. Both revertive and non-revertive switching are also supported. The $1:1$ architecture allows the protection path to be used for preemptible traffic when no failures exist. G.8031 also supports traditional manual protection switching operation.
- IEEE 802.3ag Connectivity Fault Management. Similar to the CCM of G.8031, 802.3ag specifies periodic messages to quickly detect connectivity faults in the network.

2.5.3.2 OLT Common Unit/Function Protection

Since a switch fabric failure would affect all traffic in the OLT, it should use $1:1$ unit protection.

Depending on the implementation of the control plane interaction with the fabric, control plane processing redundancy should also be considered.

2.5.3.3 Optical Line Unit (OLU) Protection

The OLU module typically contains the 1490/1550 nm laser and drivers, the optical receiver with its support circuitry, clock and data recovery circuits, and the PON MAC functions. There are multiple mechanisms available for protecting the OLUs. Since an OLU carries traffic for a limited number of subscribers, OLU protection is much more cost-sensitive than NI protection. This section discusses some of the potential OLU protection schemes. The order in which they are presented ranges from the most robust and simplest to schemes that add complexity in order to reduce the overall system cost. Note that OLU protection also provides a mechanism for upgrading an OLU without taking down all its PONs during the upgrade. The OLU protection options are:

- $1:1$ OLU Protection. With $1:1$ OLU protection, a protection OLU is dedicated to each working OLU. Only one of the two redundant OLUs (the working OLU) is transmitting data. If it fails, the other OLU (the protection OLU) transmits all the data. This option is referred to as Type B protection in [7]. Note that when an OLU contains multiple PON interfaces (each to a different set of ONUs), all PONs must be transferred to the redundant OLU in order to replace the OLU with the failure.

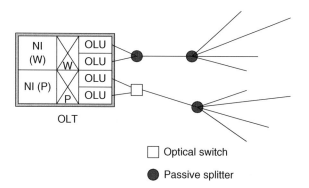

Figure 2.11 Illustration of 1 : 1 OLU protection implementations

The Type C protection illustrated in Figure 2.7, with separate working and protection PONs, assumes 1 : 1 OLU protection.[14] Figure 2.11 illustrates two different implementations of 1 : 1 OLU protection when both the working and protection OLUs are connected to the same PON. The tradeoffs between these implementations are discussed below.

- ○ 1 : 1 OLU protection using a passive splitter. One option, illustrated in the upper portion of Figure 2.11, is to use a passive splitter to connect the working and protection OLUs to the PON. The advantages of this approach are that the passive splitter is extremely reliable, and no optical switch control is required to transfer service between the working and protection OLU. The primary drawback is that the passive splitter increases the optical loss by around 3 dB. If optical splitters are used at both the OLT and ONU, FEC may be necessary; however, it is well within the range for the EPON and G-PON standard FEC. See Section 2.4.1 for a discussion of the implications of this additional optical loss.
- ○ 1 : 1 OLU protection using an optical switch. A second option, illustrated in the lower portion of Figure 2.11, is to use a 1 × 2 optical switch to connect the two OLUs to the PON. Since the optical switch is not dividing power between the branches, it only increases the optical loss by <1 dB. The main drawback to using an optical switch is that a control mechanism is required in order to activate the switch to connect the protection OLU to the PON.
- • 1:N OLU protection. In order to reduce the number of OLUs, it is possible to use a single redundant OLU to protect multiple working OLUs. The cost-effectiveness of this approach is determined by the cost of the optical switch matrix and its associated control and mechanical packaging relative to the cost of the additional OLUs and associated components for 1 : 1 OLU protection. The switch control is complex relative to 1 : 1 OLU protection. Depending on the implementation, the optical switch may need to be protected with a redundant switch unit. 1:N switches have been less common since there has never been a good application to drive them. Some potential implementations and their tradeoffs are discussed in this section.
- ○ Protecting N OLUs with an N × N optical switch. OLU protection using an N × N optical matrix is illustrated in Figure 2.12 for N = 8. The optical switch matrix functions as a bank of 1 × 8 switches that connect either the appropriate working OLU or the protection OLU to each PON. While optical switch fabrics are quite reliable, a redundant switch fabric may be required in order to achieve the desired system reliability. If a redundant switch fabric is used, additional 1 × 2 optical switches are

---

[14] With Type C PON protection, it would be possible for the protection OLU to also be transmitting data to the ONUs. For example, Ethernet Link Aggregation could be used merge the bandwidth of the two PONs. In event of a PON or OLU failure, only one of the OLUs and PONs would be available to carry data.

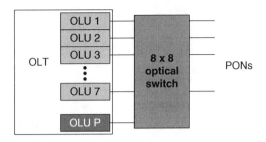

Figure 2.12   1 : N OLU protection with an N × N optical switch fabric

required on either side of the switch fabrics in order to connect the online fabric to the OLUs and the PONs.

○ Protecting N OLUs with a 1 × N optical switch. An alternative approach for 1 : N protection is to use a 1 × N optical switch that is connected to the PONs and working OLUs with passive optical splitters. This approach is illustrated in Figure 2.13. The advantage of using the passive coupling is that a switch fabric failure should not affect normal operation of the working OLUs. The disadvantage is the 3 dB additional splitter loss, as discussed in Section 2.4.1.

○ Protecting N OLUs with a M protection OLUs and an MxN optical switch. This option, illustrated in Figure 2.14, is a variation on the 1 : N protection that allows protecting multiple OLUs simultaneously. Consequently, its reliability and cost would be between that for 1 : 1 and 1 : N protection if N is less than M times the number of OLUs that could reliably be protected with 1 : N protection. However, the additional reliability inherent in an M : N protection architecture allows using a larger number of working units per protection unit than could be used with 1 : N protection.

• General Discussion about OLU Protection. Using passive splitters in order to connect the working and protection OLUs to the PON introduces a little over 3 dB additional optical link loss. This additional optical loss may necessitate FEC. Both the EPON and G-PON standards specify a shortened Reed-Solomon RS(255,239) FEC code that provides a coding gain of approximately 2.5 dB. This additional gain will be adequate for most links if the passive splitters are used only on the OLT side of the PON. If passive splitters are used for both the OLU and the optical interface at the ONU, then the RS(255,239) will not be adequate to compensate for additional optical loss.

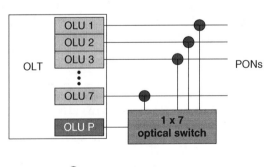

● Passive splitter

Figure 2.13   1 : N OLU protection with a 1 × N optical switch fabric and passive coupling

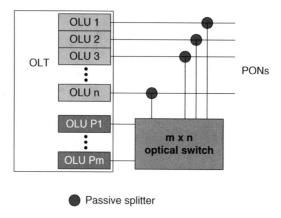

Figure 2.14 M : N OLU protection with a M × N optical switch fabric and passive coupling

There are multiple options for handling the additional loss if optical splitters are used at both the OLT and ONUs. These options are summarized in Table 2.2, with their respective pros and cons.

Using optical switches can reduce the loss to less than 1 dB. Although they are not as reliable as passive splitters, optical switches are very reliable, and complex switch components have been qualified for use in telecommunications networks for some time. Non-latching optical switches have the advantage that they default to a known position when no power is applied to them. The optical switch requirements are: (a) an integrated device so that it is simple, physically small, and cheap; (b) low power (latching has zero power draw in normal operation); and (c) reliable.

The additional optical components for protection (splitters, switches, and/or amplifiers) can be integrated into the OLT or located in a separate shelf. Using a separate shelf would allow the optical shelf to serve multiple OLTs and would also simplify the fiber management at the OLT. For applications using smaller OLTs, it could be more advantageous to integrate the additional optical components into the OLT.

- Unprotected OLUs: The final option is not to protect the OLUs. If the OLU failure rate is low enough, and the subscribers can tolerate the OLU repair time, this is clearly the least expensive option. Consequently, it is the most common option on currently deployed PONs serving residential customers. For business customers, however, it may not always be an acceptable option.

2.5.3.4   Protection of the Whole OLT or CO

Failures of an entire OLT require a redundant OLT. Full OLT protection would typically be implemented to protect against CO failures, in which case the two OLTs are located in different COs. The network topologies for redundant OLTs are illustrated in Figure 2.6 and Figure 2.8. The first illustrates protection of the entire PON, including the ONU optical interfaces. The second saves cost by protecting only the feeder portion of the PON, up to the 2 : N splitter.

For the second approach, as discussed above in Section 2.5.2.3, some type of control channel is required between the two OLTs so that only one is active at a time. In the case of CO failures, the backup OLT will need to be able to detect the failure and to become active autonomously. The CO or OLT failure could be detected by lack of communication (e.g., through periodic messages) from the working OLT or by monitoring the downstream optical signal. Monitoring the downstream signal at the backup OLT requires reflecting some of the downstream signal to it. It is difficult for the ONUs to

Table 2.2 Options for accommodating additional optical splitter loss for protection

Option	Pros	Cons
Restrict the PON in terms of distance and/ or split ratio	• Passive solution • No impact on current OLU and ONT equipment	• Can increase cost by requiring additional PONs and OLTs, and reducing the sharing of equipment and facilities among fewer subscribers
Employ a stronger FEC	• Allow full distance and split ratio capability	• Strong FEC is not standard • Stronger FEC typically requires additional overhead bandwidth, reducing the overall PON capacity
Use APD receivers	• Increases the optical budget • Works well with the RS FEC • This option has been chosen with some standard protocols	• Adds expense to all ONTs • Not backward compatible
Use optical amplifiers (note)	• Allow full and potentially increased distance and split ratio capability • No impact on current OLU and ONT equipment	• Expense of optical amplifiers and associated optical components. A circulator is required to separate the upstream and downstream signals for amplification. • The different upstream and downstream wavelengths necessitates different amplifier technology for each direction.

Note: one option is to place the optical amplifiers on the PON side of the splitter that connects the working and protection OLU to the PON. This option requires one set of amplifiers per PON, and is a good solution with 1 : 1 OLU protection. A more attractive option with 1 : N OLU protection is to place the optical amplifiers only at the input/output of the protection OLU. This option requires only one amplifier set for every N PONs. Since passive optical splitters can be manufactured with asymmetric loss when the power is split between the legs, the splitters here would provide a low loss (<1 dB) coupling to the working OLU, with a higher loss coupling to the protection OLU. The optical amplifiers compensate for the higher loss.

report the failure, since they are only allowed to transmit when the OLT grants them upstream bandwidth.

For either approach, the physical layer protection is implemented by the OLTs. Protecting the data coming from the OLTs into the network can be performed at Layer 1 through a dual-homed ring. Alternatively, the network connection protection can be performed at Layer 2, for example, through an Ethernet protection mechanism.

2.5.4 ONU Protection

There are four main options for protecting the ONUs. These options are presented here in order of the most expensive and complex to the least.

2.5.4.1 Completely Redundant ONUs, Each Connected to a Separate PON

This option, illustrated with ONUj and ONUk in Figure 2.6, is clearly the most robust, and it may be desirable for some enterprise or military customers. Each of these ONUs is connected to a separate PON and is active on that PON. A customer router/switch implements the protection at Layer 2. This option also adds cost to the CPE, since it must have a separate interface to each ONU and a means for switching between them during a failure.

2.5.4.2   Connect the ONU to Two PONs with Separate Optical Interfaces

This option, illustrated in Figure 2.6 (ONUi) and Figure 2.7, corresponds to the Type C protection in [7].

With 1 : 1 protection, the ONU is only active on one PON at a time. The ONU switches to using the other PON when the one it is currently using fails.

The Figure 2.7 architecture also supports 1 + 1 protection. With 1 + 1, the ONU is active on both PONs, with one of the PONs serving as the primary connection to the OLT. Similar to SONET/SDH, the OLT and ONU can use the bandwidth of the other (protection) PON for "Extra Traffic", i.e., data that uses the additional bandwidth available on the protection PON, as long as no failures exist on the primary PON. When a failure occurs, the ONU uses the protection PON, preempting any use of that PON for Extra Traffic. The Extra Traffic is typically regarded as lower priority, and the OLT may restrict the amount of Extra Traffic that the other ONUs can send when one or more ONUs have switched to the protection PON.

An approach similar to 1 + 1 protection is to use Ethernet Link Aggregation across the two PONs. As described above for OLT NI protection, Link Aggregation combines multiple Ethernet physical links to create a single higher bandwidth channel for the Ethernet frames. If one of the physical links fails, Link Aggregation automatically scales back its transmission rate to use just the remaining healthy link(s). The process of falling back to the healthy links is known as "fail-over". Since the bandwidth is reduced during fail-over, Link Aggregation provides service restoration for services that can tolerate the reduced bandwidth rather than providing full Layer 1 protection.

With either the 1 + 1 protection or Ethernet Link Aggregation methods, the ONU would be active on both PONs simultaneously. Consequently, the network management and traffic management becomes much more complicated. In the case of Link Aggregation, the division of the traffic flow between the two PONs would be handled as part of the standard Ethernet Link Aggregation processing. The dynamic bandwidth assignment (DBA) algorithms would become significantly more complex since they need to be aware of whether one PON or two are available for the connection to each ONU. The 1 + 1 protection architecture sends the identical data stream over both fibers, with the receiver only actively taking data from one of the streams. For the 1 + 1 PON case, the ONU must be able to take initialization, management, and provisioning data from both streams since it is logically behaving as a separate ONU for each PON.

2.5.4.3   Connect the ONU to a Single PON Using Redundant Optical Interfaces

As illustrated in Figure 2.15, this option uses redundant optical interfaces on the ONU that are connected to the fiber through either a passive splitter or an optical switch.

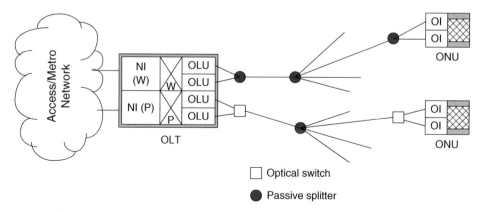

Figure 2.15 Illustration of redundant ONU optical interfaces to a single PON

As with the OLU protection case, if a passive splitter is used, then the optical budget must be able to handle the additional 3+ dB loss of the splitter. This additional loss could potentially necessitate using FEC. If the passive splitter approach is used for 1 : 1 protection at both the OLT and ONU, the total 6 dB loss may be an issue. See the discussion of passive splitter loss in Section 2.4.1 and 24.4.1.

Using an optical switch substantially reduces the optical loss, but it requires a control interface for the switch. Ideally, a non-latching switch should be used. See the discussion in Section 2.4.1.

2.5.4.4   Use a Single Optical Interface to a Single PON

This option is the cheapest and simplest, since it provides no protection. For cost reasons, it is typically used for residential customers.

With this option, it is very beneficial if the PON system can perform diagnostics during normal operation in order to detect impending failures in the optics before they occur. Detecting the degradation allows the problem to be repaired with a limited maintenance outage rather than a much longer outage due to the failure. PON devices from PMC-Sierra include integrated diagnostic features for this purpose.

2.5.5   Conclusions Regarding Protection

A wide range of protection options is available for PON systems. At one extreme, the most costly and complex of these provide full redundancy for each part of the equipment and network, including route diversity for the PON fibers. However, this level of redundancy is typically too expensive unless the subscriber requires it and pays for such a premium service level. At the other extreme, redundancy can be omitted altogether. This option may be practical for cost-sensitive non-critical services such as some video services.

PON service to residential customers typically uses redundancy for only those OLT functions that are shared by multiple PONs. Examples include the uplink from the OLT to the metro/core network and, potentially, some of the control functions. Different levels of redundancy for the PON and the interfaces to the PON may become important as differentiators for premium business services. There are a number of cost and technical issues that must be taken into account when providing this level of redundancy.

2.6   Conclusions

Due to its very high bandwidth capability, fiber is the most flexible medium for broadband service delivery to the home. After years of being a promising "next generation" technology, FTTH has finally become an economically viable option for providing residential triple-play services. The various technical and operational hurdles that have slowed large-scale deployment of FTTH have largely been resolved.

PON is the most cost-effective approach to providing FTTH broadband services. By providing a highly flexible platform for different services, and by eliminating the active electronics from the access plant, PON provides carriers with substantial ongoing OAM savings over copper-based technologies such as DSL or coaxial cable with cable modems.

Different PON protocols are favored by different carriers and different regions. The most popular PON protocols currently being deployed are the IEEE 802.3ah Gbit/s Ethernet the most popular PON and the ITU-T G.984 2.5 Gbit/s G-PON standard, which is favored by the North American and European carriers. The IEEE 802.3av 10G EPON equipment began field trials in 2010. These PON protocols are the subject of the next chapters.

Optical domain PON protocols, discussed in Chapter 5, can be used alone or in conjunction with the IEEE or ITU-T protocols. Optical domain protocols have not been cost-competitive, and are not expected to be so until possibly around the 2015 timeframe. Since optical domain techniques promise maximum flexibility for carrying different subscriber signals, and they allow the highest overall per-subscriber data rates, they will be a primary topic for further research.

Appendix 2.A: Subscriber Power Considerations

A primary factor in providing power is subscriber usage statistics. In providing power from the CO to subscribers, it can be assumed that only some fraction of the subscribers will be active ('off-hook') at a time, and that a much smaller percentage will have their phones ringing. For example, traditionally for CO-delivered voice service, fewer than 10% of subscribers could be safely assumed to be active. The probabilities associated with a large number of subscribers allows using a much smaller power and battery source than would be required if all the subscribers were off-hook or had ringing phones. Since DLC RTs connect to fewer subscribers, they must make more conservative assumptions regarding the number of simultaneously active subscribers. Here, it is only safe to assume no more than 30–50% of the subscribers are off-hook. As a result, more battery capacity is required per subscriber. With FTTC (and FTTH), it is no longer safe to assume that all of the subtended subscribers will not be simultaneously off-hook, because of data and video service usage in addition to voice.

In addition to the average power, it is also possible that multiple subscribers on a FTTC ONT will have their phones ringing simultaneously, creating a high peak power demand. In summary, the fewer subscribers that share a battery resource, the more battery capacity must be used per subscriber to provide the same level of reliable back up.

Finally, power delivery is more economical if it can re-use the existing copper wires with the existing voltages. These copper wires are small gauge (26–22 AWG). Since the loss is I^2R, these wires have substantial loss with higher currents. Higher voltages increase efficiency by lowering the current, but they quickly become considered "hazardous" and require a different type of craftsperson certification than the traditional –48 Vdc. The limit is determined again not by the average per-subscriber power, but by the peak power requirements during ringing.

References

1. Hasegawa T, Kuritani K, Makin, K, et al. Optical customer access based on digital loop carrier. *Proc. IEEE ICC'90*. 1990; 341.3.1–341.3.5.
2. Rowbotham T, Ritchie B, Hoppit, C. Plans for the Bishops Stortford (UK) fibre to the home trials. *Proc. IEEE Globecom'89*. 1989; 1320–1325.
3. ITU-TG.984.6. Gigabit-capable Passive Optical Networks (GPON): Reach Extension and its amendments; 2008.
4. ITU-TG.Sup45. GPON Power Conservation; 2009.
5. Shea D.P., Mitchell JE. A 10Gb/s 1024-Way Split 100-km long reach optical access network. *Journal of Lightwave Technology*. 2007; **25**(3): 685–693.
6. ITU-TG.665. Generic characteristics of Raman amplifiers and Raman amplified subsystems; 2005.
7. ITU-TG.983.5. A broadband optical access system with enhanced survivability; 2002.
8. ITU-TG.Sup51. PON Protection Considerations; 2012.
9. Gorshe S. Protection Strategies and Mechanisms for PON Systems, PMC-Sierra white paper, PMC-2080622; 2007.
10. IEEE1904.1. Service Interoperability in Ethernet Passive Optical Networks (SIEPON); 2013.
11. Gorshe S. Resilient Packet Ring Technology White Paper, PMC-Sierra white paper, PMC-2041096; 2005.
12. Gorshe S. Resilient Packet Ring (RPR). *China Communications* 2005; **2**(4): 91–103.
13. ITU-T G.8031. Ethernet linear protection switching, 2011.
14. ITU-T G.8032. Ethernet ring protection switching, 2012.

Further Reading

1. Van de Voorde I. et.al. The SuperPON demonstrator: an exploration of possible evolution paths for optical access networks. *IEEE Communications Magazine*. 2000; **38**(2): 74–82.

3

IEEE Passive Optical Networks

3.1 Introduction

IEEE 802.3 has developed two PON (point-to-multi-point) protocols based on their point-to-point protocols of the same rate. These include the Ethernet PON (EPON) protocol based on 1 Gigabit/s Ethernet and the 10G EPON protocol based on 10 Gbit/s Ethernet. EPON has seen extensive use, especially in Asia, with Japan taking the lead role in deploying it. Its re-use of Ethernet technology has given it some significant benefits, and 10G EPON is expected similarly to benefit from 10 Gbit/s Ethernet technology.

EPON uses 1 Gbit/s rates in both the upstream and downstream directions, and 10G EPON uses a 10 Gbit/s downstream rate with both 1 and 10 Gbit/s supported in the upstream direction. The downstream directions of both protocols are essentially the same as for point-to-point Ethernet streams of those rates, with some changes to the Ethernet frame overhead and additional management frames defined in order to support the point-to-multipoint operation. The upstream direction uses a TDMA protocol in which the ONU upstream transmissions are bursts compromised of Ethernet frames. No frame fragmentation is allowed.

In order to maximize backward compatibility and to allow co-existence on the same PON, 10G EPON is largely an extension of the EPON protocol (which is described in detail in the first section of this chapter). The second section describes 10G EPON protocol primarily in terms of how it differs from EPON. The differences between the two protocols are summarized in Table 3.2 at the end of the chapter.[1] Although both protocols are contained within the 2012 version of IEEE 802.3 [1] and its amendment IEEE 802.3bk [2], this chapter refers to the outputs of the projects under which they were developed, namely [3] and [5].

3.2 IEEE 802.3ah Ethernet-based PON (EPON)

The IEEE 802.3ah PON standard [3] was developed after the ITU-T B-PON and before the ITU-T G-PON protocol (see Chapter 4), although there was overlap in the development of the three. The EPON standard, which was developed as part of the IEEE Ethernet in the First Mile (EFM) project, was motivated by a desire

---

[1] At the time this manuscript was submitted to the publisher, the IEEE Communications Society was developing a new "Standard for Service Interoperability in Ethernet Passive Optical Networks (SIEPON)" project. The scope of the IEEE 802 standards is Layers 1 and 2. SIEPON addresses other functional aspects that are required for multi-vendor interoperability. Specifically, the scope includes "equipment functionality, traffic engineering, and service-level QoS/CoS mechanisms," and "management specifications covering: equipment management, service management, and power utilization."

Broadband Access: Wireline and Wireless – Alternatives for Internet Services, First Edition.
Steven Gorshe, Arvind Raghavan, Thomas Starr and Stefano Galli.
© 2014 John Wiley & Sons, Ltd. Published 2014 by John Wiley & Sons, Ltd.

to leverage the traditional advantages of Ethernet. These advantages included the ubiquitous presence of Ethernet at the customer premises, the relatively low cost of Ethernet UNIs, and Ethernet's potential to provide a lower cost Layer 2 technology than ATM, which was used by B-PON. EPON was developed to use the same transmission rate as the Gbit/s Ethernet interface that had recently been standardized. The B-PON protocol, in contrast, provided less bandwidth both in terms of transmission rate and the substantial ATM cell overhead. ATM is also a heavyweight Layer 2 technology that added more complexity than was required for Ethernet private line and LAN extensions through the MAN/WAN.

In contrast to the ITU-T PON protocols described in Chapter 4, which are based on a TDM frame, the EPON payload consists of Ethernet MAC frames in the upstream and downstream directions. As a consequence, EPON lacks a 125 μs reference for use with voice or other potential TDM clients and, instead, relies on VoIP for carrying voice traffic and circuit emulation service (CES) for carrying other TDM clients. While this adds complexity for TDM traffic, it reduces some complexity within the PON protocol. It also fits well with the general move to VoIP among many carriers. EPON supports auto-discovery of ONUs and FEC.

3.2.1   EPON Physical Layer

The MAC data rate of an EPON system is 1 Gbit/s. The data is encoded with the 8B/10B block code for transmission, resulting in a 1.25 Gbit/s signal transmission rate.

The data is transmitted over a single PON fiber. While a split ratio of 1 : 16 (i.e., 16 ONUs on a single PON connecting to one OLT interface) is shown in 802.3ah, actual deployments commonly use 1 : 32, with several using 1 : 64.[2] The downstream signal is transmitted with a laser using the 1490 nm wavelength window, while the upstream transmission uses the 1310 nm window to take advantage of less expensive lasers.

There are two distance options specified for EPON. One supports up to 10 km as the maximum OLT to ONU distance and the other allows up to 20 km.[3] PX refers to the optical interface options for EPON. These types are summarized as follows:

- PX10 specifies an optical channel insertion loss of ≤20 dB for ≤10 km reach with at least 1 : 16 split ratio
- PX20 specifies an optical channel insertion loss of ≤24 dB for ≤20 km reach with at least 1 : 16 split ratio
- PX30 specifies an optical channel insertion loss of ≤29 dB for ≤20 km reach with at least 1 : 32 split ratio
- PX40 specifies an optical channel insertion loss of ≤33 dB for ≤20 km reach with 1 : 64 split ratio.

3.2.2   Signal Formats

The downstream signal is simply a stream of Ethernet frames and Idle characters, as with a point-to-point Gbit/s Ethernet signal. The upstream signal is transmitted in bursts, like other TDMA protocols.

The preamble and start of frame delimiter (SFD) are modified for EPON from their normal values for Ethernet. Specifically, where the normal 8B/10B-encoded Ethernet 8-character preamble/SFD consists of /S/, 0x55, 0x55, 0x55, 0x55, 0x55, 0x55, and 0xd5, the EPON preamble/SFD consists of 0x55, 0x55, SLD, 0x55, 0x55, 2-octet LLID plus MODE bit, and CRC-8. The SLD is the Start of LLID Delimiter and has the value 0xd5. The LLID is the two-octet logical_link_ID field that uniquely identifies the ONU MAC. (Note that IEEE1904.1 allows multiple unicast LLIDs per ONU.) As discussed in Section 3.2.5, the

---

[2] The limits on the split ratio are a combination of the functions of the optical parameters (e.g., loss budget) and the desired per-ONU bandwidth.

[3] For the both options, the minimum distance between an OLT and ONU is specified as 0.5 km.

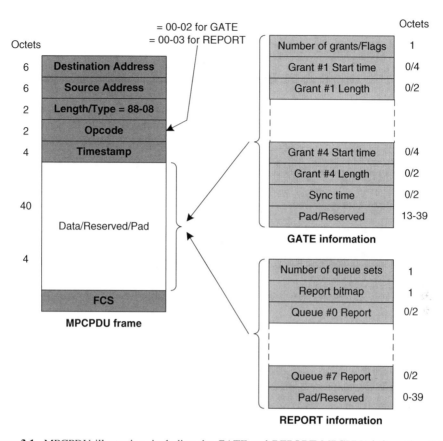

Figure 3.1 MPCPDU illustration, including the GATE and REPORT MPCPDU information fields

LLID is assigned to the ONU by the OLT during the registration phase of the discovery process. The CRC-8 covers the SLD through the LLID octets, and this uses the generator polynomial $x^8 + x^2 + x + 1$.

In addition to the one Logical Link ID (LLID) that is unique to each ONU, all ONUs respond to the Single Copy Broadcast (SCB) LLID. The SCB provides an efficient mechanism for the OLT to broadcast information to all ONUs without having to duplicate it for each. For example, the SCB is used when the OLT invites new ONUs to make their presence known during the discovery process. Multicast traffic is transmitted using the SCB LLID. An ONU can use standard L2 networking processing, such as VLAN filtering and IGMP snooping, to narrow the amount of received multicast traffic and receive only the designated multicast traffic.

Multi-Point Control Protocol PDUs (MPCPDUs) are control frames used by the ONUs to make their requests for bandwidth, and also by the OLT to assign it. As illustrated in Figure 3.1, the MPCPDU[4] frame is a basic 802.3 MAC control frame containing a four-byte timestamp and a 40-byte field filled with data and padding as needed. MPCPDU messages are also used for the discovery and ranging processes, as discussed in Sections 3.2.5 and 3.2.6. MPCPDUs are layered below the data interface, and they have higher priority than any data packet. This ensures that the bandwidth requests and grants are sent in a timely manner.

---

[4] The MPCPDU Ethertype is 0x88-08. The opcodes are assigned between 00-02 and 00-06.

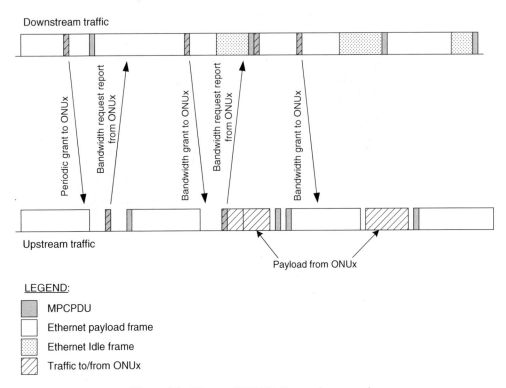

Figure 3.2 Ethernet PON MAC operation example

Note that EPON supports the use of PAUSE frames for flow control. When ONUs are a long way from the OLT, however, the delay makes PAUSE inefficient.

3.2.3   MAC Protocol

The EPON MAC uses upstream bandwidth requests from the ONUs and upstream bandwidth transmission grants from the OLT. The protocol makes use of local timers at each ONU that are synchronized to the OLTs local timer. The MAC operation discussed in this section is illustrated in Figure 3.2 with the downstream and upstream data flows.

3.2.3.1   GATE Messages for Upstream Bandwidth Grants

The OLT grants bandwidth to an ONU in the GATE message. Gating is the function that controls when the ONUs are allowed to transmit upstream data. The Gating function relies on a local timer that is synchronized to the OLT timer (see the ranging protocol description in Section 3.2.6 for a further discussion of the timers). The GATE message specifies the ONU upstream start time and transmission length relative to the ONUs local timer. The bandwidth grants from the OLT are always made for at least 1024 time quanta into the future so that the ONU has time to process the GATE message and be ready to transmit. An ONU turns its laser on when its local time matches the start time specified in the GATE message. The length field gives the length of time the ONU is allowed to transmit in that burst[5]. The start

---

[5] The transmission time is specified with respect to the number of periods of the ONU's local clock.

time is a 32-bit number (the same length as the local timer counter) and the length field is a 16-bit number. The OLT includes the time required to turn the ONU laser on and off and the time to send the upstream synchronization patterns when it assigns the grant time.

The ONU includes the inter-frame gap and FEC bit times in its requests for bandwidth. The ONU is responsible for ending its transmission long enough before the end of its time to allow its laser to turn fully off, but the grant times are specified by the OLT to be long enough to accommodate the laser on time, off time, and synchronization time.

The OLT sends GATE messages to each ONU periodically so that they can report their upstream bandwidth needs. The first (left-most) grant in Figure 3.2 is an example of such a periodic grant. The ONUs have watchdog timers that are reset whenever a GATE message is received.[6]

Up to four upstream transmission grants can be made to a given ONU in a single GATE message. As discussed in Section 3.2.5, the ONU advertises the number of outstanding grants it can accept during the discovery process. The first payload transmission from ONUx in Figure 3.2 illustrates data from two grants. The GATE message can also request reports from the ONUs corresponding to those grants. In practice, however, using multiple grants per GATE message adds considerable complexity to the OLT bandwidth assignment process. Sending a single grant in the GATE message gives much finer resolution and faster response for the upstream bandwidth assignments, and it has only a small impact on the downstream overhead bandwidth. The preferred approach, especially for per-flow dynamic bandwidth assignment (DBA), is for the OLT to send each ONU a single grant in each GATE message to service its bandwidth requests and to let the ONU decide which data should be sent in that upstream grant.

3.2.3.2 REPORT Messages for Upstream Bandwidth Requests

The ONUs communicate their upstream bandwidth requirements by sending REPORT MPCPDU messages. The OLT grants the upstream bandwidth for these REPORT messages in its GATE messages. In addition to the timestamp, the REPORT message consists of a summary of its requests for upstream bandwidth and the specific amount of bandwidth it needs. EPON supports the eight queue priority levels defined in IEEE 802.1Q. The summary field of the REPORT message indicates how many and which, if any, of these queues have data to send. The summary is followed by binary numbers to indicate the specific number of bits to be transmitted from each queue. The bit count is a 16-bit number, and this includes any required overhead bits, Inter-Packet Gap (IPG) characters, and FEC bits that it needs to send in the transmission.

Each ONU sends REPORT messages periodically, even if it has no data waiting for transmission, in order to reset a watchdog timer at the OLT. If the watchdog timer expires, the OLT deregisters that ONU from the network.

3.2.4 Encryption and Security

The EPON standard did not address the security of the downstream transmissions (i.e., protecting against ONUs listening to traffic other than their own). EPON can potentially use the 802.1ae and 802.1af link encryption standards that were subsequently developed. In the meantime, regional specifications have become common.[7]

The encryption method, described in [4], is a symmetric block code using the 128-bit key AES. A counter (CTR) mode is used, in which the output of pseudorandom counters is exclusive OR'ed with the original data to generate the cipher text. This encryption method can be implemented either for just

---

[6] For the purposes of keeping the ONU's watchdog timer alive, an OLT can also periodically send empty GATE messages when it has no pending bandwidth requests for that ONU.

[7] An encryption protocol developed by PMC-Sierra in conjunction with NTT became an early *de facto* standard in some regions [11]. As part of PMC-Sierra's agreement with NTT, no royalty fees are charged to vendors that sell to NTT carriers, and several other equipment and silicon vendors have since implemented this encryption protocol.

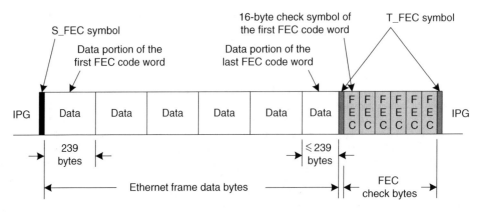

Figure 3.3 EPON FEC illustration

the downstream traffic (as is done in G-PON) or for both the upstream and downstream directions. The primary purpose of upstream encryption is to provide message authentication. If the OLT receives a transmission with the incorrect encryption key for that upstream bandwidth grant, it will discard the data. While upstream encryption is not mandatory, it is being demanded by an increasing number of carriers.

3.2.5   Forward Error Correction (FEC)

EPON allows the optional use of FEC.[8] The FEC code is the ITU-T G.975 systematic RS(255, 239, 8) code. The downstream data packets are divided into 239-octet blocks, to which 16 check code bytes are appended. The upstream bursts are similarly divided into data blocks and error check code bytes. The first upstream FEC code word of a burst begins with the /S/ character, and the last FEC code word of the burst contains the /T/ character. The additional bandwidth for the FEC check bits is taken into account in the bandwidth requests that the ONU makes to the OLT.

For both the downstream and upstream directions, the FEC is packet-based rather than stream-based. The EPON FEC code arrangement is illustrated in Figure 3.3. The FEC check bytes for all the FEC code words associated with an Ethernet frame are placed in order at the end of that Ethernet frame rather than immediately following the data portion of each code word. Since the Ethernet frame length will generally not be an integer multiple of 239 octets, typically there will be a shorter block at the end of the packet. The 16 error check bytes for the last block will be calculated over a logical block that includes the X actual data bytes followed by 239-X "0" padding bytes. The "0" padding bytes are not transmitted, but are re-inserted by the receiver when it performs its error check calculation for this block.

When the receiver FEC decoder is unable to correct a character, it replaces the uncorrectable character with a /V/ character.

In an environment noisy enough to require FEC, the preamble/SFD and end of frame (EOF) delimiters are vulnerable to transmission errors. In order to provide the desired robustness when FEC is enabled, an additional new SFD is added to the beginning of each MAC frame and an additional new EOF is added to the end. The new SFD, designated as S_FEC, is five octets long and consists of the K28.5/D6.4/K28.5/D6.4/S/character set. The new EOF is designated as T_FEC, and it has separate versions for even and odd alignment at the end of the frame. The T_FEC character sets are T_FEC = /T/R/K28.5/D10.1/T/R/ (for even alignment, positive running disparity), /T/R/K28.5/D29.5/T/R (even alignment, negative running disparity), or /T/T/R/I/T/R/(for odd alignment). The receiver is required to recognize the S_FEC or T_FEC even if they contain up to five bit errors.

---

[8] EPON FEC has the ability to be selectively activated on a per-ONU basis, thus reducing the overall FEC overhead on the PON.

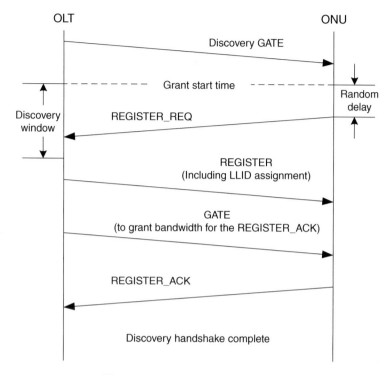

Figure 3.4 ONU discovery handshake

3.2.6 ONU Discovery and Activation

The OLT periodically opens a Discovery Time Window in order to allow new ONUs to announce themselves.[9] The discovery message, registration, and handshake flow are illustrated in Figure 3.4. The OLT opens the window by transmitting a Discovery GATE message, which includes the length of the Window and its start time.

The unregistered ONUs respond to the discovery GATE message by transmitting a REGISTER_REQ message. The REGISTER_REQ message includes the ONU's MAC address and the number of outstanding grants that it can accept (see Section 3.2.3 regarding multiple grants). A contention algorithm is used in order to minimize the chance of collision when multiple ONUs are attempting to register during the same Discovery Time Window. The contention algorithm operates by having each ONU delay its transmission by a random time relative to the beginning of the Discovery Time Window.[10] ONUs that are already registered ignore the discovery GATE message.

When the OLT receives the REGISTER_REQ, it assigns an LLID to the ONU, bonding the LLID to the ONU's MAC address. The OLT then sends a Register message to the ONU in order to communicate the ONU's LLID and the required OLT synchronization time, and to echo the maximum number of pending grants that the ONU can accept. The synchronization time is the amount of time the OLT will require in order to synchronize reliably to the ONU's upstream transmission burst. The synchronization time is

---

[9] The period of the Discovery window is implementation-dependent.

[10] The upper bound requirement on this hold-off time is that it must be short enough that the ONU can transmit its entire REGISTER_REQ before the end of the Discovery Time Window.

specified in multiples of 16-bit data patterns that the ONU sends as IDLE code pairs at the beginning of the burst.

After the ONU has processed the Register message, it sends a REGISTER_ACK message to the OLT in response to a standard GATE message from the OLT.

Note that the Discovery GATE, REGISTER_REQ, and REGISTER messages are sent on the broadcast channel since ONU doesn't know its LLID until it receives the REGISTER message. After the ONU receives its LLID, the remaining GATE and REGISTER_ACK messages are sent on the unicast channel.

Mechanisms exist in the protocol to deregister an ONU (e.g., if a watchdog timer expires) and to re-register.

3.2.7   ONU Ranging Mechanism

In order to prevent overlap in the different ONU upstream transmission bursts when they arrive at the OLT, the OLT assigns the upstream burst transmission times with enough guard band time between the bursts of successive ONUs. This guard time allows for the laser of one ONU to turn off and for the laser of the next ONU to turn on, and to cover any uncertainty in their relative signal propagation delays to the OLT.[11] Since the ONUs can be located at different physical distances from the OLT, a crude approach would be to have this uncertainty window include the round trip propagation delay difference between an ONU at the shortest fiber distance and an ONU at the longest anticipated fiber distance from the OLT. With the desired maximum ONU distance of 20 km and the 2.045 m/s speed of light through a fiber, the resulting differential delay is 200 μs. This would clearly be very bandwidth-inefficient, since only a few bit periods would be required if the ONUs were equidistant from the OLT.

The solution to this problem is to use a ranging mechanism that allows the OLT to determine the relative distances of the ONUs. The OLT can then take this range into account and assign the upstream burst times with a minimum of guard band time.

The ranging mechanism for EPON is based on local clocks and counters that are maintained at the OLT and each ONU. A counter has 32 bits and is incremented once every 16 ns.

The OLT counter is the PON master. When the OLT transmits a MPCPDU message, it loads its current counter value into the message's 32-bit timestamp field. When the ONU receives a MPCPDU, it resets its own local counter to the value contained in the MPCPDU timestamp field. When the ONU sends a MPCPDU to the OLT, the ONU loads its updated counter value into the timestamp field. The OLT then compares the offset between its current count and the value it receives in the MPCPDU timestamp field, with the difference being the round trip time (RTT) associated with that ONU. The RTT is then used to establish the ONU's range, which is taken into account when the OLT assigns the start times for upstream bandwidth grants.

Some drift may occur in the RTT over time. When the drift exceeds a provisioned threshold, a timestamp drift error condition is declared. Either the ONU or OLT can detect this condition as an offset between the expected value received in the MPCPDU and the one actually received.

3.2.8   EPON OAM

Since EPON lacks an outer transport frame structure like those used in SONET/SDH and G-PON, it has no dedicated overhead frame bits to communicate OAM data. The lack of link OAM was addressed as part of the same IEEE 802.3ah project that developed EPON for application to individual links. The 802.3ah

---

[11] If the ONUs do not turn their lasers off when they are not transmitting, spontaneous emission noise from ONUs closer to the OLT would interfere with data transmissions from ONUs further from the OLT.

Table 3.1 OAMPDU types

OAMPDU code	OAMPDU type	Comment
00	Information	To communication remote and local OAM information
01	Event notification	Alerts the remote Ethernet node of link events
02	Variable request	Request for MIB variable(s)
03	Variable response	Return of MIB variable(s)
04	Loopback control	
05-FD	Reserved	
FE	Organization-specific	Reserved for organization-specific extensions, identified by the Organizationally Unique Identifier
FF	Reserved	

group defined several Ethernet frames to communicate the link OAM information. The types of OAM information are categorized as follows:

1. Remote failure indication: Indication sent in the reverse direction by an Ethernet terminal node to indicate that it cannot properly receive messages on that link
2. Remote loopback
3. Link monitoring: Messages to support link performance notifications for diagnostic and performance monitoring purposes
4. Miscellaneous: Mechanism to provide additional OAM functions such as OAM capability discovery, or to support higher layer management applications.

All OAM PDUs share the common frame format illustrated in Figure 3.5. The different OAMPDU types are listed in Table 3.1.

3.2.9 Dynamic Bandwidth Assignment (DBA)

The most basic method of allocating upstream bandwidth is to distribute it equally among the ONUs. This method is very inefficient, especially with packet traffic, since the bandwidth needs of the ONUs will

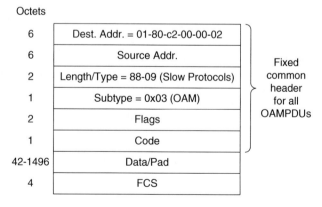

Figure 3.5 OAMPDU frame format

rarely be equal at each instant in time. Considerable overall bandwidth utilization gains can be made if the upstream bandwidth is allocated dynamically according to the current needs of the ONUs.[12] The ITU-T addressed DBA in its G.983.4 recommendation; while not specifying a particular DBA algorithm, G.983.4 specifies the framework and mechanism to implement DBA in B-PON and G-PON systems, and it is equally applicable to EPON systems. G.983.4 is discussed further in Chapter 4. Some DBA comments relevant to EPON are provided in this section.

As described above, the ONU REPORT messages inform the OLT of their current bandwidth needs. Their bandwidth requests are reported in terms of the number of characters they have in the different priority queues awaiting upstream transmission. The OLT can also take into account the service level agreements (SLAs) that have been specified for the service flows associated with an ONU. For example, an ONU with an active VoIP service will need a fixed amount of bandwidth on a regular basis. Consequently, the OLT can regularly grant the upstream bandwidth for this service flow so that the ONU does not need to waste upstream bandwidth reporting bandwidth requests for it. As another example, if the OLT receives upstream bandwidth requests from multiple ONUs, it can grant more bandwidth to ONUs that recently have been consistently requesting more bandwidth than it does to ONUs that have made few recent requests. In other words, the OLT attempts to reduce latency by anticipating the needs of the ONUs. In this example, however, the DBA algorithm needs to ensure that nodes with fewer bandwidth requests do not become starved or encounter high latency while the ONUs with more bandwidth requests are serviced.

EPON DBA has the flexibility to customize EPON network behavior to meet various carrier needs. Its flexible nature allows quick adaptation to possible carrier challenges, making the EPON infrastructure compliant with the ever-growing, ever-changing carriers' requirements. It is possible to map both user and service flows into specific containers that are managed by the DBA and to provide the QoS that is needed for every customer and service. Two straightforward adjustable parameters related to EPON DBA are latency and total system performance (upstream bandwidth utilization).

3.3   IEEE 802.3av 10Gbit/s Ethernet-based PON (10G EPON)

The IEEE 802.3av PON standard [5] was developed to increase the data rate of EPON systems from 1 Gbit/s to 10 Gbit/s, in keeping with the 10 Gbit/s Ethernet interface. 10G EPON shares much of its protocol with EPON. A combination of coarse wave division multiplexing (CWDM) and time division multiplexing (TDM) is used in order to allow EPON and 10G EPON systems to co-exist on the same PON. As with EPON, 10G EPON relies on VoIP for carrying voice traffic and circuit emulation service (CES) for carrying other TDM clients.

3.3.1   10G EPON Physical Layer

The downstream data rate of 10G EPON is 10 Gbit/s, and both 1 Gbit/s and 10 Gbit/s rates are supported in the upstream direction. The 64B/66B block line code, described in Appendix 3.A, is used for all of the 10 Gbit/s signals with a resulting signal line rate of 10.3125 Gbit/s. The 1 Gbit/s upstream uses the same 8B/10B block line code as EPON, giving a line rate of 1.25 Gbit/s.

The downstream and upstream data is transmitted over a single PON fiber, using WDM to separate the upstream and downstream signals. The wavelengths used by the different upstream and downstream signals are shown in Figure 3.6. As noted above for EPON, because there are many ONUs on the PON and only a single OLT, the wavelength bands were chosen to allow the use of less expensive lasers at the ONUs.

---

[12] DBA can be used for, or can be considered as, a form of statistical multiplexing.

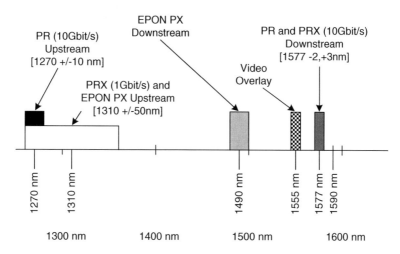

Figure 3.6 EPON and 10G EPON optical spectrum allocation

For 1 Gbit/s upstream operation, 10G EPON uses the same 1310 nm wavelength as the EPON upstream signal. This allows the OLT to use the same receiver for all 1 Gbit/s signals. The dynamic bandwidth allocation algorithm (see Section 3.3.8) allocates the bandwidth of the 1 Gbit/s upstream signal between the EPON and 10G EPON ONUs.

The 10 Gbit/s upstream signals use a separate wavelength band, but it overlaps with the 1 Gbit/s upstream wavelength band. When an OLT supports both 1 Gbit/s and 10 Gbit/s operation on the same PON it is referred to as a dual rate mode. The dual rate OLT can either separate the 10 Gbit/s and 1 Gbit/s upstream signals by dividing the signal in the optical domain or in the electrical domain. Dual-rate receiver considerations are discussed below in this section.

The advantages to allowing 10G EPON to operate over the same PON optical distribution network as EPON include:

- allowing customers to use the most cost-effective ONU for the desired service;
- allowing a network to migrate from EPON to 10G EPON by upgrading the OLT then migrating the ONUs as needed;
- continued operation of the existing network and services during the upgrade of the network.

Figure 3.7 illustrates a network where an OLT supports a mix of EPON ONUs, ONUs with 10 Gbit/s downstream and 1 Gbit/s upstream, and ONUs with 10 Gbit/s upstream and downstream. For convenience, the wavelength color key in Figure 3.7 is consistent with the key for Figure 3.6. Note that WDM is used to separate the 1 Gbit/s and 10 Gbit/s traffic in the downstream direction with the filters at the ONUs, and a combination of WDM and TDM is used in the upstream direction. The discovery and other protocol extensions to support the co-existence of EPON and 10G EPON ONUs are discussed in the appropriate sections below.

As its reference for the optical link loss budgets, the 802.3av specification uses a split ratio of either 1 : 16 (i.e., 16 ONUs on a single PON connecting to one OLT interface) or 1 : 32. In practice, larger split ratios such as 1 : 64 or 1 : 128 can be used if the other optical losses (e.g., the length of the fiber) are constrained to offset the additional 3 dB loss that is incurred when the split ratio is doubled. All of the interfaces are specified to operate at an uncorrected bit error rate no worse than (BER) 10^{-3}. After FEC

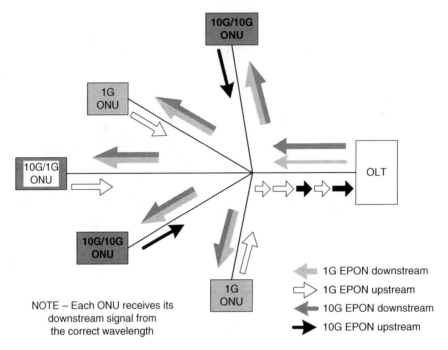

Figure 3.7 Illustration of EPON and 10G EPON ONUs sharing the same PON

correction, the bit error rate will be no worse than 10^{-12}. The nomenclature adopted to identify the different optical interface options may be summarized as follows:

- PRX interfaces use 10 Gbit/s downstream and 1 Gbit/s upstream transmission
- PR interfaces use 10 Gbit/s for both downstream and upstream transmission
- PR-Dn and PRX-Dn (n = 10, 20, 30) refer to the OLT optical interface specification
- PR-Un and PRX-Un (n = 10, 20, 30) refer to the ONU optical interface specification
- PR10 and PRX10 specifies an optical channel insertion loss of $\leq$20 dB for $\geq$10 km reach with 1 : 16 split ratio
- PR20 and PRX20 specifies an optical channel insertion loss of $\leq$24 dB for $\geq$20 km reach with a 1 : 16 split ratio or $\geq$10 km reach with a 1 : 32 split ratio
- PR30 and PRX30 specifies an optical channel insertion loss of $\leq$29 dB for $\geq$20 km reach with a 1 : 32 split ratio
- PR40 and PRX40 specifies an optical channel with a reach of at least 20 km and a split ratio of at least 1 : 64.

As with EPON, the 1550–1560 nm wavelength band is reserved for downstream video transmission.

Following the same approach as EPON, the upstream burst timing is relaxed for 10G EPON in order to allow the use of existing off-the-shelf components. The standard has mechanisms to allow for future tighter timing to be implemented with better components for increased bandwidth efficiency.

Dual-rate operation refers to an OLT that simultaneously receives upstream signals from ONUs using 1 Gbit/s and 10 Gbit/s rates. As illustrated in Figure 3.8, the received 1 Gbit/s and 10 Gbit/s streams can either be split in the optical domain or electrical domain. Since both signals time-share the

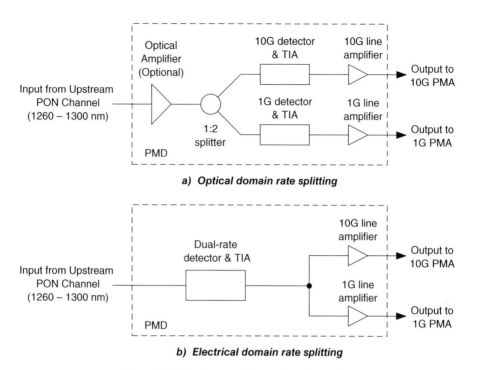

a) Optical domain rate splitting

b) Electrical domain rate splitting

Figure 3.8 Dual rate receiver option illustration

same upstream wavelength, it is not possible to use WDM filters to separate them in the optical domain.

Splitting the signals in the optical domain involves using a 1 : 2 optical splitter. Each of the two splitter outputs goes to its own photodetector followed by an electrical receiver with a filter optimized for its bandwidth in order to maximize the receiver's sensitivity. The drawback with this approach is the ≥3 dB additional optical loss introduced by the 1 : 2 optical splitter. If this additional loss cannot be tolerated, a low-gain optical amplifier must be used in the receiver.

Splitting in the electrical domain allows using a single photodetector and introduces no additional optical signal loss. In the electrical domain, one approach is to design the receiver filter as a compromise that allows reception of both the 1 Gbit/s and 10 Gbit/s signals. This means that the receiver sensitivity is not optimal for either signal, lowering it by about 1 dB for each. Alternatively, the OLT can adjust (switch) the transimpedance of its transimpedance amplifier (TIA) filter for that burst's rate. The APD bias can either be set to a compromise value or switched along with the transimpedance.[13] While the performance of an adaptable receiver is optimum, its additional complexity impacts the receiver cost. Detecting the rate of the current incoming burst must be performed fast enough to switch the receiver. The burst rate could be detected by looking for spectral energy that would only be present for a 10 Gbit/s burst. Alternatively, the OLT could exploit its knowledge of which upstream burst is scheduled to arrive. However, since this knowledge is in the MAC layer and not the PMD, requiring it would be a violation of layer stack restriction.

---

[13] Using a compromise APD bias results in a loss of around 1 dB receiver sensitivity, which is 1 dB better than using the compromise transimpedance value.

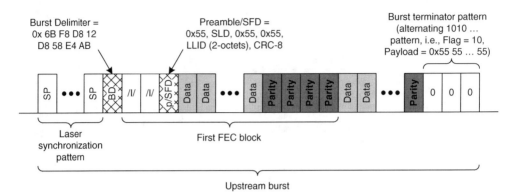

Figure 3.9 10 Gbit/s upstream burst transmission illustration

3.3.2 Signal Format

With the exception of the added forward error correction (FEC) coding, the downstream signal is simply a stream of Ethernet frames and Idle characters, as with a point-to-point 10 Gbit/s Ethernet signal.[14] The upstream signal is also essentially an Ethernet stream except that, as discussed above, a TDMA burst format is used. The upstream signal also uses FEC.

The beginning of an upstream burst is illustrated in Figure 3.9. The synchronization patterns at the beginning of an upstream transmission burst allow the OLT to synchronize its receiver to new burst from an ONU. The Burst Delimiter pattern is used by the OLT to determine the start of 66B block transmission and the FEC codeword alignment. The 66-bit value of the Burst Delimiter is 0x 6B F8 D8 12 D8 58 E4 AB (which results in a transmission bit sequence of 01 1101 0110 0001 1111 0001 1011 0100 1000 0001 1011 0001 1010 0010 0111 1101 0101, since the characters are transmitted LSB first). The FEC codeword alignment can be achieved in the presence of transmission errors. This burst delimiter is followed by two 66-bit blocks containing Idle characters. These Idle characters allow the OLT to synchronize its descrambler and delineate the start of the actual data frame. The first two blocks of Idle characters are included in the initial FEC codeword.

As discussed above in Section 3.2.2 with EPON, the preamble and start of frame delimiter (SFD) are modified for EPON and 10G EPON from their normal values for Ethernet. Specifically, the preamble bytes are replaced by the transmitting MAC's MODE and LLID variables. While the Ethernet 8-character preamble/SFD consists of /S/, 0x55, 0x55, 0x55, 0x55, 0x55, 0x55, and 0xd5, the EPON and 10G EPON preamble/SFD consists of 0x55, 0x55, SLD, 0x55, 0x55, 2-octet LLID, and CRC-8. The SLD is the Start of LLID Delimiter, and it has the value 0xd5. The LLID is the two-octet logical_link_ID field that uniquely identifies the ONU MAC. The MSB of the two octets that contain the LLID is the MODE indication bit. As discussed in Section 3.3.5, the LLID is assigned to the ONU by the OLT during the registration phase of the discovery process. The CRC-8 covers the SLD through the LLID octets, and it uses the generator polynomial $x^8 + x^2 + x + 1$.

The upstream transmission ends with a burst terminator pattern comprised of three 66-bit blocks of alternating zeros and ones (1010 . . . 10) after the last FEC codeword of the burst. The ONU turns off its laser at the beginning of the burst terminator pattern, which ensures that it will be completely off by the end of the burst.

Each ONU has one unique Logical Link Identifier (LLID) that the OLT associates to the ONU for unicast traffic. In other words, these MAC instances are used to emulate a point-to-point connection

---

[14] As explained in Section 3.3.4, both the upstream and downstream streams are encoded into FEC blocks in a manner that preserves the 64B/66B block stream format.

between and ONU and the OLT over the PON. (Note that IEEE1904.1 allows multiple unicast LLIDs per ONU.) Additionally, the OLT has two Single Copy Broadcast (SCB) MAC instances that are used as an efficient mechanism to broadcast downstream traffic to the ONUs. Such a broadcast is used for broadcast data or for when the OLT must communicate with unregistered ONUs. In the upstream direction, an SCB MAC is only used for client registration. The LLID value of 7F-FF is associated with the SCB MAC for 1 Gbit/s downstream operation and the LLID value of 7F-FE is associated with the SCB MAC for 10 Gbit/s downstream operation. An ONU can use higher layer networking processing, such as VLAN filtering and IGMP snooping, to narrow the amount of received multicast traffic that is passed to applications. It is possible that these higher layers may require addition multicast MAC instances at the OLT, in which case an OLT can have more MACS than two plus the number of ONUs.

As with EPON, MPCPDU control frames (see Figure 3.1) are used by the ONUs to make their requests for bandwidth, by the OLT to assign bandwidth, and by both ONUs and OLT during the discovery and ranging processes.

3.3.3 MAC Protocol

The 10G EPON MAC-layer control protocol is based on the protocol for EPON and includes enhancements for management of 10G FEC and inter-burst overhead. This MAC protocol operates on the basis of the ONUs informing the OLT of their upstream bandwidth requirements, and the OLT scheduling and granting bandwidth to the ONUs to transmit their upstream data (as described in Section 3.2.3 above). The details of the MAC protocol specific to 10G EPON are described in this section.

For 10G EPON, a "Sync Time" field in the GATE MPCPDU is used by the OLT to communicate to the ONU the amount of time the OLT needs at the beginning of the upstream transmission burst to synchronize its receiver to the new burst. As illustrated in Figure 3.9, each burst begins with a synchronization pattern, followed by a Burst Delimiter pattern, followed by two blocks of Idle characters. The ONU transmits the 66-bit synchronization pattern repeatedly, and then transmits the Burst Delimiter so that the duration of the entire sequence is the same as the Sync Time requested by the OLT.

Like EPON, 10G EPON supports the eight queue priority levels defined in IEEE 802.1Q. The summary field of the REPORT message indicates how many and which, if any, of these queues have data to send. Unlike EPON, the bandwidth value carried by the 10G EPON REPORT does not include burst overhead or FEC overhead. The OLT already knows this information and takes it into account.

3.3.4 Forward Error Correction

EPON and 10G EPON use different FEC approaches. FEC allows a link to function with a higher line bit error rate at the receiver. Consequently, FEC effectively increases the optical link budget, which in turn allows increased distance or split ratios. FEC becomes increasingly important as bit rate increases and, for this reason, it is mandatory in 10G EPON. Additionally, the 10G EPON FEC differs in two ways from EPON. First, 10G EPON uses a more powerful RS(255, 223) code for error correction of 16 symbols rather than the 8 symbols that can be corrected with the optional RS(255,239) code specified for EPON.[15] Second, the 10G EPON FEC is applied to fixed-length sequences of streaming data rather than Ethernet frames as illustrated in Figure 3.10. Figure 3.10 illustrates the downstream transmission direction, which is a continuous stream of FEC codewords that includes the Ethernet frames and all inter-packet information such as IPG and Ordered Set data. The upstream transmission is similar except that, as illustrated in Figure 3.9, the first FEC codeword of an upstream burst is aligned with the beginning of the burst in order to allow the OLT FEC decoder immediate codeword synchronization for each burst.

Note that when the RS(255,223) FEC parity is taken into account, the effective data rate of a 10G EPON link is approximately 8.7 Gbit/s.

---

[15] The generator polynomial $G(x) = x^8 + x^4 + x^3 + x^2 + 1$ is specified for the 10G EPON RS(255,223).

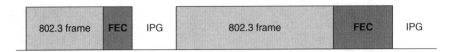

a) FEC overhead illustration for 1G EPON

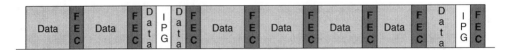

b) FEC overhead illustration for 10G EPON

Figure 3.10   FEC overhead locations (downstream)

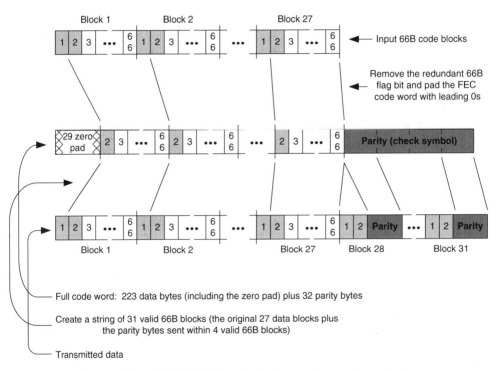

Figure 3.11   10G EPON FEC code block formatting and transmission

One of the challenges in adding FEC to the 10GE PON stream is extending the 64B/66B block code format so that a 10GbE receiver can receive and synchronize to the stream that now includes FEC parity data. The method used is illustrated in Figure 3.11, where each FEC codeword covers a group of 27 64B/66B blocks. As also shown in Figure 3.11, the first step of FEC encoding is removing the first flag bit of the 64B/66B block.[16]

---

[16] Since the two leading flag bits of the 64B/66B block are intentionally redundant, only one of them needs to be protected by the FEC.

The resulting $27 \times 65 = 1755$-bit block is padded with 29 leading zeros to get a total of 1784 bits (223 bytes). The RS(255,223) encoding produces 32 parity bytes.

In the final stage, the zero pad bits are removed, the original 27 64B/66B blocks are restored, and the parity bytes are converted into a sequence of 64B/66B blocks for transmission. Specifically, the 32 FEC parity bytes are treated a four groups of 64 bits. Each of these 64-bit parity groups is then given a pair of leading header bits in order to create 64B/66B blocks. In order to create a recognizable header pattern, the header bits for the four parity blocks are 00, 11, 11, and 00, respectively.[17] The string of 31 64B/66B characters is then transmitted.

The receiver can then synchronize to the 64B/66B character stream and extract the original data through the reverse process, performing error correction as it decodes the FEC blocks.

3.3.5   ONU Discovery and Activation

The ONU Discovery protocol for 10G EPON is the same as for EPON with the following exception. With 10G EPON, a Discovery GATE MPCPDU includes a Discovery information field that communicates to the ONUs whether the OLT is capable of receiving 1 Gbit/s upstream signals, capable of receiving 10 Gbit/s upstream signals, and whether the Discovery Window being opened is for 1 Gbit/s or 10 Gbit/s upstream signals from the ONUs. Also, as described in Section 3.3.2, the OLT uses a separate SCB LLID for Discovery messages associated with 1 Gbit/s and 10 Gbit/s upstream discovery invitations. These additions allow the OLT to communicate with and register ONUs of both upstream rate capabilities, and also allow an ONU that supports both rates to determine which upstream rate it should use.

3.3.6   ONU Ranging Mechanism

The 10G EPON ranging mechanism is identical to the EPON ranging mechanism. See Section 3.2.6 above.

3.3.7   10G EPON OAM

10G EPON also uses the 802.3ah Link OAM. See Section 3.2.7 above.

3.3.8   Dynamic Bandwidth Allocation

10G EPON DBA is similar to EPON DBA. The primary difference is that the OLT must schedule upstream traffic for both 1 and 10 Gbit/s ONUs if both types are present on the PON. As noted above, ONUs that use 10 Gbit/s upstream assume that the OLT already takes into account the required overhead bits rather than depending on the ONU to include them in its bandwidth request.

One benefit of the 10G EPON system is the ability to overcome system bottlenecks via adjustments in the EPON DBA algorithm. The DBA cycle length and bandwidth allocation per ONU can be adjusted so that the total OLT upstream transmission going into the switch will be smoother, less bursty in nature, allowing carriers to overcome blocking elements in their network topology (e.g., assigning more bandwidth to the OLT ports than the uplink ports in the switch connected to the OLT to save CAPEX). While this is also true for EPON, the higher bandwidth of 10G EPON allows additional flexibility.

3.4   Summary Comparison of EPON and 10G EPON

The essential aspects of the EPON and 10G EPON protocols are summarized in Table 3.2.

3.5   Transport of Timing and Synchronization over EPON and 10G EPON

Both EPON and 10G EPON can use the IEEE 802.1AS [6] protocol. While 802.1AS is a generalized precision timing protocol (gPTP) for use with all Ethernet applications, it includes a clause that specifies how it can be used with the Ethernet PON protocols. The 802.1AS protocol is based on a modified version

---

[17] As shown in the appendix of this chapter, the normal allowed header bit patterns are 01 and 10.

Table 3.2 Summary of EPON and 10G EPON protocols and features

Feature	EPON	10G EPON	Comment
Responsible standards body	IEEE 802.3 (IEEE 802.3ah)	IEEE 802.3 (IEEE 802.3av)	
Data rate	1 Gbit/s upstream and downstream	10 Gbit/s downstream and either 1 or 10 Gbit/s upstream	While the optical parameters were designed around 1 : 16 for EPON, and 1 : 16 or 1 : 32 for 10G EPON, in practice both support 1 : 64.
Split ratio (ONUs/PON)	1 : 64	1 : 64	
Line code	8B/10B	Down: 64B/66B Up: 8B/10B (1 Gbit/s) or 64B/66B (10 Gbit/s)	The 8B/10B line code results in a 1.25 Gbit/s line rate for the 1 Gbit/s MAC rate, and the 64B/66B line code results in a line rate of 10.3125 Gbit/s for the 10 Gbit/s MAC rate.
Number of fibers	1	1	
Wavelengths	1490 nm down and 1310 nm up	1577 nm down and 1310 nm up (1 Gbit/s) or 1270 nm up (10 Gbit/s)	Both support 1550 nm for downstream video overlay.
Maximum OLT to ONU distance	10 and 20 km (1 : 16 split)	10 km with 1 : 16 split; 20 km with 1 : 16 split; 20 km with 1 : 32 split	
Optics			Defined in regional specifications or other bodies
Protection switching	None	None	
Data format (encapsulation)	None (uses Ethernet frames directly)	None (uses Ethernet frames directly)	
TDM support	CES	CES	
Voice support	VoIP	VoIP	
Multiple QoS levels	Yes (802.1Q priority levels)	Yes (802.1Q priority levels)	All traffic types are handled using a single LLID.
FEC	RS(255, 239) – frame-oriented	RS(255, 223); – stream-oriented	
Encryption	Defined by regional standards	Not in the scope of 802.3av. Defined by regional standards	A multi-company protocol was common for EPON. IEEE 1904.1 [8] now specifies encryption for EPON and 10G EPON.
OAM	802.3ah Ethernet OAM frames	802.3ah Ethernet OAM frames	

of the IEEE 1588 [7] precision timing protocol (PTP). While a detailed description of the IEEE 1588 protocol is beyond the scope of this book, the manner in which it is used for Ethernet PON networks is summarized in this section.

Recall that EPON uses 32-bit local counters that are incremented every 16 ns (i.e., they use a time quantum of 16 ns). These counters are used for the ranging and upstream transmission synchronization processes (see Section 3.2.6). These counters are also used in the EPON timing synchronization process. Specifically, the 32-bit counter is the LocalClock entity of the time-aware system. The OLT is the clock master, and it is assumed to have an accurate synchronization time derived from a grandmaster clock source. The associated ONUs are clock slaves. A time-aware system consists of no more than one ONU, which is a clock slave to that EPON link but may contain multiple OLTs, since the ONU may have EPON links to multiple OLTs.

The PON application is different from other Ethernet links, in that the upstream direction uses a TDMA protocol that results in asymmetry between the downstream and upstream delays. (See Section 3.2.3). The use of different wavelengths for upstream and downstream transmission also impacts the directional delay asymmetry, since the fiber's index of refraction, and hence the propagation speed, are wavelength-dependent.

The 802.1AS protocol works as follows for EPON. The OLT (clock master) communicates to an ONU (the clock slave) the accurate synchronization time at the point in time when the ONU's local counter reaches a certain value. This information is communicated using an Ethernet Organization Specific Slow Protocol (OSSP) message. The specific process, which accommodates the asymmetry between the downstream and upstream transmissions delays, can be summarized as follows:

The OLT and ONU each compute their local latency factors. The ONU latency factor is the difference between the ONU's ingress latency and the scaled sum of its ingress and egress latencies. The OLT latency factor is the difference between its egress latency and the scaled sum of its ingress and egress latency. For both the ONU and OLT, there are two scaling factors. The first is the ratio of the downstream index of refraction to the sum of the upstream and downstream indices of fraction, where the effective (i.e., wavelength-dependent) index values are used. In other words, it is the ratio of the upstream propagation speed to the sum of the upstream and downstream propagation speeds. The second scaling factor is the rateRatio, which is the ratio of the grandmaster clock frequency to the local clock frequency. As part of that computation, the 802.1AS standard provides a mechanism by which the OLT clock master can measure the rateRatio.

The OLT clock master selects a timing reference that is a future value X for its local MPCP counter. The value of X is arbitrary as long as it is adequately far in the future to be communicated to the ONUs in time and is within the current MPCP counter epoch. The clock master then calculates the value of the synchronized ToD when the ONU slave MPCP counter will reach X. This time value at the ONU will be ToD at count X at the OLT clock master, plus the difference between the OLT and ONU latency factors, plus the scaled RTT (the RTT scaling factors are, once again, the rateRatio and the ratio of the upstream propagation speed to the sum of the upstream and downstream propagation speeds). The OLT then uses the TIMESYNC message to send the ONU the value X and the adjusted ToD value that its local counter should have when it reaches a count of X.

3.6 Overview of the IEEE 1904.1 Service Interoperability in Ethernet Passive Optical Networks (SIEPON)

IEEE 802.3 specified the Layer 1 and Layer 2 aspects of EPON and 10G EPON. However, additional specifications are required in order to allow equipment from multiple vendors to interoperate in a network. The ITU-T standardizes these areas for G-PON and XG-PON, but they are outside the scope of the IEEE 802 activities, so the IEEE Communications Society consequently launched the P1904.1 project to address them [8]. The reference architecture for 1904.1 is shown in Figure 3.12, which illustrates the relative scope of the 802.3 and 1904.1 standards. Service-specific functions are optional on either an OLT

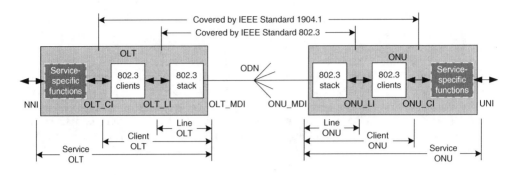

LEGEND:
OLT_MDI/ONU_MDI = Medium Independent Interface
OLT_LI/ONU_LI = Interface between the L-OLT and C-OLT / L-ONU and C-ONU
OLT_CI/ONU_CI = Client Interface
ODN = Optical Distribution Network

Figure 3.12 SIEPON reference architecture

or ONU. Since the SIEPON primarily addresses topics at higher layers than those covered in this book, this section will be restricted to an overview.

The major technical features of SIEPON include [9]:

- Management.
- QoS guarantees.
- Multicast service delivery over EPON.
- Power saving.
- VLAN modes and tunneling.
- Protection switching, including optical link monitoring.
- Data encryption.
- ONU authentication.
- ONU discovery and maintenance.
- Behavior of the MAC, MAC control and OAM clients.

Note that VLAN modes can be provisioned either for the entire ONU, or on a per-port basis.

The specifications are defined in terms of three service packages, each defining the required features for that package. The SIEPON reference model is unified to all packages but all specific features are package-specific. The feature requirements common to all packages are listed in Table 3.3, and the feature requirements that are different among the packages are shown in Table 3.4.

The ONU is further broken down into logical elements.

- Line-ONU (L-ONU), which represents the functions covered in IEEE 802.3/802.3av.
- Client-ONU (C-ONU), which represents a logical layer comprised of at least one L-ONU function, along with the associated clients (including the MAC Control, MAC, and OAM clients) that are required for proper network operation per 802.3/802.3av.
- The Service-ONU (S-ONU), which is comprised of a C-ONU, at least one UNI, and optional additional functionalities.

The OLT is similarly broken down into corresponding L-OLT, C-OLT and S-OLT elements. The additional functions that may be supported by the S-OLT include switching, POTS, and service initiation

Table 3.3 Features Required for Package A, B and C

Required feature
REPORT MPCP format
Report queue length calculation
Queue service disciple
ONU authentication (including secure provisioning)
Management (eOAM-based)
Device and capability discovery
Software update
Management entities
Power saving
VLAN support by ONU and OLT
Multicast connectivity
Multicast coexistence

Note: While these feature requirements are common to all three packages, there may be variations in how they are defined or implemented.

protocols to support delivering specific services to subscribers. Such services and solutions are typically outside the scope of 1904.1.

3.6.1 SIEPON MAC Functional Blocks

The MAC functional blocks specified by SIEPON include the Input, Classifier, Modifier, Policer/Shaper, Cross-Connect, Queue, Scheduler and Output blocks. Together, these blocks describe a unified data path architecture, which allows uniform provisioning and interoperability.

- The *Input* block is the ingress port that receives frames from the S-ONU or S-OLT (e.g., UNI, NNI, or MAC service frames).
- The *Classifier* function examines the frame headers in order to identify all frames, the EPON Service Path (ESP) to which they belong, what actions are required for that frame, and which queue should forward that frame. The Classifier operates on a set of rules that is composed of provisionable elements. The Classifier output vector specifies the actions of the Modifier, Policer/Shaper, and Cross-connect.
- The *Modifer* operates on the VLAN TAG information. Specifically, it is allowed to pass the tags, to add or remove tags, replace/alter tag fields of the outermost one or two tags, or take no action. The fields that it may modify are the TPID, PCP, CFI, DEI, or VID. The Modifier is also able to alter the IEEE 802.1ah fields.
- The *Policer/Shaper* enforces SLA conformance of the ESPs on a per-flow basis. It operates using a token bucket mechanism based on four parameters: rate, burst, action-on-conformant-frames, and action-on-non-conformant-frames. When functioning as a Policer, it deals with the coloring and discard-eligibility of frames and delaying non-conformant frames. When functioning as a Shaper, it manages the appropriate frame transmission delays.
- The *Cross-connect* routes each frame to the appropriate queue. In the case of multicast or broadcast flows, it replicates the associated packets and maps them to the appropriate set of queues.
- The *Queue* holds frames until they are polled by the Scheduler so that they can be transmitted. In addition to the data frames, the Queue inputs include control and coloring information. Its outputs include unmarked data frame, alarms, and statistics.
- A *Schedule* instance provides the multiplexing function for the frames stored within the subset of Queue block queues that are provisioned to it. In the case of the ONU upstream transmission, the OLT DBA controls its Scheduler through the mechanisms described earlier in this chapter. The Scheduler

Table 3.4 Features Requirements that are different for Package A, B and C

Feature	Package		
	A	B	C
Queue parameter discovery and configuration	N/A	Should implement discovery and configuration of queue parameters	N/A
ONU TRx status monitoring (see Note 1)	Shall implement ONU TRx status monitoring	Shall implement ONU TRx status monitoring	Shall implement ONU TRx status monitoring, associated alarms and warnings and management
OLT TRx status monitoring (see Note 2)	Shall implement OLT TRx status monitoring	Shall implement OLT TRx status monitoring	Shall implement OLT TRx status monitoring
UNI port loop detection	NA	NA	Shall implement UNI port loop detection
Events (see Note 3)	Shall implement events	Shall implement events	Shall implement events, with the event set/clear operation
Optical link protection, trunk type	NA	NA	Shall implement trunk optical link protection
Data encryption	Shall implement data encryption and integrity protection mechanism	Shall implement data encryption and integrity protection mechanism	NA
Performance monitoring	N/A	N/A	Shall implement performance monitoring
Tunneling modes	Both OLT and ONU shall support tunneling modes	N/A	N/A
MAC aging	N/A	N/A	Shall implement MAC aging function
Port selective loopback	Shall support port selective loopback	N/A	N/A

NOTES: 1. The ONU TRx status monitoring requirements are different for each package.
2. The OLT TRx status monitoring requirements are different for each package.
3. The event reporting requirements are different for each package.

polls frames within the Queue block's individual queues according to a predefined algorithm. It may shape the frame flows from either the input to individual queues or the output from the set of queues. Examples of scheduling algorithms include Weighted Fair Queue (WFQ), Strict Priority (SP), Round Robin (RR), and Weighted Round Robin (WRR). SIEPON does not specify a specific algorithm.

- Each *Output* block is associated with a single scheduler for its input, and forwards the packets to the appropriate service interface.

An ESP is the unidirectional path taken by a frame through these MAC client functional blocks, which fully determines its connectivity and QoS-related treatment. As noted above, this path is determined by the Classifier according to the rule that matches the frame.

3.6.2   VLAN Support

SIEPON supports a variety of VLAN modes, including transparent transport, translation, and filtering on port and device levels. It supports both tunneling and multicast VLAN operations. The 1904.1 standard elaborates on all the various configurations, including the use of VLAN C-Tags and S-Tags.

3.6.3   Multicast Service

SIEPON supports multicast and broadcast transmission in both directions, although it is most commonly used in the downstream direction.

In the downstream direction, at the physical layer the PON architecture naturally makes all frames available to all the ONUs subtended from an OLT. In that case, ONU filtering determines which ONU receives a multicast frame. The filtering can be performed on the basis of on one or more of the LLID, MAC address, VLAN tags, or IP address. The OLT inserts information into each multicast frame such that the ONU within its multicast group can recognize it. If the multicast group consists of both EPON and 10G EPON ONUs, then the OLT will replicate the multicast frames so that they are mapped into both the EPON and 10G EPON downstream signals. At the ONU, a multicast frame is assigned to a multicast ESP, where it is replicated to appear at the appropriate set of Output ports.

If the ONU is the source of the multicast frame, it uses a unicast channel to deliver the frame to the OLT. The OLT is provisioned to recognize the frame as multicast and to create the appropriate multicast routing.

3.6.4   SIEPON Service Management

The traffic types that an EPON system is expected to support are as follows. The equivalent DOCSIS definitions are described in Chapter 10.

- Real-time flows with periodic fixed-size data frames:
  - Using a CBR or quasi-CBR profile, in which the amount of information in a fixed time interval is relatively stable.
  - Equivalent to the DOCSIS Unsolicited Grant Service (UGS)
  - Applications include Circuit Emulation Service (CES) and mobile backhaul.
- Real-time flows with variable-sized data frames or with periodic inactivity:
  - Characterized by variable bit rate (VBR)
  - Real-time flows that generate variable-sized packets are equivalent to the DOCSIS Real-Time 1 Polling Service (rtPS)
    - Applications include IPTV service flows.
  - Real-time flows that may become inactive for substantial periods of time are equivalent to the DOCSIS Unsolicited Grant Service with Activity Detection (UGS-AD)
    - Applications include VoIP with silence suppression.

- Non-real-time flows that require throughput/frame loss guarantees:
  - Require regular variable-sized data grants.
  - Equivalent to the DOCSIS Non-Real-Time Polling Service (nrtPS).
  - Applications include tiered data services.
- Non-real-time, non-guaranteed flows:
  - Equivalent to the DOCSIS Best Effort Service.
  - Applications include non-guaranteed, non-real-time service flows.

The parameters that characterize the performance of packet-based networks include bandwidth (throughput), latency (packet delay), jitter (packet delay variation), and packet loss ratio. The QoS for each ESP is a matter of specifying and guaranteeing some or all of these parameters. The ESP can represent different constructs, depending on its configuration. Specifically, the ESP can be:

- a connection for sending frames of one or more types between a specific source and destination (e.g., a mix of IPTV and VoIP);
- a service flow consisting of all frames between a source and destination that share the same service type (e.g., multiple VoIP streams from the same ONU);
- a session consisting of all frames representing an instance of a service (e.g., and IPTV stream).

The parameters and metrics of 1904.1 follow the Metro Ethernet Forum (MEF) 10.2 specification. Specifically, they include:

- Committed Information Rate (CIR), which is the sustained bandwidth that is guaranteed for that service.
- Committed Burst Size (CBS), which is the maximum burst size that the network guarantees that it can handle for that service.
- Excess Information Rate (EIR), which is the amount of additional bandwidth beyond the CIR that a service is allowed to use in bursts when that bandwidth is available.
- Excess Burst Size (EBS), which specifies the maximum number of octets a service is allowed to send in a burst beyond its CIR/CBS.
- Peak Information Rate (PIR), which is the maximum sustained information rate allowed; data that is in excess of the PIR is dropped.
- Peak Burst Size (PBS), which is the maximum burst size allowed after the Policy/Shaper. The maximum burst size is specified in terms of the number of adjacent data frames that may be transmitted at the nominal interface rate.
- Frame delay, which includes the delays associated with the frame reception, internal processing and queuing, frame transmission and propagation. As such, they cover every link and every device through which a frame passes. The internal processing and queuing delays are a function of the QoS enforcement mechanisms, and they may be specified for each device as a single parameter called the Frame Residence Delay (FRD).
- Frame Delay Variation (FDV), which specifies the difference between the maximum and minimum samples of the FRD. It addresses the variation that inherently results from factors including the burstiness of arriving frames, queue depths, and scheduling mechanisms.
- Frame Loss Ratio (FLR), which is the ratio of the numbers of frames that are lost between the ingress and egress relative to the expected number of transmitted frames.

3.6.4.1 QoS Configurations

SIEPON supports three types of QoS configurations that can be used to satisfy the requirements of the different profiles. The first configuration supports the DOCSIS Protocol over Ethernet (DPoE) QoS model. The second configuration uses dedicated thresholds. Here, each ONU reports in a queue set the traffic it has queued up to a given threshold. The third configuration uses cumulative thresholds. Here, each ONU services its traffic in queue order up to a given threshold and reports it accordingly. All three configurations are implemented using the EPON REQUEST MPCPDU described above, which accounts for both user data frames and OAM Client frames queued at the ONU.

One of the required features to achieve the desired QoS is a queue service discipline. This is used by the L-ONU to serve its packet queues, and by the C-ONU as the rules it follows in determining the quantity of data and the order by which it serves its upstream queues.

Note that for each configuration, the queue length reports are made on the basis of queue sets, which are collections of queues. Specifically, the REPORT message communicates the occupancy of the collection of queues comprising all the queue set in a single bitmap, rather than reporting for each of the queues separately.

QoS Configuration to Support the DOCSIS Protocol over Ethernet (DPoE) QoS Model

The DPoE QoS model is followed here. Each L-ONU serves a single queue from which it transmits in FIFO order, and each L-ONU typically supports a single of service. The REPORT message carries the transmission request information for an L-ONU. Each REPORT requests bandwidth for 1–4 queue sets, where the number of queue sets depends on the thresholds of the queue fill.

A C-ONU can contain multiple L-ONUs. The OLT makes its bandwidth grant decisions based on the REPORT information from the different L-ONUs. The OLT grants each L-ONU a separate transmission window, and the C-ONU is not allowed to re-allocate the bandwidth grants among its L-ONUs.

QoS Configuration Supporting Dedicated Thresholds

The threshold-first ONU queue service discipline uses multiple queue thresholds for each given queue. The REPORT message from the ONU communicates the bandwidth required to send the whole data frames in each queue that are under each of the threshold values. A higher order queue set has a higher threshold than a lower order queue set, and thus the bandwidth it requests in the REPORT message includes the bandwidth requested by the lower order set.

When the OLT grants upstream bandwidth to an ONU using the threshold-first ONU queue service discipline, it first serves the data that is under the threshold of all its lowest queue sets, and then serves the data that is under the threshold of its next higher queue sets, and so on.

QoS Configuration Supporting Aggregated Thresholds

ONUs supporting the aggregated thresholds are also required to support the threshold-first queue discipline described above, and to report their per-queue bandwidth requirements using the same queue set threshold format. In addition to the threshold-first queue discipline, however, ONUs supporting aggregated thresholds should also support the Priority-first Queue Service Discipline described in this section.

There are two approaches to implementing the Priority-first Queue Service Discipline. For both approaches, the ONU uses queue priority rather than queue thresholds for its upstream transmission decisions. The first is the reported-first approach, in which the ONU sends all the data it had reported in higher priority queues before serving lower-priority queues, regardless of the queue thresholds. The second approach is the strict-priority approach. Here, ONU preempts lower-priority frames that it has reported, in order to send higher-priority frames that have arrived at the ONU since it sent its REPORT.

3.6.5 Performance Monitoring and Verification

SIEPON specifies performance monitoring requirements for the ONU and OLT. The performance associated with each LLID and each L-OLT is monitored by the C-OLT. The performance associated with each UNI port and each L-ONU is monitored by the C-ONU. All of the performance monitoring data processing is performed by the OLT or Element Management System (EMS). The ONU only collects data and consequently only retains data history from the last performance monitoring period. The default performance monitoring period is 15 minutes.

The performance monitoring parameters for both the downstream and upstream directions include counts of drop events, octets, frames, broadcast frames, multicast frames, CRC-errored frames, undersized frames, oversized frames, fragments, jabbers, discards, errors, and frames within specific size bins (64, 65–127, 128–255, 256–511, 512–1023, and 1024–1518 octets). Status change timers are also monitored.

3.6.6   SIEPON Service Availability

SIEPON addresses service availability by supporting device and transceiver monitoring, with alarms and warnings, and supporting optical link protection and control of the ONU power supply.

The monitored parameters are those defined by IEEE 802.3, including received errored symbols and frames, errored symbol and error frame periods, and errored frame seconds. They also include counts of the number of corrected FEC blocks or codewords.

In addition to performance parameters, the ONU and OLT monitor their optical transceiver's parameters, including internal temperature, bias current and power supply voltage. The optical transmitter's output power and the optical receiver's input power are also monitored.

The mandatory ONU alarms are the equipment alarm, power alarm, battery missing, battery failure, low battery voltage, intrusion alarm, self-test failure, high and low temperature alarms, and a sleep status update (i.e., to indicate that the ONU has returned from power saving mode to active mode and the reason for becoming active). An Integrated Access Device (IAD) connection failure alarm is required if the ONU supports an internal VoIP client. A PON interface switch alarm is required if the ONU supports optical link protection.

SIEPON also supports the signal failure and degradation events defined by ITU-T G.984 for G-PON. If the ONU UNI participates in the client's LAN Spanning Tree Protocol (STP), it indicates an alarm if it detects a STP loop condition.

Loopbacks are used at the UNI ports for diagnostics. More than one UNI port can be looped back simultaneously and loopbacks can be set up between two or more UNI ports. The loopbacks are requested by the OLT, using the IEEE 802.3 loopback function.

3.6.7   SIEPON Optical Link Protection

SIEPON specifies optical link protection for both the trunk side and tree side. It supports a variety of alternatives, such as those described in Chapter 2. It also supports protecting the circuit functions described above in Section 3.6.1.

3.6.8   SIEPON Power Savings

SIEPON supports a range of power savings methods, similar to those supported by G-PON. Power savings methods are discussed in Chapter 2. The EPON power savings requirements include:

- ONUs that support power savings and those that do not must co-exist on the same PON.
- The packet loss rate, delay, and jitter resulting from power saving must not increase beyond the limit specified in the SLA for that service.
- The ONU data load, user activity cycles, and both the configured and active services on an ONU should be taken into account by the power savings mechanism.

- The power savings mechanism should not interfere with the MCPC and higher layer protocols (for example, the sleep time must be short enough that the ONU does not become deregistered due to an MCPC timeout)[18].

SIEPON specifies power savings methods that are initiated by either the OLT or the ONU. The ONU can either support a sleep mode in which just its transmitter is shut down, or a sleep mode in which both the transmitter and receiver are shut down.

3.6.9 SIEPON Security Mechanisms

SIEPON expands on the encryption and authentication specifications. EPON, as discussed above, did not fully specify link encryption and ONU authentication.

3.6.10 SIEPON Management

As discussed above, EPON provided all the required message types for management. SIEPON builds on these to specify management approaches and messages. One approach is optimized for SNMP-based management, and the other for non-SNMP-based management. A full set of management entities is specified for both. It also elaborates on device and capability discovery and software updates.

3.7 ITU-T G.9801 Ethernet Passive Optical Networks using OMCI

Carriers interested in using EPON want to be able to integrate it seamlessly into their network management systems. Furthermore, with the standardization of 10G EPON, it is expected that some carriers will deploy a mix of G-PON, 10G EPON, and XG-PON. The 10G EPON would primarily be used to serve enterprise customers who desire symmetrical 10 Gbit/s capacity.[19] For these reasons, ITU-T extended its ONU management and control interface (OMCI) to support EPON. As additional higher layer functionalities have been added to EPON, including Quality of Service, protection switching and ONU power management, it became desirable to have a standard that covered the whole system specification. These extensions are specified in ITU-T G.9801 [10].

3.8 Conclusions

Ethernet-based PON protocols have the advantage of being able to use the substantial Ethernet eco-system, which extends from physical layer devices through application and OAM software. Ethernet PON hardware is typically less complex than the FSAN/ITU-T protocols discussed in the next chapter. The primary complexity reduction is due using the more relaxed specifications of 1000BASE and 10GBASE Ethernet physical layer components. Complexity is also reduced by restricting the protocol to Ethernet frames rather than adding additional modes for constant bit rate clients. As IP becomes the common protocol for voice (VoIP) and video (IPTV) in addition to data services, Ethernet is a good fit for Layer 1 and 2 functions. EPON has seen extensive deployment, especially in countries like Japan and South Korea. Ethernet-based PON is also being deployed by Cable Television network providers, especially to provide services to enterprise customers. For that reason, the cable network operators have been very involved in the SIEPON standard development. Also, as described in Chapter 6, the DOCSIS protocol has been extended to operate over Ethernet PON systems.

A primary feature of 10G EPON is its ability to interoperate in a backwards compatible manner with EPON equipment on the same PON. This feature allows a network provider a straightforward solution for upgrading PON bandwidth on an as-needed basis without rendering already deployed ONUs obsolete.

---

[18] Note, however, that a timestamp drift error resulting from the sleep period could force the deregistration of the ONU.

[19] Since 10G EPON was defined prior to XG-PON, ITU-T chose not to define a 10 Gbit/s symmetric option for XG-PON. The assumption was that there was no need to have both protocols target these applications.

Once the OLT has been upgraded, an ONU on the PON can be upgraded from 1G to 10G without disrupting other ONUs.

The G.9801 further expands the potential applications for Ethernet-based PON to carriers that otherwise deploy the FSAN/ITU-T PON systems.

In the next chapter, we examine the FSAN/ITU-T PON protocols that were optimized for protocol and service compatibility with the existing telecommunications networks.

Appendix 3.A: 64B/66B Line Code

There are two primary requirements for any line code. The first is to ensure an adequate number of line code level transitions within a given time window so that the receiver can align the frequency and phase of its clock and data recovery circuit to the incoming signal. This alignment is critical for making a decision about the value of the received bit or symbol at the point with the least impact from noise, distortion, and inter-symbol interference. The second requirement is that there must be a method of determining the boundaries of the line code characters.

Gigabit/s Ethernet typically uses the 8B/10B line code in which eight data bits are mapped into ten bits for transmission. The 256 8-bit data values are mapped into pairs of 1024 possible 10-bit characters in a manner that maintains a running balance between the transmitted ones and zeros (the running disparity). Some of the 10-bit codes not used for data encoding are used to communicate control information. Special Idle control characters are transmitted between packets that allow the receiver to synchronize the 8B/10B characters to the alignment boundaries. The 25% bandwidth overhead of the 8B/10B code becomes a problem with high-speed signals. Consequently, the more bandwidth-efficient 64B/66B line code was developed for 10GE.

The 64B/66B block code consists of eight octets (64 bits) of data and/or control characters preceded by two flag bits that indicate whether the block contains control characters or only data characters. The mapping is performed in a manner that allows mapping all the 8B/10B special control characters into the 64B/66B code. The two leading code bits also provide the means for the receiver to synchronize to the code block boundaries. The code maintains an average balance between transmitted ones and zeros through a scrambler, although the balance is not as strictly maintained in the short term as with the 8B/10B code.

Figure 3.13 illustrates the 64B/66B code construction, and the following bit and byte label conventions are used. Hexadecimal values are used to represent the values of the block type fields, control characters, and data octets. In Figure 3.13, the information, including any binary values, is transmitted from left to right with the LSB (bit 0) of the hexadecimal values transmitted first. For example, a block type value of $0x87 = 10\,001\,110$ is transmitted from left to right as $01\,110\,001$. The transmitted (received) block bits are designated as TxB < 65:0> (RxB < 65:0 >).

The 64B/66B code can either be a data block or a control block. A synchronization header (sync header) value of 01 indicates a data block where the payload of the 64B/66B code contains only Ethernet frame data characters. A sync header of 10 indicates a control block in which the 64B/66B block contains control information such as Ethernet frame Start or Termination characters, or control characters. A sync header of 10 is immediately followed in the payload area by an 8-bit field (the Block Type Field or BTF) that specifies the structure of the 64B/66B code payload. The receiver can achieve block code alignment by searching for an adjacent pair of bits that always have complementary values (i.e., a transition between their values). The payload of the 64B/66B code is scrambled, which makes it highly unlikely that any two payload adjacent bit positions will retain complementary values for a significant period of time.

Data blocks contain eight data characters, and control blocks contain eight characters that are either all control characters or a combination of control and data characters, as specified by the BTF. Two different techniques are used to fit eight characters into the 56 block payload bits after the BTF. First, control characters are encoded as 7-bit values or 4-bit O-codes. Second, the locations of the Ethernet frame Start and Termination characters are implied by the BTF, and hence these characters do not need to be carried explicitly within the block.

Block Type	Input Data (Block Format)	Sync	Block Payload							
Data	$D_0D_1D_2D_3D_4D_5D_6D_7$	01	D_0	D_1	D_2	D_3	D_4	D_5	D_6	D_7
			B.T.F.							
Control	$S_0D_1D_2D_3D_4D_5D_6D_7$	10	0x78	D_1	D_2	D_3	D_4	D_5	D_6	D_7
	$C_0C_1C_2C_3S_4D_5D_6D_7$	10	0x33	C_0	C_1	C_2	C_3		D_5	D_6	D_7
	$O_0D_1D_2D_3S_4D_5D_6D_7$	10	0x66	D_1	D_2	D_3	O_0		D_5	D_6	D_7
	$O_0D_1D_2D_3O_4D_5D_6D_7$	10	0x55	D_1	D_2	D_3	O_0	O_4	D_5	D_6	D_7
	$C_0C_1C_2C_3O_4D_5D_6D_7$	10	0x2D	C_0	C_1	C_2	C_3	O_4	D_5	D_6	D_7
	$O_0D_1D_2D_3C_4C_5C_6C_7$	10	0x4B	D_1	D_2	D_3	O_0	C_4	C_5	C_6	C_7
	$C_0C_1C_2C_3C_4C_5C_6C_7$	10	0x1E	C_0	C_1	C_2	C_3	C_4	C_5	C_6	C_7
	$T_0C_1C_2C_3C_4C_5C_6C_7$	10	0x87		C_1	C_2	C_3	C_4	C_5	C_6	C_7
	$D_0T_1C_2C_3C_4C_5C_6C_7$	10	0x99	D_0		C_2	C_3	C_4	C_5	C_6	C_7
	$D_0D_1T_2C_3C_4C_5C_6C_7$	10	0xAA	D_0	D_1		C_3	C_4	C_5	C_6	C_7
	$D_0D_1D_2T_3C_4C_5C_6C_7$	10	0xB4	D_0	D_1	D_2		C_4	C_5	C_6	C_7
	$D_0D_1D_2D_3T_4C_5C_6C_7$	10	0xCC	D_0	D_1	D_2	D_3		C_5	C_6	C_7
	$D_0D_1D_2D_3D_4T_5C_6C_7$	10	0xD2	D_0	D_1	D_2	D_3	D_4		C_6	C_7
	$D_0D_1D_2D_3D_4D_5T_6C_7$	10	0xE1	D_0	D_1	D_2	D_3	D_4	D_5		C_7
	$D_0D_1D_2D_3D_4D_5D_6T_7$	10	0xFF	D_0	D_1	D_2	D_3	D_4	D_5	D_6	
Bit Position		0 1 2								65

B.T.F. = Block Type Field

Figure 3.13 64B/66B block code structure

The mapping of data and control characters into the 64-bit code payload is designed around the 32-bit XGMII interface. Specifically, data blocks are filled with two four-octet transfers from the XGMII. The data octets and control characters are designated according to their position within the code block, which also corresponds to their position within the eight characters of the XGMII bus transfer. For example, if the block contains eight data octets, these octets are designated D0 to D7, with D0 located immediately after the leading flag (synchronization) bits. The control octets are encoded as 8-bit values on the XGMII interface but, as discussed above, they are encoded into 7-bit or 4-bit values when encoded into a 64B/66B control block.

The set of legal control code encodings is shown in Tables 3.5 and 3.6. Any other control codes are regarded as errors if they are received. The control codes for start of packet (/S/) are designated S0 or S4, and the start of an ordered set control character (/O/) is designated O0 or O4 since these are only valid on

Table 3.5 Encoding for non-ordered set control codes

Control character	Notation	XGMII control code	10G-BASE-R 7-bit control code	8B/10B code
Idle	/I/	0x07	0x00	K28.0, K28.3 or K28.5
Start	/S/	0XFB	Implied by BTF	K27.7
Terminate	/T/	0xFD	Implied by BTF	K29.7
Error	/E/	0xFE	0x1E	K30.7
reserved0	/R/	0x1C	0x2D	K28.0
reserved1			0x33	K28.1
reserved2	/A/	0x7C	0x4B	K23.3
reserved3	/K/	0xBC	0x55	K28.5
reserved4		0xDC	0x66	K28.6
reserved5		0xF7	0x78	K28.7

Note: The/A/,/R/, and/K/codes are XAUI interface codes that can only appear on the XGMII due to bit errors.

Table 3.6 Encoding for ordered set control codes

Control character	Notation	XGMII control code	10G-BASE-R 4-bit O-code	8B/10B code
Sequence ordered_set	/Q/	0x9C	0x0	K28.4
Signal ordered_set	/Fsig/	0x5C	0xF	K28.2

the first octet or the fourth-octet XGMII position (note that while the start and end of packet information is communicated across the XGMII, the ordered set information is typically not communicated over the XGMII). The end of the packet control code (/T/) can occur anywhere within the code block, and thus can be designated T0 to T7. Similarly, control codes other than /S/, /O/, and /T/ can be located in any position within the block and are therefore designated C0 to C7.

When the O0 O-codes are encoded as 4-bit values they are placed in the first nibble of the fifth octet of the 64B/66B block payload, and the 4-bit value representing the O4 O-codes are placed in the last nibble of the fifth octet of the 64B/66B block payload. O-codes represent the start of a 4-octet ordered set.

The narrow rectangles in Figure 3.13 represent single pad bits within the control block payload area. These pad bits contain a value of 0 and are used to achieve the desired alignment of the control and data characters when their encoding leads to fewer than 56 total bits following the BTF.

A self-synchronous scrambler is used on the block payload in order to help to randomize the payload bits. The sync header bits bypass the scrambler. The resulting effects of the scrambler are to improve DC

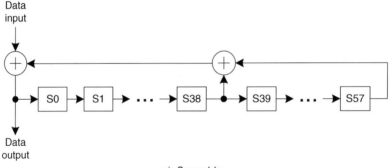

a) Scrambler

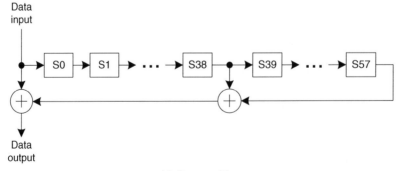

b) Descrambler

Figure 3.14 64B/66B block code self-synchronous scrambler

balance, to reduce the likelihood of payload bits mimicking the sync header bits, and to increase the average number of data value transitions, which aids the receiver data recovery circuit. Self-synchronous scramblers avoid the need for initialization or periodic explicit synchronization by feeding the data back on itself through exclusive-OR gates, as illustrated in Figure 3.14. The descrambler uses a circuit similar to the scrambler in order to reverse the process. The scrambler polynomial is: $G(x) = 1 + x^{39} + x^{58}$.

References

1. IEEE802.3. IEEE Standard for Ethernet; 2012.
2. IEEE802.3bk. Amendment 1: Physical Layer Specifications and Management Parameters for Extended Ethernet Passive Optical Networks; 2013.
3. IEEE802.3ah. Amendment to - Information technology - Telecommunications and information exchange between systems - Local and metropolitan area networks – Specific requirements - Part 3: Carrier sense multiple access with collision detection (CSMA/CD) access method and physical layer specifications - Media Access Control Parameters, Physical Layers and Management Parameters for subscriber access networks; 2004.
4. Federal Information Processing Standard 197 Advanced Encryption Standard, National Institute of Standards and Technology, U.S. Department of Commerce; November 26 2001.
5. IEEE802.3av. Amendment: Physical Layer Specifications and Management Parameters for 10Gb/s Passive Optical Networks; 2009.
6. IEEE802.1as. IEEE Standard for Local and Metropolitan Area Networks – Timing and Synchronization for Time-Sensitive Applications in Bridged Local Area Networks; 2011.
7. IEEE1588. Standard for Precision Clock Synchronization Protocol for Networked Measurement and Control Systems; 2008.
8. IEEE1904.1. Service Interoperability in Ethernet Passive Optical Networks (SIEPON); 2013.
9. Kramer G. *et al.* The IEEE 1904.1 Standard: SIEPON Architecture and Model. *IEEE Communications Magazine.* 2012; **50**(9):98–107.
10. ITU-TG.9801. Ethernet Passive Optical Networks using OMCI; 2013.
11. Haran, O. and Hiironen, O. CTR Mode for Encryption. 2002; http://grouper.ieee.org/groups/802/3/efm/public/jul02/p2mp/haran_p2mp_2_0702.pdf.

Further Readings

1. Gorshe S. FTTH/FTTH technology and standards. *China Communications.* 2006; **3**(6):104–114.
2. Gorshe S. Overview and comparison of the IEEE EPON and ITU-T GPON protocols. *China Communications.* 2007; **4**(2):69–78.
3. Gorshe S. *Introduction to IEEE 802.3av 10Gbit/s Ethernet Passive Optical Networks (10G EPON)*, PMC-Sierra, PMC-2081212; 2010.
4. IEEE802.1AV. Standard for Local and metropolitan area networks – Timing and Synchronization for Time-Sensitive Applications in Bridged Local Area Networks; 2011.
5. ITU-TG.983.4. A broadband optical access system with increased service capability using dynamic bandwidth assignment (DBA); 2001.
6. ITU-TG.983.7. ONT management and control interface specification for dynamic bandwidth assignment (DBA) B-PON systems; 2001.

4

ITU-T/FSAN PON Protocols

4.1 Introduction

The Full Service Access Network (FSAN) consortium was formed under the leadership of several telephone network providers (carriers) with the goal of researching, standardizing, and promoting broadband access solutions and services. FSAN settled on PON technology as the best long range network solution, and the first FSAN PON protocol was called Broadband-PON (B-PON). FSAN chose to have the ITU-T publish and maintain the protocols it develops – a logical choice, since the carriers were already very active in ITU-T, and ITU-T also had expertise in the various areas of technology, including optical link physical layers. The working model between the two bodies has continued to be for FSAN to define the requirements of new PON protocols and functions. The high level descriptions of the protocol are developed within FSAN, and the details are developed within, and published by, ITU-T.

The B-PON protocol, specified in the ITU-T G.983 Recommendation series, provided data rates in the range of hundreds of Mbit/s on the PON (tens of Mbit/s per ONU).[1] The protocol was based on ATM technology, which was then ubiquitous in DSL systems. B-PON has seen some significant deployments, most notably the initial phase of the Verizon FiOS[TM][2] FTTH network in the USA. B-PON is described briefly in Section 4.2 of this chapter.

The second generation of FSAN/ITU-T protocols was called Gigabit-capable PON (G-PON), and this provided data rates in the Gbit/s range (up to 100 Mbit/s per ONU on average). G-PON was developed to take advantage of the decreasing price for optical components, and also to compete with EPON, which had higher bandwidth than B-PON. G-PON also moved from ATM to a more bandwidth-efficient adaptation method optimized for carrying Ethernet frames. G-PON is specified in the ITU-T G.984 series, although it re-uses portions of the G.983 series standards as appropriate (e.g., for the protection and dynamic bandwidth algorithm descriptions). G-PON is described in Section 4.3 of this chapter.

FSAN is currently developing next generation protocols, both for the relatively near term (XG-PON) and also for the longer term (NG-PON). The details of XG-PON are described in Section 4.4. The NG-PON work is just beginning, so it cannot be covered in any level of detail in this book.

---

[1] The per-ONU rates here are very approximate, since there are typically multiple PON rates specified for an FSAN protocol, different numbers of ONUs can be deployed on a PON, and different service rates can be offered to different subscribers.

[2] FiOS (Fiber Optic Service) is Verizon's (and in some regions, Frontier's) FTTH network and service offering for combined support of data, voice, and video.

Broadband Access: Wireline and Wireless – Alternatives for Internet Services, First Edition.
Steven Gorshe, Arvind Raghavan, Thomas Starr, and Stefano Galli.
© 2014 John Wiley & Sons, Ltd. Published 2014 by John Wiley & Sons, Ltd.

4.2   ITU-T G.983 Series B-PON (Broadband PON)

Most of the PON systems deployed in North America and Europe prior to 2008, including initial phases of Verizon's ambitious FiOS^{TM} project, used the ITU-T G.983 series B-PON. Since most carriers that deployed B-PON are now moving to G-PON, the B-PON discussion will not have the same level of detail as G-PON or the IEEE PON protocol descriptions.

The G.983 series includes specifications of the ONT and OLT functional blocks, the upstream and downstream frame rates and formats, the TDMA upstream access protocol, physical interfaces, ONT management and control interfaces, survivability enhancements, and the framework for a dynamic bandwidth allocation algorithm (DBA) [1–10]. The B-PON features are summarized in Table 4.1.

The downstream transmission is a stream of ATM cells. A downstream frame consists of 56 53-byte cell slots for 155 Mbit/s and $4 \times 56 = 224$ cell slots for 622 Mbit/s, with a physical layer OAM (PLOAM) cell inserted every 28 cell slots. The PLOAM contains a framing bit to identify the PLOAM cells. Otherwise, the PLOAM cells are programmable and contain information such as upstream bandwidth grants and OAM messages. The ONTs use the ATM cell VPI/VCI addresses to identify their data in the downstream signal.

The upstream frame consists of 53 56-byte time slots. Each time slot is comprised of an ATM/PLOAM cell and 24 bits of overhead. The overhead consists of guard time, a preamble to allow timing and signal level recovery by the OLT, and a delimiter to indicate the end of the overhead. The overhead field lengths and contents are programmable by the OLT. ONTs transmit PLOAM cells when they are requested by the OLT.

One advantage of using ATM as the underlying Layer 1+ protocol in B-PON is that the adaptation methods for most clients into ATM were already specified. B-PON uses these standard ATM adaptation techniques in same way as DSL systems.

The bandwidth grant information from the OLT tells each ONT which upstream time slots it may use for its upstream data. The B-PON DBA protocol allows the OLT to learn the ONT bandwidth needs either through explicit reports from the ONTs, and/or by observing the number of ATM Idle cells that the ONTs transmit. The OLT can decrease the bandwidth of an ONT sending Idles and increase the bandwidth of an ONT that is filling all its upstream transmission slots with data.

The OLT periodically halts upstream transmissions so that it can invite any new ONTs to announce themselves. The new ONTs transmit a response during this window, using a random time delay in order to

Table 4.1 B-PON specification summary

Feature	B-PON
Responsible standards body	FSAN and ITU-T SG15 (G.983 series)
Data rate	155.52 Mbit/s upstream and 155.52 or 622.08 Mbit/s downstream
Split ratio (ONUs/PON)	1 : 64
Line code	Scrambled NRZ
Number of fibers	1 or 2
Wavelengths	1310 nm up and down or 1490 nm down and 1310 nm up
Maximum OLT to ONU distance	20 km
Protection switching	Supports multiple protection configurations
Data format (encapsulation)	ATM
TDM support	via ATM
Voice support	via ATM
Multiple QoS levels	Yes (mix of fixed, assured, and best effort bandwidth assignments)
FEC	None
Encryption	Churning
OAM	PLOAM and ATM

minimize the risk of collisions if there are multiple new ONTs. The OLT determines the distance to each new ONT by sending it a ranging message and measuring the time until it receives the response. The OLT then sends the ONT an equalization delay time value such that the sum of the round trip and equalization delays is the same for each ONT. This allows the upstream transmissions from the ONTs to arrive at the OLT with a minimum of guard time.

4.3 ITU-T G.984 Series G-PON (Gigabit-capable PON)

As discussed in Section 2.2 of Chapter 2, G-PON is a third generation PON protocol, and the second generation developed through the FSAN consortium. G-PON was developed to support higher data rates that were becoming achievable through technology advances. While the EPON standard was available at the time, the G-PON standard was able to add two improvements from a telephone carrier perspective. The first was that carriers desired a payload mapping method that simplified the process of carrying voice and traditional TDM telecom interface data. The second was that downstream rates of roughly 2.5 Gbit/s had become feasible.

ATM, which was used in B-PON, requires a substantial amount of adaptation processing for mapping packets such as Ethernet frames. In addition, since each 53-byte ATM cell has at least five overhead bytes, ATM is inefficient in its bandwidth use. For this reason, a new encapsulation method known as the G-PON Encapsulation Method (GEM) was adopted for G-PON. GEM was based in large part on the Generic Framing Procedure (GFP) encapsulation method of ITU-T G.7041. [11] GEM, including its relationship to GFP, is discussed in detail in Section 4.3.3.[3]

This section describes the full range of options for G-PON. A group of access carriers in FSAN agreed to a Common Technical Specification (CTS), which specifies their common subset of options. The goal of the CTS was to help vendors expedite the delivery of G-PON equipment to the market. The CTS subset of options has largely become the only set of options supported in more recent versions of the G.984 series standards revisions [12–18].

Similar to the EPON protocol, G-PON supports auto-discovery of ONUs and FEC. The similarities and differences between EPON and G-PON are summarized in the appendix of this chapter.

4.3.1 G-PON Physical Layer

The G-PON physical layer is defined in G.984.2 [13]. G-PON supports split ratios of 1 : 64 (i.e., up to 64 ONUs on the fiber from a single OLT interface).[4]

The optical parameters are specified in G.983.1 [1] and Table 2 of G.984.2 [13], with three classes of path loss specified in G.982 [19]. The optical loss budgets for the three classes are summarized in Table 4.2. It has become common for implementations to use an overall loss budget between Class B and Class C, referred to as Class B+. Longer reach applications can use the enhanced C+ class that can allow up to 60 km distance between the OLT and ONU.[5] The Class C+ requires enhanced receiver sensitivity through the use of the FEC option at both the ONU and OLT and something like a Semiconductor Optical Amplifier (SOA) preamplifier at the OLT. The protocol supports a maximum OLT/ONT fiber length difference between ONTs on the same fiber tree of 20 km.

Note that the G-PON reference models assume the use of an APD receiver. Improvements in transimpedance amplifiers (TIA) have since made it possible to support the B+ class even without FEC.

---

[3] The first release of the G-PON standard supported using ATM in addition to GEM. When the OLT or an ONU supported both ATM and GEM, the ATM cells were transmitted at the beginning of the payload frame followed by the GEM frames. The ATM option was not implemented and was subsequently removed from the G-PON standards in early 2008.

[4] The 1:64 split ratio is based on the optical power budgets. From logical protocol standpoint, G-PON can support up to a 1:128 split ratio.

[5] From a protocol (logical) standpoint, 60km is the longest reach that could be supported by G-PON. The distance limitation comes from the ONU distance delay compensation discussed below.

Table 4.2 G-PON optical link class optical attenuation ranges

Class	Optical loss budget (min; max. attenuation range)
A	5–20 dB
B	10–25 dB
B+	10–28 dB
C	15–30 dB
C+	17–32 dB

In contrast, the EPON reference models are based on PIN receivers that are significantly less expensive than the G-PON APD reference receivers. However, the EPON reference models only support 4 dB less link loss.

Another significant difference between the EPON and G-PON physical layer specifications is the time allocated for the ONU lasers to turn on and off for an upstream burst transmission. The G-PON guard time between bursts of different ONUs is a maximum of 12.86 ns. The time allocated for an ONU laser to turn on and turn off must be less than the combination of the guard time and any uncertainty in the signal flight time. In contrast, EPON specifies a maximum of 512 ns for the ONU lasers to turn on or turn off. Consequently, the laser control circuitry for G-PON is more complex than for EPON. This had a small negative effect on G-PONs early availability but, in the present day, both G-PON and EPON operate with the same short transmitter guard times.

While the initial G-PON standard supported multiple physical layer options, the standard eventually settled on a single approach, which uses a single fiber with wave division multiplexing (WDM) to achieve the upstream/downstream data separation. The OLT lasers use the 1480–1500 nm band, while the ONTs use less expensive lasers in the 1260–1360 nm band.

In all cases, the NRZ line code is used with a scrambler in the Transmission Convergence (TC) layer to guarantee an adequate density of transitions on the line. The scrambler is frame synchronous with the polynomial $x^7 + x^6 + 1$. Each bit to which the scrambling is applied is Exclusive-OR'ed with the output stage of the scrambler. The scrambler is reset to all ones at the first bit following the physical layer synchronization field and runs through the rest of the frame.

The G-PON upstream/downstream data rate options are listed in Table 4.3. While the original G-PON standards included several lower speed options, the industry has settled on using the 2.48832 Gbit/s downstream rate and either 1.24416 Gbit/s or 2.48832 Gbit/s upstream.[6] Consequently, the 2008 revision to the G-PON standards includes only these options. The predominant option being deployed and planned is the one using 2.488 Gbit/s downstream and 1.244 Gbit/s upstream.[7]

Table 4.3 G-PON downstream and upstream data rates

Upstream	Downstream
1.24416 Gbit/s	2.48832 Gbit/s
2.48832 Gbit/s	2.48832 Gbit/s

---

[6] Note that the 2.48832 Gbit/s rate is convenient from a carrier perspective since it is identical to the rate of the SONET STS-48 (SDH STM-16) transport signal rate. The B-PON rates were similarly identical to lower SONET/SDH transport signal rates.

[7] This is the option specified in the CTS.

4.3.2 G-PON Frame Formats

Data is mapped onto the G-PON physical layer through the G-PON Transmission Convergence (GTC) layer, which is defined in G.984.3 [14]. In both the downstream and downstream directions, there is a GTC frame, compromised of physical layer overhead (PLO) fields and payload areas. Several of the downstream and upstream fields are covered by a CRC-8 error check code. Each CRC-8 uses a $g(x) = x^8 + x^2 + x + 1$ polynomial.

4.3.2.1 Downstream Frame Format

Figure 4.1 illustrates the GTC downstream frame format. The frame begins with the downstream Physical Control Block (PCBd), which contains various fields associated with physical layer parameters, the

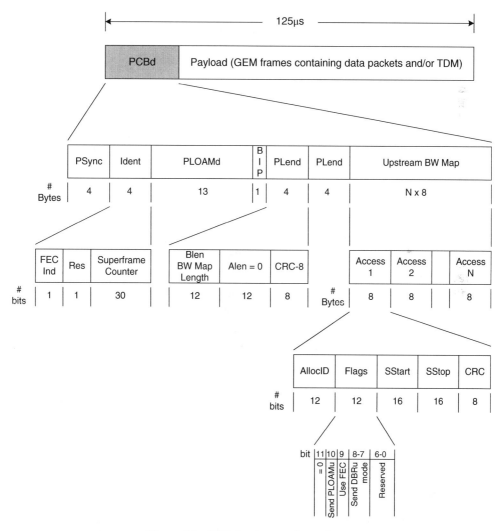

Figure 4.1 GTC downstream frame format

downstream frame payload format, and the upstream bandwidth assignments that the ONTs will use during their next upstream frame. Table 4.4 summarizes these fields.

The PCBd begins with a 32-bit Physical synchronization (Psync) field. The ONUs use the Psync field as a framing pattern to locate the beginning of the downstream frame.

The downstream payload data follows the PCBd and consists of GEM frames. The total length of the payload area is specified by the BLen field. Each ONU recognizes its data based on the Port ID field in the GEM frame.

As discussed in the next section, the OLT uses the upstream bandwidth map (US BW Map) fields to specify the transmission times granted to ONUs for the next upstream frame.

Table 4.4 Description of PCBd fields

Field	(Sub-field) description		
Psync	The frame alignment pattern used by the ONUs. The pattern is 0xB6AB31E0, and is not scrambled.		
Ident	FEC Ind.	Indicates whether the downstream frame is using FEC.	
	Superframe counter	30-bit wrapping counter, incremented once per GTC frame, that can be used for superframe structures such as data encryption and lower rate reference signals.	
PLOAMd	13-byte field to carry the downstream physical layer OAM message		
BIP	BIP-8 error check over the frame (i.e., all the bits between it and the last BIP)		
PLend	Downstream payload length field. The PLend field is sent in duplicate, with the ONU choosing the least errored copy.		
	BLen	Length of the upstream bandwidth map field (BLen value times 8 = the number of bytes in the field)	
	ALen	This field originally indicated the length of the ATM portion of the payload area. Since ATM payloads are no longer supported, this field has a fixed value of 0.	
	CRC-8	CRC check over the PLend field. It allows single error correction.	
US BW Map	Upstream bandwidth map field – each Alloc. field is the bandwidth allocation to a particular T-CONT (N is specified by the BLen field)		
	AllocID	T-CONT to which the bandwidth allocation applies (By convention, the 8 LSBs identify the ONU with ID = 254 reserved for the ONT Activation ID for discovering unknown ONUs and ID = 255 reserved as the unassigned AllocID.)	
	Flags	Send PLOAMu	Request for the ONU to send its PLOAM information
		Use FEC	Request for the ONU to use FEC for its upstream transmissions of this allocation
		Send DBRu mode	00 Don't send DBRu
			01 Send DBRu mode 0 (two bytes)
			10 Send DBRu mode 1 (three bytes)
			11 Reserved (Note 1)
	SStart	StartTime field – The number of bytes after the first upstream frame byte in which the allocated transmission of valid data begins. (In order to keep the pointer meaning independent of the position within the burst, it does not include the PLO time.)	
	SStop	StopTime field – The byte number of the last valid data byte transmission for this allocation.	
	CRC	CRC-8 error check over the allocation structure.	

Note – The value 11 was originally assigned for DBRu mode 2 (five bytes), but this mode is no longer supported.

4.3.2.2 Upstream Frame Format

The GTC upstream frame format is illustrated in Figure 4.2, and its fields are defined in Table 4.5. The ONUs receive their upstream bandwidth grants in the upstream bandwidth map (US BW Map) fields, which specify the start and stop times for which the ONT can transmit its data in the next upstream frame. By specifying the ONT start and stop times, the OLT can prevent signals from multiple ONTs from arriving simultaneously at the OLT receiver. These start and stop times are specified by the OLT such that they take into account the physical layer overhead (PLOu) transmitted by the ONU at the beginning of its upstream transmission burst. Note that the PLOAMu and/or DBRu fields follow the PLOs only if the preceding downstream frame requested them in its US BW Map Flags field. The OLT scheduler takes the bandwidth of these messages into account when it schedules the upstream bandwidth map for frames that include them.

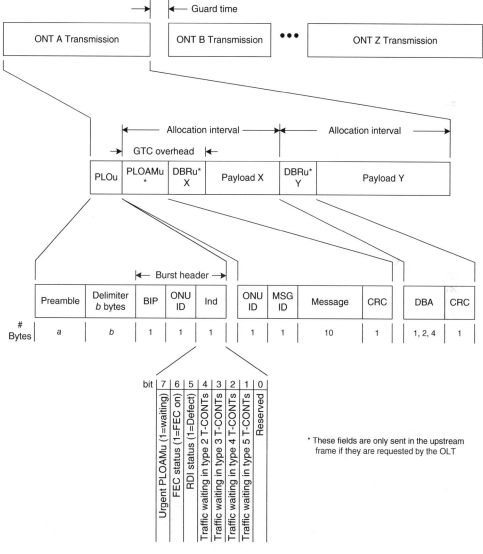

Figure 4.2 GTC upstream frame format

Table 4.5 Description of PLOu fields

Field	(Sub-field) description
Preamble	Data string transmitted at the beginning of the upstream burst to allow the OLT to lock onto the ONT's signal for clock and data recovery of the remainder of the burst. Its format and length are specified by the OLT.
Delimiter	A *b*-byte data field to indicate the end of the preamble field so that the OLT can detect the start of the actual upstream data. Its format is specified by the OLT.
BIP	8-bit bit-interleaved parity over all the bytes the ONU has transmitted since the last BIP
ONU ID	The unique ONU_ID that identifies which ONU is sending this data burst
Ind field	Upstream indicator field for real time ONT status reporting. (bit 7 is the MSB)
	bit 7 =1 when an urgent PLOAMu is waiting to be sent
	bit 6 =1 when FEC is on
	bit 5 Remote Defect Indicator (RDI) =1 when there is a defect
	bit 4 =1 when traffic is waiting in type 2 T-CONT
	bit 3 =1 when traffic is waiting in type 3 T-CONT
	bit 2 =1 when traffic is waiting in type 4 T-CONT
	bit 1 =1 when traffic is waiting in type 5 T-CONT
	bit 0 Reserved
PLOAMu	Field for sending the 13-byte upstream PLOAM message when requested by the OLT.
	ONU ID The unique ONU_ID that identifies which ONU is sending this PLOAM.
	MSG ID Indicates the type of PLOAMu message
	Message The actual physical layer OAM message (10 bytes)
	CRC CRC-8 check over the PLOAMu
DBRu	Field for upstream Dynamic bandwidth report when requested by the OLT
	DBA The dynamic bandwidth assignment (DBA) traffic status for that T-CONT. The message format is specified by the OLT in the downstream frame's Flags field.
	CRC CRC 8 check over the DBRu

Each upstream data burst is scrambled with a frame synchronous scrambler. The scrambler is reset to all ones in the first bit following the preamble field. This reset occurs only once during the upstream transmission, and it is not repeated on internal boundaries if there are multiple contiguous allocations for that ONT.

Physical Layer Overhead
The PLO time, which includes the guard time between the beginning of the transmission of that ONT and the end of the previous ONT's burst, is specified to accommodate the physical layer processes that are required for the OLT to recover the data from the ONTs. The five processes affecting the PLO time are the turn-on time for the laser, the laser turn-off time, the timing drift tolerance, the clock recovery time, and delimiting of the start of the burst information. The guard time between the upstream burst transmissions of two ONTs must obviously be longer than the laser on/off-time and the peak-to-peak uncertainty in the propagation delay. The preamble field is used by the OLT to adapt to the signal level of the new ONT's burst and to lock its clock recovery circuit to the data burst. The delimiter allows the ONT to complete the data recovery synchronization process and reliably detect the beginning of the rest of the upstream frame information. The preamble and delimiter fields are specified by the OLT in an upstream overhead message. The durations of the guard time and the preamble and delimiter patterns depend on the upstream bit rate, and their recommended times are shown in Table 4.6.

The preamble consists of some number of 1 bits (type 1 preamble) followed by some number of 0 bits (type 2 preamble) followed by some number of repetitions of an 8-bit type 3 preamble (The 8-bit pattern of the type 3 preamble is specified in the upstream overhead message). The duration of the type 1 and type 2

Table 4.6 Recommended duration of guard and PLO times

Upstream data rate (Mbit/s)	Tx enable (bits)	Tx disable (bits)	Total PLO time (bits)	Guard time (bits)	Preamble time (bits)	Delimiter time (bits)
155.52	2	2	32	6	10	16
622.08	8	8	64	16	28	20
1244.16	16	16	96	32	44	20
2488.32	32	32	192	64	108	20
Notes	Maximum	Maximum	Mandatory	Minimum	Suggested	Suggested
	The 1244.16 Mbit/s rate is only one of current interest in the industry.					

preamble bits can be 0–255 bits. The type 3 preamble pattern can be repeated as many times as deemed necessary.

The delimiter field consists of the three bytes after the preamble, although the actual delimiter function may use only a portion of this field. Specifically, the delimiter function will typically use the 16 or 20 LSBs of the 24-bit field, with the remaining delimiter field MSBs becoming part of the preamble function. Proposed values for 16-bit delimiters are 0x85B3, 0x8C5B, 0xB433, 0xB670 and 0xE6D0, while 0xB5983 is a value proposed for the 20-bit delimiter.

The status indicator field in the PLOu gives the OLT a quick alert that the ONU needs attention from the dynamic bandwidth allocation (DBA) algorithm. The DBA controller can then query the ONU to determine which T-CONT(s) needs how much bandwidth (a T-CONT is a transmission container, discussed below). In order to avoid initial delays in providing bandwidth to the ONUs, the OLT can grant bandwidth to an ONU without having to wait for the status indication.

The ONU initially uses an ONU-ID = 255 to indicate it is unassigned. The ONU's ONU-ID is assigned by the OLT during the ranging process.

When the OLT receives the PLOAMu message-waiting indicator, it should respond within 5 ms with a request for the ONU to send the PLOAMu message (the OLT request is indicated in the downstream frame Flags field). This mechanism allows fast reporting of ONT conditions requiring OAM-related action by the OLT.

Upstream Payload Structure
As illustrated in Figure 4.3, the upstream payload area is divided into transmission containers (T-CONTs). The T-CONT is used as a tool for managing upstream bandwidth allocations. It enables improved upstream bandwidth efficiency by grouping traffic with similar bandwidth characteristics together. The full T-CONT definition is given in ITU-T G.983.4 (which covers dynamic bandwidth allocation) and summarized here. Some characteristics of T-CONTs include:

- The T-CONT is defined as a unit for bundling traffic flows.
- Each T-CONT contains GEM frames.
- Multiple input queues with the same traffic characteristics can be multiplexed into a single T-CONT (the Port-ID is used for multiplexing flows of GEM frames).
- A T-CONT is identified by its Alloc-ID.
- A T-CONT can also report its ingress buffer status to the OLT.
- The OLT makes its bandwidth assignments to the ONU on a per-Alloc-ID basis (each ONU can have one or more Alloc-IDs).
- The OLT controls the bandwidth and QoS of each Alloc-ID by the number of upstream time slots it assigns to the T-CONT.

The five types of T-CONTs are summarized in Table 4.7.

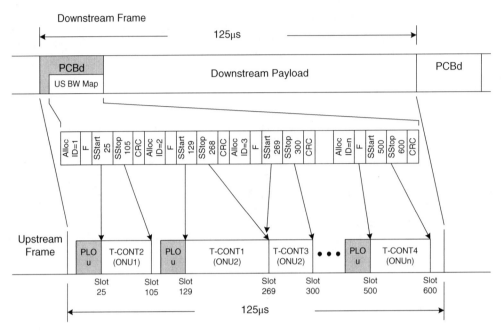

Figure 4.3 Relationship between upstream and downstream frames

Table 4.7 Summary of T-CONT types

T-CONT type	Bandwidth type	Delay sensitive	Assignment type	Comment
1	Fixed	Yes	Provisioned	The bandwidth is assigned whether it is used or not. Used for CBR or real-time services requiring a fixed rate and controlled delay.
2	Assured	No	Provisioned	Assured bandwidth is fixed average bandwidth over some unit of time. Can be used for any non-real time service.
3	Assured and non-assured	No	Provisioned or dynamic	Bandwidth is assigned up to its assured rate, but only as needed. Non-assured bandwidth is assigned across all T-CONTs requesting it, proportionate to their Assured bandwidth, up to a maximum value. Used for variable bit rate traffic.
4	Best-effort	No	Dynamic	No guaranteed bandwidth. Bandwidth is assigned equally across all Type 4 T-CONTs.
5	Fixed	—	Provisioned	The Type 5 T-CONT is a superset of the other four types, and can be downgraded to one or more of them.
	Assured	—	Provisioned	
	Non-assured	—	Dynamic	
	Best-effort	—	Dynamic	

GEM Hdr	Frame Fragment	GEM Hdr	Full Frame	GEM Hdr	Frame Fragment

Figure 4.4 GTC upstream payload field format example

Upstream Payload Transmission

The relationship between the US BW Map in the downstream frame and ONT upstream frame transmissions is illustrated in Figure 4.3. As discussed above, the US BW Map specifies the start and stop times of the Alloc-IDs in the bursts within the 125 µs upstream frame window, taking the PLOu time into account. The start and stop times are typically specified in units of bytes, although larger granularities are also used in some systems. Whatever granularity is chosen for the start/stop times, the OLT can use dynamic scheduling to achieve finer bandwidth control. For example, a single byte allocated in every 125 µs upstream frame window gives 64 kbit/s bandwidth. Allocating 48 bytes once every 6 ms (48 125 µs upstream frame windows) gives the same bandwidth.

The upstream payload data follows the upstream overhead fields in the burst. As illustrated in Figure 4.3, the payload area consists of the data assigned by the OLT in the downstream frame's US BW Map fields. The Map assignments are made per Alloc-ID, so a given ONT may be given multiple allocations for the same upstream frame. The multiple allocations can be sent within the same burst in order to avoid repeating the PLOu information between the allocations (i.e., the PLOu is just sent at the beginning of the burst). The bandwidth map from the OLT indicates the contiguous allocations by setting the StopTime of one to be 1 less than the StartTime of the next one, as shown for ONU 2 in Figure 4.3. However, sending multiple allocations adds complexity.

The upstream payload area, as illustrated in Figure 4.4, consists of GEM frames containing whole packets, packet fragments, or data blocks. In this figure, the payload begins with the final fragment of a fragmented GEM frame, followed by a GEM frame that completely encapsulates its client frame (i.e., no fragmenting), and concludes with the first fragment of another GEM frame. See Section 4.3.3.3 for a further discussion of GEM fragmentation. As noted above, each ONT can be granted bandwidth for multiple Alloc-IDs.

4.3.3   G-PON Encapsulation Method (GEM)

The primary drawbacks to the ATM used by B-PON are its inefficient bandwidth utilization, the derived protocol translation, and the cost involved in implementing a high bandwidth switching PON chassis that can switch up to 160 Gbit/s. Achieving this in ATM is more costly than with Ethernet. ATM uses five overhead bytes for every 48 payload bytes. In addition, since ATM cells must always be 53 bytes long, there is typically some wasted padding at the end of the ATM cell that contains the end of an encapsulated packet. While this degree of bandwidth inefficiency is acceptable in a core network with ample bandwidth availability, it is unacceptable in the access network where bandwidth is very expensive. For this reason, G-PON looked to the then new ITU-T G.7041 Generic Framing Procedure (GFP) [11] as the model for its encapsulation method. GFP allows direct encapsulation of variable length packets without repeating the encapsulation overhead on a regular basis.

The GFP frame overhead, however, was optimized for point-to-point, NE-to-NE links rather than for the multi-point PON application. The GFP overhead contains fields to communicate the type of encapsulated payload frame, whether a frame check sequence (FCS) is used over the payload, and what type of Extension header (if any) is used. Applications using GFP were typically expected to rely on higher layer protocols for packet/frame multiplexing, so G.7041 defines a second, optional payload header (i.e., the Extension header) for communicating channel/port number information in applications where it is needed.

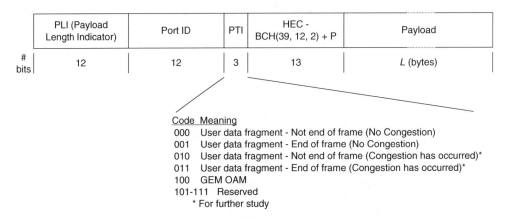

Figure 4.5 GEM frame format

In contrast, since the ONTs and OLT of a PON system can be considered conceptually as a single NE with the PON fiber functioning as a backplane extension, much of this information is already known through provisioning. Since the PON uplink multiplexes multiple client streams, it typically needs a port identifier (Port-ID) for each GEM frame. Due to provisioning, the port identifier gives the OLT and ONT adequate knowledge about the type of encapsulated frame. Consequently, the G-PON standard re-used the basic concepts of GFP but redefined the overhead fields so that their information and bandwidth are optimized for the PON application. Figure 4.5 illustrates the resulting GEM frame format.

4.3.3.1 GEM Frame Header

The Payload Length Indicator (PLI) is the primary mechanism for delineating GEM frames, and it is functionally equivalent to the GFP PLI. The value of the PLI, L, is the number of bytes in the fragment payload area of the GEM frame. The PLI of the first GEM frame in the downstream GTC frame or upstream burst is located at the beginning of the payload area. The next GEM frame will begin L bytes after the header of the current GEM frame. This GEM frame also contains a PLI, and its value can likewise be used to find the beginning of the next GEM frame, and so forth through the remainder of the G-PON payload area.

Note that since each GEM partition or payload begins with a GEM header, GEM frames can be delineated immediately at the start of each partition or payload. This feature is especially valuable for the OLT, since it removes the need for the OLT to keep track of the delineation alignment between upstream frames of each ONU separately.

The Port ID field allows multiplexing up to 4096 traffic streams per PON.

In contrast to GFP, the GEM payload type indicator (PTI) does not identify the type of client data frame encapsulated into the fragment payload area. It only indicates whether this fragment contains the end of the client data frame or if this GEM frame is carrying a GEM OAM message.

The header error check (HEC) comprises a 12-bit BCH-2 code followed by a parity bit that allows detection and correction of transmission errors within the GEM header. The BCH code is a double error correcting BCH(39, 12, 2) code that covers all the header bits except the parity bit. The BCH generator polynomial is $x^{12} + x^{10} + x^8 + x^5 + x^4 + x^3 + 1$. The parity bit is set to provide an even number of 1 s within the header.

In order to provide good transition density to correctly delineate a series of idle frames, the 40-bit header is exclusive OR'ed with the pattern 0xB6AB31E055 prior to transmitting the frame.

4.3.3.2 GEM Idle Frames

There will typically be cases when the ONT or OLT has no data to send during a portion of its frame. In these cases, the unused bandwidth is filled with GEM idle frames. A GEM idle frame is a frame with zero payload length, consisting of an all-zero GEM header. Since the header is exclusive OR'ed prior to transmission, the GEM idle header is actually transmitted as 0xB6AB31E055. The zero PLI keeps the frame delineation state machine locked onto the stream.

It is possible that at the end of a partition or payload, there will be bandwidth needing idle fill that is smaller than the 5-byte GEM idle frame. In these cases, the GEM idle is transmitted, but its end is pre-empted (i.e., only as much of the idle is sent as payload/partition bytes exist to carry it). The receiver can recognize and discard the pre-empted header. Since each partition or payload begins with a GEM header, the GEM frame delineation will be restored immediately at the start of the next partition or payload.

4.3.3.3 GEM Payload Mapping

GEM supports two types of payload mappings, as illustrated in Figure 4.6. GEM was optimized to carry Ethernet frames, but the encapsulation was also designed to readily accommodate higher-priority traffic such as TDM data. This property of GEM distinguishes G-PON from pure packet-based PON protocols such as IEEE EPON.

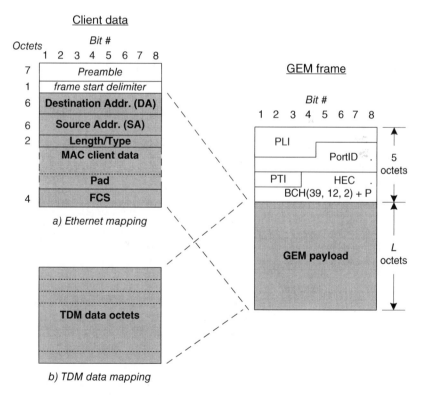

Figure 4.6 GEM payload mapping examples

Packet-Based Data Mappings

As shown in Figure 4.6, the characteristic information of the Ethernet frame (i.e., the Destination Address through the FCS, but not the preamble or frame start delimiter) is mapped directly, octet-wise, into the GEM payload area.[8]

For all packet data clients, the frame/packet can be fragmented and sent in multiple GEM frames. Each GEM frame, however, can only carry information from a single client frame/packet. Fragmentation is discussed further below.

TDM-Based Data Mappings

In order to support legacy TDM signal transport, a group of consecutive TDM data samples is gathered in the ingress buffer and mapped as a block into the GEM payload area. Due to the 125 µs GTC frame repetition rate, sending a GEM payload byte once per GTC frame corresponds to 64 kbit/s of payload bandwidth.

In order to accommodate differences in reference clock frequency, the amount of data mapped into each GEM frame (i.e., the payload length) can vary. Typically, the same number of data bytes will be sent in each GTC frame. If the client data rate is fast relative to the GTC frame rate, the ONT or OLT will need periodically to send an additional byte. Likewise, if the client data rate is slow relative to the GTC frame rate, the ONT or OLT will need to periodically send one fewer bytes in that frame. In this manner, the ingress buffer will keep from overflow or underflow. The PLI tells the receiver when the number of bytes in the GEM frame has changed. The Transmission Convergence (TC) function re-assembles these bursts into a constant rate TDM signal.

TDM is not popular among G-PON implementations. TDM encapsulations have been proposed to FSAN, such as direct mapping of SONET/SDH tributaries using the embedded signaling bits to mark the number of data bytes. This mechanism eases the CO side of implementation. CO complexity was the main obstacle for TDM implementation; with SONET/SDH direct mapping it should be solved.

The 125 µs GTC frame rate was chosen to provide a convenient reference clock for voice, for mapping signals such as DS1/E1, and for interconnecting to SONET/SDH at the OLT. Since the development of the G-PON standards, however, voice-over-IP (VoIP) has become an increasingly popular option for sending voice traffic over broadband access connections. VoIP has already become more typical for voice over G-PON than the TDM-based mapping into GEM. Similarly, it is also possible to use circuit emulation over Ethernet for other TDM services as an alternative to the TDM-based GEM mappings.

Fragmentation

In order to accommodate traffic from multiple ONTs, it is often desirable to fragment an Ethernet frame. A longer data packet can be fragmented to allow the transmission of an urgent frame (e.g., one carrying data for clients using T-CONT type 1) in the middle of transmitting a longer, lower priority frame. It also allows more flexibility in dividing the transmission of long packets in order to better serve the bandwidth requirements across the various ONUs. In addition, since the 12-bit PLI only allows payloads up to 4096 bytes, fragmentation is required for larger client frames such as 9600-byte Ethernet "jumbo" frames.

As illustrated in Figure 4.7, the fragments are sent in shorter GEM frames across multiple GTC frames. While this fragmentation applies to both the upstream and downstream directions, it is primarily intended to increase the upstream bandwidth utilization.

Each GEM payload or partition must begin with a GEM header (i.e., a fragmented frame is not allowed to continue across GTC frame boundaries without a GEM header at the beginning of each payload or partition). As shown in Figure 4.7, the first and middle fragments (if any) are transmitted using a PTI code that indicates that this GEM frame does not contain the end of the client data frame. The PON system

---

[8] Support was also added in 2008 for a direct mapping of IP packets into GEM frames rather than first encapsulating them into an Ethernet frame. The IP packets are inserted directly into the GEM payload. It is not clear whether this new mapping will gain significant acceptance in the market.

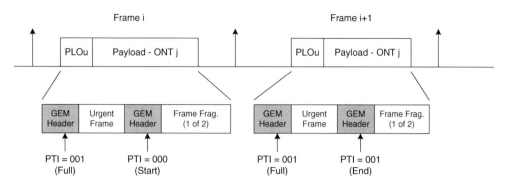

Figure 4.7 Fragmentation example for the upstream direction

inherently delivers the fragments in the order in which they were transmitted. The OLT uses the PTI, Alloc-ID, and Port-ID information to re-assemble the client data frame. Each Alloc-ID can have up to two open fragments.

Fortunately, most high-priority real-time data does not need to be sent more often than every 125 μs. This allows a substantial simplification of the process for inserting urgent real-time data. Rather than inserting it into the middle of an upstream transmission (e.g., the example of Figure 4.4), it can always be sent at the beginning of each partition or payload, as illustrated in Figure 4.7. Typically, however, TDM traffic will have a dedicated Alloc-ID and be transmitted in separate bursts.

4.3.4   G-PON Multiplexing

G-PON upstream multiplexing is illustrated in Figure 4.8. Each service flow's logical connection between the ONU and OLT is associated with a port at the ONU. As described in Section 4.3.3, the GEM frame Port-ID field identifies the ONU port associated with the data in that GEM frame and, as described in

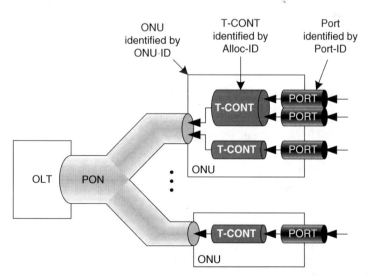

Figure 4.8 G-PON multiplexing illustration

Section 4.3.2.2, services with the same transmission characteristics can be multiplexed into the same T-CONT. The T-CONT is smallest entity for which bandwidth is assigned, and it is identified in the upstream bandwidth allocation map by its Alloc-ID. Each ONU recognizes its bandwidth grants from the Alloc-ID. The ONU-ID in the upstream overhead is only used to identify the ONU that is the source of an upstream transmission burst. Note that in addition to its use with T-CONTs, the Alloc-ID is also used to assign bandwidth for ONU Management and Control Channel (OMCC) messages.

The result is a distributed multiplexing and bandwidth assignment approach for upstream traffic. The OLT determines how much bandwidth to allocate to each ONU T-CONT for each upstream GTC frame. The ONU is responsible for choosing which port or ports to service with the bandwidth allocation for that T-CONT, and for multiplexing the data from the different ports into the T-CONT. In summary, service flows (associated with the ports) are multiplexed by the ONU as GEM frames into the appropriate T-CONTs (identified by the Alloc-ID), and the ONU upstream transmissions are time multiplexed onto the fiber according to the OLT's upstream bandwidth assignments (made per Alloc-ID).

In the downstream direction, the OLT multiplexes the GEM frames from the different service flows into the GTC payload. As with the upstream traffic, each of the logical connections for a service flow between the OLT and an ONU are identified by the Port-ID, which is included in the GEM frame overhead. Each ONU identifies and processes its own frames on the basis of the GEM frame Port-ID value, and it ignores all other payload frames.

4.3.5   Encryption and Security

Security is an inherent issue in any network in which a shared medium is potentially visible at the customer premises. Although each PON ONT should only be able to access the data destined to it, the potential exists for someone to modify an ONT such that all the downstream data is visible. Like other networks, G-PON uses an encryption process to protect against unwanted listening. Unlike wireless access or cable modems, however, the directional nature of the fiber transmission prevents an ONT from seeing upstream traffic from other ONTs, which allows some simplification of the security process. First, encryption only needs to be applied to the downstream data. Second, the upstream data can carry encryption keys in the clear.

G-PON uses the Advanced Encryption Standard (AES), which is a block cipher operating on 16 byte data blocks [20]. Specifically, the Counter Mode is used, in which a stream of 16-byte pseudorandom cipher blocks is generated and exclusive OR'ed with the input data to produce the cipher text at the OLT. The inverse process is used at the ONT to recreate the clear text data. Only the GEM payload is encrypted. The crypto-counter is aligned with the data payload rather than with the GTC downstream frame.

The OLT initiates the key exchanges by sending a message to the ONT through the PLOAM channel. The ONT is then responsible for generating the key and sending it back to the OLT. Support for key sizes of 128 bytes is required, but keys of 192 and 256 bytes are also allowed as options. After receiving the key from the ONU, the OLT sends the ONU a message with the time at which the ONU will begin to use the encryption.

Encryption has also been added for the upstream signals in order to support robust ONU authentication.

4.3.6   Forward Error Correction

A systematic forward error correction (FEC) code was chosen so that FEC decoding is optional at the ONT or OLT receivers. The FEC code is the Reed Solomon code RS(255, 239), which takes 239 data bytes and adds 16 error check code (parity) bytes after the data bytes to create a 255-byte block. The RS (255, 239) code allows correction and detection of multiple byte errors, which effectively results in a 3–4 db link loss budget gain. For both the upstream and downstream directions, no additional bandwidth is added for the error check bytes.

As illustrated in Figure 4.9, the downstream frames and upstream bursts are a string of 255-byte blocks consisting of the 239 data bytes followed by the 16 error check bytes, with the error check bytes occupying

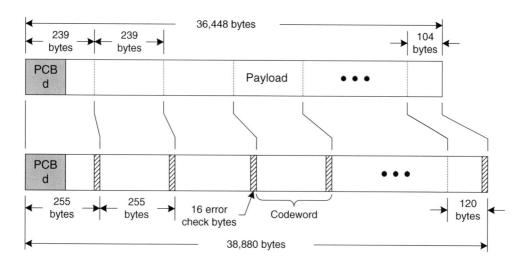

a) Downstream frame FEC illustration

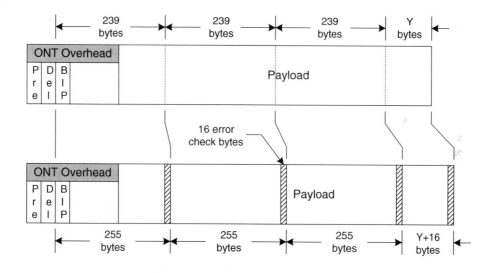

b) Upstream burst example

Figure 4.9   FEC block code examples

time slots that would be used for payload data if no FEC were present. Hence, using FEC decreases the payload bandwidth of the signal in the hope of increasing the overall bandwidth efficiency by reducing the number of client data packets that have to be re-transmitted due to transmission errors. Also, FEC is desirable for constant bit rate TDM clients since re-transmission is not an option for them.

Figure 4.9a illustrates the division of the downstream frame into 255-byte blocks. The PCBd is included in the first 255-byte code word. Since the downstream frame is not divisible by 255, the last block is shorter than 255 bytes. Specifically, at 2.488 Gbit/s the GTC frame length is 38 880 bytes, which leads to 152 255-byte code blocks and a last block with 120 bytes. The 16 error check bytes for the last block are calculated over a logical block that includes the 104 actual data bytes preceded by 135 "0" pad bytes to create the 16-byte error check. The "0" padding bytes are not transmitted, but are re-inserted by the receiver when it performs its error check calculation for this block.

The upstream bursts, illustrated in Figure 4.9b, are similar to the downstream frame except that the first codeword does not include the entire physical layer overhead. Rather, the first codeword begins with the BIP and does not include the preamble and delimiter fields. For this reason, the OLT must be able to tolerate up to three bit errors in a 16-bit delimiter and up to four bit errors in a 20-bit delimiter field. The burst is divided into 239-byte payload blocks that are mapped into the 255-byte codeword blocks. The last codeword will typically contain less than 239 payload bytes, so it uses the same "0" padding mechanism for this codeword as is used with the downstream signal. Note that when FEC is enabled at an ONT it is applied to all of the upstream transmissions from that ONT. Note also that FEC encoding occurs prior to scrambling.

4.3.7 Protection Switching

Protection switching is an optional feature for G-PON since the need for protection is a function of both cost and service reliability requirements. Protection switching can be performed as either an automatic protection switch (e.g., due to detection of a loss of signal (LOS) or loss of frame (LOF) condition at either the ONU or OLT) or as a craftsperson-initiated forced protection switch. The protocol is more complicated than for point-to-point systems. See Chapter 2 for an expanded discussion of different PON protection options that can be used with either G-PON or EPON.

4.3.8 ONU Activation

There are several scenarios for the installation and activation of ONUs. This section provides an overview rather than a detailed look at each scenario.

There are three triggers for initiating the activation of an ONU:

1. Manual initiation by a craftsperson after a new ONU is known to have been connected.
2. Automatic polling by the OLT to see whether a "missing" but previously active ONU has returned to service.
3. Automatic polling by the OLT to detect any new ONUs connected to the PON.

For options 2 and 3, the polling rates are programmable.

ONUs begin in an Initial-state after power-up or reset due to a long alarm condition. Once all alarms are cleared and the ONU receives the downstream signal, it moves to the Standby-state. When the ONU is in the Standby-state, it waits until it sees the default transmit power setting specified in the Upstream_Over-head message from the OLT and moves to the Power-Setup-state. The ONU then enters the Serial-Number-state and waits for the OLT to send a Serial_Number request message.

The Serial_Number request message is the mechanism by which the OLT searches for new ONUs or ONUs that had previously been activated, but which had been "missing." The ONU responds by sending its serial number. This process is repeated until all new ONUs have responded successfully to two Serial_Number request messages. When the OLT sees two successful responses from an ONU, it assigns the ONU an ONU-ID and communicates it through the Assign_ONU-ID message. This whole process repeats until the OLT sees no new responses for at least two cycles.

Since the new ONU's distance from the OLT is not known when the Serial_Number request message is sent, there is a danger that the new ONU's response message will collide with traffic from other ONUs.

The OLT and ONU both have mechanisms to avoid such collisions. The OLT first opens a window for the responses by halting the transmissions from the active ONUs for a period of time before and after it sends the Serial_Number request message. The halting is accomplished by sending zero length upstream bandwidth allocations to the active ONUs. This halt period is typically two frames. Each new ONU waits a random delay time before transmitting its response message, which contains its serial number and the random delay value it used. If a collision does occur, the random delay should reduce the probability of another collision from occurring in the next serial number request cycle.

Once the ONU has been assigned its ONU-ID, it enters the ranging-state. The ranging mechanism is described in Section 4.3.9. After the ranging process is complete, the ONU enters the Operation-state and proceeds with normal operation.

4.3.9   Ranging Mechanism

The OLT assigns the burst transmission times for the different ONTs such that there is no overlap between these bursts when they arrive at the OLT receiver. As discussed in Section 4.3.2.2, these time assignments leave some guardband time between the bursts of successive ONTs in order to allow for the laser of the one ONT to turn off and the laser of the next ONT to turn on, as well as to cover any uncertainty in their relative signal propagation delays to the OLT. As discussed in Chapter 3, since the ONTs can be located at different physical distances from the OLT, a crude approach would be to have this uncertainty window include the round trip propagation delay difference between an ONT at the shortest fiber distance and an ONT at the longest anticipated fiber distance from the OLT. With the desired maximum ONT distance of $20\,\text{km}$ and the 2×10^8 m/s speed of light through a fiber, the resulting differential delay is $200\,\mu\text{s}$. Since the upstream frame duration is $125\,\mu\text{s}$, this crude approach will not work.

The solution to this propagation delay difference problem is to use a ranging mechanism that allows the OLT to determine the relative distances of the ONUs. The OLT then assigns an Equalization_Delay time to each ONU. The ONU uses the Equalization_Delay for the phase (bit) offset of its upstream GTC frame alignment relative the downstream GTC frame. Specifically, an ONU uses the Equalization_Delay as the offset between the start of the received downstream frame and the start of its upstream frame. The upstream transmission grant start and stop times it receives from the OLU are relative to the start of this normalized upstream frame. As a result, each of the ONTs appears to be equidistant from the OLT (i.e., each has the same equalized round trip delay from the OLU), and the guardband time can be reduced to just a few bit periods. (See Table 4.6.)

The ranging mechanism for G-PON is somewhat different from the approach taken for EPON. Rather than using local counters, the G-PON ranging mechanism uses the $125\,\mu\text{s}$ GTC frame boundaries as a reference point. The mechanism is based on a round trip delay (RTD) measurement procedure. The RTD includes the bidirectional propagation delay through the fiber, the delay through the electro-optical and opto-electrical circuits at the ONU and OLT, and any signal processing delay at the ONU. If the ONU Equalization_Delay defaults to a value other than zero, it must also be added to the RTD. Specifically, the RTD is measured as the time from when the OLT PON processing circuit launches the first bit/byte of the Ranging-request in the downstream frame until it receives the last bit/byte of the Ranging-transmission from the ONU. The Ranging-request is sent by setting the ONU-ID to Ranged ONU-ID and requesting a PLOAM from the ONU. The Ranging-transmission from the ONU is a PLOAM message that contains the ONU's serial number.

The trigger for the RTD calculation can either be the connection of a new ONU to the PON or a "missing" ONU being detected through the periodic OLT serial number polling described in Section 4.3.8. The ranging protocol can also be triggered by a protection switch, since the switch can change the length of the fiber. This is especially true of type C and X : N protection switches. (See Section 4.3.7 and also Chapter 2.) The OLT continually calculates the RTD of active ONUs.

When calculating the RTD, the OLT checks several conditions to confirm whether the RTD information is valid (e.g., ONU ID and serial number, and whether the RTD is within an expected/

allowed range). In order to increase the accuracy of the RTD measurement, it may be performed multiple times and the average RTD value used.

As discussed in Section 4.3.8, when a new ONT is connected to the PON it proceeds to the Ranging-state after its ONU-ID is assigned. At this point, the OLU needs to open a ranging window that allows the new ONU to perform its RTD without any danger of its upstream transmissions colliding with the data from the other ONUs. To accomplish this, the OLT sends a Halt grant message (or zero bandwidth pointers or no allocations) to the other ONUs for multiple frames in order to prevent them from transmitting upstream data. If the new ONU distance is unknown, the "quiet zone" in the upstream transmission can include all the ONUs. If, however, the OLT knows the new ONU's approximate distance (e.g., through provisioning), it can halt the upstream transmissions of only those ONUs having upstream bursts that would arrive within a window around when the new ONU's burst is expected to arrive. The time window within which other ONU upstream transmissions must be halted is referred to as the "quiet zone."

Since data in the input queues of working ONUs within the quiet zone continues to accumulate during the Halt cycle, the OLT must wait for at least some minimum time between Halt cycles to let these ONUs catch up to normal queue fill levels.

The effects of things like temperature and component aging can cause the RTD to drift over time. The OLT should detect this drift between the expected and actual burst arrival times, and update its Equalization_Delay value for an ONU accordingly. The new Equalization_Delay value can then be sent directly to the ONU, thus avoiding the need for a new ranging operation.

4.3.10   Dynamic Bandwidth Assignment (DBA)

The most basic method of allocating upstream bandwidth is to distribute it equally among the ONUs. This method is very inefficient, especially with packet traffic, since the bandwidth needs of the ONUs will rarely be equal at any given time. Considerable overall bandwidth utilization gains can be made if the upstream bandwidth is allocated dynamically according to the current needs of the ONUs. While the ITU-T does not specify a specific DBA algorithm, G.983.4 [4] specifies the framework and mechanism to implement DBA in B-PON and G-PON systems.

G.983.4 specifies two different DBA mechanisms:

With the first method, the ONUs take a passive role. The OLT monitors how much bandwidth each of the ONUs is using, based on the number of GEM idle frames it receives in the upstream GTC frames. For this reason, this technique is referred to as "idle cell adjustment." While this method was originally referred to as the Non-Status Reporting (NSR) strategy, it is now commonly referred to as Traffic Monitoring DBA (TM-DBA). More bandwidth is assigned to an ONU if its bandwidth utilization exceeds a predefined threshold. The advantages to this method are simplicity for the ONUs and avoiding using upstream bandwidth for reporting bandwidth needs. The disadvantage is a potentially slower response to the ONU bandwidth needs.

With the second method, the ONUs report their buffer status to the OLT; hence, it is called the "buffer status reporting" method or Status Reporting (SR) strategy. The quick indication of a need for bandwidth in a T-CONT type is communicated in the PLOu Indicator field. The more detailed report of the per T-CONT buffer status is communicated in upstream dynamic bandwidth reports (DBRu), as illustrated in Figure 4.2. The OLT uses the status report information to determine the appropriate bandwidth allocation for each Alloc-ID.

An OLT can also use a hybrid of the NSR and SR strategies.

As noted above, DBA can also handle multiple Alloc-IDs in an ONU, with each Alloc-ID operating separately. The T-CONT types, discussed in Section 4.3.2.2 are used in determining the bandwidth grants.

The priority order for bandwidth types, from highest to lowest priority, is fixed bandwidth, assured bandwidth, non-assured bandwidth, and best effort bandwidth. The first two are types of guaranteed bandwidth, and the sum of the latter two is referred to as "additional bandwidth". "Surplus bandwidth" is the additional bandwidth minus any bandwidth reserved for other uses such as OAM. It is the surplus bandwidth that is available for dynamic DBA allocation.

The OLT is responsible for traffic management of the downstream flows based on their service specifications, bandwidth and memory resource availability, and the dynamic traffic conditions. This traffic management can include traffic policing and scheduling per Port-ID. In the upstream direction, the OLT and ONT share the traffic management responsibilities. The OLT is responsible for assigning bandwidth to each T-CONT based on the service specifications for the flows that are multiplexed into that T-CONT, as well as the upstream bandwidth availability. The ONT is responsible for traffic management of the flows that are aggregated into the T-CONT. The ONT traffic management functions may include policing and traffic shaping per Port-ID, and scheduling and traffic shaping per T-CONT.

4.3.11   OAM Communication

G-PON has three different mechanisms for communicating OAM information. As illustrated in Figures 4.1 and 4.2, the embedded OAM channels and PLOAM are sent as part of the upstream and downstream frame structures. The embedded OAM channels are the real-time data information fields that communicate bit errors, RDI, and DBA status.

The PLOAM fields provide an efficient mechanism to communicate additional OAM information. The PLOAM field consists of 13 bytes including the ONU ID (1 byte), Message ID (1 byte), OAM data (10 bytes), and a CRC-8. For bandwidth efficiency, the upstream PLOAM messages are only sent when the OLT requests them. One of the flag bits in the PLOu allows the ONT to notify the OLT that it has an OAM message that it needs to send (see Figure 4.2).

The channel for carrying this OAM communication between the OLT and the ONTs within the payload area is called the ONU Management and Control Channel (OMCC) and the specification of the communications on this channel is called the ONU Management and Control Interface (OMCI). The OLT uses the OMCI to control the ONT. Specifically, it allows the OLT to establish and release connections going through the ONT, to manage an ONT's UNIs, to request performance statistics and configuration information from an ONT, and to inform the system operator of events detected by the OLT, such as link failures. In other words, the OMCI is used for configuration, fault, performance, and security management. The OMCI and associated MIBs are defined in G.984.4 [15].

The OMCC packet format is illustrated in Figure 4.10. When the OLT sends a request message to an ONT, it assigns a Transaction correlation identifier so that it can correlate the ONT response with the proper request. The choice of the identifier is up to the OLT; however, the MSB is used to indicate the priority of the transaction (1 = high priority, 0 = low priority). The message types are shown in Tables 4.8–4.10. More recently, an extended OMCC message was adopted that allows message contents up to 1500 bytes. This extension allows more efficient data intensive operations, such as software download and MIB upload.

4.3.12   Time of Day Distribution

G-PON supports time of day (ToD) distribution to the ONUs with an accuracy of ±1 μs. An OLT distributes the ToD information for each ONU based on a real-time clock available to the OLT. The ONUs then adjust the received ToD value to accommodate the transmission and processing delays. The concept

GEM header	Transaction correlation identifier	Message type	Device identifier	Message identifier	Message contents	OMCI trailer
5	2	1	1	4	32	8

(first column label: Bytes)

Figure 4.10 OMCC packet format

Table 4.8 Downstream OAM message types

Type	Code	Description
Upstream_Overhead	1	Informs the ONU of its upstream delay, preamble, and optical power level parameters
Assign_ONU_ID	2	ONU_ID assignment (tying it to the serial number indicated in the message)
Ranging_Time	3	Tells the ONU its equalization delay (in number of bits)
Deactivate_ONU_ID	4	Instructs the ONU to stop upstream transmissions and reset itself
Disable_serial_number	5	Disables the ONU having this serial number
Encrypted_Port_ID	6	Tells the ONU which channels are encrypted
Request_password	7	Asks the ONU for a password to verify its identify
Assign_Alloc_ID	8	Tells the ONU it has been assigned the specified allocation ID
No message	9	
POPUP	10	Forces ONUs in the POPUP state to move to the ranging state, unless they are in LOS/LOF state, or commands a specific ONU to go to the operation state
Request_Key	11	Instructs the ONU to generate and send a new encryption key
Configure_Port-ID	12	Links an ONU Port-ID to its internally processed OMCI channel, allowing the OMCI to be routed over a GEM channel
PEE	13	Physical equipment error to inform the ONU that the OLT cannot send GEM and OMCC frames
Change-Power-Level	14	Instructs the ONU to change its transmitted optical power level
PST message	15	For connectivity check and APS when the PON is configured for survivability
BER interval	16	The interval (number of downstream frames) over which the ONU accumulates the downstream BER
Key switching Time	17	Instructs the ONU when to switch to a new encryption key
Extended_Burst_Length	18	Tells the ONU how many upstream type 3 preamble bytes to use

makes use of a hypothetical reference ONU that has zero equalization delay and response time. The OLT tells the ONU the ToD at which a certain GTC downstream frame (identified by its frame number within the superframe counter) would arrive at this hypothetical reference ONU. The ToD information is communicated over the OMCI channel, and the use of a future frame as the reference point removes the need communicate in real-time. A more detailed description follows, using Figure 4.11 to illustrate the relationships between the different timing parameters at the OLT and ONUs.

Table 4.9 Upstream OAM message types

Type	Code	Description
Serial_number_ONU	1	Communicates the ONU's serial number
Password	2	Password to verify the ONU to the OLT
Dying_gasp	3	Tells the OLT that a local power condition exists (e.g., low power or battery conservation mode) that may prevent the ONU from responding to upstream bandwidth grants. (note: the ONU does not commit itself to power off, and the OLT can continue to send bandwidth allocations. The OLT does not regard a failure to transmit as an alarm condition to be reported).
No message	4	
Encryption key	5	Sends the OLT a fragment of the ONU's encryption key
PEE	6	Tells the OLT that the ONU cannot send both GEM and OMCC frames
PST message	7	For connectivity check and APS when the PON is configured for survivability
REI (Remote Error Indication)	8	The number of downstream BIP errors in the BER interval
Acknowledge	9	Acknowledgment that the ONU has received a downstream message

Table 4.10 Management message types

Type	Code	Description
Create	4	Create a managed entity instance and associated attributes
Delete	6	Delete a managed entity instance
Set	8	Set one or more managed entity attributes
Get	9	Get one or more managed entity attributes
Get all alarms	12	Get a managed entity's alarm active alarm status
MIB upload	13	Latch the MIB
MIB upload next	14	Get a managed entity's latch attributes
MIB reset	15	Clear the MIB, initializing it to its default, and reset the MIB data sync. counter to 0
Alarm	16	Alarm notification
Attribute value change	17	Notification of an autonomous attribute value change
Test	18	Request a test on a specific managed entity
Start software download	19	Begin the download of software
Download section	20	Download a software image section
End software download	21	End the software download
Activate software	22	Activate the downloaded software image
Commit software	23	Commit the downloaded software image
Synchronize time	24	Synchronize the ONT time to the OLT
Reboot	25	Reboot the ONT, subscriber line card, or PON interface line card
Get next	26	Get the managed entity's latched attribute values within the current snapshot
Test result	27	Notification of the test results that were initiated by the "Test"
Get current data	28	Get the current counter value associated with, or more of a managed entity's attributes

An OLT possessing an accurate real time clock schedules a ToD event to occur at the beginning of a downstream frame far enough in the future that the ONU can prepare for it, and informs the ONU of that frame number through a management message. Let this frame number be N. Recall that the frame numbers are specified by the superframe counter in the PCBd. The times at which the first bit of this GTC frame cross the optical connectors/splices leaving the OLT and arriving at the ONU are used as the time reference points.

Recall also that the OLT had measured the round trip delay to each ONU during the ranging procedure, including the signal propagation time and the ONU processing delay. This information was communicated to the ONU as the Equalization_Delay time, which is used as the offset between the downstream and upstream frame alignment at the ONU. The ToD information sent by the OLT is with reference to a hypothetical ideal ONU that has no processing response time and an Equalization_Delay time of zero. In other words, the hypothetical ONU has a combined upstream and downstream optical propagation delay equal to the Equalization_Delay that the OLT would assign to an ONU with zero distance and processing delay (called the zero distance equalization delay).[9]

The timestamp is the exact ToD when the first bit of the GTC frame N would arrive at this hypothetical ONU (i.e., cross the optical connector/splice between the ODN and the ONU). Specifically, it is calculated as the exact ToD at which the first bit of transmitted downstream GTC frame N crossed the optical connect/splice connecting the OLT to the ODN plus the wavelength velocity adjusted zero distance equalization delay time. The zero distance equalization delay corresponds to the difference at the OLT

---

[9] Remember that the closer an ONU is to the OLT, the greater the Equalization_Delay it must have in order to make all ONUs appear to be the same functional distance from the OLT.

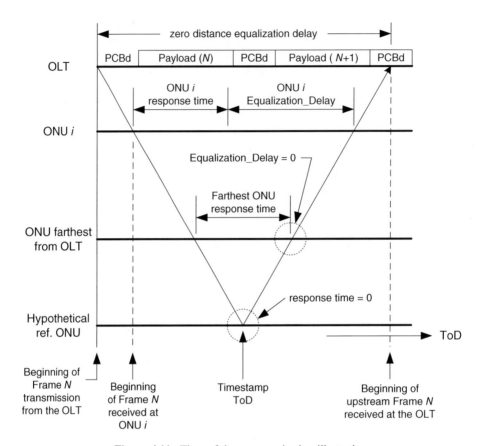

Figure 4.11 Time of day communication illustration

between the start times of a transmitted downstream frame and the corresponding received upstream frame at the OLT (i.e., it is the Equalization_Delay value that would be associated with an ONU that has no propagation distance from the OLT).

$$\text{Timestamp} = \text{Send time} + (\text{Zero distance equalization delay}) \, (\lambda \, \text{velocity adjustment})$$

The wavelength velocity adjustment for the Equalization_Delay time corrects it so that it reflects the downstream signal propagation time of the round trip propagation delay used to determine the Equalization_Delay. It is the ratio of the downstream wavelength propagation velocity to the combined velocities of the downstream and upstream wavelengths.[10]

The OLT pre-computes and stores the timestamp value associated with frame N and uses a management message to send both the frame number and the associated timestamp value to an ONU.[11]

---

[10] The wavelength affects the propagation delay of light through the fiber.

[11] The superframe counter rolls over after roughly 37 hours. Consequently, the frame selected for the ToD can be scheduled hours in advance. The ToD is sent to newly activated ONUs, and sending the ToD information at least once every 24 hours is recommended for all ONUs.

An ONU uses this timestamp value to calculate the expected ToD when the first bit of frame N will arrive at its reference point. Specifically, this ToD is the timestamp value minus the wavelength velocity adjusted sum of that ONU's equalization delay and response time.

$$\text{Receive time} = \text{Timestamp} - (\text{Equalization\_Delay} + \text{ONU resp.time})\,(\lambda\,\text{vel.adjust.})$$

In other words, the resulting receive time is:

Receive time = Send time
   + [(Difference between the downstream and upstream frame offsets at the OLT and ONU_i)
   − (ONU_i's response time)](λ velocity adjustment)

When the first bit of frame N actually arrives at the ONU, it sets its ToD to the pre-computed value, with appropriate adjustment to compensate for its internal delays.

4.3.13   G-PON Enhancements

In the 2008–2009 timeframe, two important enhancements were made to G-PON. One was a specification for reach extension, and the other was a description of optional methods for ONU power savings. Both are discussed in Chapter 2.

4.4   Next Generation PON (NG-PON)

In 2007, FSAN and ITU-T begin work on the next generation of PON protocols beyond G-PON. The primary motivation for this work is to increase the per-subscriber bandwidth capabilities in a cost effective manner. Minimizing cost implies increasing the number of subscribers served by each PON. The end of this section describes the protocols that were developed just before this book was published, and discusses the expected directions of the work that was still in progress.

Due to differing requirements for these protocols, the work was divided into multiple categories. NG-PON refers to Next Generation PON, beyond G-PON. XG-PON refers to an extension of G-PON for 10 Gbit/s rates, with the "X" chosen since it is the Roman numeral designation for "10".

- NG-PON1:
  - Evolution of G-PON, allowing both G-PON and NG-PON1 to share the same optical distribution network.
  - XG-PON1: G-PON extension using 10 Gbit/s downstream with 2.4 Gbit/s upstream[12].
- NG-PON2:
  - New PON protocol that may be disruptive in terms of potentially not being restricted to backwards compatibility with G-PON.
  - A downstream bandwidth of 40 Gbit/s will be supported.

NG-PON will also support IPv6 for all services in addition to IPv4.

This section first introduces some of the technology considerations and constraints on next generation PON protocols. It then describes the NG-PON1 protocols, followed by a summary of the initial agreements and expected directions of the different NG-PON2 protocols.

---

[12] FSAN/ITU-T has also considered an XG-PON2 with 10 Gbit/s upstream and downstream, but chose not standardize it. One motivation for this decision is that the applications for it could also be served with 10G EPON.

4.4.1   Introduction to G.987 series XG-PON (NG-PON1 – 10Gbit-capable PON)

XG-PON is defined in the ITU-T G.987 series [21–25]. It provides an evolutionary path for increasing the bandwidth per subscriber that is "backwards compatible" with G-PON. The backwards compatibility is primarily achieved through allocating wavelengths for XG-PON that do not interfere with G-PON operating on the same ODN. This approach is the same one used for 10G EPON and EPON, as discussed in Chapter 3. As long as the G-PON ONUs have the appropriate wavelength blocking filters (WBFs), wavelength separation allows both G-PON and XG-PON to co-exist independently on the ODN.

The service requirements for XG-PON include:

- FTTH – telephony, television and high-speed Internet access for residential users;
- FTTO (Fiber to the Office) – leased line, L2 VPN (e.g., Ethernet services) and IP services (e.g., VoIP and L3 VPN) for enterprise customers; and
- FTTCell (Fiber to the Cell site) – for wireless backhaul.

Power savings are very important for XG-PON. The equipment should support "full service" and "sleep" nodes of operation, and ONUs should also support power shedding when operating on their backup batteries. See Chapter 2 for a discussion of power saving techniques.

4.4.2   XG-PON Physical Layer

The XG-PON physical layer is defined in G.987.2 [23]. Achieving the desired XG-PON 10 Gbit/s downstream transmission is complicated by dispersion, which creates inter-symbol interference (ISI), especially in the presence of chirp.[13] ISI can be reduced by reducing chirp (e.g., external laser modulation), using wavelengths in the low dispersion region of the fiber (around 1310 nm for the most common fibers), electronic dispersion-compensation, or adding dispersion compensating fiber.

Due to the technical difficulties of 10 Gbit/s serial transmission, FSAN considered multiple alternatives for increasing the downstream bandwidth per subscriber. Approaches that were considered but rejected included using the existing G-PON with reduced split ratios, adding more G-PON systems to an ODN with WDM, and using separate wavelengths for four G-PON downstream signals and a single G-PON 1.2 Gbit/s upstream signal with a wavelength filter that behaves like a splitter in the upstream direction. FSAN also considered using a new 4×2.5 Gbit/s downstream signal, which would also have the advantage of reducing power consumption due to the lower processing speeds. Ultimately, a single 10 Gbit/s serial stream was chosen for downstream.

At the physical layer, XG-PON specifies two "Nominal" loss budgets (N1 and N2) that use no optical amplification, and two "Extended" loss budgets (E1 and E2) based on optical pre/post amplifiers to gain a 4 dB increase over the Nominal loss budget. The Nominal1 (N1) class has a minimum and maximum loss budget of 14 dB and 29 dB, respectively. The Nominal2 (N2) class, which supports a loss budget of 16–31 dB, is divided into two subcategories, N2a and N2b for the downstream direction. The minimum and maximum mean launch power for N2a are 4.0 and 8.0 dB, respectively, while the minimum and maximum mean launch power for N2b are 10.5 and 12.5 dB. The E1 and E2 Extended class minimum and maximum loss parameters are each 4 dB higher than the respective N1 and N2 loss budgets. Class E2 is also subdivided into E2a and E2b for the downstream direction. The minimum and maximum mean launch power for E2a are 6.0 and 10.0 dB, respectively, while the minimum and maximum mean launch power for E2b are 14.5 and 16.5 dB. Note that in order to avoid damage to receivers, optical attenuators must be used on a path with less than 5 dB loss.

XG-PON also specifies two categories of differential fiber difference (i.e., the maximum difference in the distances between the ONU nearest to the OLT and the ONU farthest from the OLT). The first

---

[13] Chirp refers to the variations that occur in a laser's output wavelength as it transitions between its on and off states.

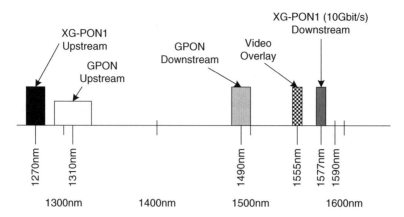

Figure 4.12 XG-PON1 wavelength allocation

category is 20 km differential distance (DD20), and the second category is 40 km differential distance (DD40).

The minimum split ratio requirement is 1 : 64, with logical support for at least 1 : 256. The minimum physical reach is 20 km, and the protocol provides logical support for at least 60 km reach. It will be possible to overlay multiple XG-PON systems on an ODN, but this will require colored ONUs (see Chapter 5 for more on colored ONUs).

The downstream signal uses a wavelength in the 1575–1580 nm range, as illustrated in Figure 4.12. The wavelength range is extended up to 1581 nm for outdoor applications.

Although the 64B/66B line code used by 10G EPON had been considered for the 10 Gbit/s XG-PON signals, the decision was made to use the NRZ line code with scrambling for both the upstream and downstream XG-PON signals, similar to G-PON. The 10G rate for XG-PON is actually 9.95328 Gbit/s, which is the same as SONET STS-192 (SDH STM-16) and, thus, simplifies integrating the PON systems into the network clock schemes.

The clock accuracy of the downstream signal is determined by the timing source available to the OLT. An OLT timing must free run at no worse than Stratum-4 (32 ppm) accuracy, although higher accuracy free-running clocks may be required if the OLT is supporting services requiring it (e.g., Stratum-3). An OLT will typically have a Stratum 1 source (0.0001 ppm) available to it, either from an external physical interface or from a packet-based timing source. Since the ONT derives its timing from the downstream signal, the timing accuracy of its upstream data transmissions will be the same as that of the downstream signal.

4.4.2.1 XG-PON1 Physical Layer Specifics

The key issue for PON systems was optical burst mode transmission and reception, which becomes increasingly difficult at higher data rates. Practical OLT burst mode receivers become more difficult to implement beginning at rates around 5 Gbit/s. The significant technical barriers encountered at these rates pertain to the speed and accuracy at which the receiver can adapt its thresholds to each burst and to achieve timing recovery. Eventually these barriers will be overcome, but the timeframe is not clear. Consequently, FSAN considered different options for upstream transmission including $N \times 2.5$ Gbit/s with WDM, and a single serial 5 Gbit/s stream. Using multiple wavelengths creates additional spectrum compatibility issues with G-PON. It also creates additional optical access complexity and requires more filter elements, which increases the optical loss. Since 2.5 Gbit/s components are already available, FSAN decided that using a single 2.5 Gbit/s upstream channel provided the best technical tradeoffs and was adequate for the target applications.

The specific upstream line rates for XG-PON1 upstream rate is 2.48832 Gbit/s, which is exactly one-quarter of the rate of the downstream signal. It is also the same as the SONET STS-48 (SDH STM-64), and again simplifies integrating the systems into the network clock schemes. As noted above, the physical layer adaptation sublayer uses scrambled NRZ rather than going to a new upstream line code.

As illustrated in Figure 4.12, the upstream wavelength band for the 2.5G signal is 1260–1280 nm.[14] This wavelength band is known as the O-minus band, which has the advantages of having optical component compatibility with 10G EPON (and hence faster expected optical component availability), and that laser chirp is easier to manage for this band than with longer wavelengths.

Note that, unlike with EPON and 10G EPON, there is no overlap between the upstream wavelength bands of G-PON and XG-PON. The lack of overlap significantly simplifies the OLT receiver, eliminating much of the dual-mode OLT receiver complexity discussed in Chapter 3.

"Enhancement" bands have also been reserved for XG-PON1. Potential applications for these bands include support for G-PON and/or video services. The three enhancement band options for XG-PON1 are:

- Option 1: 1290–1330 nm for upstream use;
- Option 2: 1360–1480 nm for future use;
- Option 3: 1480–1560 nm for G-PON downstream (1480-1500 nm) and/or video distribution (1550–1560 nm);
- Option 4: xx-1625 nm for future use, where "xx" is to be determined.

An extended wavelength plan (extended practice recommendation) is being considered that utilizes dense WDM (DWDM) to increase the PON capacity. See Section 4.4.2.2.

Upstream bursts include physical layer overhead that allows the receiver to accommodate the laser on/off time, timing drift, signal level recovery, clock recovery and delineating the start of the upstream burst. The overhead consists of three sections. The first is the guard time, in which the ONU transmits at its nominal zero level. The second is the preamble pattern, which is a pattern with an adequate transition density to enable fast recovery of the signal level and clock. The third is the delimiter pattern, which allows the OLT to identify the beginning of the burst. The duration of each section is implementation-dependent; however G.987.2 provides recommended constraint values, resulting in recommended times of 64 bits for the guard time, 160 bits for the preamble, and 32 bits for the delimiter.

4.4.2.2 Hybrid XG-PON and DWDM

As indicated in the previous section, FSAN/ITU is also studying the use of DWDM to combine multiple XG PON systems on the same ODN in order to increase the number of subscribers and/or capacity per subscriber. The key assumptions here are:

- compatibility with the characteristics of the existing G-PON ODN;
- colorless ONUs;
- a combination of WDM and power splitters in the ODN.

This extended wavelength plan would use 0.5 nm wavelength windows based on the ITU-T dense WDM (DWDM) grid. For example, it could use wavelengths centered at 189.7, 189.9, 190.1, 190.3, 234.3, 234.5, 234.7, and 234.9 THz. One application for this would be to use a single wavelength with optically-pre-amplified receivers so that narrowband filters can be used for increased receiver sensitivity. Another application would be multiplexing four logical XG-PON1 systems onto a single ODN by using four wavelength pairs. These wavelengths could be used for the entire PON section or just the feeder fiber.

---

[14] The potential upstream wavelength bands considered for XG-PON1 were the L-band (1595–1615 nm), C-band (1539–1559 nm), video compatible C-band (1530–1540 nm), O-plus band (1340–1360 nm), and O-minus band (1260–1280 nm). Each band posed different cost and complexity issues with respect to achieving the necessary optical isolation from adjacent G-PON or video overlay channels, impact on loss budgets, and the cost and complexity of the ONU filters.

As discussed in Chapter 5, WDM filters have the advantage of lower loss than power splitters. Chapter 5 also discusses the issues associated with colorless ONUs.

The downstream wavelength region for DWDM operation was under study at the time this book was written. While the 1539–1559 nm C-band is ideal for the loss sensitive DWDM application, it conflicts with the 1555 nm video overlay wavelength assignment. The 1575–1582 nm region is another possibility. An 8-channel DWDM system using 100 GHz spacing or a 16-channel system using 50 GHz (7 nm) spacing are also being considered as options.

4.4.3   XG-PON Transmission Convergence Layer and Frame Structures

The three sublayers that make up the XG-PON Transmission Convergence (XGTC) layer in both the upstream and downstream directions are the XGTC service adaptation, XGTC framing, and XGTC PHY adaptation sublayers [24].

The service adaptation sublayer uses the XG-PON Encapsulation Method (XGEM) to map service data units (SDUs) into XGTC frames. SDUs include user data and OMCI messages, with each of the two message types having its own TC adapter. The XGTC service adaptation supports the XGEM frame delineation and XGEM Port-ID filtering as well as SDU frame fragmentation and reassembly.

The XGTC framing sublayer handles functions associated with the XGTC frame. In addition to the basic XGTC frame delineation function, the three types of functions it performs are:

- Multiplexing/demultiplexing the XGTC and PLOAM information into the upstream and downstream frames.
- Header creation/decoding, including OAM insertion and extraction.
- The internal Alloc-ID based routing function for data to/from the XGTC adapter

In the upstream direction, the OLT uses the BW map that it transmitted to the ONU to identify and demultiplex the contents of the upstream frame.

The XGTC PHY adaptation sublayer deals with physical layer overhead, line codes and FEC. Specifically, its functions are:

- Physical layer synchronization, including inserting and delineating the physical layer synchronization blocks (PSBd for downstream and PSBu for upstream) that are used for signal timing recovery.
- FEC, which must be supported in both the upstream and downstream directions.
- Scrambling, using a frame synchronous scrambler with the $x^{58} + x^{39} + 1$ scrambler polynomial. The scrambler is reset at the first bit after the PSB by preloading the scrambler shift register. In the downstream direction, the 51-bit superframe counter value in the PSBd is used as the 51 MSBs of the preload pattern, with the 7 LSBs set to 1. In the upstream direction, the 51-bit superframe counter of the PSBd is used as the 51 MSBs of the preload pattern, with the 7 LSBs set to 1.
- Applying the NRZ line code.

Next the frame formats and their overhead fields will be described.

4.4.3.1   XGTC Frame Structures

The XG-PON frame structures are very similar to those for G-PON, except that they have been optimized for use with the higher transmission rates. In some cases, this optimization was achieved by defining fields in a granularity of 32-bits to better support 32-bit data paths within systems and devices. In other cases, items were added, removed, or modified based on experiences with G-PON deployments. Consequently, this section will refer to the corresponding G-PON sections above wherever the field definitions or functions are unchanged, and will highlight the changes where appropriate.

As with G-PON, the upstream and downstream frame durations are 125 μs.

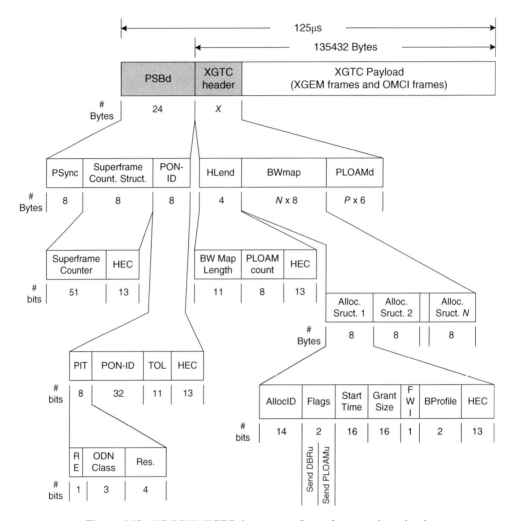

Figure 4.13 XG-PON1 XGTC downstream frame format and overhead

Downstream Frame

The downstream XGTC frame format is illustrated in Figure 4.13. The PSBd functions have been described in the previous section. The PON-ID structure contains fields for the PON ID, the Transmit Optical Level (TOL), and the PON ID Type (TIP), which contains the ODN class identifier (loss budget class) and an indicator of whether the TOL pertains to the launch power of an OLT or to a reach extender (RE).

The remainder of the frame is comprised of the XGTC header and the payload areas. The header begins with the fixed length HLend field, which defines the length of the other two header fields.

The BWmap is the upstream bandwidth map field that contains the bandwidth grants to the ONUs. Each bandwidth grant is an 8-byte Allocation Structure that communicates, on a per-Alloc-ID basis, the allocation start time, grant size. It also communicates any requests/allocations for the ONU to send either a

DBRu or PLOAMu in the next frame, a Forced Wakeup Indication (FWI), and an error check/correction field over the structure.

This format differs from the one illustrated in Figures 4.1 and 4.3 for G-PON in several ways. The first difference is the use of grant size instead of a stop time to establish the end of the burst allocation. The second is that the granularity of the start time and grant sizes are units of 4-byte words rather than single bytes. The resulting bandwidth granularity is (32 bits/125 μs) = 256 Kbit/s.

It is still possible for the OLT to achieve finer bandwidth granularity by allocating bandwidth in a fixed fraction of the upstream frames rather than in every frame. For upstream transmission efficiency, the OLT can arrange for an ONU to transmit its data for multiple Alloc-IDs in a single, contiguous upstream burst. It does this by sending a burst allocation series, which is a sequence of allocation structures associated with those Alloc-IDs.

The BWmap flag field has been simplified relative to that for G-PON, and the structure includes the Forced Wakeup Indication (FWI) bit, which can be used to speed the waking up of an ONU in the sleeping or dozing power saving states. The structure also adds the Burst Profile (BProfile) field, which is a 2-bit index to the set of valid burst profiles for that ONU to use in forming its PHY burst. The OLT uses the burst profile to adjust the ONU's optical signal level, plus preamble and delimiter patterns values to optimize them for the OLT receiver. It communicates the burst profile and associated index to the ONUs through the PLOAM message channel.

The PLOAMd header field is used to send PLOAM messages from the OLT to the ONUs. The number of 48-byte PLOAM messages sent in that frame is specified as 0–255 in the PLOAM Count field of the HLend.

Upstream Frame

The upstream XGTC frame format is illustrated in Figure 4.14. The upstream frame length is 125 μs (38880 byte periods) long, and this is comprised of the upstream transmission bursts of all the ONUs granted bandwidth to transmit in that frame. Each upstream burst begins with the PSBu physical layer overhead described above. The remainder of the upstream burst is described in this section.

The XGTC upstream header contains the ID of the transmitting ONU and status indicators, protected by an error check code. Two indicators are specified at this time. One indicates that the ONU has a PLOAMu waiting to transmit and, thus, needs a bandwidth grant to send it. When the OLT subsequently grants the bandwidth to send the PLOAMu, the PLOAMu is sent as a 32-byte message in an optional field of the XGTC upstream header. The other indicator is referred to as a "Dying Gasp" (DG) that the ONU uses to tell the OLT that it is dropping off the PON due to a local condition. For example, it could indicate that the ONU has run out of battery power after a prolonged local power outage. Note that the ONU is not obligated to leave to PON after signaling DG, and the OLT can continue to send bandwidth grants to that ONU. The OLT will not regard a lack of response from the ONU as an alarm condition to be reported.

The next part of the upstream burst is an optional 4-byte field used for reporting the ONU buffer status associated with a specific Alloc-ID. The OLT controls whether this DBRu structure is sent by setting the "Send DBRu" flag in the downstream Allocation Structure associated with that Alloc-ID. The buffer occupancy (BufOcc) indicates the number of 4-byte words queued in all the buffers corresponding to that Alloc-ID. This Allocation Overhead field includes error protection.

The XGTC payload follows the burst overhead. The amount of data sent is determined by the upstream bandwidth grant from the OLT. The upstream data is encapsulated using XGEM, as described in Section 4.4.5.

The upstream burst concludes with a 4-byte trailer that is a BIP error check over the entire XGTC burst. In the absence of FEC, the OLT uses the BIP information to estimate the upstream error rates. When FEC is present, its correction results are used for estimating the bit error rates.

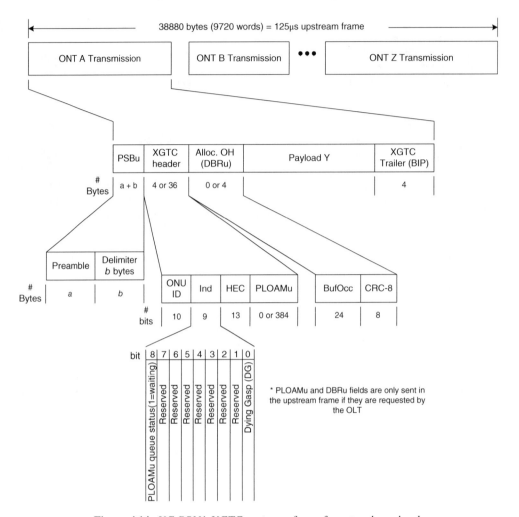

Figure 4.14   XG-PON1 XGTC upstream frame format and overhead

4.4.4   Forward Error Correction

FEC implementation is required for both the upstream and downstream directions. The downstream FEC is a RS(248, 216) code and is always enabled at the OLT. The upstream FEC is a RS(248,232) code and is used by the ONUs under dynamic control of the OLT. The RS(248,232) is a truncated form of the RS (255,239) code, which is capable of correcting eight symbols, and the RS(248, 216) is a truncated form of the more powerful RS(255, 223) code, which is capable of correcting 16 symbols.[15] FEC overhead is added prior to scrambling.

In the downstream frame, 32 bytes of FEC check overhead are added after every 216 payload bytes (starting with the XGTC header, but excluding the PSBd). Specifically, the 135456 bytes of the downstream frame are divided into 216-byte blocks, which become 672 248-byte FEC code words after the parity bytes are added. The resulting transmitted downstream frame is 155520 bytes long.

---

[15] The higher bit rate of the downstream signal relative to the upstream signal increases the need for FEC on the downstream signal. This is the reason why a more powerful FEC code was specified for the downstream signal than for the upstream signal.

PLI (Payload Length Indicator)	Key Ind.	XGEM Port ID	Options	L F	HEC - BCH(63, 12, 2) + P	Payload
14	2	16	18 (For Further Study)	1	13	P = L + pad (bytes)

\# bits

Figure 4.15 XGEM frame format

Similar to the downstream frame, the first upstream FEC codeword begins with the burst XGTC overhead and excludes the PBSu. The upstream uses RS(248, 232), which adds 16-bytes of FEC check overhead after every 232 bytes of the upstream burst to create a string of 248-byte FEC codewords. Since the pre-FEC upstream burst length is typically not an integer multiple of 232 bytes, the last FEC code typically contains less than 232 bytes of upstream data. A last code word containing X bytes of upstream data is handled by computing the FEC based on filling the $232\text{-}X$ leading bytes with all zeros. These all-zero bytes are then removed so that only a $X + 16$ byte codeword is transmitted at the end of the burst. The OLT receiver re-inserts the all-zero padding when it decodes the last FEC codeword.

Note that when an ONU is using upstream FEC, the OLT should account for this by assigning the upstream bandwidth grant such that the last FEC codeword does not require padding. In other words, the upstream burst should be an integer number of 232-byte blocks.

4.4.5 XG-PON Encapsulation Method (XGEM)

XG-PON uses an encapsulation similar to G-PON GEM, but optimized for the XG-PON rates and XGTC layer. As with GEM, XGEM is used in both the downstream and upstream directions to encapsulate user data and protocol signaling SDUs, and it supports SDU fragmentation and delineation.

The XGEM frame is comprised of the XGEM frame header and payload area. The 64-bit XGEM header is illustrated in Figure 4.15. The field definitions are as follows:

The PLI field gives the length L in bytes of the SDU contained in the XGEM frame payload. Since the payload area length is always an integer number of 4-byte words, there will typically be some number of pad bytes (0x55) after the end of the SDU. Specifically, the padding rules are:

- The payload length P must be at least 8 bytes, so up to 7 pad bytes are added if $L < 8$.
- Otherwise, 0–3 pad bytes are added so that payload length P is $N \times 32$ bits long, where N is an integer.

The two-bit Key Index indicates which encryption key is being used for the XGEM payload data, with the XGEM Port-ID indicating whether it pertains to a unicast or broadcast key type. Each of these key types can have up to two active keys at a time. A Key Index of 00 indicates that no encryption is used on that XGEM frame, a Key Index of 01 indicates the first key, and a value of 10 indicates the second. The 11 value is reserved for future use. A frame is discarded if the received Key Index is 11 or points to an invalid key.

Since the higher bandwidth of XG-PON makes it practical to have more ports, the XGEM Port-ID has been expanded to 16 bits.

The Options field is undefined at this time and is left for potential future definition.

When SDUs are fragmented, it is important to have an indicator of whether the current XGEM frame contains the end of the SDU or an earlier fragment. The LF (Last Fragment) bit is set to 1 whenever the XGEM frame contains either the complete SDU or the last fragment of a fragmented SDU. Otherwise, it is set to 0.

XGEM uses the same type of 13-bit header error check (HEC) as GEM (i.e., a 12-bit 2-error correcting BCH code plus a parity bit), except that it now covers 63 bits instead of 39 bits due to the longer header. A correct HEC confirms that there was proper frame delineation and interpretation of the PLI in the last XGEM frame.

XGEM SDU fragmentation works essentially the same as described above for GEM, subject to the $N \times 32$ bit payload size constraint and padding rules of XGEM. No padding should be used for fragments, except for the last. The main difference from GEM is that downstream SDU preemption is not allowed. In other words, the fragments of a downstream SDU must be transmitted one after another without any other SDU transmitted between the fragments.

When there is no data (SDU or SDU fragment) to be sent in an XGEM frame, XGEM Idle frames are inserted into the XGTC payload area to fill the unused space. This includes cases in which an SDU is too large for the available XGTC space in a given frame, but the fragmentation rules prevent fragmenting that frame. Unlike the GEM Idle frames, which all have the same zero length payload area, the XGEM Idle frames can have any payload length $4k$, (k ranges from 0 to the maximum supported SDU size in 4-byte words), as specified by the PLI. An XGEM Port-ID value of 0xFFFF is used to tell the receiver that this is an Idle frame. The payload area of the XGEM Idle frame is arbitrarily defined by the transmitter[16] and is ignored by the receiver. No encryption is used on the Idle frames. Note that there is also a special XGEM Short Idle frame consisting of four all-zero bytes. The Short Idle frame is sent whenever the space available in the XGTC payload is too small to carry a normal Idle frame (i.e., the space is less than the XGEM header size).

4.4.6 XG-PON Management

Like G-PON, XG-PON uses embedded OAM, PLOAM, and OMCI for control and management communications. The OAM channel uses fields embedded within the XGTC frame overhead. PLOAM provides a message-based communications mechanism where the messages are sent in a predetermined portion of the XGTC frame as required. A uniform system for managing higher layers that define the services is provided by the OMCI.

4.4.7 XG-PON Security

XG-PON supports multiple authentication methods. All XG-PON systems must support a registration-based ID authentication in which there is a management-level assignment of the ONU registration ID to a subscriber. This ID is provisioned into the OLT and communicated to the subscriber or field personnel so that it can be entered into the ONU. This method only authenticates the ONU to the OLT. A second option uses an OMCI message exchange, and a third option is based on IEEE 802.1X authentication. Support for the latter two options is mandatory at the component level, but not at the equipment level. Both of these options provide mutual authentication for the ONU and OLT to each other.

XG-PON supports encryption at both the Transmission Convergence layer, and optionally for the XGEM payloads also. The XGEM encryption uses the AES-128 cipher. The downstream encryption can either be unicast to each specific ONU, or multicast to multiple ONUs.

The XG-PON standard also allows using a reduced strength encryption. This option is implemented by setting some of the bits in the 128-bit AES encryption key to a well-defined bit pattern. The remaining number of randomly generated bits forms the effective key length.

4.4.8 NG-PON2 40 Gbit/s Capable PON

ITU-T SG15 consented the NG-PON2 requirements in recommendation G.989.1 in September 2012 [26]. These requirements and the anticipated directions for the resulting standard are discussed in this section.

The key requirements for NG-PON2 can be summarized as follows:

- At least 40 Gbit/s downstream aggregate capacity, with a target capacity of 160 Gbit/s.
- At least 10 Gbit/s upstream aggregate capacity, with the target potential for rates up to 80 Gbit/s.

---

[16] The patterns must not create problematic patterns in the physical layer, such as long strings of bits with the same value.

- An ONU capable of supporting up to 10 Gbit/s service capacity (symmetrical or asymmetrical).
- At least 40 km reach over a passive ODN, with a goal of 60 km, and with the potential for up to 100 km using reach extension techniques.
- Maximum differential fiber distance configurable for up to 20 km and 40 km.
- Split ratio support up to at least 1 : 256.
- Support for at least four TWDM (concurrent time and wavelength division multiplexing) channels per direction, although not all of them may be populated.
- ONUs must be colorless.

Some additional desired characteristics include:

- Downstream/upstream asymmetry should allow a range from at least 1 : 1 to 100 : 1 at an ONU.
- Achieving 60 km to 100 km reach through reach extension
- Reach extension technology may also be used to achieve greater split ratios
- Flexible capacity through access to different wavelength counts (e.g., 4, 8, 16)

4.4.8.1 Required System Configurations

Since it will not be possible to achieve all these goals simultaneously in an individual system, tradeoffs will be required. G.989.1 also specifies some specific configurations that NG-PON systems must be capable of supporting:

- Required configurations:
  - 40 Gbit/s downstream (10 Gbit/s per downstream channel) with at least a 1 : 64 split and 20 km reach.
  - 10 Gbit/s upstream with at least a 1 : 64 split and 20 km reach.
  - 10/2.5 Gbit/s downstream/upstream access to peak rates.
  - the ability to achieve longer distances with lower split ratios.
- Desired configuration capabilities:
  - 40 Gbit/s upstream with at least a 1 : 64 split and 20 km reach.
  - 10/10 Gbit/s downstream/upstream access to peak rates.
  - extension to support eight TWDM channels.
  - tunable point-to-point WDM.

4.4.8.2 Coexistence and Migration Considerations

Coexistence of G-PON, XG-PON1, RF video overlay, and NG-PON2 systems on the same ODN is a requirement. The coexistence is achieved through wavelength assignment, and it requires that NG-PON2 be capable of using the legacy power budgets, within the same legacy spectrum. This includes not requiring wavelength filtering within the ODN itself.

Seamless migration to NG-PON2 from G-PON and XG-PON (and also EPON and 10G EPON) systems is also a key requirement. While the migration ideally occurs with no service outage, the requirement is that it occurs without prolonged service outage. In any case, a legacy ONU or OLT that is not part of the upgrade must be able to remain unchanged.

Achieving flexible capacity assumes that an NG-PON2 system will be capable of accessing multiple wavelengths or wavelength groups, or multiple wavelength bands. These wavelengths can be separated logically or physically. Furthermore, they can be driven independently by a single OLT or independently by multiple OLTs that are fully operationally independent. It is expected that either the ONU transmitter, receiver, or both will need to be tunable.

4.4.8.3   Physical Media Dependent (PMD) Sublayer

The specifics of the PMD were still being defined as this book went to print. However, the material in this section reflects the most likely agreements and the rationale behind them.

Line Rates

As indicated in the required configuration discussion, one of the typical configurations will use nominally 40 Gbit/s downstream transmission that consists of four 10 Gbit/s streams, each on a different wavelength. This combination of concurrent time division and wavelength division multiplexing for NG-PON2 is referred to as TWDM. The upstream signal rate will optionally either be nominally 2.5 Gbit/s or 10 Gbit/s. Other rate combinations are under consideration.

As with G-PON and XG-PON, the actual rates will be the same as the respective SONET/SDH rates. In other words, the 10 Gbit/s and 2.5 Gbit/s rates are actually 9.95328 and 2.48832 Gbit/s, respectively. Choosing these rates preserves an 8 KHz factor, which greatly simplifies network timing for carrying voice and other legacy signals.

Since XG-PON already defines FEC encoding for the 9.95328 Gbit/s downstream, the 9.95328 Gbit/s upstream, and the 2.48832 Gbit/s upstream signals, NG-PON will use the same FEC coding for those respective signals.

Note that the NRZ line code will continue to be used for both upstream and downstream signals, although other line codes (e.g., DQPSK) are under consideration for their potential to reduce crosstalk into overlaid video signals.

4.4.8.4   Resiliency

Due to its high capacity, and its intended use for supporting business customers, resiliency is critical to NG-PON2. Consequently, NG-PON2 will be required to support duplex and dual parenting, in addition to other extensions described in the Appendices to G.984,1 [12]. See also Chapter 2 for more on PON protection options.

Appendix 4.A: Summary Comparison of EPON and G-PON

Feature	G-PON	EPON	Comment
Responsible standards body	FSAN and ITU-T SG15	IEEE 802.3	
Data rate	2.488 Gbit/s downstream and either 1.244 or 2.488 Gbit/s upstream	1 Gbit/s upstream and downstream	The CTS specifies 2.488 Gbit/s downstream and 1.244 Gbit/s upstream for G-PON.
Split ratio (ONUs/PON)	1 : 64	1 : 64	While the EPON optical parameters were initially designed around a 1 : 16 split ratio, current implementations support 1 : 64.
Line code	Scrambled NRZ	8B/10B	Due to its line code, the EPON transmission signal rate is 1.25 Gbit/s
Number of fibers	1	1	
Wavelengths	1490 nm down and 1310 nm up	1490 nm down and 1310 nm up	Both G-PON and EPON support 1550 nm for downstream video overlay.

Maximum OLT to ONU distance	10, 20, and up to 60 km	10 and 20 km	The CTS specifies the 20 km reach for G-PON.
Optics	G-PON timing budget is very strict, requiring purpose-built implementations.	EPON timing budget is loose, allowing the use of off-the-shelf components	Currently, both G-PON and EPON achieve tight timing performance.
Protection switching	Supports multiple protection configurations	None	IEEE 1904.1 specifies preferred EPON protection options
Data format (encapsulation)	GEM and/or ATM	None (uses Ethernet frames directly)	ATM support was removed from the G-PON standard
TDM support	Direct via GEM, or CES	CES	G-PON supports CES through an intermediate mapping into Ethernet
Voice support	Via TDM or VoIP	VoIP	
Multiple QoS levels	Yes (mix of fixed, assured, and best effort bandwidth assignments)	Yes (802.1Q priority levels)	For G-PON, a dedicated Alloc-ID should be assigned to each traffic type. For EPON, all traffic types are typically handled using a single LLID.
FEC	RS(255, 239)	RS(255, 239)	Both use ITU-T G.975
Encryption	AES – 128 bit key required, 192 and 256 optional	None	A multi-company protocol using 128-bit key AES has become the de facto standard for EPON. Other regional options also exist.
OAM	GTC frame fields and GEM OAM	802.3ah Ethernet OAM frames	

References

1. ITU-TG.983.1. Broadband optical access systems based on Passive Optical Networks (PON); 2005.
2. ITU-T G.983.2. ONT management and control interface specification for B-PON; 2005.
3. ITU-T G.983.3. A broadband optical access system with increased service capability by wavelength allocation; 2001.
4. ITU-T G.983.4. A broadband optical access system with increased service capability using dynamic bandwidth assignment (DBA); 2001.
5. ITU-T G.983.5. A broadband optical access system with enhanced survivability; 2002.
6. ITU-T G.983.6. ONT management and control interface specifications for B-PON system with protection features; 2002.
7. ITU-T G.983.7. ONT management and control interface specification for dynamic bandwidth assignment (DBA) B-PON systems; 2001.
8. ITU-T G.983.8. B-PON OMCI support for IP, ISDN, video, VLAN tagging, VC cross-connections and other select functions; 2003.
9. ITU-T G.983.9. B-PON OMCI management and control interface (OMCI) support for wireless Local Area Network interfaces; 2004.
10. ITU-T G.983.10. B-PON ONT management and control interface (OMCI) support for Digital Subscriber Line interfaces; 2004.
11. ITU-T G.7041/Y.1303. Generic Framing Procedure; 2011.
12. ITU-T G.984.1. Gigabit-capable Passive Optical Networks (G-PON): General Characteristics; 2008.
13. ITU-T G.984.2. Gigabit-capable Passive Optical Networks (G-PON): Physical Media Dependent (PMD) layer specification; 2003.

14. ITU-T G.984.3. Gigabit-capable Passive Optical Networks (G-PON): Transmission convergence layer specification; 2008.
15. ITU-T G.984.4. Gigabit-capable Passive Optical Networks (G-PON): ONT Management and control interface specification; 2008.
16. ITU-T G.984.5. Gigabit-capable Passive Optical Networks (G-PON): Enhancement band; 2007.
17. ITU-T G.984.6. Gigabit-capable Passive Optical Networks (G-PON): Reach Extension; 2008.
18. ITU-T G.984.7. Gigabit-capable Passive Optical Networks (G-PON): Long Reach; 2010.
19. ITU-T G.982. Optical access networks to support services up to the ISDN primary rate or equivalent bit rates; 1996.
20. Federal Information Processing Standard 197, Advanced Encryption Standard, National Institute of Standards and Technology, U.S. Department of Commerce, November 26, 2001.
21. ITU-T G.987. 10-Gigabit-capable passive optical network (XG-PON) systems: Definitions, abbreviations and acronyms; 2012.
22. ITU-T G.987.1. 10 Gigabit-capable passive optical network (XG-PON): General Requirements; 2010.
23. ITU-T G.987.2. 10-Gigabit-capable passive optical networks (XG-PON): Physical media dependent (PMD) layer specification; 2010.
24. ITU-T G.987.3. 10 Gigabit-capable Passive Optical Network (XG-PON) systems: Transmission Convergence layer specification; 2010.
25. ITU-T G.987.4. 10-Gigabit-capable passive optical network (XG-PON) systems: Reach Extension; 2012.
26. ITU-T G.989.1. 40-Gigabit-capable passive optical networks (NG-PON2): General requirements; 2012.
27. ITU-T G.988. ONU management and control interface specification (OMCI); 2010.

Further Readings

1. Effenberger F. The XG-PON system: Cost effective 10Gb/s access. *IEEE/OSA Journal of Lightwave Technology.* 2010 **29**:403–409.
2. Effenberger F, Mukai H, Soojin P, Pfeiffer T. Next generation PON – part II: Candidate systems for next generation PON. *IEEE Communications Magazine.* 2009 **47**(11):50–57.
3. Effenberger F, Mukai H, Kani J, Rasztovits-Wiech M. Next generation PON – part III: System specifications for XG-PON. *IEEE Communications Magazine.* 2009; **47**(11):58–64.
4. Effenberger F, Ichibangase H, Yamashita H. Advances in broadband passive optical networking technologies. *IEEE Communications Magazine.* 2001; **39**(12):118–124.
5. ITU-T G.Imp983.2. Implementers' Guide to G.983.2 (2002); 2006.
6. ITU-T G.Imp984.3. Implementers' Guide for ITU-T Rec. G.984.3; 2004.
7. Gorshe S. FTTH/FTTH technology and standards. *China Communications* 2006; **3**(6):104–114.
8. Gorshe S. Overview and comparison of the IEEE EPON and ITU-T G-PON protocols. *China Communications.* 2007; **4**(2):69–78.
9. ITU-T G.Sup39. Optical system design and engineering considerations; 2008.
10. Jingjing Z, Ansari N, Yuanqiu L, Effenberger F. Next-generation PONs: A performance investigation of candidate architectures for next-generation access stage 1. *IEEE Communications Magazine.* 2009; **47**(8):49–57.
11. ITU-T G.Sup45. GPON power conservation; 2009.

5

Optical Domain PON Technologies

5.1 Introduction

Optical domain PON technologies (i.e., PONs whose media access is defined by optical rather than electronic means) are ideal from the standpoint that they can be largely agnostic to the rate and format of the client signals they carry over the PON. For example, a purely optical domain PON could potentially carry each subscriber's UNI signal in its native format (e.g., DS1/DS3, 10/100/1000BASE Ethernet, etc.), allowing simple service upgrades without needing to upgrade other remote ONUs on the same PON. It also allows the highest potential data rates per subscriber. However, the optical domain technologies have been more expensive than TDM technologies, and will probably continue to have a cost disadvantage at least until around the 2015 time frame. Another drawback to all the optical domain technologies relative to TDMA is that they do not allow statistical multiplexing on the PON for more bandwidth efficiency at the OLT PON interface.

An important application for optical domain PON technologies in the meantime is to carry TDMA PON protocols in order to increase the number of ONUs and/or the per-subscriber service rates on a PON. This WDM application was introduced in Chapters 2 and 4 in the discussion of next generation FSAN/ITU PON protocols. This chapter focuses on the optical domain technologies, since their extension to hybrid combinations with TDMA PON is relatively straightforward.

5.2 WDMA (Wavelength Division Multiple Access) PON

5.2.1 Overview

With WDMA PON, the OLT uses a separate pair of wavelengths – one upstream and one downstream – to communicate with each ONU as a logical point-to-point connection.[1] In other words, WDM is used instead of TDMA for multiplexing each subscriber's signal onto the PON. The ODN is constructed using WDM multiplexers instead of passive power splitters, so that each ONU only receives the wavelength designated for it. Since bandwidth for each wavelength is not shared, WDM PON allows the highest per-subscriber bandwidth, and it is largely independent of the subscribers' signal formats at the ONU.

WDM wavelength grids have been standardized by the ITU-T, primarily for use in metro and core optical networks. However, the same grids are generally expected to be used by WDM PON systems. These grids are illustrated in Figure 5.1. The Coarse WDM (CWDM) grid spacing is 20 nm between

---

[1] As described below, in some cases the ONU uses the downstream wavelength as a 'seed' for its optical source, so that the ONU uses the same wavelength for upstream transmission as it received.

Broadband Access: Wireless and Wireless – Alternatives for Internet Services, First Edition.
Steven Gorshe, Arvind Raghavan, Thomas Starr and Stefano Galli.
© 2014 John Wiley & Sons, Ltd. Published 2014 by John Wiley & Sons, Ltd.

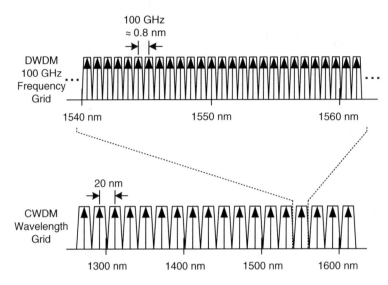

Figure 5.1 ITU-T WDM wavelength/frequency grid

channels [1]. The Dense WDM (DWDM) spacing is expressed in terms of frequency rather than wavelength [2]. G.694.1 specifies both 50 GHz channel spacing and the 100 GHz channel spacing shown in Figure 5.1. If the entire range is used with 50 GHz channel spacing, 1000 channels could be accommodated. However, the 1539–1582 nm region is best suited to upstream transmission due to its lower loss, laser technology maturity and compatibility with EDFAs.

A laser's wavelength will drift with temperature. In order to maintain a laser's output wavelength within the channel boundaries over an extended environmental temperature range, a mechanism such as a thermo-electric cooler is required to stabilize the laser's operating temperature. Temperature stabilization is less of a problem at the OLT since it will typically reside in a CO with relatively little environmental temperature range. The ONU, however, is subject to the -40 to $+85\,°C$ temperature range of outside plant equipment. Since higher downstream bandwidth is typically required, and since temperature stabilization is less difficult at the CO, the DWDM grid may be practical for downstream signals.

Since temperature stabilization circuits would add considerable cost to an ONU, it is much more cost-effective for the upstream signals to use the CWDM grid where the channel spacing allows operation without temperature stabilization. However, the CWDM grid, as defined in ITU-T G.694.2, only provides 18 wavelengths, ranging from 1271 nm to 1611 nm with 20 nm spacing. Hence, only 18 ONUs could be supported with CWDM for the upstream transmission.

5.2.2   Technologies

The primary drawback of WDMA PON is the cost of the optical components to create and filter the different wavelengths. At the OLT, some component sharing is possible in order to spread the cost over multiple ONU connections. The ONU- related costs are the most critical, since they cannot be shared among multiple subscribers. Having a separate ONU type for each possible upstream wavelength is impractical in terms of the added operation and the inventory costs associated with stocking and deploying the different ONUs. Using different fixed wavelength ONUs also works against the volume production cost savings that are possible if all ONUs are identical. Ideally, each ONU is "colorless", in that it does not have a specific fixed upstream wavelength and it is flexible enough to allow any of a number of different wavelengths to be used. Multiple alternatives have been explored for generating the different colorless ONU wavelengths, including the following:

- Field installable optical modules to select the ONU wavelength.
- Tunable lasers at the ONUs.
- Spectrum slicing at the ONUs or in the PON
- Reflective approaches in which the OLT provides the optical carrier signals to the ONUs and each ONU modulates the carrier in some manner as it reflects it back to the OLT.
- Using the downstream signal to control the output wavelength of the ONU laser.

Field installable ONU optical modules are not a very flexible approach. Although this alternative has somewhat more flexibility and lower inventory cost than dedicated wavelength ONUs, it still has similar problems. Adding new ONUs to a PON requires determining which wavelengths are already in use and selecting an optical module with an available wavelength. Stocking and tracking the different optical modules adds considerable operations expense.

Tunable lasers at the ONUs are very flexible, but they are not yet cost-effective for this application.

With spectrum slicing, a light source with a reasonably broad optical spectrum is used at the ONUs. For example, the light source could be a spectrum-sliced light-emitting diode (SSLED). Filters are used to select the ONU transmission carrier wavelength, but placing the wavelength selection filters at the ONU would still require different components for each ONU. A more practical approach is to use a wavelength filter in the PON to combine the signals from the different ONUs. The wavelength filter function selects which wavelength is passed through to the OLT for each ONU. The main problem with spectrum slicing is that it is difficult to achieve enough optical power for upstream transmission at each wavelength.

For reflective approaches, the OLT provides the optical carrier signals to each ONU. The ONU then reflects and modulates this signal to create the upstream signal. There are two basic methods for the OLT to provide the upstream carrier signal to the ONU. One method is for the OLT send a continuous wave signal to the ONU at the upstream wavelength in addition to the downstream signal on its own wavelength. The second approach is for the ONU to modulate the downstream data signal that it receives in order to create the upstream signal.

The latter approach is feasible when the upstream data rate is lower than the downstream signal rate by some multiple, and has the advantage of requiring half as many lasers at the OLT and half as many wavelengths to filter within the network. The drawbacks are the upstream rate restrictions, and the additional noise associated with modulating a data signal rather than a continuous wave signal. For all reflective approaches, the optical network must be carefully constructed to minimize reflections of the downstream signal to be used by the ONU for upstream transmission. Any reflections of these signals add to the noise at the OLT receiver.

Reflective Semiconductor Amplifiers (RSOA) are one way to implement the reflective approach. There are multiple ways to implement an RSOA and modulator. In general, an RSOA is an SOA in which one or both endfaces of the optical chamber are non-reflective. This allows the RSOA to amplify the signal based on the electrical pump input without lasing. Since the RSOA does not need to operate as a laser, it can handle the amplification of a large enough range of wavelengths to cover the WDM upstream signal band.

One type of implementation is to modulate the RSOA directly. This approach is relatively simple, although the performance has been limited by the long optical carrier lifetime to rates of 2.5 Gbit/s or lower. Another implementation is to have an optical absorption modulator at one end of the RSOA chamber and the reflective surface at the other end. This implementation is illustrated conceptually in Figure 5.2. In order to send an upstream pulse, the modulator allows the carrier signal into the RSOA where it is amplified by passing back and forth once through the active amplification region. RSOA implementations have been recently reported that can achieve 2.5 Gbit/s upstream transmission with an 8 dB extinction ratio and power penalty of 2 dB for up to 20 km of fiber at 1550 nm [3].

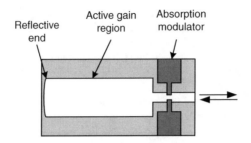

Figure 5.2 Basic example of an RSOA and modulator

Using the downstream signal to control the output wavelength of the ONU laser is similar to the reflective approach. For example, it has been shown that inserting some of the downstream signal into a vertical-cavity surface-emitting laser (VCSEL) will cause the VCSELs output to lock to the same wavelength as the downstream signal. This technique is known as optical injection locking.

The other key technology requirement is inexpensive wavelength filters. The leading candidate is Arrayed Waveguide Grating (AWG) technology. An AWG is a passive device that can be constructed so as to be sufficiently athermal to allow for use in outside plant environments. Since they can be implemented with the silica-on-silicon technology that is used for other optical integrated circuits, they have the potential to become reasonably cost-effective. AWG filters can be constructed to filter up to 80 wavelengths, depending on the wavelength spacing. Additional wavelengths or dense wavelength spacing complicates the AWG design.

The concept behind the AWG is illustrated in Figure 5.3. The light on the input fiber passes through a free space region and couples into multiple waveguides. The waveguides have a constant incremental length difference relative to each other. Specifically, the optical path length of adjacent waveguides differs by an integer multiple of the central wavelength of the filter. The difference in waveguide lengths introduces a relative phase shift in the optical signals of different wavelengths passing through the waveguides. The output of the waveguides re-combines in a second free space region. The light that

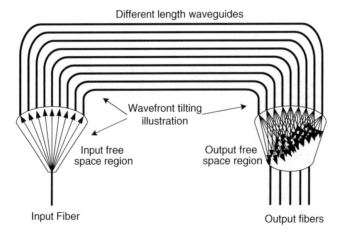

Figure 5.3 Arrayed Waveguide Grating (AWG) illustration

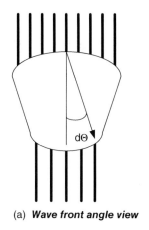

(a) *Wave front angle view*

Waveguides from the array

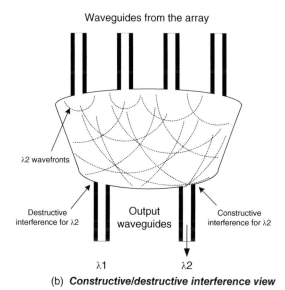

(b) *Constructive/destructive interference view*

Figure 5.4 AWG output free space wavelength filter illustration

couples into the output fibers is wavelength selective, with a different wavelength coupling into each output fiber.

The resulting filtering phenomenon can be described in terms of the different waveguide lengths "tilting" the wave front of the optical signal. Due to this tilt, each wavelength exits the array at a different angle, with the output ports located at the point where each wavelength focuses. This view is illustrated in Figure 5.4a, where $d\Theta$ is the output tilting angle for a given wavelength relative to the central wavelength of the filter.

An alternative way to describe the phenomenon is illustrated in Figure 5.4b. The wave fronts coming out of each waveguide are offset in phase. As a result, the wave fronts for each wavelength combine such that they interfere constructively at one output port but interfere destructively at all other output ports.

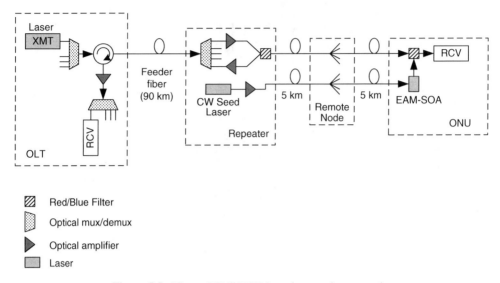

Red/Blue Filter

Optical mux/demux

Optical amplifier

Laser

Figure 5.5 "Super-PON" WDM reach extension example

5.2.3 Applications

Another important advantage to WDMA PON is that optical filters have much less optical loss than the passive power splitters used with TDMA PON. For example, a 1 : 32 optical power splitter introduces a loss of about 18.4 dB, while a 1 : 32 wavelength multiplexer introduces about 3.5 dB of loss. The nearly 15 dB difference can be used for additional distance/reach or to support additional ONUs on the same ODN.

While WDMA PON has some flexibility advantages with respect to carrying different client signals and higher data rates per subscriber, the high number and relative complexity of the optical components have kept it less cost-effective than the TDMA-based PON systems such as EPON and GPON, or point-to-point fiber connections using media converters.

The one significant deployment of WDMA PON to date is in South Korea. This system appears to have used AWG receiver filters with spectrum slicing at the ONUs. The spectrum slicing is performed with acousto-optic tunable filters (AOTFs). With AOTF, an acoustic wave is used to create a long-period diffraction grating that acts as a notch filter for the desired wavelength.

One potential approach to building a "Super-PON" (SPON) with 1000 ONUs and 100 km reach has been demonstrated [4]. This approach is illustrated in Figure 5.5.

Hybrid DWDM/TDMA PON is another possibility, and this is under study for use with the NG-PON systems discussed in Chapter 4. One recent demonstration system uses a hybrid DWDM-TDM approach to allow bit rates of up to 10 Gbit/s per wavelength, with each wavelength shared by 256 ONUs [5]. This system uses the same concepts as illustrated in Figure 5.5 and is a follow-on to the SPON project. The XG-PON1 is considering using DWDM in order to overlay multiple XG-PON1 systems onto the same ODN. A 200 GHz channel spacing is being considered in order to reduce the cost of the optical components. In its basic form, a hybrid WDM/TDMA PON system would use WDM downstream with TDMA upstream in order to avoid the problems associated with colorless ONU upstream capabilities. This technique is referred to as "composite" PON in [6].

5.3 CDMA PON

Code-division multiple access (CDMA) technology can also be applied to PON applications. One of the advantages to CDMA PON is that each ONU can use a different signal rate and format

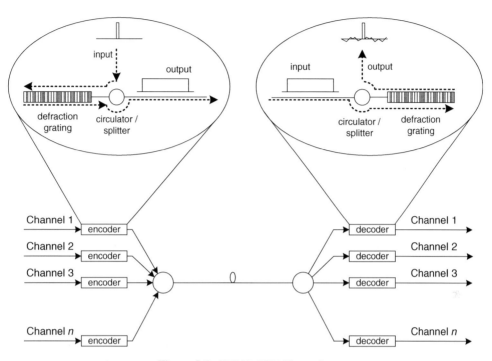

Figure 5.6 CDMA PON illustration

corresponding to the native client signal from the subscriber, without requiring separate wavelengths per ONU. Optical CDMA can also be used in conjunction with WDM for increased bandwidth capabilities. A basic, classical implementation, as described in [7] and illustrated in Figure 5.6, is described here.

The concept behind CDMA is to carry multiple client signals with their transmission spectrum spread over the same channel. The symbols from the constituent signals are encoded in a manner that is recognizable by the decoder. In wireless systems, one bandwidth-spreading technique is to use pseudo-random frequency hopping among different carrier frequencies. Another technique is to encode the symbols (e.g., 0 and 1) of each client signal encoded as a longer string of symbols at a higher rate, with each client using a different string value for their symbols. The latter approach is referred to as direct sequence spread spectrum, and this is the one discussed here due to its amenability to optical CDMA. Wavelength hopping optical CDMA systems have also been explored, especially for applications where additional security is required [8,9].

Fortunately, optical direct sequence CDMA can be implemented with passive diffraction filters. One implementation uses Bragg diffraction grating, although other grating types could also be used. A fiber Bragg grating can be constructed using UV exposure of standard single mode fibers through a mask with the desired pattern.

As illustrated in Figure 5.6, the encoder and decoder can use the same basic implementation. The signal is launched into one end of the filter. As the signal propagates through the filter, the grating pattern creates interference patterns as the light is reflected. The resulting signal that reflects back out of the filter is thus modified in both amplitude and phase, with the symbol spread a function of the propagation time through the filter. At the encoder, the input symbol is thus converted into a spread spectrum symbol. At the receiver, the inverse operation converts the received spread spectrum symbol back into the original symbol.

In the upstream direction each CDMA PON ONU will use a unique grating pattern to encode its transmitted symbols. Due to the linearity of the Bragg filter, the spread spectrum bandwidth is proportional to number of ONUs. As illustrated in Figure 5.6, the OLT splits the received optical signal to multiple diffraction filters in order to recover the data from the different ONUs. In a classical optical CDMA implementation, the transmitter and receiver use identical Bragg gratings. More sophisticated receivers can make use of different gratings at the transmitter and receiver, and can use a combination of optical and electrical domain processing. Proper design of the grating patterns results in a system where the effects of crosstalk can be eliminated at the receiver's symbol sample time.

Note that temperature control of the grating is important, since physical expansion or contraction of the filter changes the effective pattern. This feature can also be exploited, however, to achieve tunable filters (other tuning mechanisms also exist). The lasers in an optical CDMA system, however, do not require frequency stability.

Other phase coding technologies that have been demonstrated include arrayed wave guides (AWG), imaged phased array (VIPA) filters, micro-ring resonators (MRR), and holographic Bragg reflectors (HBR). The MRR solutions are promising due to allowing integration with other photonic elements, offering high spectral resolution, and allowing reconfigurable phase codes. The HBR is a promising technology for implementing wavelength hopping CDMA.

One major drawback to optical CDMA PON is that optical amplifiers are typically required to achieve an adequate signal-to-noise ratio. The second bank of passive splitters at the receiver give at least 3db of attenuation per splitter level, and the circulators and filters also introduce some attenuation. As a result, ONU/OLT splitter ratios without amplifiers are only in the 2 : 1 to 8 : 1 range. Another factor limiting the split ratio is that a limited number of adequately orthogonal codes are available unless more complex codes are used, which increases the number of coding elements. Codes that are not orthogonal create crosstalk. Also, the receivers are complex relative to other PON technologies. Consequently, they have not proven to be as cost-effective as some of the other alternatives.

Readers interested in an in-depth discussion of optical CDMA can refer to [10].

5.4 Point-to-Point Ethernet

For some applications, point-to-point Ethernet architectures are attractive. As the name implies this architecture consists of a direct, dedicated fiber connection from the CO to each subscriber. The primary benefit to this architecture is that it is the most flexible for providing different services per subscriber and scaling to higher bandwidth. It is frequently used to serve enterprise subscribers. The ITU-T has defined the interfaces for the access application of 100 Mbit/s and 1 Gbit/s Ethernet signals in G.985 [15] and G.986 [16], respectively.

The main drawback to this approach is the cost associated with not sharing the fiber plant or CO terminal port interfaces. The lack of fiber sharing adds cost for the distribution fiber and also adds cost associated with the increased amount of fiber management at the CO. The lack of CO terminal port sharing means that, unlike with a TDMA PON OLT, each subscriber has a dedicated laser and optical receiver at the terminal. These added costs are partially offset by being able to use less expensive enterprise network-type Ethernet switches instead of OLTs. Due to their much higher deployment volume than OLTs, these Ethernet switches are significantly less expensive, but they need to be capable of supporting carrier grade Ethernet services. Also, their optical interfaces must be capable of covering the longer distance to the subscribers. One study shows the cost premium of using point-to-point Ethernet instead of point-to-multipoint PON to be in the range of 5–25% [11].

5.5 Subcarrier Multiplexing and OFDM

5.5.1 Introduction

Another multiplexing technique is to modulate the optical carrier signal with different frequencies to create subcarriers. For multiple access, all the ONUs use the same nominal carrier wavelength, but each ONU uses a different modulation frequency to create its own subcarrier for communication with the OLT. In other words, the traffic between the OLT and different ONUs is separated in the frequency domain rather than using different time slots or wavelengths.

The primary advantage of this approach to multiple access is that each ONU has its own independent channel to the OLT, just as with WDMA or CDMA. There are several drawbacks, however. One major drawback is that having independent channels for each ONU means that no statistical multiplexing is performed on the PON. The main technical drawback is the implementation complexity, as relatively expensive analog RF components are required at the ONUs and OLT. Also, since the carrier wavelengths used by each ONU will not be precisely the same, beat noise can be generated in the received electrical domain signal at the OLT. Methods to combat this problem (including using greater, intentional wavelength separation of the carriers from each ONU) further increase the complexity.

Due to the issues associated with upstream transmissions, this method has not seen significant use for multiple access. The one partial exception is RFoG, discussed in Chapter 6.

For downstream transmission, however, many of these problems go away. Subcarrier multiplexing is commonly used for delivering video signals over a separate wavelength from the data traffic. Each video signal is modulated into a separate frequency channel (subcarrier), typically in the same format used for cable television signals. In some applications, discussed in the next section, data signals are also carried over the subcarrier channels.

5.5.2 OFDMA PON

The most promising subcarrier multiplexing approach for providing ultra-high bandwidth makes use of orthogonal frequency division multiplexing (OFDM) techniques. The resulting orthogonal frequency division multiple access (OFDMA) systems can be used independent from, or layered onto TDMA and WDMA PON techniques. A key technical advantage of OFDMA is that most of the processing can be handled in the electrical domain through digital signal processing (DSP) rather than in the optical domain. Electrical domain processing is typically more cost effective than optical domain processing, and Moore's Law tends to reduce its costs more rapidly over time. Also, OFDMA allows using a single optical source, with the DSP used to generate the different optical WDM channels (frequencies) of the OFDM signal.

The basic principle behind OFDM is to divide the available bandwidth into multiple partially-overlapping frequency slots. The center frequency of each slot is a subcarrier, and the partial overlap is chosen such that the overlapping does not cause interference between the frequency slots. Specifically, the center frequencies of the slots (i.e., the subcarriers) are chosen such that:

$$f_n = n/T, (n = 1, 2, \ldots N) \tag{5.1}$$

where T is the symbol period for each subcarrier.

The resulting spectrum is illustrated in Figure 5.7, where it can be seen that the frequency spectrum of each slot has nulls at the center frequencies all other slots. Because the frequencies do not interfere with each other, they are orthogonal.

The OFDM signal is robust to optical domain impairments such as chromatic dispersion (CD) and polarization mode dispersion (PMD). Another key advantage of OFDM is that it can be implemented at the transmitter based on Fast Fourier Transform (FFT) techniques using DSP. The FFT method allows a straightforward modulation format adaptation for each subcarrier, including the use of multi-level

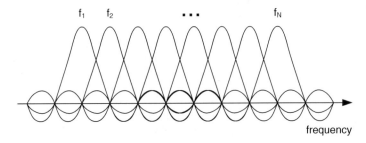

Figure 5.7 OFDM signal spectrum illustration

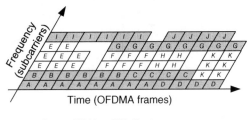

Gray = TDM or RF clients
White = packet clients

Figure 5.8 OFDMA bandwidth assignment illustration

signaling for greater bandwidth efficiency on each subcarrier. For example, M-ary quadrature amplitude modulation (QAM) can be used on a subcarrier where the value of M can be chosen based on the quality of that channel slot. Multiple techniques exist for creating the complex QAM signal.

As discussed above, each subcarrier provides an independent channel between the OLT and ONU. When combined with TDMA, a subcarrier could be shared by multiple ONUs, allowing the potential for normal TDM or statistical multiplexing gain. Further, bandwidth efficiency can be increased through the dynamic assignment of TDMA time slots and OFDM carriers among the ONUs on a PON. An example of such allocation across time and subcarrier slots is illustrated in Figure 5.8. In this example, it can be seen that TDMA or radio frequency (RF) client signals (shown in gray) occupy a single frequency. Packet multiplexed clients (shown in white) can occupy a combination of time and frequency slots, with the amount of bandwidth allocated to each client assigned dynamically. Figure 5.9 illustrates how this OFDMA time and frequency allocation could be mapped to a set of ONUs within an OFDMA frame.

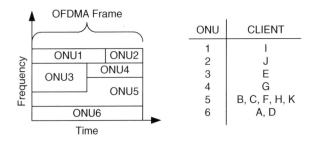

Figure 5.9 OFDMA bandwidth assignment to ONUs within an OFDMA frame

As with other PON protocols, the ONU time and frequency assignments are communicated to them by the OLT within a downstream overhead channel. For upstream transmission, each ONU generates an OFDM signal using the subcarriers assigned to it and nulling all other subcarriers.

Since OFDMA signals are immune to PMD effects, they can be combined with optical domain techniques such as polarization multiplexing for additional bandwidth gains. Using polarization multiplexing, though, creates additional complexities for the ONU. Metro and longhaul systems typically use coherent optical receivers for optimal reception of polarized signals. However, this technology is too expensive for ONUs, since it involves using an optical local oscillator for a down conversion in the optical domain to produce the baseband IQ signals.

Direct detection is possible at the ONUs with relatively simple, low-cost receivers based on a single photodiode. However, using direct detection is complicated by the loss of polarization information in the direct detection process, which can lead to destructive interference between the incoming signals. DSP-based techniques have been proposed to solve this cross-polarization interference problem [12].

Another advantage of OFDMA is that the ONUs can use the same wavelength, yet remain effectively "colorless" due to using different subcarriers or time slots. Using a single wavelength eliminates the problems with "colorless" WDMA ONUs. Using a common upstream wavelength, however, requires using a coherent receiver at the OLT in order to handle the interference between the upstream ONU signals. Note that coherent reception requires accurate source wavelength control that is not possible at an ONU without temperature compensation.[2] It is also possible to have the OLT transmit the upstream wavelength to the ONUs, as described above in Section 5.2.2, so that they are optical source-free.

Ultra-high-capacity OFDMA PON systems have been described that can achieve, for example 100 Gbit/s capacity [12] or up to 1 Gbit/s per user with extended range [13], while supporting a variety of applications.

OFDMA PON systems have been cost-prohibitive so far. The ADC, DAC, and DSP-based components are the main cost drivers for ultra-high-capacity OFDMA systems. The DAC is complicated by the OFDM being complex, containing both in-phase and quadrature components. This requires that the DAC has two analog outputs to the optical modulator, each operating a sufficiently high sampling rate. The typical rule of thumb is to use a sampling rate that is 1.5 times the OFDM subcarrier frequency ($f_{sDAC} = (1.5)f_N$). Likewise, the ADC must have separate channels to handle the in-phase and quadrature components. The ADC sampling clock should be at least 2.5 times the OFDM subcarrier frequency ($f_{sADC} \geq (2.5)f_N$). It is not clear whether the cost of these components will decrease faster than the alternative WDM approaches (e.g., using tunable optics).

It should be noted that DSP techniques can also be used to create multiple non-orthogonal frequencies from a single optical source. These frequencies would be spaced further apart than with OFDM (i.e., non-overlapping), making each a separate WDM channel [14]. DSP techniques can also be used for functions such as implementing coherent receivers and chromatic dispersion compensation. The biggest drawback to DSP-based techniques is power consumption for the DSP computations.

5.6 Conclusions

A variety of optical-domain technologies exist for broadband access. Some, such as OFDM, are useful in specific applications where they provide a physical layer for carrying broadband signals that would normally use a medium other than all fiber. Technologies that provide multiple access within the optical domain, such as WDMA and CMDA PON, are feasible and have great potential. WDMA PON appears to be the best long-term technology. However, due to the cost of the WDM optical components, TDMA PON technologies will continue to be more cost-effective for at least the next several years. In the

---

[2] For cost reasons, EPON, 10G EPON, G-PON and XG-PON do not use temperature compensation at the ONUs. If DWDM techniques are used with ITU-T NG-PON, they will require temperature compensated optical sources.

2015 timeframe, it is possible that WDMA PON will become more attractive due to either the decreasing cost of the optical component or the increasing per-subscriber bandwidth demands.

References

1. ITU-T G.694.2. Spectral grids for WDM applications: CWDM frequency grid; 2003.
2. ITU-T G.694.1. Spectral grids for WDM applications: DWDM frequency grid; 2012.
3. Kim H-S, Choi BS, Kim KS, Kim DC, Kwon OK, Oh, DK. Improvement of modulation bandwidth in multisection RSOA for colorless WDM-PON. *OSA Optics Express*. 2009; **17**(19): 16372–16378.
4. Talli G, Townsend PD. Feasibility Demonstration of 100km Reach DWDM SuperPON with Upstream Bit Rates of 2.5Gb/s and 10Gb/s *Proceedings of OFC*; 2005.
5. Talli G, Townsend PD. Hybrid DWDM-TDM long-reach PON for next-generation optical access. *IEEE/OSA Journal of Lightwave Technology*. 2006; **24**(7): 2827–2834.
6. Feldman RD, Harstead EE, Jiang S *et al.* An evaluation of architectures incorporating wavelength division multiplexing broad-band fiber access. *IEEE/OSA Journal of Lightwave Technology*. 1998; **16**: 1546–1558.
7. Grunnet-Jepsen A *et al.* Fibre Bragg grating based spectral encoder/decoder for lightwave CDMA. *Electronic Letters*. 1999; **35**(13): 1096–1097.
8. Bres C-S, Glesk I, Prucnal PR. Demonstration of an eight-user 115-Gchip/s incoherent OCDMA system using supercontinuum generation and optical time grating. *IEEE Photonics Technology Letters*. 2006; **18**(7): 889–891.
9. Tranceski L, Andonovic I. Wavelength hopping/time spreading code division multiple access systems. *Electronic Letters*. 1994; **30**(9): 721–723.
10. Prucnal PR. *Optical Code Division Multiple Access: Fundamentals and Applications*. Boca Raton: Taylor and Francis; 2006.
11. Medcalf R, Mitchell S, Nicosia M. Fiber to the Home: Technology Wars. Cisco IBSG Economic Insight, on www.cisco.com; 2007
12. Cvijetic N, Qian D, Hu J. 100Gb/s optical access based on optical orthogonal frequency-division multiplexing. *IEEE Communications Magazine*. 2010; **48**(7): 70–77.
13. Kanonakis K *et al.* An OFDMA-based optical access network architecture exhibiting ultra-high capacity and wireline-wireless convergence. *IEEE Communications Magazine*. 2012; **50**(8): 71–78.
14. Yoshimoto N, Kani J, Kim S-Y, Iiyama N, Terada J. DSP-based optical access approaches for enhancing NG-PON2 systems. *IEEE Communications Magazine*. 2013; **51**(3): 58–64.
15. ITU-T G.985. 100 Mbit/s point-to-point Ethernet based optical access system; 2003.
16. ITU-T G.986. 1 Gbit/s point-to-point Ethernet-based optical access system; 2010.

Further Readings

1. Chowdhury P, Sarkar S, Mukherjee B. Building a green wireless-optical broadband access network (WOBAN). *IEEE/OSA Journal of Lightwave Technology*. 2010; **28**(6): 2219–2229.
2. Sarkar S. Dixit S, Mukherjee B. Hybrid wireless-optical broadband-access network (WOBAN): A review of relevant challenges. *IEEE/OSA Journal of Lightwave Technology*. 2007; **25**(11): 3329–3340.
3. Shami A, Maier M, Assi C. (eds) *Broadband Access Networks*. New York: Springer; 2009.
4. Shea DP, Mitchell JE. A 10Gb/s 1024-way split 100-km long reach optical access network. *IEEE/OSA Journal of Lightwave Technology*. 2007; **25**(3): 685–693.
5. Van de Voorde I. *et al.* The SuperPON demonstrator: An exploration of possible evolution paths for optical access networks. *IEEE Communications Magazine*. 2000; **7**(2): 74–82.
6. Yin H, Richardson D. *Optical Code Division Multiple Access Communication Networks*. Berlin: Springer; 2009.

6

Hybrid Fiber Access Technologies

6.1 Introduction and Background

The high bandwidth potential of fiber access networks make them attractive for carrying protocols and signal types that are normally transmitted over different types of media. This is especially true for PON, which is the most economical fiber access topology. This chapter addresses two such hybrid uses of PON. The first section discusses the evolution of the DOCSIS protocol to using PON instead of, or in addition to its traditional coaxial cable medium. The second section covers various technologies for transporting radio frequency signals over PON, both for television signals and for mobile radio signals. The opposite situation is emerging, with the new IEEE 802.3bn standards project for carrying the EPON protocol over a coaxial cable network. Unfortunately the project had not progressed enough to provide any description beyond some early technical agreements by the time this book was released for publication.

6.2 Evolution of DOCSIS (Data-Over-Cable Service Interface Specification) to Passive Optical Networks

6.2.1 Introduction and Background

Cable television companies, also known as Multiple System Operators (MSOs), use the DOCSIS (Data Over Cable Service Interface Specification) protocol to provide broadband access over their hybrid fiber-coaxial cable (HFC) networks. The DOCSIS protocol is discussed in detail in Chapter 11. While DOCSIS, especially DOCSIS 3, enables high bandwidth connections over coaxial cable to the home, it is clear that FTTH provides the most flexible and highest performance access network. Consequently, the MSOs are also interested in PON technology.

Most MSOs have upgraded their HFC networks for DOCSIS 2 and DOCSIS 3 relatively recently, and hence they have not been anxious to launch another major network upgrade. However, there are multiple applications where PON deployment is very attractive. The first is to serve enterprise customers. EPON or 10G EPON network [1] to reach enterprise customers can potentially be employed directly, without being integrated into the DOCSIS systems. Another application is greenfield residential deployments such as new housing developments where builders desire an FTTH infrastructure as a selling point for the neighborhood. Since it requires less ongoing maintenance than coaxial cable, fiber is also attractive in some rehab applications. In addition, there is increasing competitive pressure on the MSOs from telephone companies that are deploying their own FTTH networks in order to compete for new services.

Broadband Access: Wireline and Wireless – Alternatives for Internet Services, First Edition.
Steven Gorshe, Arvind Raghavan, Thomas Starr and Stefano Galli.
© 2014 John Wiley & Sons, Ltd. Published 2014 by John Wiley & Sons, Ltd.

Two FTTH approaches are available to MSOs. The first is the SCTE (Society of Cable Tele-communications Engineers) "RF over Glass" (RFoG) protocol, as described below in Section 6.2.3. RFoG essentially replaces the coax portion of the HFC network with fiber and otherwise uses the same DOCSIS signal formats and protocols. The second approach is to deploy a PON protocol such as EPON or 10G EPON for Layer 1, and to integrate the upper layers of the PON system into their DOCSIS management systems. The advantage to the RFoG approach is that it can be deployed in the near term, using the same Cable Modem (CM), DOCSIS Cable Modem Terminating System (CMTS), and DOCSIS management infrastructure as their current HFC networks. The drawback to RFoG is that its throughput is limited to the DOCSIS 3.0 rates, and it cannot support the much higher potential data rates of EPON or GPON, much less their upgrades to 10G EPON or XG-PON.

MSOs have also been pursuing a second approach through a combination of two standards projects. One of these is the development of the new DOCSIS Provisioning of EPON (DPoE) specification recently released by CableLabs ([1], [2], [3], [4], [9], [15], [17]), and the second is the active participation of CableLabs and MSOs in the IEEE SIEPON effort (see Chapter 3). The DPoE concepts have been explored in recent years and discussed in papers such as [13]. DPoE is a natural extension of the DOCSIS network to a FTTH-based physical topology.

EPON and 10G EPON were chosen as being better suited for MSO PON applications than the ITU protocols. One reason is that the EPON MAC, especially with the multiple-LLID option extension supported by the emerging SIEPON standards, operates more like the DOCSIS MAC than the GPON MAC does. A second reason is that the EPON physical layer parameters typically allow for lower cost components. Yet another reason is that, since the telephone carriers are mostly committed to GPON, there could be an appearance of a late "me-too" reaction if the MSOs were also to deploy GPON.

The basic concept behind DPoE is that the EPON network is made to emulate a DOCSIS HFC network. This section describes the manner in which this emulation and functional division is accomplished.

6.2.2 DOCSIS Provisioning of EPON (DPoE)

6.2.2.1 DPoE Overview

DPoE exploits the existing DOCSIS and EPON protocols, together with their respective ecosystems, by combining them. The EPON protocol is used for the Layer 1 network connectivity, while the DOCSIS protocols are emulated by the EPON network such that it appears to be a normal DOCSIS HFC network to both the DOCSIS management systems and CPE. A DOCSIS ONU (D-ONU) behaves in the same way as a CM (Cable Modem) in terms of service capabilities.

In order for the D-ONU to behave like a CM, it must have a connection to the DOCSIS management system. For example, the management system needs to identify its capabilities and configure it to forward the appropriate traffic. Since the EPON system does not carry the DOCSIS channels, and hence cannot be reached directly by the DOCSIS management system, it was necessary to develop a standard method for translating this information for transport over the native EPON protocol. As described below, the DPoE System, which includes the PON OLT, provides the translation service between the native DOCSIS protocol for interfacing with the DOCSIS management system, and the native EPON protocols for communicating with an ONU.

DPoE provides both voice and data (e.g., internet access) services for both enterprise and residential customers. The data services for residential users are the same as those supported by DOCSIS (see Chapter 11).

The DPoE enterprise data services are built on the MEF (Metro Ethernet Forum) service definitions. These include the MEF service definitions for E-Line, E-LAN, and E-Tree[1]. As discussed below, Ethernet

---

[1] E-Line services pertain to point-to-point circuit or virtual connections. E-LAN services emulate a LAN connection between the various service end-points. E-Tree services provide a one-to-many (i.e., tree-like) service connectivity. A more detailed description of MEF services is beyond the scope of this book.

VLAN (Virtual LAN) technology is used for routing the Ethernet traffic, thus requiring the D-ONU to be capable of VLAN processing.

DPoE implements DOCSIS IP services through Ethernet Virtual Circuits (EVCs) between an Ethernet port on a standalone DPoE ONU (S-ONU) and an IP router function within the DPoE System. These services are sometimes called IP high-speed data service (IP-HSD), which is defined in DOCSIS 3.0.

Voice services for residential users are provided through Voice over Internet Protocol (VoIP), in the same manner as with DOCSIS. This is also an option for enterprise customers. Alternatively, an enterprise customer can get voice service through a DS1 connectivity (e.g., from a PABX) to its CPE. In this case, the DS1 is carried over the DPoE network by a Circuit Emulation Service (CES) over Ethernet.

Video services can either by provided as packetized traffic (either as video over IP or video directly over Ethernet), or as an RF signal over a separate wavelength of the PON. The IPTV approach transmits video to an ONU as an IP-based data service, using the appropriate QoS parameters. This approach has no direct impact on the PON or DOCSIS protocols. It is also much better suited to video on demand and virtual VCR (Video Cassette Recorder) functionality than the video overlay approach.

The service discussion here makes it apparent that the high-speed packet data capability of EPON fits well with the DOCSIS approach of providing all service types over IP. However, this is a key difference between DOCSIS and DPoE networks. Where DOCSIS uses IP as the common denominator for all services, DPoE uses Ethernet. Since DOCSIS services are built on IP transport, they must perform Ethernet emulation in order to provide Ethernet services. In contrast, DPoE uses native Ethernet, which can either provide Ethernet services directly or serve as the Layer 2 protocol for providing IP services.

6.2.2.2 DPoE Network Architecture

The DPoE network architecture is specified in [1]. The network architecture model is illustrated in Figure 6.1. This figure includes the lower layer portions of the architecture model most relevant to this book. The figure also indicates which portions of the DPoE network are covered by the different DPoE standards sections, [2], [3], [4], [9], [15] and [17]. Note that in the following DPoE discussions, EPON is used to indicate both EPON and 10G EPON unless a distinction is explicitly stated.

With respect to terminology, the DPoE Network is illustrated in Figure 6.1 as the entire network between interfaces to the routers on the left side of the DPoE System and interfaces on the right side of the D-ONUs to the CPE, CMCI (Cable Modem CPE Interface) or DEMARC points. Hence, it includes the DPoE System, a PON, and all the D-ONUs connected to that PON. It can also include the portion of a DEMARC defined in the DPoE specifications.

The DPoE System provides the functions needed to imitate the DOCSIS Cable Modem Terminating System (CMTS) (see Chapter 11). Physically, it can include one or more devices. The DPoE System functions include the EPON OLT, the DOCSIS service functions for interfacing to the IP network, the MEF service functions for the customer side interfaces from the D-ONU, IP NE management, routing, and forwarding functions. See [2] for the IP NE specification.

As noted above, a D-ONU is a DPoE-capable ONU. The taxonomy of the D-ONU types is shown in Table 6.1. A D-ONU can contain one or more embedded Service/Application Functional Entities (eSAFE) in order to support a particular service to a customer. The eSAFE types include interfaces to routers for customer home networks, telephone service support, interfaces to sensors (e.g., power monitors) or keypads, and interfaces to set-top boxes.

Since the DPoE System provides the connectivity for all OAMP traffic to the D-ONUs, the system must translate all the messages between the EPON and DOCSIS protocol domains. This domain translation is illustrated at a high level in Figure 6.2. As shown in Figure 6.1, a DPoE System includes a virtual Cable Modem (vCM) function for providing this translation. A unique vCM is instantiated within the DPoE system for each registered D-ONU, but this only applies to the D-ONU functions and not to other potential embedded functions in that D-ONU. Specifically, the vCM handles all the DOCSIS functions specified in DPoE-MULPI [17] and DPoE-OSSI [3], serving as the proxy between the DOCSIS management system

D	DOCSIS IP INNI
TU	Interface between D-ONU and DPoE system
TUL	Virtual interface representing a MAC domain as a logical EPON on the ODN

LCI	Logical CPE (virtual) Interface
CMCI	DPoE equivalent of the DOCSIS Cable Modem CPE Interface
S	IEEE 802.3 interface (S_1 for an S-ONU and S_2 for interfaces on a DPoE system DEMARC)

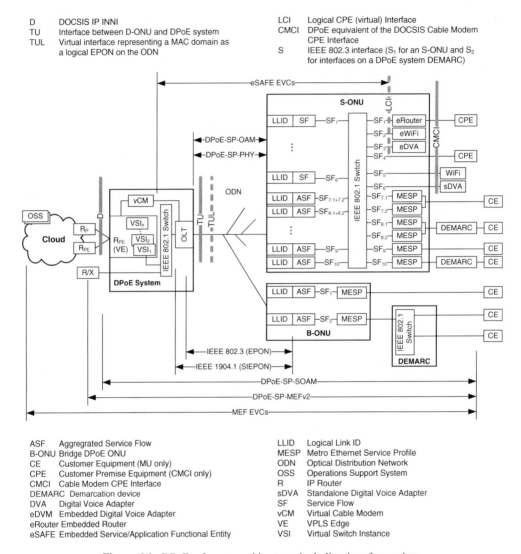

Figure 6.1 DPoE reference architecture, including interface points

ASF	Aggregrated Service Flow
B-ONU	Bridge DPoE ONU
CE	Customer Equipment (MU only)
CPE	Customer Premise Equipment (CMCI only)
CMCI	Cable Modem CPE Interface
DEMARC	Demarcation device
DVA	Digital Voice Adapter
eDVM	Embedded Digital Voice Adapter
eRouter	Embedded Router
eSAFE	Embedded Service/Application Functional Entity

LLID	Logical Link ID
MESP	Metro Ethernet Service Profile
ODN	Optical Distribution Network
OSS	Operations Support System
R	IP Router
sDVA	Standalone Digital Voice Adapter
SF	Service Flow
vCM	Virtual Cable Modem
VE	VPLS Edge
VSI	Virtual Switch Instance

Table 6.1 DPoE ONU taxonomy

Specified logical ONU types (normative)			Comments
D-ONU	S-ONU		Standalone ONU
	B-ONU	BB-ONU	Bridge broadband ONU
		BP-ONU	Bridge pluggable ONU • Can be embedded into SFP (1G), SFP+ (10G), or XFP (10G) modules

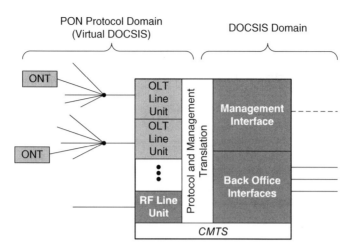

Figure 6.2 Simplified illustration of a DPoE system and its protocol domain translation function

and the D-ONU. These DOCSIS messages for the D-ONU, including requests and signaling, are communicated over the EPON network using EPON OAM messages, as defined in [9]. The customer traffic does not go through the vCM.

DOCSIS and DPoE both classify the traffic from each customer and treat the traffic associated with each service from each customer as a separate service flow (SF). Each SF is identified by a unique service ID (SID) (see Chapter 11). As with DOCSIS, DPoE uses a centralized scheduler for all the service flows. EPON, however, does its own scheduling for each service. Consequently, the scheduler in the OLT must translate the DPoE scheduler information such that the PON network provides the equivalent scheduling of the packets from each service flow.

As originally specified, EPON lacked a convenient method to distinguish the different service flows within the PON. This issue has been addressed by extending EPON beyond using a single unicast Logical Link Identifier (LLID) per-ONU to assigning a unique LLID to each service flow or aggregated service flow. This extended EPON functionality is also defined in IEEE 1904.1 Service Interoperability in Ethernet Passive Optical Networks (SIEPON). See Chapter 3 and [5]. The use of LLIDs is discussed further in Section 6.2.2.5 below.

DPoE follows the DOCSIS model in supporting the concept of an optional embedded DOCSIS device called an eDOCSIS. As illustrated in Figure 6.3, the eDOCSIS consists of an embedded Cable Modem (eCM) and one or more embedded Service/Application Functional Entities (eSAFE), and may also include a physically exposed external CMCI. Each eCM is connected to an eSAFE through a logical CPE interface (LCI). As illustrated in Figure 6.3, some of the eDOCSIS functions for DPoE are split between DPoE System and the S-ONU. For the purposes of management, the DPoE vCM must function as the eDOCSIS eCM.

6.2.2.3 DPoE Physical Layer Considerations

As described above, DPoE uses EPON or 10G EPON for its physical layer. However, DPoE specifies the requirements of these standards that must be explicitly supported for DPoE, as well as how these must be supported [4].

Forward Error Correction (FEC) support is a requirement for DPoE. The requirements differ, depending on whether EPON or 10G EPON is being used. See Chapter 3 for the respective EPON and 10G EPON FEC specifications. When EPON systems are used, FEC must be supported on a per logical link basis, independently in the upstream and downstream directions. This is possible because

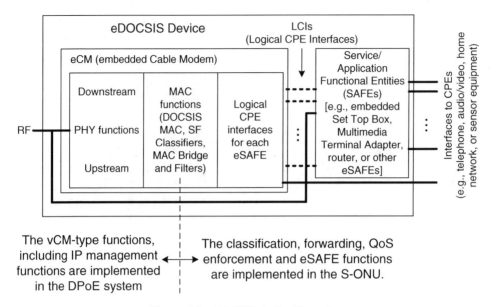

Figure 6.3 eDOCSIS device illustration

EPON applies FEC on a per Ethernet frame basis. When a D-ONU prepares to register, it must automatically detect whether the downstream frames use FEC, then default to using the same initial FEC state for its upstream signal. Since 10G EPON applies FEC to the entire data stream, it cannot be selected per logical link for DPoE. While 802.3 specifies that FEC use is mandatory for 10G EPON, DPoE allows the option of disabling it for all ONUs. Since that possibility exists, a 10G ONU may look for FEC on the downstream signal and set its FEC state to match.

DPoE also addresses the effects of Raman interference. Raman scattering is a phenomenon that can occur when a high-powered laser output is coupled into a single mode fiber [6]. The Raman scattering occurs as the light travels through the fiber, and the energy from the scattering can create crosstalk into other wavelengths. Analog video signals require low carrier to noise ratios, and hence are especially vulnerable to Raman interference. Unfortunately, Raman crosstalk from the 1490 nm downstream wavelength used by EPON is particularly strong at 1550 nm, which is the center of the wavelength range used for downstream analog video (see Chapter 3 for the EPON wavelength usage).

Modulation (e.g., the NRZ format used with Ethernet PON) increases the Raman scattering crosstalk. To make matters worse, the Ethernet IDLE fill pattern, which is a sequence of alternating 1 s and 0 s, exacerbates the Raman crosstalk by making strong power spikes associated with the frequency of this alternating pattern. For that reason, DPoE specifies a modified idle pattern that mitigates the Raman interference. Instead of sending the Ethernet IDLE pattern, the idle periods are filled by transmitting Ethernet frames containing a random pattern. Using a random pattern spreads the optical power over a wider spectrum so that the peak interference at other wavelengths is reduced. Specifically if an idle period is detected lasting at least 84 bytes (i.e., the duration of the 64-byte minimum Ethernet frame length plus eight preamble and 12 IPG bytes), the DPoE must replace the Ethernet IDLE. These replacement frames are internally generated, and use an LLID that it is not being used by any of the D-ONUs on that PON.

DPoE also requires support for monitoring at both the DPoE system and the D-ONU for optical parameters such as transmitted and received power levels, transmit bias current, and temperature.

DPoE addresses the potential mode of a D-ONU that continues to send an optical signal outside its allocated times. For example, the laser driver may have failed, such that the laser will not turn off. The

DPoE solution is to require the D-ONU to have the ability to disconnect the optical module from its power source through a software command. This disconnect is triggered by the DPoE system sending a special DPoE OAM message to the D-ONU when it detects the problem. Once the D-ONU has disconnected the optical module power, it must remain disconnected (i.e., neither D-ONU power cycling nor software reboot can be allowed to restore the optical module power).

6.2.2.4 D-ONU Registration and Initialization

PON protocols specify registration and initialization for the physical layer and the PON MAC. In contrast, the DOCSIS protocol also specifies registration and initialization for higher layers, including Dynamic Host Configuration Protocol (DHCP), Time of Day (TOD), and Trivial File Transfer Protocol (TFTP). DPoE uses the EPON protocol to handle its normal registration and initialization layers, and has the DOCSIS protocol handle the data link and network layers once the lower layers are established. This approach provides a clean functional partition between the protocols.

The next step after the EPON physical layer and MAC initialization is the initialization of the D-ONU authentication and encryption.

The D-ONU is not directly addressable using IP, due to its lack of an IP stack. A vCM represents a management entity within the DPoE System that is IP-addressable and is used to provide all the IP-based management functions on behalf its unique corresponding D-ONU. In other words, the vCM allows the DOCSIS management system to send management requests to a given D-ONU. Hence, during the third step of the registration process, the D-ONU relies on the vCM to obtain the IP address and CM configuration file from the DOCSIS management system on its behalf. Here, the IP initialization assigns an IPv4 and/or IPv6 address to each vCM through the appropriate DHCP.

After the initialization comes the Registration stage, in which the DPoE System processes the CM configuration file. After its contents are validated, the CM configuration file service provisioning information is used to configure the DPoE System vCM and the D-ONU. The vCM stores the D-ONU configuration and registration state.

Once the DPoE System is initialized, the vCM becomes a manageable NE within the MSO's IP network, using SNMP to communicate with the management system. The vCM handles locally those D-ONU management requests that do not require interaction with the D-ONU. If D-ONU interaction is required, the management message is converted to the appropriate extended OAM (eOAM) message to be sent to the D-ONU over the PON.

If the state of either a D-ONU or its associated vCM changes, such that it is no longer fully operational (e.g., due to an operator request or loss of power), the DPoE System is responsible for making sure the other element experiences a similar transition.

Since DOCSIS supports using DHCP for CPE device provisioning, the DPoE System contains a DHCP Relay Agent for associating the D-ONU MAC address with a CPE IP address DHCP request. Note that this mechanism is also used to prevent IP address spoofing.

6.2.2.5 DPoE MAC Layer Considerations

There are several DOCSIS functions that are not supported by EPON, including DOCSIS Dynamic QoS establishment and its two-phase activation process, DOCSIS-specific load balancing, DOCSIS upstream and downstream channel bonding, and transport layer frame fragmentation (see Chapter 11). DPoE effectively accommodates these DOCSIS functions through the EPON Dynamic Bandwidth Allocation (DBA). Since the entire PON bandwidth can be used in spurts by individual D-ONUs, DOCSIS-type channel bonding is not applicable. Similarly, EPON DBA allows it to function without the need for frame fragmentation.

DPoE supports the concept of having different virtual EPON networks sharing the same optical distribution network, as illustrated by the TUL interface in Figure 6.1. The MAC Domain of a DPoE Network consists of a shared group of upstream and downstream channels on the same logical EPON that share the same DBA for all D-ONUs within that MAC Domain. Each D-ONU belongs uniquely to one MAC Domain. While a MAC Domain can support multiple upstream and downstream channels, each downstream and each upstream channel also belongs uniquely to one MAC Domain.

The DPoE System performs both IP Network Layer routing and Ethernet MAC Layer bridging for data traffic. The D-ONU, which does not contain an IP stack, only performs MAC layer bridging. Both the DPoE System and D-ONU, however, are required to be aware of the transport and network layers so that they can support classification of user traffic for QoS and packet filtering. The classification is based on criteria configured by the network operator.

The DPoE MAC uses an EPON LLID as the identifier for each logical link between the DPoE System and a D-ONU. The DPoE system assigns at least one LLID to a D-ONU during D-ONU registration. When the D-ONU receives a downstream frame containing an LLID assigned to it, the frame is forwarded to the D-ONU's MAC layer. Otherwise, the frame is rejected. For upstream frames, the DPoE uses the LLID to forward the incoming frame to the appropriate MAC entity. Hence, the LLID assignments allow point-to-point communication over the physical PON tree. As described in Chapter 3, EPON also uses separate LLIDs for broadcast and multicast services.

With DPoE, an LLID also represents a traffic-bearing entity requiring upstream bandwidth allocation from the DPoE System. A LLID can be assigned to an individual SF, or multiple SFs can be aggregated for transmission using a single LLID. For example, the multiple MEF services can be aggregated by the D-ONU onto the same LLID. Since both DOCSIS and EPON use TDMA bandwidth assignment/scheduling mechanisms, the OLT can perform the LLID scheduling under the same criteria used by the DOCSIS system for the associated SF(s). See Chapter 3 for additional information on LLID and EPON bandwidth allocation. A D-ONU must support at least eight LLIDs.

As indicated in the previous paragraph, DPoE supports the concept of an aggregated Service Flow (ASF), in which multiple SFs, in either the upstream or downstream direction, are associated with a single ASF. Note that an ASF containing no SFs can be provisioned, allowing the possibility of adding SFs to it in the future. The aggregated SF traffic then uses the LLID associated with the ASF when it is transmitted, rather than using individual LLIDs for each of the SFs.

DPoE supports both the DOCSIS any-source and source-specific IP multicast services to the D-ONUs, including for IP-HSD. The DPoE System handles all the processing and management functionality for multicast groups so that the D-ONU does not need to do any Layer 3 processing such as snooping or proxy. The DPoE System assigns a multicast LLID (mLLID) to a set of multicast sessions, such that the mLLID is unique per MAC Domain and identifies the multicast sessions that a set of D-ONUs may receive. The DPoE System then uses that mLLID to forward all downstream packets associated with that multicast session set. As with the other LLIDs, the D-ONU is provisioned to receive and perform the appropriate forwarding of information frames using the mLLIDs. The DPoE system further associates a unique mLLID with each group Service Flow (GSF), and a given GSF can carry single or multiple multicast sessions. While DOCSIS assigns a unique Downstream Service ID (DSID) to each IP multicast session, DPoE assigns a unique mLLID to each group Service Flow (GSF). One or more downstream IP multicast sessions can be forwarded downstream using the GSF and the assigned mLLID.

The EPON DBA mechanisms are described in Chapter 3. DPoE supports the same SF scheduling types provided over DOCSIS networks. These are Best Effort (BE), Real-time Polling Service (RTPS), Non-Real-time Polling Service (NRTPS), Unsolicited Grant Service, and Unsolicited Grant Service-Activity Detection (UGS-AD). See Chapter 11 for further definition of these service types. Support for BE and RTPS is required for the DPoE System. Support for the other scheduling types is optional. The DPoE will not allow the registration of a vCM if its configuration file calls for support of a scheduling type that the DPoE system does not support.

As noted above, DPoE uses MEF service definitions as the basis for Ethernet services [7]. Such services make use of the Ethernet VLAN technology of IEEE 802.1ad "Provider Bridge" (PB) [14] or 802.1ah "Provider Backbone Bridge" (PBB) [18] to establish data connectivity between the DPoE System and a D-ONU. VLAN tag-based switching is used, with an S-VID (Service VLAN ID) for the outer tag and C-

VID (Customer VLAN ID) for the inner tag. Note that this tag usage is independent of the IEEE 802.1ad tag usage. Each service is granted a dedicated Ethernet Virtual Circuit (EVC), identified by its VLAN tags. The S-VLAN is used to organize C-VLANs into manageable Service groups. Frames belonging to the same EVC follow the same path through the DPoE Network. To establish the appropriate data forwarding, a unique combination of S-Tag and C-Tag headers is assigned to each D-ONU port during DPoE System provisioning. The D-ONU port information is associated with the Cable Modem Interface Mask used locally in the D-ONU.

For D-ONU timing and synchronization, including frequency, phase and time of day distribution, DPoE uses the EPON mechanisms described in Chapter 3.

6.2.2.6 D-ONU Provisioning

Each D-ONU has a slave relationship with its DPoE system. After the completion of all the D-ONU discovery, registration, and OAM capability discovery within the EPON domain, a D-ONU is provisioned by the DPoE system. The provisioning is performed through the CM configuration file that the management system writes into the vCM for its corresponding D-ONU. As discussed above, the vCM uses eOAM messages over the EPON network to communicate with the D-ONU.

DPoE Security
Security is an inherent issue in shared medium networks since the traffic from all customers is potentially visible to each customer. Like other shared medium networks, PON systems use an encryption process to protect against unwanted listening.

EPON now uses the Advanced Encryption Standard (AES), which is a block cipher operating on 16-byte data blocks. Specifically, it uses the Counter Mode, in which a stream of 16-byte pseudo-random cipher blocks are generated and exclusive OR'ed with the input data to produce the cipher text. The inverse process is used at the receiver to re-create the clear text data. EPON uses a 128-byte key (see Chapter 3 for a further discussion of the encryption and authentication available for EPON). DOCSIS 3.0 also uses AES, however, as discussed in Chapter 11, it uses CBC mode (cipher block chaining) where the residual of the encryption of a block is the "seed" for the encryption of the next block. The strong link layer security used for DPoE is specified in [14].

6.2.3   Conclusions for DPoE

The MSO infrastructure is widely deployed in much of the world. Due to its use of a combination of fiber for the access distribution plant and coaxial cable to the home, it has many advantages over copper telephone line connections for providing broadband data services to home users. The channel bonding capabilities introduced with DOCSIS 3.0 allow it to compete with at least some FTTH deployments for downstream data traffic.

However, the nature of the legacy video spectrum assignments leaves relatively little bandwidth available for upstream data transmission. It is also more limited in the maximum potential overall throughput it can achieve in both the downstream and upstream directions relative to an FTTH protocol such as those described in Chapters 3 and 4. Since telephone carriers have used FTTH to leapfrog the bandwidth capabilities of the MSOs, the MSOs have needed to respond with a tailored FTTH approach of their own. DPoE provides the means for maintaining the DOCSIS management infrastructure while taking advantage of an existing PON technology and ecosystem with only minor modifications.

6.3   Radio and Radio Frequency Signals over Fiber

There are two types of applications for sending radio frequency or radio information over fiber access networks. The first application is for wireless access networks that connect cellular base stations with serving central offices. The second is for carrying the video channels of a cable television network. Both are addressed in this section. While the connection between a radio base station (BTS) and the network is commonly referred to as a backhaul link, connections between a BTS and a subtended radio are often referred to as "fronthaul".

In order to increase the per-user capacity of wireless access systems and reduce the per-transmitter radio signal power, it is important to move more radio antennas as close to the subscriber as possible (i.e., create smaller cells). As noted in the previous section, it is possible with linear lasers to carrier analog RF signals over a fiber. The linearity requirements for wireless access are only slightly less stringent. Carrying the RF cellular signal as an optical signal is known as Radio over Fiber (RoF). The other approach uses a baseband digital signal rather than an analog RF signal.

The cable television application extends the fiber portion of the CATV network to the subscriber, rather than using a coaxial copper cable interface to the subscriber as discussed in Chapter 11. This approach is known as Radio Frequency over Glass (RFoG).

6.3.1   Radio over Fiber (RoF)

The complexity and power consumption of equipment at the radio sites can be significantly reduced if an RF signal is delivered to the site. Fiber is the most practical medium to carry these RF signals without creating or encountering electromagnetic interference problems. The fibers are connected to a central node that modulates the data onto the high frequency carrier signals. This approach is known as Radio over Fiber (RoF). Another application for RoF is delivering video signals over a PON.

Of course, transmitting analog signals over a fiber with minimum distortion requires highly linear sources and receivers. Note that either an RF-band signal or an intermediate frequency (IF)-band signal can be used to modulate the optical carrier. When IF-band modulation is used, sometimes an RFreference frequency signal is transmitted in order to simplify the receiver.

Intensity modulation of the carrier wavelength is relatively straightforward. However, dispersion in the fiber can cause interaction of the resulting sidebands such that fading occurs. Various techniques exist to address this fading issue, including heterodyne and single sideband transmission techniques, but these techniques add complexity (See [16] for more a more detailed discussion). Otherwise, transmission over fiber yields fewer impairments than with transmission over copper cable.

A related technique is known as microwave-over-fiber (MoF), in which a microwave carrier signal is modulated onto an optical carrier wavelength in order to carry the signals for WiMAX base stations.[2] This microwave signal is the analog signal between the base station and the antenna. With WDMA PON, the signal for each antenna uses a separate wavelength. Overlay onto a TDMA PON requires frequency-shifting the microwave signals so that they create unique optical subcarriers on the PON (See Chapter 5 for more on optical subcarriers).

When TDMA PON is used to carry signals to an integrated base station, such as one for WiMAX, the base station is commonly known as a wireless gateway. The potential exists with this architecture for the integration of the wireless and PON bandwidth allocation algorithms. For example, if a TDMA PON ONU and a WiMAX base station are integrated, the ONU can take advantage of knowing how much bandwidth the base station has granted for its WiMAX users when it makes its bandwidth requests to the OLT. The result is more efficient bandwidth allocation and shorter latency, due to avoiding a sequential, independent request for the base station and the ONU.

6.3.2   Baseband Digital Radio Fiber Interfaces

In contrast to RoF, two industry specifications have been developed for transmitting a digital baseband signal between the base transceiver station (BTS) and a remote RF head (RRH) rather than using an RF signal. One specification is the Common Public Radio Interface (CPRI). CPRI was developed by a

---

[2] While in principle MoF could be used for other applications, it was initially defined for use with WiMAX.

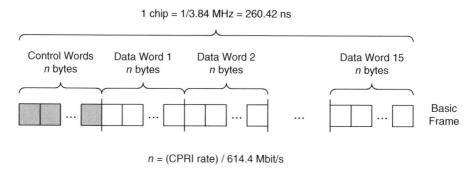

$$n = (CPRI\ rate) / 614.4\ Mbit/s$$

Figure 6.4 CPRI protocol Basic Frame structure

cooperative agreement between five large BTS manufacturers,[3] and the specification has been made publically available. The other specification is a part of the Open Base Station Architecture Initiative (OBSAI). OBSAI was also developed by an industry consortium that includes many members, with its Promoter companies leading the effort.[4] The goal of OBSAI covered most aspects of the BTS architecture; however its Reference Point 3 (RP3) specification has a scope similar to CPRI.

There are two main advantages to transmitting a digital baseband signal. Most importantly, it avoids the various challenges associated with transmitting an analog RF signal. For example, the component linearity requirements are not a factor for digital signals. Likewise, signal fading due to dispersion is not an issue. The second advantage is that it allows embedded diagnostic channels for the OAM&P of the RRH and fiber link. A secondary advantage is that embedded overhead allows an accurate measurement of the link delay, which can be used by a BTS, for example, for better handoffs between cell sites.

The one drawback to using a digital baseband signal is that additional remote electronics are required at the antenna to convert between digital baseband signals and the RF signals of the antenna. Note that the bit rates chosen for the baseband digital signals are ones that allow the RRH to derive a highly stable RF clock for the air interface from the recovered clock of the baseband digital signal.

There are a number of similarities between CPRI and OBSAI RP3. For example, both were designed to support either optical or electrical media and to support network topologies that include point-to-point, chain, tree-and-branch, and rings. As will be discussed below, CPRI has received the most industry attention with respect to expanding its applications over fiber access networks. For that reason, it will be used to illustrate the digital baseband BTS to RRH link concepts. A more detailed discussion and comparison of these interfaces can be found in [10].

CPRI supports the UMTS, LTE, and WiMAX radio signals [11] (see Chapters 14–17 for the discussion of these wireless protocols). As illustrated in Figure 6.4, the CPRI frame structure is constructed from 16-word sub-frames called Base Frames (BF), in which user payload data is carried in 15 words and one word is used for control information. The CPRI signal transmits BFs at the 3.84 MHz UMTS radio chip rate. Since each data word for the base rate is a byte that is encoded with the 8B/10B line code, the resulting base rate is 614.4 Mbit/s. CPRI also supports multiples of this base rate (1.2288 Gbit/s, 2.4576 Gbit/s, 3.072 Gbit/s, 4.9152 Gbit/s, 6.144 Gbit/s and 9.8304 Gbit/s) for growth to increased bandwidth.[5] Each of these higher rates still transmits BFs at the same 3.84 MHz rate, but the word size is increased to 2, 4, 5, 8, 10 and 16 bytes, respectively.

---

[3] Ericsson AB, Huawei Technologies Co. Ltd., NEC Corporation, Nortel Networks, and Siemens AG.
[4] OBSAI Promoter companies included Hyundai Syscomm, LG Electronics, Nokia, Samsung Electronics, and ZTE.
[5] A 9.8304 Gbit/s rate is under consideration for addition to the CPRI standard.

A Hyperframe (HF) consists of 256 BFs and uses the 8B/10B K28.5 comma character as the control word in the first BF as the HF delimiter. HFs can be combined into larger radio frames. For example, the UMTS radio frame consists of 150 hyperframes. The control words in the other 255 BFs are structured as time multiplexed sub-channels allocated for carrying control, management, vendor-specific, and air interface synchronization information. The multiframe structure is used by CPRI to support different mappings for each type of application data.

The payload portion of the BF consists of In-phase/Quadrature-phase (IQ) baseband samples. The I and Q samples of each IQ sample are first bit-interleaved, then the bit-interleaved samples are multiplexed into the BF payload into fixed time slots associated with each supported antenna carrier. The baseband data is aligned with its associated air interface frame period by using data from the control word synchronization sub-channel.

The CPRI electrical interface uses either the Ethernet XAUI low voltage interface for all rates, or the 1000Base-CX for higher voltage. The optical interface typically makes use of Ethernet or Fiber Channel transceivers.

The ITU-T SG15 has added a mapping for native CPRI signals into G.709 Optical Transport Network (OTN) payloads. See [19] for more on OTN. This mapping was requested by multiple carriers, especially China Mobile. There are two main applications driving the interest in this mapping. One application is to allow a wireless carrier to lease a connection through a wireline carrier's metro or access network to provide the CPRI link. By leasing this connection, the wireless carrier can avoid or postpone the need to build its own fiber infrastructure in those areas.

The second application for carrying CPRI over OTN is to better support WDM access where the CPRI signals share a PON infrastructure that provides both CPRI links to RRH locations and EPON/GPON wireline links to other users. Examples of where the EPON/GPON links connect include enterprise customers and ONUs in multiple-dwelling units that serve wireless LANs. This application allows a wireless carrier to build a single, flexible fiber access network for different applications that uses the same OTN technology that they plan to use for their backhaul and core network applications. Another advantage of using OTN in this application is that it provides a TDM method for combining multiple CPRI signals onto the same wavelength, thus avoiding some of the expense associated with using separate wavelengths for each CPRI signal.

A 614.4 Mbit/s or 1.2288 Gbit/s CPRI signal would be mapped as a CBR signal into a level 0 Optical Data Unit (ODU0), and the 2.4576 Gbit/s CPRI signal would be mapped into the level 1 ODU (ODU1). The higher rate CPRI signals would be mapped into a new flexible ODU for CBR signal transport, referred to as an ODUflex(CBR). The ODU0, ODU1, or ODUflex(CBR) signals could then be multiplexed into a higher rate OTN transport signal.

The main obstacle to using OTN is that CPRI has extremely stringent timing requirements that are driven by the need to support mobile handoffs between antenna sites. In terms of frequency accuracy, the CPRI downlink must be accurate to within ±0.002 ppm. In order to coordinate the signals between antenna sites, CPRI includes a delay measurement protocol that must also be extremely accurate. For example, for UMTS applications, the delay measurement must be accurate to within ±0.03 125 chip for the downlink and to ±0.0625 chip for the round trip. Supporting this level of frequency and delay accuracy over OTN is very challenging, since other OTN clients allow for a greater degree of jitter and wander. It is not yet clear whether it will be cost effective to use OTN for the transport of CPRI signals.

6.3.3 Radio Frequency over Glass (RFoG)

The RFoG standard (SCTE IPS SP 910, [8]) has been developed by the Engineering Committee of the Society of Cable Telecommunications Engineers (SCTE) in order to provide a FTTH infrastructure option for DOCSIS and CATV video signals. (See Chapter 11 for a complete discussion of DOCSIS.) The RFoG system is a passive optical network that allows cable television operators to use the same equipment at

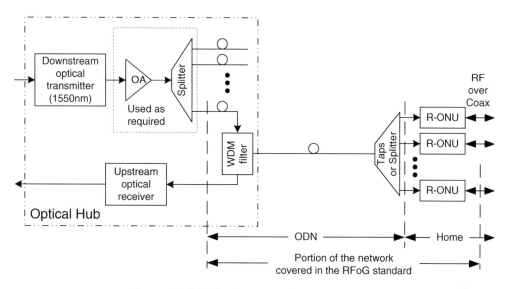

Figure 6.5 RFoG reference architecture illustration

both the home and head end[6] as are used with their legacy hybrid fiber-coax (HFC) networks, with only the coax access portion now replaced with fiber. As such, it covers the portion of the plant between the last active element and the home interface.

As illustrated in Figure 6.5, the last active element is called the Optical Hub, and the subscriber terminal at the home is called the RFoG Optical Network Unit (R-ONU). The RFoG standard covers the ODN between the Optical Hub and R-ONU, specifying the distance and optical loss of this portion of the network, and the R-ONU up to its coax RF interface to the subscriber equipment. The structure of the Optical Hub is application-specific, and hence it is not specified by the RFoG standard. By focusing on the R-ONU, the RFoG standard allows interoperability with R-ONUs from multiple equipment vendors on an ODN.

The R-ONU consists of:

- A WDM filter to separate the downstream and upstream wavelengths.
- The downstream optical receiver and upstream optical transmitter.
- An optical diplexer with high- and low-pass filters to interface between the bidirectional coax interface and the downstream OE and upstream EO modules (see Chapter 11 for the upstream and downstream frequency allocations).
- A signal detector for upstream transmission that enables the laser when it detects an upstream signal coming from the coax interface.

Two upstream wavelengths are allowed with RFoG, 1310 nm and 1610 nm. The 1310 nm wavelength provides for using relatively inexpensive lasers at the R-ONU when the ODN only carries DOCSIS traffic upstream. Using the 1610 nm wavelength allows the RFoG upstream signals to co-exist on the same ODN with either EPON, 10G EPON, GPON, or XG-PON, all of which use 1310 nm for their upstream transmissions (see Chapter 3 for EPON and 10G EPON, and Chapter 4 for G-PON and XG-PON

---

[6] The cable operator head end is effectively equivalent to the telephone carrier central office.

details). However, it requires a notch filter (either separate or an integrated version) at the R-ONU to filter out the 1577 nm carrier.

The RFoG specification supports a split ratio of up to 1: 32 and a range of 0–20 km, with a maximum optical loss budget of 25 dB. Tradeoffs can be made between split ratio and distance to support more R-ONUs or longer distances, as long as the 25 dB link loss budget is maintained. Implementations could potentially support higher link loss budgets, however this would complicate any migration attempt to share the ODN with EPON or GPON systems.

The RFoG standard also specifies the RF coaxial interface parameters at the R-ONU, including the allowed RF band separation points between the upstream and downstream RF signals. Both FM and AM modulation formats are supported on the upstream signal.

The R-ONU portion of the RFoG standard includes the system power input, physical, and environmental requirements for both indoor and outdoor deployments.

6.4 IEEE 802.3bn Ethernet Protocol over Coaxial Cable (EPoC)

While this protocol is the opposite of what this chapter's title suggests, its hybrid nature and extension of a PON protocol makes it fit here better than in any other chapter. The Ethernet protocol used here is an extension of EPON in order to provide Ethernet access over a shared coaxial cable medium. The work on this new standard was in its very early stages when this book was written so, while no details were available, the project is mentioned for completeness. The early technical agreements include:

- OFDM will be used for the downstream signal, with OFDMA for the upstream signal (see Chapter 5 for a description of OFDMA as it pertains to PON).
- Multiple modulation orders will be supported, with up to 4096 QAM downstream and 1024 QAM upstream.
- The MAC/PLS rates will scale with the number of OFDM channels
- A 16 QAM PHY Link channel will be used to carry PHY Link information
- Both 25 KHz and 50 KHz sub-channel spacings will be supported
- The PHY will use the same shortened header version of the 64B/66B line code used by 10G EPON.
- The upstream channel bandwidth will be 192 MHz
- Both Frequency Division Duplexing (FDD) and Time-Division Duplexing (TDD) will be used.
- An LDPC(384,288) FEC will be used.

6.5 Conclusions

Fiber connectivity to users offers the highest potential performance for broadband data access, and the PON optical distribution network is the lowest cost method of building the fiber access infrastructure. As can be seen in this chapter, the PON infrastructure is also highly versatile and there are multiple different ways to use it for broadband access. Some of these alternatives involve using different types of signal formats, such as analog or digitized RF signals over the PON. Some, like DPoE, use a standard TDMA PON protocol to emulate a different legacy physical layer network in order to support the higher layers of that network.

The choice of how the PON is used depends on the service provider's legacy network and services. The applications include providing wireline subscriber access and providing connectivity to wireless antennas to support wireless broadband service. Of course, as discussed elsewhere in this book, PON can also be used for connections to FTTC terminals from which xDSL is used for the final subscriber connection. In short, the PON network is a very compelling as an access infrastructure that can be used either on its own or in conjunction with nearly all the other broadband access technologies discussed in this book.

References

1. DPoE-SP-ARCH DOCSIS Provisioning of EPON Specifications – DPoE Architecture Specification; 2012.
2. DPoE-SP-IPNE DOCSIS Provisioning of EPON Specifications – DPoE IP Network Element Requirements; 2012.
3. DPoE-OSSI DOCSIS Provisioning of EPON, Operations and Support System Interface Specification; 2012.
4. DPoE-SP-PHY DOCSIS Provisioning of EPON Specifications – DPoE Physical Layer Specification; 2012.
5. IEEEP1904.1. Service Interoperability in Ethernet Passive Optical Networks (SIEPON); 2013.
6. Mahlein H F. Crosstalk due to stimulated Raman scattering in single-mode fibres for optical communication in wavelength division multiplex systems. *Optical and Quantum Electronics.* 1984; **16**: 409–425.
7. MEF6.1. Ethernet Services Definitions – Phase 2; 2008.
8. SCTE IPSSP 910. Radio Frequency over Glass Fiber to the Home Specification; 2010.
9. DPoE-OAM DOCSIS Provisioning of EPON, OAM Extensions Specification; 2012.
10. Leavey G, Gorshe S.Open Standards for Cellular Base Stations. In Furht B. (ed.) *Encyclopedia of Wireless and Mobile Communications.* Boca Raton; CRC Press: 2007.
11. Common Public Radio Interface (CPRI). Interface Specification, v5.0; 2011.
12. IEEE802.3. IEEE Standard for Ethernet; 2012.
13. Bernstein A, Gorshe S. A proposal for DOCSIS 4.0: The best of both worlds, DOCSIS and PON. Proc. of the Emerging Tech. conference; 2008.
14. IEEE802.1ad. Provider Bridges; 2005.
15. DPoE-SP-SECv2.0 DPoE Security and Certificate Specification; 2012.
16. Koonen AMJ, Ng'Oma A. *Integrated Broadband Optical Fibre/Wireless Lan Access Networks, In Broadband Optical Access Networks and Fiber-To-The-Home: System Technologies and Deployment Strategies*, Chap. 11. New York: John Wiley and Sons; 2006.
17. DPoE-SP-MULPI DOCSIS Provisioning of EPON Specifications – DPoE MAC and Upper Layer Protocols Specification; 2012.
18. IEEE (2008) 802.1ah. Provider Backbone Bridges.
19. Gorshe S. *A Tutorial on G*.709 *Optical Transport Networks (OTN).* PMC-Sierra white paper PMC-2081250 available at pmcs.com/; 2011.

Further Readings

1. IEEE802.1Q. IEEE Standard for local and metropolitan area networks – Virtual Bridged Local Area Networks; 2011.
2. MEF10.2. Technical Specification MEF 10.1 – Ethernet Services Attributes – Phase 2; 2009.
3. MEF12.1. Carrier Ethernet Network Architecture Framework Part 2: Ethernet Services Layer - Basic Elements; 2010.
4. CM-SP-MULPIv3.0-I03-070223 Data-Over-Cable Service Interface Specifications 3.0 – MAC and Upper Layer Protocols Interface Specification; 2007.
5. CM-SP-OSSIv3.0-I02-070223 Data-Over-Cable Service Interface Specifications 3.0 – Operations Support System Interface Specification; 2007.
6. CM-SP-PHYv3.0-I03-070223 Data-Over-Cable Service Interface Specifications 3.0 – Physical Layer Specification; 2007.
7. CM-SP-SECv3.0-I03-070223 Data-Over-Cable Service Interface Specifications 3.0 – Security Specification; 2007.

7

DSL Technology – Broadband via Telephone Lines

7.1 Introduction to DSL

Digital subscriber line (DSL) technology transmits broadband data over telephone lines. There are nearly a billion telephone lines connecting homes and businesses in the developed world, and most of these connect to network equipment with DSL capability. A telephone line consists of a dedicated twisted pair of wires that runs from one or more rooms in the customer premises to a telephone network serving node. In the past, the line ended at a local central office (CO, also known as an exchange). Today, a telephone line more commonly ends at a remote equipment cabinet containing a multiplexer serving a local neighborhood. As a result, the length of the line is reduced, and this enables the DSL technology to convey higher bit rates. With approximately 480 million DSL lines in service at the end of 2012, DSL technology is the most popular form of broadband access in the world.

Figure 7.1 is a high-level diagram of common DSL systems. The DSLAM (digital subscriber line access multiplexer) is the network end for each DSL. The DSLAM contains a DSL modem dedicated for each line, and it also multiplexes the traffic from all DSL lines into a high-speed trunk to the core network. Different types of DSL technology have different attributes:

- ADSL (asymmetric digital subscriber line) provides up to 8 Mbits/s downstream (towards the customer) and 1 Mbits/s upstream (from the customer) simultaneously with POTS (plain old telephone service). ADSL operation is possible for lines up to 6 km (20 000 feet), but the bit-rate capacity is less for longer lines. Most often, ADSL lines connect directly from the central office to the customer, as shown in Figure 7.1.
- VDSL (very high bit-rate digital subscriber line) technology provides much higher bit rates simultaneously with POTS over relatively short lines. For example, a 1 km (3280 feet) line can provide 25 Mbits/s downstream and 3.5 Mbits/s upstream. Most often, VDSL lines connect from a DSLAM at a remote terminal (RT) location to the customer, as shown in the lower portion of Figure 7.1.
- SHDSL (single-pair high-speed digital subscriber line) provides the same bit rate for both upstream and downstream, up to 5.7 Mbits/s in both directions. SHDSL technology is often thought of as "symmetric high bit-rate DSL". Unlike ADSL and VDSL, SHDSL technology does not support simultaneous POTS on the same line. While most SHDSL lines do not have mid-span repeaters, Figure 7.1 shows the case with one mid-span repeater. The repeater would enable high bit rates to be carried over a longer distance.

Broadband Access: Wireline and Wireless – Alternatives for Internet Services, First Edition.
Steven Gorshe, Arvind Raghavan, Thomas Starr, and Stefano Galli.
© 2014 John Wiley & Sons, Ltd. Published 2014 by John Wiley & Sons, Ltd.

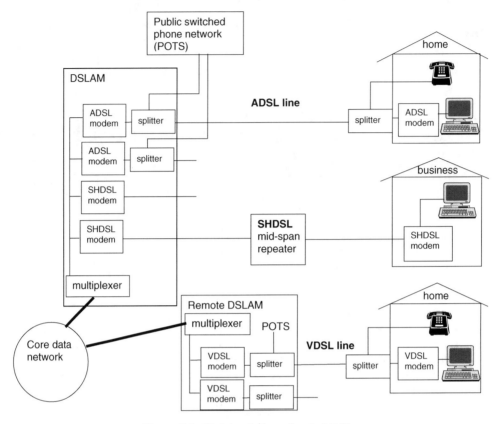

Figure 7.1 High-level view of typical DSLs

In Figure 7.1, splitters are shown at both ends of the ADSL and VDSL lines to combine the analog voice signals (POTS) with the digital DSL signals.

7.2 DSL Compared to Other Access Technologies

7.2.1 Security and Reliability

Each telephone line connects to only one customer and, thus, it provides a dedicated, point-to-point physical connection. This provides the advantages of greater physical security, better reliability, and more constant data throughput than shared-media access such as coax, radio, and broadband powerline access.

While it is still possible for hackers to attack via higher-layer access through the Internet, or attack locally via a WiFi local network in the home, the DSL link from the home to the network provides effective physical security. Even if an attacker were to connect physically to the telephone line, it would be very difficult to intercept the traffic on the telephone line without disrupting the signal transmission. Shared access systems such as coax have large numbers of users connected to the same media and have effective data security protocols to provide security. However, hackers have proved their abilities to defeat many security protocols.

A shared access system is vulnerable to service disruption due to accidental or intentional rogue transmissions, and it works well only if all the equipment sharing the media obeys the access etiquette. To a large degree, DSL transmission is immune to disruption due to improper transmissions by other customers. However, DSL can be affected somewhat by signals from other sources. This will be discussed in Section 7.4.7 under crosstalk, spectral compatibility, and noise.

7.2.2   Point-to-Point Versus Shared Access

A shared access system is susceptible to a few users consuming a large portion of the total available capacity and, thus, leaving little capacity for other users. This is referred to as "bandwidth hogging". Some network service providers have attempted to enforce limits on bandwidth hogging by restricting peer-to-peer type traffic, but this practice has been widely criticized. The capacity of a DSL line is dedicated to a single customer, so usage by other customers does not affect the traffic on other DSL lines. However, the traffic from many DSL lines (typically 100 to 1000 lines) are multiplexed together at a DSLAM (digital subscriber line access multiplexer), and then the traffic for all customers on a DSLAM share the capacity of the link from the DSLAM to the transport network. As a result, bandwidth hogging is possible with DSL systems, but it is less of a concern because of the large DSLAM uplink capacity (typically 150 Mbits/s to 1 Gbits/s). Also, the traffic management within the DSLAM can be more effective than a shared access system. Figure 7.2 shows a DSLAM located in a central office.

Figure 7.2   DSLAM

7.2.3   Common Facilities for Voice and DSL

The use of existing telephone lines for DSL technology provides several benefits. There are economic benefits from using existing facilities between the network and the home, as well as sharing these facilities with the existing voice/fax services. Also, the shared use of DSL transmission over the same lines as voice service helps to assure that the lines used for DSL are in good operating condition, because the lines have been maintained to provide voice service. In areas with existing telephone lines, the cost for the facilities to reach the customer is small because the facilities already exist, and the cost of the lines has been largely recovered by the existing voice services. However, in recent years, a growing number of customers do not subscribe to traditional voice services because they use wireless and VoIP for their voice services. Coax facilities provide similar benefits, since they have been widely deployed to provide video services. Broadband wireless often lacks this advantage, due to the high cost of deploying new broadband base station equipment and the acquisition of additional licensed radio spectrum.

7.2.4   Bit-rate Capacity

Generally speaking, fiber and coaxial cable technologies can provide higher service bit rates than DSL. However, VDSL2 technology served from a network node less than 100 meters from the customer can achieve up to 100 Mbits/s, and multi-pair bonding (serving a customer with multiple lines) can achieve even higher bit rates. Also, the cancellation of crosstalk by vectoring techniques greatly improves DSL performance above 40 Mb/s.

Access bit rates in excess of 25 Mb/s provide benefit for applications such as IPTV where multiple HDTV streams are provided by specialized local servers with cached content. However, access bit-rates above 25 Mb/s solely for internet access have little benefit today because servers connected to the internet and the internet itself rarely provide sustained per-flow rates above 25 Mb/s.

7.2.5   Hybrid Access

Unlike wide-area broadband wireless services, DSL based services do not provide access mobility, except to the limited extent that a DSL-fed WiFi gateway or a femtocell base station permits wireless access within a small area.

What may appear to be wireless or fiber access may really be hybrid access. For many locations, wide-area wireless systems use DSL technology for the backhaul from the wireless cell site to the core network.

Another common example is access comprised of fiber, copper, and wireless used in tandem. Broadband access provided by telephone companies often consists of several kilometers of fiber from a central office to a neighborhood serving node with VDSL2 or ADSL over copper telephone lines from the serving node to the customer's home. The journey then continues within the home, possibly via wireless WiFi distribution of the service throughout the home.

7.2.6   Future Trends for DSL Access

DSL provides broadband access to nearly two-thirds of broadband homes and many businesses today. This proportion will decline as fiber to the home (FTTH) and ever faster wireless broadband continues to serve more customers. However, the capabilities of DSL technology will continue to improve and new DSL lines will continue to be deployed for many years.

The future industry trend for DSL will be higher bit rates and shorter loops. Both the downstream and upstream bit-rate will increase, initially with ADSL2plus providing up to 20 Mb/s downstream and 1 Mb/s upstream. The deployment of VDSL2 profiles 8 and 12 are already underway, providing bit rates up to 40 Mb/s downstream and 5 Mb/s (20 Mb/s for profile 12) upstream. The next stage of the trend will be VDSL2 profile 17a providing bit rates up to 60 Mb/s downstream and 20 Mb/s upstream. Following this will be VDSL2 profile 30a providing bit rates up to 120 Mb/s downstream and 100 Mb/s upstream.

There will be increasing use of multi-pair bonding where 2-pair bonding can provide 1.6 times greater loop reach for a set bit rate, or at the same loop reach the bit-rate will nearly double. For business customers, bonding of up to eight pairs will be employed to provide nearly eight times the bit-rate of a single line.

During the next several years, level one and two dynamic spectrum management (DSM) will be employed with increasing sophistication to reduce the amount of generated crosstalk and improve service quality. Starting in 2011, vectoring (also known as level three DSM) has been used to cancel crosstalk. For loops shorter than 3000 feet, where cable pairs are suitably managed, vectoring will increase the bit rates by 50% to 100%.

The DSL transmission quality will also improve with the application of more powerful error correcting techniques, including erasure decoding and PHY-retransmission.

7.3   DSL Overview

DSL technology was derived from the preceding voice-band modem technology. The transmission techniques used by both DSL modems and voice-band modems are similar. Raw binary data can not be directly transmitted over a telephone line because the resulting signal would have energy at frequencies beyond those that will pass through the telephone line. A modem (modulator/demodulator) is a device that creates a signal that represents the data and is confined to the available channel bandwidth.

7.3.1   Voice-band Modems

Voice-band modems were introduced in the early 1950s to convey data through the PSTN (public switched telephone network), with the Bell 103 modem sending and receiving 300 bits/s. Voice-band modem technology evolved to ITU-T V.90 modems in the late 1990s, sending 56 kbits/s downstream (towards the customer) and 33.6 kbits/s upstream. At the time, this was considered "screaming fast". Voice-band modems are also called dial-up modems.

Historically, when the telephone voice network was the only ubiquitous communications network, the only way to send data anywhere was to transmit the data though the voice telecommunications network. Voice-band modems converted data into signals that used the same range of frequencies as voice. The voice frequency band is from 300 Hz to about 3.4 kHz.

As shown by the dotted line in Figure 7.3, the transmission path consists of a telephone line (A) from a customer to a telephone central office, a voice switching system, a trunk to another voice switch, and then another phone line (B) to another customer's home or an Internet service provider's (ISP's) bank of modems. The voice-band modem transmission is limited by the available bandwidth of the entire end-to-end transmission path.

Voice-band modems are still widely used today, mainly in facsimile (fax) machines. It is interesting to note that these modems disguise data to resemble a voice signal so that it can be conveyed through a voice communications network, while voice-over-Internet-protocol (VoIP) does the opposite. VoIP disguises voice to look like data to be conveyed by an IP data network. In the 1950s, the PSTN was the best available ubiquitous network, and now the best available ubiquitous network is the packet switched IP network. Now, all types of information are converted to fit the IP network.

7.3.2   The DSL Concept

As shown in Figure 7.4, the transmission path between the corresponding pair of DSL modems consists of one telephone line. Thus, DSL transmission occurs over one segment of the total end-to-end path shown in Figure 7.3 for voice-band modems. While the data flow end to end, the DSL signals are confined a single telephone line. The fundamental advantage of DSL technology over voice-band modem technology is the expanded frequency bandwidth of the transmission channel. Specifically, the analog-to-digital conversions at the local digital telephone switches are excluded from the DSL transmission channel. Also, the cumulative degradation due to transmitting through two tandem local telephone lines is eliminated. Each

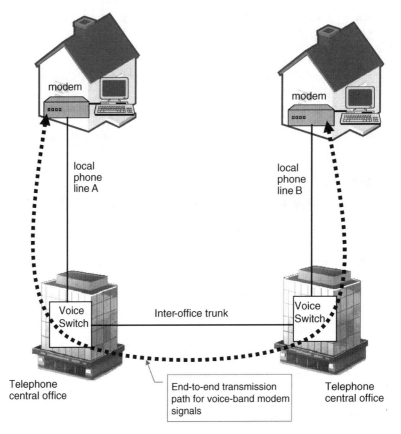

Figure 7.3 Voice-band modem end-to-end transmission path

pair of DSL modems deal with the transmission impairments of one telephone line, but not the remainder of the network.

DSL signals become weaker and more distorted as they traverse the telephone line; this is called signal attenuation. The degree of signal attenuation is greater for longer lines and at higher frequencies. As a result, shorter lines have greater bit-rate capacity than longer lines. Since DSL signals traverse only one telephone line, and not two tandem phone lines, the attenuation of the line at any frequency is less than the end-to-end path experienced by voice-band modems. However, the attenuation across the entire DSL frequency band is quite high because DSL systems transmit signal energy at much higher frequencies than voice-band modems.

Since higher bit rates can be achieved through shorter telephone lines, there is a trend to shorten the DSL transmission path. This is shown on the right-hand side of Figure 7.4 for telephone line "B." Whereas DSL line "A" transmits over the entire distance from the central office to the customer's home, line "B" has been shortened by placing the DSLAM closer to the customer. The DSLAM has been placed at a remote terminal (RT) cabinet in the customer's vicinity. A few years ago, the practice was to place remote DSLAM up to two kilometers from the customer, with nearly a thousand customers served from the remote DSLAM. Now, the trend is to make the subscriber line even shorter to achieve yet higher bit rates with VDSL2 technology. The extreme of this trend towards shorter lines is the fiber-to-the-curb (FTTC) and the fiber-to-the-building (FTTB) architectures, where the DSLAM is, at most, 100 meters from the

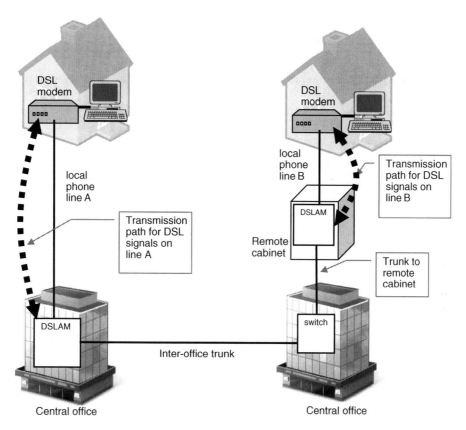

Figure 7.4 The DSL transmission channel consists of one telephone line

customer and the DSLAM serves up to 12 customers. The remote DSLAMs are usually connected to the core network via a fiber but, in a few cases, the trunk to the DSLAM is achieved by bonding multiple DSL lines from the remote terminal (RT) to the central office.

Figure 7.5 shows an RT cabinet that houses a remote DSLAM. A wiring pedestal is shown in the background.

7.3.3 DSL Terminology

The DSL industry has some unique jargon, and some of the terms are not used consistently throughout the industry. For example, the letter L in DSL stands for "line", and this book generally uses the term "line". However, the term "loop" is frequently used in the industry to refer to a telephone line, e.g., the term DLC, which is an abbreviation for Digital Loop Carrier.

In most of the world, the length of telephone lines is measured in meters and the diameter of the copper wire is measured in millimeters. However, in the United States, the length of telephone lines is measured in kilofeet (1 kft is approximately 305 meters), and the diameter of the wire is measured in AWG (American wire gage). A 24 AWG wire has a diameter of $1/24^{th}$ of an inch and equals 0.5 mm, and a 26 AWG wire has a diameter of $1/26^{th}$ of an inch and equals 0.4 mm.

The term POTS (plain old telephone service) refers to traditional switched telephone service using analog telephones. Downstream data flows from the network to the customer, upstream data flows in the opposite direction, and symmetric transmission refers to the same data rate flowing in both directions.

Figure 7.5 Remote terminal

The DSL transmission bit rates presented in this book are net bit rates unless noted otherwise. The net bit rate is the rate available to convey the user's traffic; in effect, it is the payload rate. The gross bit rate is the net bit rate plus the overhead such as framing, error correction coding, and maintenance information.

A modem is a modulator/demodulator; it is a device that transforms digital data into a signal well suited for transmission through a line.

Figure 7.6 shows the generic DSL protocol reference model based on the open systems interconnection (OSI) standard protocol reference model. The physical layer functions (also known as layer one) consist of

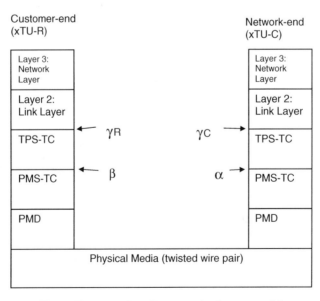

Figure 7.6 Generic DSL protocol reference model

the PMD (physical media dependent sublayer), PMS-TC (physical media specific transmission convergence sublayer), and TPS-TC (transport protocol specific transmission convergence sublayer) functions. Above this is layer two, the link layer, which uses protocols such as HDLC to perform message delineation and error control between network nodes. Layer three, the network layer, performs addressing, switching and routing functions.

DSL modems usually perform the layer-one functions and may also perform some layer-two and layer-three functions. The PMD functions include line coupling (also known as AFE-analog front end), filtering, symbol timing, equalization, echo canceling, modulation, and protection against overvoltage. The PMS-TC functions include framing, scrambling, and error monitoring. The TPS-TC performs the mapping of the bearer channels used to carry the user payload. As shown in Figure 7.6, the α and β interfaces exist between the PMS-TC and the TPS-TC, and the γR and γC interfaces exist between the TPS-TC and the link layer.

7.3.4 Introduction to DSL Types

Several types of DSL technology have evolved, with a trend toward higher bit rates. Each type of DSL technology has distinct abilities. The standardized and widely deployed types of DSL technology are described here. There have been many proprietary (nonstandard) types of DSL technology, but they are not described here because the number of deployed lines is comparatively small.

7.3.4.1 BRI-ISDN

The first DSL technology was BRI-ISDN (basic rate interface, integrated services digital network), which was introduced to service in 1985. It provided fixed, symmetric bit rate service at 144 kbits/s organized as two 64 kbits/s circuit switched B channels and one 16 kbits/s packet switched D channel. Lines up to 5.5 km (18 kft) are supported. ISDN-BRI does not support POTS on the same line.

More than ten million lines of ISDN-BRI have been deployed, but they have been largely replaced by newer technologies supporting higher bit rates and better suited for data traffic. ISDN-BRI chips are also used for DAML (digital added main line) equipment that digitally multiplexes two telephone circuits onto one line to provide additional telephone service areas without spare lines.

See Section 8.3 for more information about BRI-ISDN.

7.3.4.2 HDSL

HDSL (high bit rate digital subscriber line) service was introduced in 1992 as a lower-cost alternative to repeatered T1 and E1 lines. T1 lines (1.544 Mbits/s) and E1 lines (2.048 Mbits/s) required repeaters for virtually all lines. The T1 and E1 repeaters with waterproof cases and cable splicing labor were expensive. Many HDSL lines required no repeaters, and the longer HDSL lines require about one-third the number of repeaters in comparison to the older T1 and E1 technologies. HDSL technology provides fixed-rate, symmetric transmission of 1.544 Mbits/s or 2.048 Mbits/s, with no POTS on the same line.

The HDSL technology evolved to HDSL2 and HDSL4. At 1.544 Mbits/s, the non-repeatered reaches are:

- HDSL: up to 3.7 km using two pairs of wires.
- HDSL2: up to 3.7 km using one pair of wires.
- HDSL4: up to 4.6 km using two pairs of wires.

With one repeater the reach is nearly doubled, and multiple repeaters can be used to reach further. Section 8.4 provides more information about HDSL.

7.3.4.3 SHDSL

SHDSL is officially called "single-pair high-speed digital subscriber line", but it is often thought of as "symmetric high bit rate DSL". SHDSL service was introduced about 1999 to provide symmetric bit rate service ranging from192 kbits/s (at 6 km) to 5.7 Mbits/s (at 600 m) via one pair of wires with no POTS. SHDSL is best for high upstream bit rates on medium-to-long lines. Section 8.5 provides more information about SHDSL technology.

7.3.4.4 ADSL

ADSL (asymmetric digital subscriber line) service was introduced in 1995, and approximately 365 million lines were in service in 2012. ADSL is the most widely deployed form of broadband access in the world. It is most often used for high-speed Internet access, but it is also used for some video services. Compared to earlier forms of DSL technology, it embodied several new ideas:

- Asymmetry: the downstream bit rate was nearly ten times faster than the upstream.
- POTS and data transmission on the same line simultaneously with in-line filters, making customer self-installation feasible.
- Frequency division multiplexing: crosstalk is minimized by placing downstream and upstream transmission at different frequencies.
- Rate adaptive operation: the transmission bit rate automatically adapts to the line capacity (but the bit rate can also be fixed to a set rate).
- DMT modulation: discrete multi-tone modulation provides higher performance.
- Several versions of ADSL evolved:
  - ADSL: up to 8 Mbits/s downstream and 1 Mbits/s upstream.
  - ADSL2: slightly better performance, improved robustness, and additional diagnostic functions (Section 3.8 of [1] provides more details).
  - ADSL2plus: builds upon ADSL2, supporting up to 24 Mbits/s downstream and 1 Mbits/s upstream.
  - G.lite: a version of ADSL limited to about 1.5 Mbits/s which was standardized, but its deployment was negligible – see Section 8.1.1.

See Section 8.1 for more about ADSL.

7.3.4.5 VDSL

VDSL service was introduced in 2003, and it is widely used to provide triple-play service (video, voice, and data). VDSL supports POTS on the same line as asymmetric bit-rate transmission. The VDSL2 version of the technology supports symmetric bit-rate transmission up to 100 Mbits/s (100 Mbits/s downstream plus 100 Mbits/s upstream) on short lines. At 1 km (3.2 kft), VDSL2 provides 25 Mbits/s downstream and 3.5 Mbits/s upstream. See Section 8.2 for more about VDSL.

7.3.5 DSL Performance Improvement, Repeaters, and Bonding

The electrical properties of telephone lines and the electrical noise on the lines limit the achievable DSL bit rate. Advances in coding and modulation techniques over the past 20 years have provided large improvements for DSL performance, but the current technology is approaching the limits of classical transmission efficiency. However, other approaches are being used to gain higher performance. Higher bit rates can be achieved by:

- Making the line shorter, e.g., VDSL2 fed from a remote node.
- Reducing the noise. Chapter 9 describes DSM (dynamic spectrum management) methods to reduce the noise generated by other lines.
- Vectoring cancels the noise by learning how the noise couples between the lines.
- Placing a repeater near the middle of the line to regenerate and amplify the signal. In effect, repeaters make the line shorter by transforming the line in two (or more) shorter segments. Chapter 9 discusses repeaters.

- Bonding is a term for using several telephone lines to send data to and from a customer. Each line has its own pair of DSL modems and a portion of the total data payload for one customer is sent via each line. This is much like a multi-lane highway, and it is discussed Chapter 9.

Except for the first technique (making the line shorter), any of these techniques may be employed to send higher bit rates over a given line, or to send a given bit rate over a longer line.

7.3.6   Splitters and Filters for Voice and Data

ADSL and VDSL have the ability to convey traditional analog voice telephone service (POTS) simultaneously with data. This is accomplished by placing the DSL signals above 26 kHz to allow the voice signals to reside at the usual 300 Hz to 3400 Hz frequency range. The guard band of 3.4 to 26 kHz is needed for filters to separate the data and voice signals. Some service providers require customers to subscribe to both voice and data service for business reasons, but other service providers permit customers to subscribe to data-only service.

Carrying voice as a traditional analog voice signal assures the voice service will continue to function if the commercial power fails. The cost to provide the voice service is small, since the existing telephone network and customer telephones are used. All of the services based on POTS will work in a way familiar to customers; this includes facsimile transmission, alarm reporting, caller ID, and voicemail.

There are two configurations for ADSL and VDSL to combine POTS and data at the customer's premises: splitters and in-line filters. In both cases, a splitting-type filter is located at the network end of the line to separate the DSLAMs DSL signals from the voice signals for the voice switch. At the central office splitter, the path from the line to the DSLAM may contain a high-pass filter (HPF), which consists of a pair of series capacitors to block the flow of dc current. This optional dc blocking function is useful when the line is shared by different service providers for voice and data because it prevents the data service provider from intentionally or accidentally affecting the voice service.

The splitter or in-line filters perform three functions:

- High frequency noise resulting from POTS ringing signals and switch-hook transitions are isolated from the DSL modem.
- The high-frequency DSL signals are isolated from the telephone transmission because these signals would cause audible noise in some telephones due to intermodulation distortion.
- The DSL modems are isolated from the wideband impedance load of the telephones.

Figure 7.7 shows the customer-end splitter located at the point of entry to the premises. In North America, the splitter is usually located within the NID (network interface device) on the outside wall of the customer's house. A low-pass, passive filter in the splitter permits only frequencies below about 4 kHz to pass through to the inside wires connecting to the phones. The path through the splitter from the telephone line to the DSL modem permits the high frequencies to pass and, in most customer-premises splitters, this is an "all pass" path, i.e. there is no filter in the path from the telephone line to the DSL modem. The customer-premises splitter is sometimes called a centralized filter. The standard for customer-end splitters in the United States is ATIS-0600016 [2].

The advantages of the splitter configuration are:

1. isolation of the DSL modem from most of the house wiring; and
2. elimination of the additional shunt impedance resulting from placing multiple in-line filters on the line.

For homes with multiple phones, higher DSL bit rates can usually be achieved using a centralized splitter instead of in-line filters. For ADSL and ADSL2plus operating below 2 MHz, this difference is usually small but, for the higher frequencies used by VDSL, the difference is so large that splitters are often used.

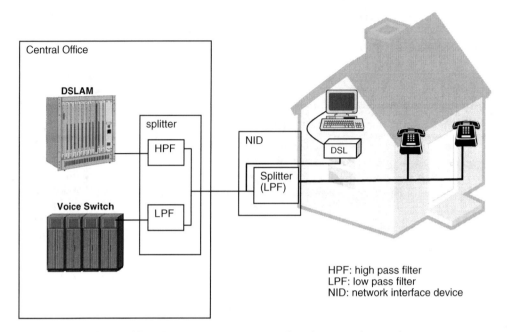

Central Office

DSLAM

splitter

HPF

Voice Switch

LPF

NID

DSL

Splitter
(LPF)

HPF: high pass filter
LPF: low pass filter
NID: network interface device

Figure 7.7   DSL/POTS Splitter at the point of entry to the premises

Figure 7.8 shows a network interface device (NID) and a splitter designed to fit in the NID. The NID is usually mounted to the wall outside a customer's home, and it provides the point of demarcation between the telephone network and the customer's premises. Technical requirements for splitters at the network end of the line are provided in ATIS T1.TRQ.10-2003.

For ADSL and ADSL2plus, the in-line filter configuration shown in Figure 7.9 is the most popular method to combine the POTS and DSL signals at the customer premises. The advantage of using in-line filters is that it easy for customers to install. An in-line filter (also known as a micro-filter) is plugged into the telephone wall jack so that it is in series with the telephone cord leading to each phone and other voice-band devices (answering machine, facsimile machine) in the house. Nearly all customers can install the ADSL modem and the in-line filters, which avoids the large cost of having a trained technician visit each customer's home. Furthermore, it also avoids the inconvenience of the customer having to wait for the technician to turn up.

In contrast, very few customers have the technical skill to correctly connect a centralized splitter and install the new wire usually needed to connect to the DSL modem. Thus, a splitter results in a large labor cost for a technician to visit the premises.

One drawback of in-line filters is a reduction of DSL performance due to the loading affect of the bridged taps in the house wiring and the impedance resulting from connecting multiple in-line filters. For the high-frequency signal used by VDSL, these affects are so large that only the splitter configuration is used. Also, homes with alarm reporting systems may have problems, because the customer will probably not know where to install the in-line filter for the alarm unit.

The technical requirements for in-line filters are provided in ATIS T1.421 and [1] describes in-line filters in detail.

The in-line filter is plugged into the telephone wall jack, and then the cord to the telephone is plugged into the socket in the in-line filter. One of the in-line filters shown in Figure 7.10 has two sockets; one is for

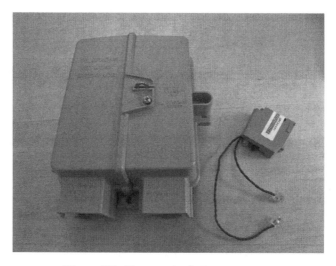

Figure 7.8 Photograph of NID and splitter

the telephone cord, and the other socket for the cord to the DSL modem. This figure also shows an in-line filter packaged to fit between a wall-mounted telephone and a wall jack.

7.3.7   Other Ways to Convey Voice and Data

Splitters and in-line filters combine baseband analog voice signals with DSL signals residing at higher frequencies. There are also all-digital methods for DSL to simultaneously carry voice and data. These all-

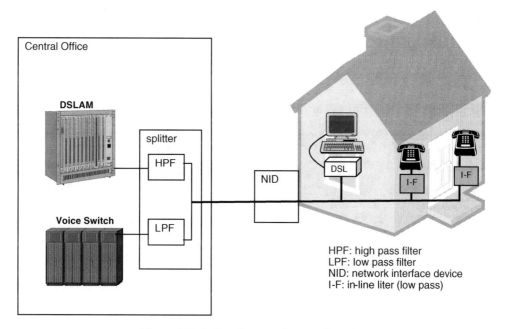

Figure 7.9   In-line filter premises configuration

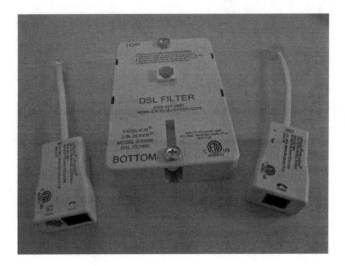

Figure 7.10 Photograph of in-line filters

digital methods lack the reliability advantage of network-powered POTS, but they have the advantage of easily supporting multiple voice calls. Also, the all-digital methods avoid the transmission impairments from analog filters.

DSL modems are often integrated within a residential gateway (RG) device that also contains voice-over-Internet-protocol (VoIP) functions. In this case, some of the IP packets carried by the DSL are used for one or more voice calls. The RG converts the VoIP packets to analog telephone signals that connect to the telephones in the house via the existing inside telephone wiring.

Alternatively, using a femtocell premises architecture, the digitized voice and the data IP packets can be converted to mobile phone radio signals which are then received by a mobile phone in or near the home.

Lastly, DSL can carry digital voice and data using a channelized voice configuration, with a fixed-bit rate, digital channel reserved for voice. This is similar to ISDN, but it has seen little use.

7.4   Transmission Channel and Impairments

A transmission system can only be as good as its channel. Telephone cables were designed to carry analog voice signals, but DSL requires the twisted copper wire pairs to carry a thousand times more bandwidth. Channel imperfections that were imperceptible for voice can be devastating for broadband digital transmission. DSL modem designers primarily work to undo the degradation of transmitted signals caused by telephone line impairments.

The impairments to DSL transmission over telephone lines include:

- signal attenuation;
- bridged taps;
- loading coils;
- nonlinear behavior of electrical protectors;
- wire gage changes;
- corroded and loose splices;
- pair imbalance;
- water intrusion;

- flat wire (not twisted pair);
- split pair (a line comprised of wires from different twisted pairs);
- impedance mismatch and return loss;
- intersymbol interference;
- crosstalk;
  - from like systems;
  - from other types of DSL;
  - near-end (NEXT);
  - far-end (FEXT);
  - from sources at an intermediate location;
- external stationary noise, such as radio frequency interference (RFI);
- impulse noise;
  - repetitive (REIN)
  - isolated (SHINE).

Since several of the impairments are simultaneously present to some degree on most lines, the DSL modem designer faces a daunting gantlet of challenges.

A telephone line consists of two insulated, solid-copper wires that are twisted together. Figure 7.11 shows a 1500-pair feeder cable and a three-pair drop wire. The twisting of the two wires is clearly visible in the drop wire. The pair of wires is twisted for two reasons. Firstly, twisting the wires helps to reduce the coupling of signals between the wire pairs. This signal coupling is known as crosstalk. Secondly, twisting the pair of wires makes it easier for network technicians to identify the two wires associated with a customer's line. The idea of twisting the pair of wires was introduced by Alexander Graham Bell.

Before 1990 in North America, the average length of a telephone line was about 2.4 km (8 kft), and it ran from the central office in a feeder cable having about 1500–4000 pairs of wires to a cross-connection cabinet (called the serving area interface, SAI) where the wire pairs were connected to distribution cables having 25–200 pairs of wires. The feeder cable was often placed underground (the photograph in Figure 7.12 shows a SAI with the doors open to illustrate the cross-connection panel). The distribution cable then ran to wiring pedestals, where the wire pairs were cross-connected to individual drop wires to

Figure 7.11 Photograph of 1500-pair, 26 AWG feeder cable, and 3-pair drop wire

Figure 7.12 Serving Area Interface (SAI)

each customer's home. The drop wires typically had two to four pairs of wires. The distribution and drop wires were typically aerial wires hung from wooden poles.

For both feeder and distribution cables, the wire pairs are organized into binder groups, usually having 25 pairs of wires. The wire pairs within a binder group remain adjacent to each other for the length of a cable segment. The grouping of the wire pairs is often preserved when cable segments are spliced together, although this is not always true. Furthermore, due to cross-connections at the SAI, the grouping of pairs from the feeder cable may differ from the grouping in the distribution cable.

Today, many telephone lines are still served from the central office but, more often, the telephone line is served from equipment placed in a remote cabinet. As a result, the average North American telephone line is about 1.5 km (5 kft) long at the time this was written, and the trend will continue towards about half this length in the future.

In some areas outside North America, four wires are twisted together into units call quads. The crosstalk between the wires within the quads is very high.

7.4.1   Signal Attenuation

Figure 7.13 is a graph of the signal attenuation for frequencies up to 18 MHz for 26 AWG twisted pair wires 100 m, 500 m, and 2000 m long.[1] In this graph, attenuation is shown as negative decibels to illustrate that signals become weaker on longer telephone lines. The topmost line shows that a 100 m line has relatively little attenuation, and the attenuation increases only slightly with frequency. The middle line shows that the 500 m line has more attenuation, especially at higher frequencies. The bottom line shows that the 2000 m line has much more attenuation, and the increased attenuation at high frequencies is yet more pronounced. At higher frequencies, there is an enormous difference between the attenuation for a 100 m and a 2000 m line – more than 200 dB of difference. Since a wider range of frequencies (e.g.

---
[1] Figures 7.13, 7.15, and 7.16 were reproduced by permission of Craig Schelp of PMC-Sierra

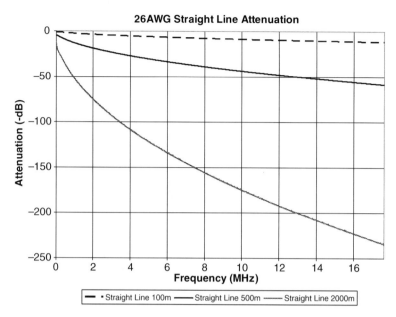

Figure 7.13 Signal attenuation of twisted pair wires. (Source: [Shelp 2009]. Reproduced with permission of Craig Shelp, PMC Sierra.)

bandwidth) is needed to transmit higher bit rates, this illustrates why the bit rate capacity is less for longer lines.

Under low noise conditions, DSL modems can operate with up to approximately 65 dB of attenuation at the frequencies used for transmission. As a result, the approximate maximum usable frequencies for the following line lengths (with no bridged taps) are:

- 1.5 MHz at 2000 m.
- 5 MHz at 1000 m.
- 18 MHz at 500 m.
- Over 30 MHz at 100 m.

Figure 7.13 shows the signal attenuation for "straight lines." The term "straight line" does not refer to the wire having no bends; rather, a straight line may have any amount of bends, but it does not have any branches (also called bridged taps).

Signal attenuation is also affected by the thickness of the copper wires, especially at lower frequencies. A 0.5 mm (24 AWG) wire will have less signal attenuation per meter than a 0.4 mm (26 AWG) wire.

The temperature of the wires also changes signal attenuation, with higher temperatures causing more signal attenuation. As a result, DSL modems must track the changes in signal attenuation. For aerial cables exposed to the sun in desert conditions, cable temperature can change rapidly at sunrise and sunset.

7.4.2 Bridged Taps

Signal attenuation is affected by the presence of bridged taps. A bridged tap is a segment of a wire pair connected to the pair of wires serving a customer. This can also be described as an unterminated wire stub. As shown in Figure 7.14, the end of the wire tap does not connected to anything.

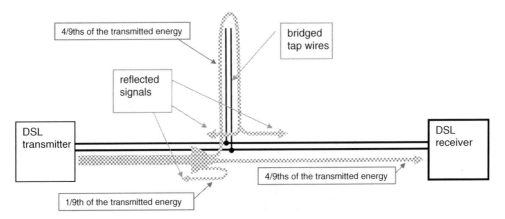

Figure 7.14 Bridged tap

In this figure, the thin black lines are the wires that comprise a twisted pair. For simplicity, the twisting of the wires is not shown. The term "working line" in used below to describe the direct path between the pair of modems. The bridged taps are not part of the "working line".

The checkered lines depict the signals sent from the transmitter on one end of the line to the receiver at the other end. The checkered line leaving the transmitter is wide, showing that it is a strong signal. This signal travels along the twisted pair of wires, eventually encountering the bridged tap. Four-ninths of the transmitted signal energy continues along the working line to the DSL receiver. One-ninth of the signal energy is immediately reflected back towards the transmitter. The other four ninths follows the bridged tap wires until the signal reaches the unterminated end. The impedance mismatch at the end of the tap causes the signal energy to be reflected back and travel along the bridged tap wires towards the working line. When the reflected signal reaches the working line, the signal energy splits yet again, with part of the signal energy returning to the transmitter, part of the reflected signal energy traveling towards the receiver, and part of the energy reflecting back along the bridged tap recursively. The reflected signal traveling towards the receiver is delayed and will arrive at the receiver later than the signal that traveled on the direct path.

Thus, bridged taps impair transmission in two ways. First, the bridged tap attenuates the signal by diverting some of the transmitted signal. Second, the bridged tap causes a delayed echo to arrive at both ends of the line. This echo will arrive when the receiver is attempting to receive a symbol sent at a later time, and thus the echo will interfere with the reception of the signal following the direct path.

Clearly, bridged taps are a severe impairment for DSL transmission, and the effects are even worse for lines that have multiple bridged taps. The degree of impairment is greater at higher frequencies. Below 1 Mhz, the effect is small enough that HDSL and ADSL systems are usually able to operate with only a small reduction in bit rate, so bridged taps are rarely removed for HDSL and ADSL lines. However, since VDSL operates at much higher frequencies, bridged taps are almost always removed for VDSL lines. As a result, the prevalence of bridged taps is gradually diminishing.

The large majority of telephone lines in North America have bridged taps, but they are rarely found outside of North America. There can be more than one bridged tap on a line, and the taps may be present at any point along the line. However, the most common location is 20 m to 800 m from the customer's end of the line. Bridged taps exist because they have no effect on traditional telephone service, and they enable cabling practices that reduce the operating cost for the telephone network.

A simplified example for the telephone network in North America consists of a 200-pair distribution cable serving approximately 150 homes. All 200 pairs pass all 150 homes. At the wiring pedestal, the drop wires to a home can connect to any of the 200 pairs. This permits every home to be served with one line, and the remaining 50 pairs are available to provide second and third line service to any of the 150 homes until the pairs are exhausted. When a customer's drop wire is connected to the distribution cable, the remaining unused length of the distribution cable remains connected to the wire pair. This preserves the future utility of that wire pair to serve any other customer after the current customer stops using it. However, the section of distribution wire beyond the wiring pedestal becomes a bridged tap. There are other reasons why bridged taps sometimes are present near the central office end of the cable and near the middle of the line.

Only one direction of transmission is shown in Figure 7.14, whereas DSLs actually transmit signals in both directions. The bidirectional transmission on the same pair of wires is accomplished by various duplexing methods. Frequency division duplexing (FDD) places the "eastbound" and "westbound" signals in different frequency bands. Time division duplexing (TDD) alternates short bursts of unidirectional transmission in each direction. Echo canceled hybrid (ECH) transmission simultaneously sends signals in both directions in the same frequency band, with the receiver subtracting the locally transmitted signal to extract the signal sent by the transmitter at the other end of the wire.

Figure 7.15 shows the frequency domain signal attenuation of 200 m lines with and without a bridged tap. The dashed line at the top of the graph shows the signal attenuation versus frequency for a "straight line", i.e., a 200 m line with no bridged taps. The solid, black "wiggly" line shows the affect of a 200 m bridged tap located at the customer's end of a 200 m line; the signal attenuation averages 3 dB greater than the line with no bridged tap. The grey "bouncing" line shows the affect of a 25 m bridged tap at the customer's end of a 200 m line. The average attenuation with the 25 m tap is again 3 dB greater than the line with no bridged tap, but the variation of attenuation with frequency is much greater than the case with the 200 m bridged tap. The 25 m tap causes up to 15 dB more attenuation at some frequencies but, at other

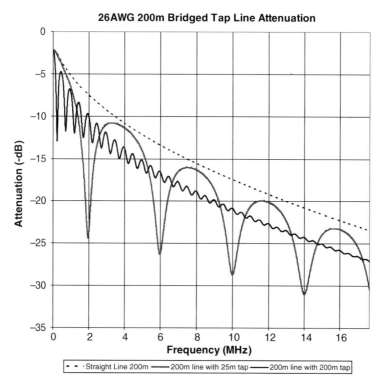

Figure 7.15 Attenuation versus frequency for 200 m line with bridged tap. (Source: [Shelp 2009]. Reproduced with permission of Craig Shelp, PMC Sierra.)

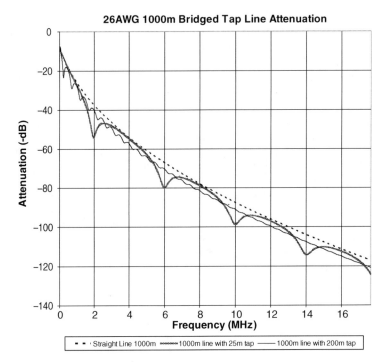

Figure 7.16 Attenuation versus frequency for 1000 m line with bridged tap. (Source: [Shelp 2009]. Reproduced with permission of Craig Shelp, PMC Sierra.)

frequencies, the 25 m tap has almost no affect. Notice that both bridged taps have relatively little effect at frequencies below 1 MHz.

Figure 7.16 shows a similar graph for a longer line (1000 m) with and without bridged taps. Since the maximum signal attenuation is much greater for the longer line, a larger scale is used for this graph. This makes the frequency variability appear smaller than the previous graph, but the effects are nearly the same. Note that the frequencies where attenuation changes for the line with the 25 m bridged tap are the same as the shorter line. This is because the length of the bridged tap is the same for both graphs. Further information regarding bridged taps and wire gage changes is provided in [3].

7.4.3 Loading Coils

In North American telephone networks, loading coils are spliced into cables longer than 5.5 km (18 kft) to improve the perceived speech quality of telephone service. As shown in Figure 7.13, signal attenuation for long twisted pair copper lines increases with frequency. As a result, the quality of analog voice transmission is affected by the loss of the high-frequency components of the human speech.

The frequency response of a long twisted pair line between 200 Hz and 4 kHz is flattened when inductors are wired in series with the line. The attenuation in the frequency band is reduced by inserting loading coils. The common practice is to insert a pair of 88 mH inductors in series with the pair of wires at approximately 1.8 km intervals. Depending on the length of the line, three or more loading coils may be present on the line.

While the frequency response below 4 kHz is flattened by loading coils, the attenuation above 4 kHz increases very rapidly with frequency. A loaded line has no usable capacity above 10 kHz. DSL transmission is not possible on loaded lines.

Loading coils are not used outside of North America. In North America, loading coils are present only on lines longer than 5.5 km (except for an occasional provisioning mistake). In 1990, about 20% of lines had loading coils. With the advent of ISDN and the subsequent generations of DSL technology, loading coils have been removed from many lines. At this time, the proportion of North American lines with loading coils is estimated to be less than 5%. Most of the lines with remaining loading coils are so long that they would not support high DSL rates even if the loading coils were removed.

7.4.4   Return Loss and Insertion Loss

Ideally, all of the signal energy transmitted by a DSL modem would be conveyed by the twisted pair line to the receiver. However, in reality, the signal is attenuated by the resistance, capacitance, and inductance of the line. Also, when the signal encounters an impedance change, a portion of the transmitted energy is reflected back towards the transmitter and, thus, this portion of the energy does not reach the receiver. Examples of impedance changes include bridged taps, wire-gauge changes, connectors, and the difference between the impedance of the modem and the line. Modem designers attempt to match the modem impedance to the line and, in practice, the modem impedance is usually similar to line but not exactly the same.

Return loss and insertion loss are measured in decibels (dB). They may pertain to an individual element, such as a splitter, or to the entire system, comprised of the modems and everything in the transmission path. Return loss is the ratio of the signal reflected back towards the transmitter to the transmitted signal. When this is expressed as echo return loss, then a higher number indicates better performance. For example, DSL splitter specifications require the echo return loss to be greater than 8 dB.

Insertion loss is the ratio of the signal energy not reaching the output measurement point to the energy transmitted. Thus, a low insertion loss results in good performance. Insertion loss is the combined effects of signal attenuation and reflected energy.

Section 3.9 of [4] provides further discussion of return loss and insertion loss.

7.4.5   Balance

The balance of a twisted pair line is measured in decibels (dB), and it represents the logarithm of the proportion of a longitudinal signal applied to a line to the resulting differential signal. A differential signal, also known as a metallic signal, is the voltage difference between the two wires. A longitudinal signal, also known as a common mode signal, is the average voltage of the two wires with respect to ground. Balance is also known as LCL (longitudinal conversion loss).

Ideally, balance should be high so that a transmitted signal remains differential and very little of the signal transforms into a longitudinal signal. The high longitudinal signal resulting from poor balance causes electromagnetic radiation which can cause radio frequency interference (RFI). Also, the radiated energy can couple into nearby pairs of wire, causing crosstalk. Conversely, a line with poor balance is more susceptible to picking up ambient radio frequency energy, including impulse noise and crosstalk from other wires. As a result, a cable with two lines with poor balance will have high crosstalk between the poorly balanced pairs.

Poor balance results from impedance differences between each wire in a twisted pair. Telephone cables are designed to have high balance, and twisting the two wires of a pair helps to improve balance. The balance of telephone lines tends to be less at higher frequency. The balance of a typical telephone line is greater than 60 dB at 3 kHz, but it is typically 40 dB at 1 MHz and may drop to 15 dB at 30 MHz.

7.4.6   Intersymbol Interference (ISI)

DSL modems modulate the data to be transmitted into short symbols. The symbols are pulses of precisely adjusted phase, frequency, and amplitude. For ADSL, 4000 symbols are sent each second, and the duration of each symbol is 250 microseconds. Different frequencies propagate at different speeds along a

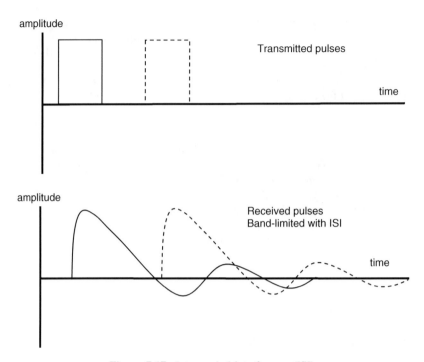

Figure 7.17 Intersymbol interference (ISI)

twisted pair line. This is sometimes called "delay distortion", and it causes the symbols to disperse in time. Some of the frequency components in a symbol arrive later than others causing a portion of the symbol to spread into the next symbol.

Intersymbol interference (ISI) is the effect of transmitted symbols spreading into each other, and this is illustrated in Figure 7.17. ISI is one of the foremost factors limiting DSL performance and it is particularly detrimental on long lines and for high symbol rates. The effects of ISI are worse for lines with bridged taps, but ISI effects all lines to some degree.

DSL modems mitigate ISI with techniques including receiver decision feedback equalization (DFE), transmitter precoding, and maximum-likelihood detection (Viterbi algorithm). Also, discrete multi-tone transmission (DMT) inserts a short cyclic prefix to each symbol to help to combat ISI. The cyclic prefix is a copy of the last few microseconds of the symbol, and it is placed at the start of the symbol. As a result, the first and last few microseconds of the DMT symbol are identical, and it makes the channel look periodic to the receiver.

The length of the cyclic prefix is designed to be as long as the anticipated duration of the channel response. In addition to acting as a guard interval during which the ISI from the prior symbol will occur, the cyclic prefix improves the effectiveness of the FFT (Fast Fourier Transform) and the IFFT (inverse FFT) used for DMT modulation. When combined with time domain equalization, the length of the cyclic prefix may be shorter. See [5] for more information about DMT modulation and the FFT.

7.4.7 Noise

In addition to the low-level thermal noise that results from the flow of electrons though a circuit, there are many external sources of noise that electromagnetically couple into a twisted pair line. Lines with good balance are less susceptible, but even lines with good balance are affected by noise.

Noise is characterized by its power spectral density (PSD). The PSD describes the noise amplitude as a function of frequency. Impulse noise exists for only a few seconds or less, whereas continuous noise lasts for a much longer duration. Stationary noise maintains a nearly constant PSD for more than a few minutes, whereas non-stationary noise exhibits large changes of PSD every few minutes or sooner. To some extent, DSL modems can adapt to stationary noise by moving transmission to frequencies with less noise, or by reducing the total bit rate. Adaptation to non-stationary noise is more difficult.

7.4.7.1 Crosstalk

Telephone cables typically have between 25 to a few thousand pairs of copper wires. Even in cables with well-balanced pairs, some of the signal energy from each pair of wires radiates into the other pairs of wires in the cable. The energy electromagnetically coupled into a wire pair appears as noise to the DSL receivers connected to that line. The electromagnetic coupling consists of mutual capacitance and mutual inductance between the respective pairs of wires. This electromagnetic coupling works both ways, i.e., the signal on wire pair A produces crosstalk noise into all of the other pairs in the cable, while the radiated signal energy from all of the other wire pairs in the cable produces a composite crosstalk noise into wire pair A.

Crosstalk between wire pairs carrying the same type of DSL signal is called "self crosstalk" (e.g., crosstalk from one HDSL line to another HDSL line). "Foreign crosstalk" is crosstalk between different types of DSLs or from a non-DSL source to a DSL. For example, a T1-line or an ADSL would be a foreign crosstalker for an HDSL line.

Near-end Crosstalk (NEXT)
Near-end crosstalk (NEXT) is crosstalk noise from a transmitting modem located at the same end of the cable as the victim receiver (see Figure 7.18).

NEXT coupling between pairs increases with signal frequency, and it is a major factor limiting the performance of DSL systems that use the same frequencies for upstream and downstream transmission (e.g., SHDSL, and ISDN). ADSL and VDSL virtually eliminate the effect of NEXT by placing downstream and upstream transmission in separate frequency bands, which is known as frequency division multiplexing (FDM). For FDM-type DSLs, the NEXT appears outside of the frequency band used by the near-end receiver. The standard 99% worst-case model for received NEXT at frequency f is defined by Equation 7.1:

$$NEXT(f, N) = 10^{-13} \cdot (N/49)^{0.6} \cdot f^{1.5} \cdot S(f) \qquad (7.1)$$

where N is the number of disturbing wire pairs (for a 50 pair binder group), and $S(f)$ is the signal power input to one of the disturbing wire pairs at frequency f. Note that the NEXT is independent of line length.

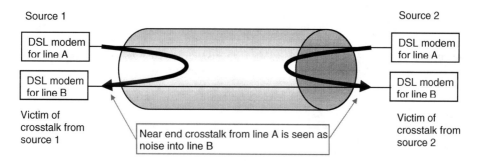

Figure 7.18 Near-end crosstalk (NEXT) path

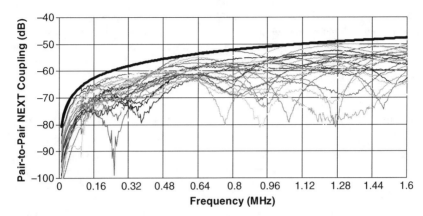

Figure 7.19   Near-end crosstalk (NEXT) graph. (Source: [Kerpez 2009]. Reproduced with permission of Ken Kerpez, ASSIA Inc.)

This equation is based on empirical analysis of laboratory measurements of telephone cables. NEXT increases with frequency and with the number of disturbers (line causing crosstalk). The total amount of NEXT produced by two disturbing lines is about 50% greater than the NEXT produced by one disturbing line, but the total amount of NEXT produced by 24 disturbers is only about 10% more than the total NEXT produced by 20 disturbers. The effect of NEXT is dominated by the first few disturbers.

Figure 7.19[2] shows the NEXT coupling between a victim wire pair and the other disturbing wire pairs in the cable.

The bold line at the top of the set of curves is the 99% worst case model defined in ANSI Standard T1.417 for spectrum management of DSL systems [8]. The pair-to-pair NEXT coupling at each frequency empirically results in 99 out of 100 measured NEXT couplings being equal to or less (better) than the NEXT model. At some frequencies, there is a 30 dB difference between the pair with the most NEXT coupling and the pair with the least NEXT coupling. For a given wire pair, the crosstalk coupling may change radically with frequency, such that it is nearly the greatest disturber at one frequency and nearly the least disturber at a different frequency. Some of the pair-to-pair coupling lines are generally higher or lower than the other lines. The lines with greater NEXT coupling represent wire pairs that are located more closely within the cable. NEXT drops rapidly below 100 kHz. The NEXT at high frequencies is more than 30 dB greater than at low frequencies.

The wire pairs in telephone cables are organized in binder groups of 10, 25 or 50 wire pairs, with 25-pair groups being the most common. A binder group is a set of wires that remain close to each other for the length of a cable segment. The NEXT equation (Equation 7.1) and graph above describe the NEXT coupling between wire pairs within the same binder group of 50 pairs. Wire pairs in different binder groups have greater physical distance in the cable than wire pairs in the same binder group. As a rule of thumb, the NEXT coupling between pairs in different binder groups is 10 dB less than pairs in the same binder group. When cable segments are spliced together, the binder group mapping is usually preserved, but not always.

Far-end Crosstalk (FEXT)

Far-end crosstalk (FEXT) is crosstalk noise from a transmitting modem located at one end of the cable into a victim receiver connected to a different pair of wires at the other end of the cable (see Figure 7.20).

---

[2] Figure 7.19 was reproduced by permission of Ken Kerpez of Telcordia.

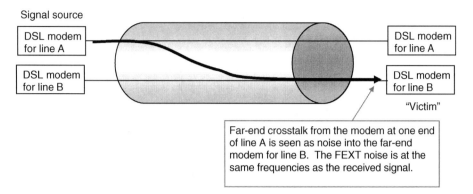

Figure 7.20   Far-end crosstalk (FEXT) path

FEXT coupling between wire pairs increases with signal frequency to a greater degree than NEXT. While frequency division multiplexing (FDM) effectively defends against NEXT, FDM provides no benefit against FEXT because the FEXT noise is located at frequencies used by the victim receiver.

FEXT has a greater impact on shorter lines for two reasons. First, the short lines are capable of conveying higher frequency signals, so DSLs often use this higher frequency capacity, and FEXT coupling is much greater at higher frequencies. Second, the FEXT couples over the entire length of the line, but most of the FEXT coupling occurs near the transmitter, where the transmitted signal is strongest. Also, the FEXT noise is attenuated by the remaining length of the line.

The FEXT noise seen by the victim receiver is defined by Equation 7.2.

$$\text{FEXT}(f, L, N) = 9 \times 10^{-20} \cdot (N/49)^{0.6} \cdot f^2 \cdot L \cdot H(f, L)^2 \cdot S(f) \tag{7.2}$$

The degree of FEXT is a function of the number of disturbers (N), frequency (f), length of the line (L), and the power of the disturbing signal source S(f). The "L" represents the length of wire over which the crosstalk will couple. Thus, as the wire length increases, more FEXT energy is coupled from the disturbing wire pair to the victim wire pair. However, the "H(f,L)^2" factor is the channel response that attenuates the FEXT energy coupled into the victim wire pair. The combined effect of these two factors plus the f^2 factor for frequency results in the FEXT noise seen by the victim receiver being greater for shorter lines, primarily due to higher frequency signals typically present on shorter lines.

Figure 7.21[3] illustrates the pair-to-pair FEXT coupling between some of the pairs within a 25-pair binder group. The thick, black line is the 99% worst case model described in Equation 7.2. The light gray lines show the FEXT coupling between pairs of lines, but not all combinations of pairs within the binder group are shown in order to make lines easier to see. If all the combinations had been shown, three more light gray lines would have been shown above the 99% worst case model for a portion of the graph. This graph shows that a few of the combinations of line-to-line FEXT are much worse than most of the combinations. Also, some frequency-dependent variability from the FEXT model can be seen for each combination.

Midpoint Crosstalk
When DSL is served from a remote terminal (RT) located between the central office (CO) and the customers, the signals from the DSL transmitters at the RT can cause crosstalk into other lines served from the CO. This midpoint crosstalk is also known as the near-far problem because the disturber is nearer to the receiver than the source of the desired signal. Midpoint crosstalk is illustrated in Figure 7.22.

---
[3] Figure 7.21 was reproduced by permission of Alynn Wilson of Adtran.

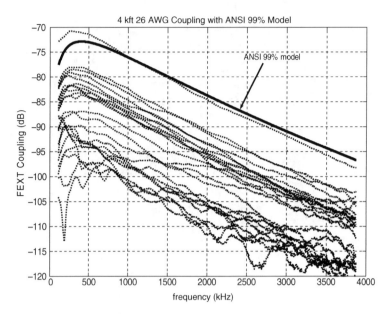

Figure 7.21 FEXT for individual line-to-line combinations. (Source: [Wilson 2009]. Reproduced with permission of Arlynn Wilson, Adtran Inc.)

Midpoint crosstalk can also be caused by mid-span repeaters. However, it can be avoided by serving all customers in a distribution area from the RT, or by having the RT-fed DSLs transmit at higher frequencies than those used by the CO-fed lines.

7.4.7.2 Narrowband Noise

Narrowband noise affects a small portion of the frequency band used by the DSL system, unlike crosstalk which often affects most of the used frequency band. Common sources of narrowband noise include

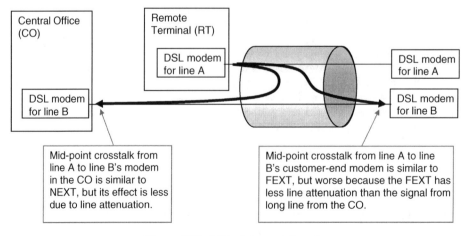

Figure 7.22 Midpoint crosstalk path

broadcast radio, amateur radio, and switching power supplies with faulty radio frequency (RF) noise suppression. The amplitude of narrowband noise can be so high that the DSL system will have no usable transmission capacity within the affected frequencies. DSL systems using DMT modulation can usually limit the impact to the frequency range of the narrow band noise plus a few adjacent DMT subcarriers. However, in some cases, high-level narrowband noise can overwhelm the analog front end (AFE) such that DSL operation is not possible. Sometimes, an external RF filter inserted at the customer-end modem can reduce the effects of narrowband noise.

The high power of AM broadcast radio can be very damaging, especially if the AM broadcast antenna is within two kilometers of the DSL equipment. In North America, AM radio transmits in narrow frequency bands from 550 kHz to 1.6 MHz. Television is broadcast at frequencies above 54 MHz, so it should not affect DSL.

Amateur radio (also known as HAM radio) uses the following frequency bands: 1.8–2 MHz, 3.5–4 MHz, 5.332–5.4 MHz, 7.025–7.3 MHz, 10.1–10.15 MHz, 14.025–14.35 MHz, and 18–18.168 MHz, 21.025–21.45 MHz, 24.89–24.99 MHz, 28–29.7 MHz.

Interference to and from amateur radio is highly localized. To reduce interference of amateur radio operation due to emission from DSL signals, most DSL systems have the ability to reduce the transmitted power in the amateur radio frequency band. This is also known as "notching".

7.4.7.3 Impulse Noise

Impulse noise occurs in relatively short bursts that sometimes have such high amplitude that it overwhelms the signal on DSLs operating with high signal-to-noise margin. The temporal behavior of impulse noise depends on the source of noise. The following types of impulse noise are based on commonly observed noise in the field.

- PEIN (prolonged electrical impulse noise) are randomly occurring impulses with duration of less than ten milliseconds. This type of impulse noise is often correctable by interleaved Reed Solomon forward error control (FEC). PEIN may be caused by analog telephone signaling transients on the same line or an adjacent line. For example, when a ringing telephone is taken off-hook ("ring trip") an electrical transient may be produced.
- SHINE (short high amplitude impulse noise event) are randomly occurring impulses with duration of ten milliseconds or more. This type of impulse noise is often uncorrectable by FEC, but may be correctable by packet retransmission. SHINE may be caused by an electrical motor starting or stopping. For example, a voltage spike may be caused when the relay controlling a refrigerator motor opens. The word "short" in the acronym is misleading since SHINE impulses are longer than PEIN.
- REIN (repetitive electrical impulse noise) are periodic impulses typically having a duration of less than one millisecond. REIN impulses usually occur every 10 ms in countries with 50 Hz AC power or 8.3 ms in countries with 60 Hz AC power. A REIN impulse is produced by some AC-powered equipment at a certain voltage of the AC power phase and is usually produced at both the positive and negative portions of the AC cycle. REIN may be produced by AC light dimmers, which turn the current on and off during part of the AC cycle. REIN may also be produced by florescent lights with a ballast inductor.

A fourth type of impairment, termed a *microinterruption*, has affects similar to impulse noise, but it is not caused by noise. A microinterruption is a brief open-circuit condition, typically having duration less than 10 ms. This may be the result of a loose splice in the wire or a loose connector. For example, the vibrations caused by a heavy truck driving by a cable splice could cause the wires in a cable splice to separate momentarily. A microinterruption causes a brief loss of signal, which has similar affects to a brief noise impulse.

Increasing the amplitude of the DSL signal is a poor defense against high amplitude impulse noise. The most effective mitigation is redundancy coding combined with interleaving. For example, ADSL and VDSL employ Reed Solomon forward error control (FEC) which can correct a limited number of bit errors, and the data is interleaved so that a long impulse is spread across multiple FEC blocks, enabling the

FEC to correct longer bursts of errors. This works well up to a point but, if the error burst exceeds the correction ability of the interleaved FEC, then the interleaving may result in several packets being affected.

Packet retransmission is also an effective defense against impulse noise. Retransmission is sometimes called ARQ (automatic repeat request). Retransmission is typically performed at layer 3 (IP) and the application layer, but it is under discussion for possible addition to the layer-1 ADSL and VDSL standards. Performing retransmission at layer 1 would reduce the delay jitter resulting from a packet retransmission. The ITU-T has developed draft standard G.998.4 (G.inp), specifying a retransmission technique for ADSL2 and VDSL2.

7.4.8 Transmission Channel Models

The twisted pair of wires that DSL signals traverse may be represented as a transmission line model where the distributed characteristics are equivalent to a lumped electrical circuit. This is called the RLGC transmission line model where R represents the series resistance, L represents the series inductance, G represents the parallel conductance (resistance between the tip and ring wires), and C represents the parallel capacitance between the tip and ring wires. The characteristic impedance (Z_0) of the cable can be derived from the primary constants R, L, G, and C, as shown in Equation 7.3.

$$Z_0 = \sqrt{\frac{R + j\omega L}{G + j\omega C}} \qquad (7.3)$$

where:

$j = \sqrt{-1}$
$\omega = 2\pi f$
f = frequency.

The twisted pair of wires may also be represented by a two-port, equivalent circuit defined by ABCD parameters based on the input and output voltages and currents. Telephone lines consisting of differing segments of wire are modeled as a cascaded series of models with each segment represented by a model using the transmission characteristics for that segment.

ABCD network models and RLGC line models are discussed in depth in Chapter 3 of [4] and Chapter 3 of [6].

7.5 DSL Transmission Techniques

DSL modems are designed to overcome the many imperfections of the twisted pair transmission channel. Modulation converts the digital signal to an analog signal which is better suited for transmission over the wires because the signal energy is placed at frequencies with better channel characteristics. The modulation techniques include 2B1Q (2 binary 1 quaternary), PAM (pulse amplitude modulation), and DMT (discrete multi-tone). These and other modulation techniques are discussed in [4,5].

7.5.1 Duplexing

Duplexing techniques are used to send data to the customer (downstream) and from the customer (upstream) over the same pair of wires.

ECH (echo canceled hybrid) is the duplexing techniques used for Basic-rate ISDN, HDSL, and SHDSL systems. With ECH, data is simultaneously transmitted in both directions in the same frequency band. The signals in both directions are superimposed upon the same pair of wires. The receivers at both ends recover the desired signal by subtracting the portion of the signal sent by the local transmitter. Since the signal sent from the far end is attenuated by the line, the amplitude of the signal from the local transmitter is much greater than the desired signal send from the far end. The near-end signal can be removed with high precision, because the locally transmitted data pattern is exactly known, and the DSL modem adapts to the reflection characteristics of the channel. The term *training* is used to describe the process of the modem learning the channel characteristics.

The advantage of the ECH duplexing method is that the utilized frequency band does not extend as high as would be needed for the FDM technique. Disadvantages are the high levels of NEXT (near end crosstalk) and the limited precision for cancelling the signal from the local transmitter.

FDM (frequency division multiplexing) is the duplexing technique used for ADSL and VDSL systems. With FDM, downstream data is placed in one or more frequency bands that are separate from the frequencies used for upstream transmission. Interference from the local transmitter is simply removed by filtering out the frequencies used by the local transmitter. NEXT is also simply removed by the same filtering, provided that all other lines use the same frequency plan (i.e., all customer-end modems transmit upstream data in the same frequency bands, and all network-end modems transmit downstream data in the same frequency bands).

ADSL and VDSL systems are asymmetric, that is, the downstream frequency bands are larger than the upstream bands. To avoid excessive NEXT, all lines must transmit in the same direction at each frequency. Thus, it is not possible to operate ADSL or VDSL systems with reverse asymmetry (upstream bandwidth greater than downstream).

The advantages of FDM are the simplicity and effectiveness of avoiding NEXT and interference from the local transmitter. The disadvantage of FDM is that it is necessary to use higher frequencies, because frequencies are not reused to send data in both directions.

Time Compression Multiplexing (TCM) is the duplexing technique used for some of the basic-rate ISDN systems in Japan. This technique is also known as Time Division Duplexing (TDD) and Ping-Pong. Data is transmitted in short unidirectional bursts in alternating directions, about one millisecond in duration. Thus, data is transmitted downstream only for 1 ms and then transmitted upstream only for 1 ms, with a brief idle period (a guard interval) between the bursts to reduce interference from echoes.

TCM is a simple and highly effective method of avoiding NEXT and interference from the local transmitter. However, the effectiveness of the NEXT avoidance depends on the transmitted bursts from all modems connected to a cable being precisely synchronized. The required synchronization can become difficult when some of the pairs in the cable are fed from different DSLAMs, possibly at different locations. Also, synchronization may not be precise when the lines in the cable have substantially different lengths.

Lastly, a two-pair duplexing technique is used by old T1 and E1 transmission systems. Data is transmitted in only one direction on one pair of wires, and data in the opposite direction is transmitted on a separate pair of wires. To reduce crosstalk, the wire pairs are specially selected to ensure that they are in separate binder groups. One binder group is dedicated for downstream transmission, while another binder group is dedicated for upstream transmission. High-cost, special network engineering is required to ensure that the correct wire pairs are used, and the dedicated binder groups result in inefficient usage of wire pairs.

7.5.2 Channel Equalization and Related Techniques

DSL signals are distorted by variations in attenuation and group delay at different frequencies. This distortion impairs transmission by causing intersymbol interference (ISI, see Section 7.4.6) and by altering the modulation constellation. Equalization methods attempt to achieve the same attenuation and group delay at all used frequencies by either adjusting the transmitted signal (precoding) or adjusting the received signal. When the modems start up, special signals are sent to learn the distortion characteristics of the transmission channel; this process is called training. Since the transmission channel characteristics may change over time (due to the temperature of the wires changing or a change in the moisture content within the cable), most DSL modems attempt to track the transmission channel characteristics during operation. Many DSL modems use decision feedback equalization (DFE) method in the receiver to compensate for the distortion of the transmission channel.

Another technique used in DSL receivers is the Viterbi algorithm with trellis coding. This technique helps to protect against transmission errors by adding some extra (redundant) constellation points to the transmitted signal constellation, and then the receiver decides on the value of received symbols based on observing a sequence of several received symbols.

ADSL and VDSL modems improve channel equalization by dividing the utilized range of frequencies into a large number of small sub-bands. The channel variation within each 4 kHz sub-band is small compared to the variation across the entire range (1.104 MHz for ADSL per ITU-T Rec.G.992.1) with discrete multi-tone (DMT) and orthogonal frequency division multiplexing (OFDM). In effect, DMT is similar to each sub-band having an independently modulated signal that carries a portion of the total payload. DMT modulation utilizes a Fast Fourier Transform (FFT) algorithm.

Channel equalization methods, the Viterbi algorithm, trellis coding, and DMT modulation are discussed in [4,7].

7.5.3 Coding

There are two aspects to coding: error detection and error correction. To detect errors, a cyclic redundancy block code (CRC) checksum is appended to each transmitted block. CRC codes can detect errors with very high confidence, even if there are multiple bit errors. A received block where the CRC checksum indicates an error is called a *code violation* (CV).

DSL systems have performance monitoring (PM) registers that maintain an error performance history for the line. As specified in ITU-T Rec. G.997.1 (PLOAM), a record of code violations (CV) for each direction of transmission, Error Seconds (ES), Severely Error Seconds (SES, more than 18 CRC errors in a second), and Unavailable Seconds (UAS, ten or more contiguous SES) is maintained for each 15-minute interval for the previous four hours, and also ES, SES, and UAS count are recorded for the current and previous 24-hour period. This data is useful for diagnosing a DSL line with an intermittent problem.

The two strategies for error correction are forward error control (FEC) and automatic repeat request (ARQ, also known as retransmission). ARQ is typically performed by upper layer protocols such as TCP-IP, where the receiver detects if the received packets are correct by examining the CRC checksum appended to each packet. If the receiver detects an error, it sends a NAK message to the transmitter, which then retransmits the packet. ARQ error correction is effective for correcting large bursts of errors if they occur infrequently, but it can be overwhelmed by frequent errors. Also, the delay to retransmit the packet results in additional delay (i.e., jitter) for the corrected packet.

A DSL with FEC sends additional check bits (i.e., redundant bits) with the data that the receiver uses to correct transmission errors. A crude example would be to send each block of data three times, and then the receiver could correct errors by choosing the two identical received copies as the correct data. This majority-vote error correction is used for the low bit rate embedded operations channel (eoc) for some types of DSL, but DSLs use much more powerful error correcting codes for the high bit rate payload data. ADSL and VDSL systems use a Reed Solomon (RS) forward error correcting code.

The RS code takes a block of digital data (i.e., K bytes of data) from the output of the scrambler,[4] and appends some redundancy data (i.e. R bytes of redundancy) to form a codeword of $N = K + R$ bytes for transmission on the line. The RS codeword structure is shown in Figure 7.23. Since the data bytes are enclosed in their original form and the redundancy bytes are appended, this structure is referred to as a systematic (N, K) code. Each byte within a codeword is sometimes referred to as a data symbol, which is distinct from a DMT symbol. When the codewords are transmitted through the channel, the noise in the channel may cause errors to occur in one or more byte locations within the codeword. The RS decoder examines the received data and redundancy in each of the received codewords and tries to optimize the probability in recovering the original data correctly.

---

[4] The scrambler converts long series of all zeros or all ones into a sequence of zeros and ones because a long series of bits with no change of value can cause problems for modulation and timing.

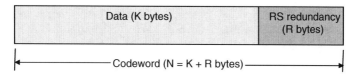

Figure 7.23 RS codeword structure

For ADSL2 and VDSL2, the maximum number of bytes in a RS codeword is 255. The Reed-Solomon decoder can correct up to $t = R/2$ bytes that contain errors within a codeword. In VDSL2 and ADSL2 systems, the valid number of bytes N per codeword ranges from 32 to 255 inclusive, and the valid number of redundancy bytes R are the even integers ranging from 0 to 16 inclusive. Hence, the RS code is configurable to correct up to eight bytes with errors inclusive per codeword. Even if only a few error bits are present in each byte, the maximum number of correctable bytes is the same.

Given the above mentioned error correction capability in each codeword, the RS code is suitable for correcting bursts of errors. The length of correctable error burst may be extended beyond that of each codeword through the use of interleaving. As showed in Figure 7.24, the interleaver shuffles the order to the transmitted bytes, and the de-interleaver un-shuffles the bytes to regain the order of the original data.

The bytes of numerous sequential RS codewords may be interleaved to a specified interleave depth of D codewords prior to being transmitted on the channel. When an impulse noise event hits the line, a sequence of interleaved bytes will be corrupted (i.e., cause a burst of errored bytes). Figure 7.24 shows the reference model for the DSL transmission system showing the interleaver located between the RS encoder output and the DMT transmitter input. The DMT receiver reconstructs the received RS codewords (which may contain errors), the de-interleaving process spreads the location of errored data bytes across the span of the interleave depth D as it reconstructs the original codewords. In effect, the interleaving transforms one long series of errors into many shorter series of errors spread across several codewords. Thus, there are fewer byte errors per codeword which is hopefully within the correction capacity of the RS decoder.

Interleaving has two drawbacks. First, it increases transmission latency. Also, if the amount of errors still exceeds the capacity of the interleaved RS FEC, the interleaving may cause several packets to be corrupted, whereas a non-interleaved operation would have had only a single corrupted packet. Interleaving is a brittle technique; it protects well up to a degree, and then it fails catastrophically.

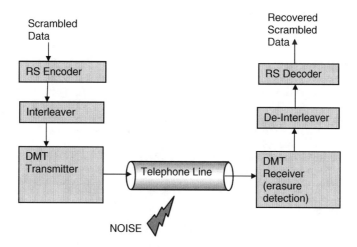

Figure 7.24 DSL system reference model

An RS decoder can correct both errors and erasures. An erasure occurs when the position of an errored byte is known, i.e. the demodulator flags a data byte as being "likely to contain errors". A data byte flagged as being "likely to contain errors" may be referred to as a marked error (or an erasure), whereas an unidentified errored data byte may be simply referred to as an unmarked error. Various proprietary techniques are used to decide which bytes are marked; these techniques may use analog measures of the received DMT symbol such as the deviation of the instantaneous signal-to-noise ratio (SNR) from the average SNR. Also, a Trellis code analysis may be employed in deciding if a DMT symbol may have an error. When a received DMT symbol is suspected of having an error, all the data bytes within the DMT symbol may be marked.

The RS decoder may correct a combination of u unmarked errors and m marked errors (i.e., erasures) if the following relation is held in each codeword:

$$2u + m \leq 2t = R \tag{7.4}$$

Within the bounds of Equation 7.4, the transmitted codeword will always be recovered correctly. Otherwise, the decoder will either detect that it cannot recover the original codeword, or it will incorrectly decode a received codeword without any kind of indication of an error occurring. A receiver with well implemented erasure decoding may be capable of correcting twice as many byte errors as a RS decoder without erasure decoding. If bytes are incorrectly marked or incorrectly not marked the direct result is not an error in the decoded data, but it does reduce the RS correction capacity.

Recently, erasure decoding has been applied to DSL modems. Compared to a conventional RS decoder, there is no additional latency or transmission overhead for erasure decoding. Since erasure decoding is performed entirely in the receiver, without coordination with the transmitter, a new generation receiver may perform erasure decoding while it is receiving a signal sent by a modem that does not implement erasure decoding in its receiver.

Turbo codes and low density parity check (LDPC) codes provide better FEC performance than RS FEC, but they generally require more complex circuitry. In general, FEC techniques are effective in correcting frequent, but small error bursts. A long series of bit errors will exceed the correction capacity of even the best FEC. FEC adds a constant delay regardless of the number of bit errors, and the additional delay is very small if there is no interleaving.

DSLs generally employ FEC at the physical layer and ARQ at the data-link layer. Thus, the small error bursts are corrected at the physical layer FEC, so that relatively few large error bursts remain to be corrected by the ARQ at the data-link layer. The combined result is very effective at correcting errors, but it still has a limit.

Further details on DSL error control are provided in [1,4,7].

References

1. Starr T. Sorbara M. Cioffi JM, Silverman PJ. *DSL Advances*. Prentice Hall: 2003.
2. ATIS Dynamic Spectrum Management (DSM) Technical Report, 0600007. ATIS; 2008.
3. Galli S, Waring DL. Loop makeup identification via single ended testing: beyond mere loop qualification. *IEEE Journal on Selected Areas in Communications*. 2002; **20**(5): 923–935.
4. Starr T, Cioffi JM, Silverman PJ. *Understanding Digital Subscriber Line Technology*. Prentice Hall; 1999.
5. Bingham JAC. *ADSL, VDSL, and Multicarrier Modulation*. John Wiley & Sons; 2000.
6. Chen WY. *DSL Simulation Techniques and Standards*. Macmillan; 1998.
7. Golden P, Dedieu H, Jacobsen K. *Fundamentals of DSL Technology*. Auerbach Publications; 2006.
8. ANSI T1.417-2003. Spectrum Management for Loop Transmission Systems. ANSI; 2003.

Further Readings

1. IETFRFC 1157. A Simple Network Management Protocol (SNMP). IETF; 1990.
2. Oksman V, Galli S. G.hn: The New ITU-T Home Networking Standard. *IEEE Communications Magazine*. 2009; **47**(10): 138–145.

8

The Family of DSL Technologies

8.1 ADSL

ADSL (asymmetric digital subscriber line) and later VDSL (very high bit rate DSL) were designed to transport higher bit rates downstream (towards the customer) than upstream. This was achieved by allocating most of the frequency band for downstream transmission. The transport asymmetry was chosen based on the principal applications of web browsing, email, and video delivery from a server to a customer. These applications require higher downstream bit rates than upstream. At the end of 2012, ADSL was the most popular form of DSL technology, with approximately 365 million lines in service. Figure 8.1 shows a customer-end ADSL modem.

In 1994, the T1E1.4 standards working group in the United States completed the first ADSL standard, and it was published in 1995 as ANSI T1.413. The ADSL standard borrowed technology from voice-band modems (ITU-T Rec. V.34) and basic rate ISDN DSL (ITU-T Rec. G.961), but it contained many new advances, including discrete multi-tone (DMT) modulation. During the development of T1.413, a fierce debate ("the second line code war") took place in T1E1.4. The first line code war took place during the development of the BRI-ISDN standard in the 1980s.

The pros and cons of DMT and single carrier modulation (SCM in the form of QAM and CAP) were discussed for nearly two years. SCM had the advantage of lower circuit complexity and slightly less latency. In theory and simulations, DMT provided higher performance, especially against certain types of impairments such as narrow band noise and bridged taps.

In laboratory tests using the T1E1.4 test plan, DMT modems demonstrated much better performance than the SCM modems. The differences in the laboratory results were largely a reflection of the SCM modems being products optimized for high volume production, whereas the DMT modems were prototypes optimized and hand-tuned for performance. Eventually, the T1E1.4 working group decided that only one modulation method would be chosen for the ADSL standard, and that method was DMT. In hindsight, it seems either SCM or DMT could have served well, but it is clear that it was best for the T1E1.4 standards committee to produce a standard that fostered interoperability between modems of different manufactures.

Astonishingly, six years later VDSL standards followed the same scenario with a lengthy line-code debate between SCM and DMT. Laboratory tests once again had hand-tuned "hot rod" DMT modems demonstrating better performance than mass-produced SCM modems, and DMT was again chosen for the VDSL standard. This facilitated the production of multi-mode ADSL/VDSL modems.

In 1995, field trials of ADSL-based service began. The T1.413 ADSL standard was later followed by an Issue 2 T1.413 standard that included both ATM (asynchronous transfer mode) and STM (synchronous

Broadband Access: Wireline and Wireless – Alternatives for Internet Services, First Edition.
Steven Gorshe, Arvind Raghavan, Thomas Starr, and Stefano Galli.
© 2014 John Wiley & Sons, Ltd. Published 2014 by John Wiley & Sons, Ltd.

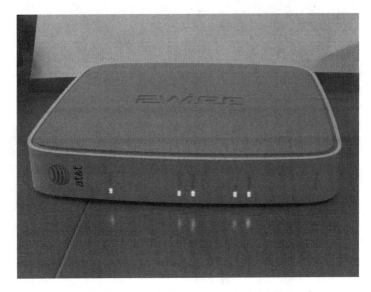

Figure 8.1 Photograph of a customer-end ADSL modem

transfer mode) and reduced overhead framing. The T1.413 standard was then further refined to become the ITU-T Rec. G.992.1. All of these standards are now known as ADSL1.

ADSL1 specifies simultaneous transmission of analog voice below 4 kHz (POTS, or "plain old telephone service"), upstream transmission up to 640 kbits/s in the frequency band between 25.8 kHz and 138 kHz with a peak PSD (power spectral density) of −34.5 dBm/Hz, and downstream transmission up to 7 Mbits/s in the frequency band between 138 kHz and 1.104 MHz with a peak PSD of −36.5 dBm/Hz. DMT modulation is defined using 256 tones of 4.3125 kHz each. The DMT technique loads (modulates) between 0 and 15 bits into each tone depending on each tone's capacity and the total required bit rate. Every DMT tone is sent 4000 times per second as a transmitted symbol. Thus, each symbol's duration is 250 microseconds, and each bit loaded into one DMT tone conveys 4 kbits/s.

The ADSL standards species FEC using trellis coding, Reed Solomon block coding, and interleaving to correct transmission errors. In combination, the coding provides about 7 dB of gain. A CRC block code is specified to detect transmission errors.

To reduce power consumption when little data is being transmitted, an L2 mode is defined with a reduced transmitted power (up to 40 dB less). However, the L2 mode is not used because of the delay to exit the L2 mode when traffic resumes. Also, the frequent changes in transmitted power due to transitions between the L2 (low power) and L0 (full power) modes could cause fluctuating crosstalk that would be disruptive to other lines.

ADSL and VDSL standards specify the ability to greatly reduce the transmitted power in narrow-frequency bands used for amateur radio (HAM) transmission; this is known as RFI notching. Steep PSD notches are possible, reducing the transmitted power in a few DMT tones. A technique called windowing helps to improve the effectiveness of the RFI notching.

8.1.1 G.lite

In 2002, the ITU-T produced the G.992.2 ADSL Recommendation; this is also known as G.lite. Interestingly, G.lite was a commercial failure, but it was largely responsible for ADSLs future mass-market success. The development of G.lite focused on reducing the cost of ADSL-based service by reducing the complexity of ADSL modems and enabling self installation of service by customers.

Before G.lite, ADSL1 service installation required a service provider technician visit to the customer's premises to install a POTS/DSL splitter and often to run a new inside wire to the customer's computer. G. lite introduced the concept of using in-line filters (also known as micro-filters) instead of the centralized POTS/DSL splitter. Both the POTS and DSL signals traversed throughout the premises on the existing LINE 1 inside twisted-pair wire. At every phone (and other POTS devices such as Caller ID devices) the customer inserted a series in-line filter at the wall jack (see Figures 7.9 and 7.10 in Chapter 7). The low-pass characteristic of the in-line filter prevented DSL signals from causing audible noise in the phones, and it also prevented the phone's impedance and noise from impairing operation at the DSL frequency band. The splitterless configuration enabled customers to install ADSL service themselves and thereby to avoid the large cost of a technician visit to the customer's premises to install service.

To reduce the complexity of ADSL modems, G.lite reduced the number of DMT tones to 128, the number of bits-per-tone to 8, and the maximum frequency to 512 kHz. As a result, the maximum downstream bit rate was 1.5 Mbits/s. The hope was that this simplification would enable the implementation of software modems operating in the microprocessor of the customer's computer. However, it was discovered that typical computers did not have sufficient processing power to perform an effective ADSL modem while simultaneously supporting the necessary computer applications. Furthermore, it was found that the potential cost savings of the G.lite modem for hardware-based implementations was trivial compared to 7 Mbits/s capable ADSL1 modems.

The big benefit of G.lite was splitterless self-installation by customers. However, it was found that 7 Mbits/s capable ADSL1 modems could function well in a splitterless configuration with in-line filters. As a result, the industry borrowed the splitterless concept from G.lite and applied it to the higher speed ADSL1 technology.

8.1.2 ADSL2 and ADSL2plus

In 2002, the ITU-T completed Rec. G.992.3 (ADSL2) which provided rate/reach performance improvements over ADSL1. ADSL2 provides 8 Mbits/s downstream and 800 kbits/s upstream, tone reordering and pilot tone repositioning for improved robustness, enhanced line diagnostics, transmit power spectra shaping, and seamless rate adaptation. Also, ADSL2 was the first DSL designed to use ITU-T Rec. G.994.1 (G.hs, "handshake") for negotiation of capabilities between modems during start up.

In 2003, the ITU-T completed Rec. G.992.5 (ADSL2plus), increasing the maximum frequency to 2.208 MHz with 512 DMT tones. ADSL2plus provides up to 24 Mbits/s downstream and up to 1 Mbits/s upstream when the upstream/downstream frequency boundary is set at 138 kHz.

G.992.3 and G.992.5 specify support of up to four frame bearers numbered 0, 1, 2, and 3. A frame bearer (often called a bearer channel) is a data stream that conveys the bits for one instance of a TPS-TC. Each frame bearer can be placed in either the "fast path" or an "interleaved path." The fast path has no interleaving, so its one-way latency is typically 1–2 milliseconds through both ADSL modems. An interleaved path may have interleaving of a depth determined by a MIB parameter. Depending on the interleaver setting, the one-way latency of the interleaved path may be 3 ms to 20 ms.

Annex A specifies ADSL operation above the POTS band, with the lowest used DMT tone starting at 25.8 kHz. Both overlapped (echo canceled) and FDM (non-overlapped) type spectral modes are specified but, in practice, FDM is widely used to minimize NEXT.

Annex B specifies ADSL operation above Basic Rate ISDN (ITU-T Rec. G.961), with the lowest upstream DMT tone starting at 120 kHz.

Annex C specifies ADSL operation optimized for operation in the same cable where other lines are conveying TCM-ISDN (for use in Japan).

Annex E specifies POTS/ADSL splitters. For North America, this has been superseded by ATIS T1. TRQ.10 for the network-end splitter, and ATIS-0600016 for the customer-end splitter.

Annexes F and G for North American and European performance requirements have been superseded by the Broadband Forum TR-100.

Annex I specifies the "all digital mode" where the frequencies below 25 kHz are used for upstream transmission. This permits upstream data rates up to 320 kbits/s greater than the Annex A mode, but no POTS transmission is possible on a line operating in the Annex I mode. Annex I is especially helpful when operating on a long line.

Annex J specifies all digital operation for improved spectral compatibility in cables where other lines are operating in the Annex B mode (ADSL over ISDN).

Annex K specifies operation for STM (synchronous transfer mode) transport.

Annex L specifies operation with transmit spectra optimized for higher performance on lines longer than 4.5 km (15 kft). This is also known as READSL for reach-extended ADSL.

Annex M specifies operation expanding the number of DMT tones used for upstream transmission from 32 (up to 138 kHz, like ADSL1) to as high as 64 tones (276 kHz). This increases the maximum upstream bit rate up to nearly 2 Mbits/s at the cost of reducing the available downstream bandwidth slightly. Special care is needed to avoid excessive crosstalk if ADSL lines with different upstream/downstream frequency boundaries are present in the same cable.

Annex N, added in 2005, specifies a 64/65 octet encapsulation method for packet transport mode (PTM) based on IEEE standard 802.3.

Annex P, added in 2008, specifies ADSL operation with constant reduced transmit power to reduce power consumption. This is useful for shorter lines.

8.1.3   ADSL1 and ADSL2plus Performance

As with all forms of DSL, the performance of ADSL technology varies greatly depending on the length of the line, wire gage, crosstalk noise, and other impairments such as bridged taps. The DSL performance shown in this book is based on commonly used assumptions that approximate slightly conservative, real-world conditions, but they are far from worst-case conditions.

Figures 8.2 and 8.3 show the downstream and upstream net bit rates for ADSL1 and ADSL2plus modems. The measurement conditions include simulated crosstalk from other lines of the same type. As a

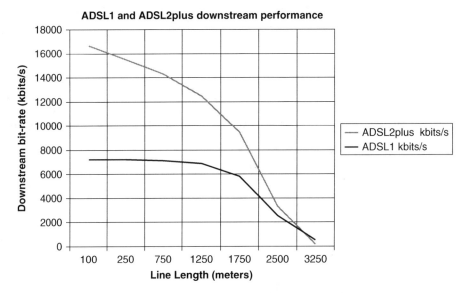

Figure 8.2   ADSL1 and ADSL2plus downstream performance

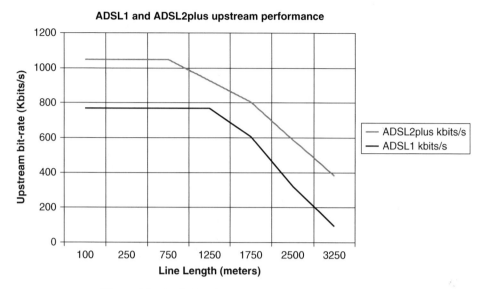

Figure 8.3 ADSL1 and ADSL2plus upstream performance

result, the upstream and downstream bit rates shown in these graphs are less than could be achieved for operation with no noise on the line, but the bit rates shown are more than would be achieved under worst-case conditions such as a line with multiple bridged taps and multiple types of noise.

These graphs show net bit rate which, is sometimes also called payload data rate. The net bit rate is the bit rate available to carry layer-two traffic such as IP packets or ATM cells, including the overhead associated with the layers two and above. The total transmitted bit rate (the "gross" rate) includes physical-layer overhead bits in addition to the net bit rate. ADSL overhead bits include framing, RS FEC bits, and the embedded operations channel (eoc). ADSLs typically operate with 10–25% overhead, depending on the degree of RS FEC redundancy.

ADSL1, with downstream transmission starting at 138 kHz, has up to 224 DMT tones available in the downstream direction. One tone is reserved for use as the timing pilot tone. If all 223 tones were fully loaded with 15 bits each, then the maximum total gross downstream bit rate for ADSL1 would be 13.38 Mbits/s. As discussed in the preceding paragraph, the ADSL performance graphs shown below contain net bit rates, not the gross bit rate. Furthermore, even in ideal conditions, it is never possible to load 15 bits onto every DMT tone. In practice, ADSL1 modems have a maximum downstream net bit rate of approximately 8 Mbits/s. The maximum ADSL1 downstream net bit rate shown in Figure 8.2 is 7.2 Mbits/s, because this was measured with crosstalk noise on the line.

Similarly, ADSL2plus has 479 downstream tones to carry data. If all 479 tones were loaded with 15 bits each, the total gross downstream bit rate would be 28.74 Mbits/s. In practice, ADSL2plus modems have a maximum downstream net bit rate of approximately 24 Mbits/s, and the highest downstream net bit rate shown in Figure 8.2 is 16.6 Mbs/s, because this was measured with crosstalk noise on the line.

8.2 VDSL

Work on very-high-bit rate digital subscriber line (VDSL) started in the T1E1.4 and ETSI TM6 standards groups in 1994. The basic concept was to apply ADSL techniques to a wider frequency band, so that much higher bit rates could be transmitted downstream and upstream. Due to the high line attenuation at the higher frequencies, VDSL is designed for operation on relatively short lines, typically less than 1.5 km

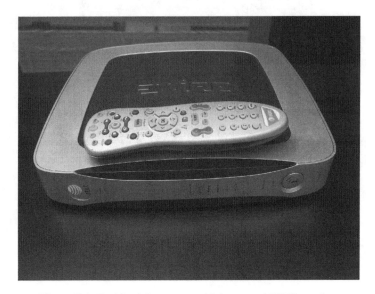

Figure 8.4 Residential gateway with integrated VDSL modem

(5000 feet). Since the length of the twisted pair lines must be short, most VDSLs are served from remote terminals (RTs) which use fiber optic links to the central office (CO). Figure 8.4 shows a residential gateway which includes the customer-end VDSL modem.

Borrowing from ADSL, VDSL specifies asymmetric digital transmission at frequencies above the POTS frequency band to enable simultaneous POTS and VDSL operation on the same pair of wires.

Like the development of the basic rate ISDN and ADSL standards, a lengthy debate took place regarding the choice of modulation method (the "third line code war"). This time, the debate delayed the creation of the standard by three years. Despite the ample experience with ADSL, some people wanted to delay a choice until the results of large scale field deployments were known.

In 2001, the ITU-T produced a generic VDSL foundation specification defining the VDSL frequency bands but little else. In 2002, as an interim step, T1E1.4, and ETSI TM6 produced the VDSL1 standards, containing two separate specifications – one being based on DMT, and the other on SCM.

In 2003, T1E1.4 organized laboratory tests of VDSL modems; this was informally called the "VDSL Olympics". Similar to the ADSL scenario, DMT again demonstrated superior performance in part because the SCM modems were optimized for high volume production, whereas the DMT modems were optimized for performance. As a result, later in 2003, Committee T1 produced a VDSL standard (T1.424) specifying only DMT and, as a political concession, Committee T1 simultaneously produced a separate technical requirements document (not a standard) specifying an SCM version of VDSL.

The VDSL1 standards produced during 2002 and 2003 specified transmission at frequencies up to 12 MHz, supporting downstream bit rates up to 52 Mbits/s and upstream bit rates up to 26 Mbits/s. Two different FDM frequency plans are specified for VDSL1, which are named the 998 and 997 frequency plans. These names come from the number of computer-generated frequency plans studied to meet a set of requirements to deliver asymmetric video services and symmetric data services, as well as to avoid transmission in frequency bands used for amateur radio (HAM bands). The standardized VDSL frequency plans were the 997th and 998th versions of analyzed frequency plans. Both plans specify downstream transmission starting at 138 kHz. There is no downstream transmission below 138 kHz for spectral compatibility with ADSL, and upstream transmission below 138 kHz is usually not implemented in order to simplify the design of the analog front end (AFE) of VDSL1 modems.

The VDSL1 998 frequency plan is used in North America and some other parts of the world. Downstream transmission is located at 138–3.75 MHz and from 5.2–8.5 MHz. Upstream transmission is located at 3.75–5.2 MHz and from 8.5–12 MHz.

The VDSL1 997 frequency plan, used in some parts of Europe, provides slightly more upstream capacity and less downstream capacity. Downstream transmission is located at 138 kHz to 3.25 MHz and from 5.1–8.1 MHz. Upstream transmission is located at 3.25–5.1 MHz and from 8.1–12 MHz.

Both the 998 and 997 VDSL1 frequency plans specify four non-overlapping frequency bands that progress from low frequencies to high frequencies in the pattern: down, up, down, up. This enables VDSL operating on relatively long lines (more than 1 km) to have at least one useful upstream band and one useful downstream band. Also, VDSL operating on intermediate length lines (less than 1 km) will have two useful downstream bands, plus at least one useful upstream band.

A third VDSL1 frequency plan, called Fx, specifies flexible (adjustable) frequency bands, but it has seen virtually no use.

VDSL1-based services were introduced in 1998 with more than 100 000 lines in service by the year 2001. However, the deployment of millions of lines occurred after the ITU-T completed the VDSL2 standard in 2005.

8.2.1   VDSL2

The ITU-T completed the VDSL2 standard (G.993.2) in 2005, and it was published in 2006. This standard specifies only DMT as the modulation technique, and it contains eight profiles to permit lower-cost modems when comparatively lower bit rates are sufficient. The VDSL2 standard borrows much from ADSL2plus, including its framing technique, mandatory trellis coding, PSD shaping, built-in line diagnostics, packet transport mode (PTM) encapsulation, and an all-digital mode. VDSL2 specifies many other advances beyond VDSL1, including:

- transmission at frequencies up to 30 MHz;
- time domain equalizer training and cyclic extension to enable operation on longer lines;
- a mandatory U0 frequency band (25 kHz to 138 kHz) for some profiles;
- mandatory support of modulation constellations from 1–15 bits;
- receiver directed tone ordering;
- expanded RS FEC capability;
- an improved interleaver.

Like the VDSL1 standards, VDSL2 specifies HAM band notching and an upstream power back off (UPBO) function to automatically reduce the upstream transmit power on very short lines. This helps to reduce the effects of far-end crosstalk (FEXT).

In effect, the VDSL2 standard is eight standards in one. There are four profiles with transmission up to 8.5 MHz:

- the 8a profile specifies a maximum aggregate downstream transmit power of +17.5 dBm;
- the 8b profile specifies +20.5 dBm;
- 8c specifies +11.5 dBm;
- 8d specifies +14.5 dBm.

The higher power profiles can operate on somewhat longer lines, whereas the lower power profiles consume less power and, thus, enable more lines to be packaged in a given space. Support of the U0 frequency band (25 kHz to 138 kHz) is required for profiles 8a, 8b, 8c, 8d, and 12a. Table 8.1 shows the characteristic of each of the VDSL2 profiles based on frequency plan 998. The maximum bit rates in this

Table 8.1 VDSL2 Profiles

Profiles	Highest frequency (MHz)	Maximum downstream bit rate (Mbits/s)	Maximum upstream bit rate (Mbits/s)	Longest suitable line length (meters)
8a, 8b, 8c, 8d	8.5	48	10	2300
12a, 12b	12	48	30	1500
17a	17	60	30	1000
30a	30	120	100	400

table are achievable only on very short lines with little noise. The downstream and upstream bit rates for the longest suitable line length are much lower.

The upstream bit rates in Table 8.1 show that, in addition to asymmetric bit rate services, VDSL2 can support high bit rate symmetric data services on relatively short lines.

VDSL2 extends the available frequency range beyond what was used for VDSL1. Frequency plan 998 of VDSL2 specifies:

- U0: 25 kHz to 138 kHz upstream for profiles 8a-d and 12a.
- D1: 138 kHz to 3.75 MHz downstream for all profiles.
- U1: 3.75 MHz to 5.2 MHz upstream for all profiles.
- D2: 5.2 MHz to 8.5 MHz downstream for all profiles.
- U2: 8.5 MHz to 12 MHz upstream for profiles 12a, 17a, and 30a.
- D3: 12 MHz to 23 MHz downstream for profiles 17a (partial) and 30a.
- U3: 23 MHz to 30 MHz upstream for profile 30a.

The VDSL2 standard includes specifications for transport of synchronous transport mode (STM), asynchronous transport mode (ATM), and packet transport mode (PTM, based on IEEE standard 802.3)

8.2.2   VDSL2 Performance

As with the ADSL performance graphs shown in Section 8.1, the following graphs show the net downstream and upstream bit rates for VDSL2 profile 8d, 12a, and 17a systems operating on lines with no bridged taps, but with typical simulated crosstalk. The shortest line length graphed is 150 meters. VDSL2 systems using these profiles would achieve higher bit rates for lines shorter than 150 meters. Higher bit rates would also result if there was less crosstalk from other lines.

The downstream performance shown in Figure 8.5 shows virtually the same performance for profiles 8d and 12a because they use the same downstream frequency bands, but the performance of profile 17a is higher for shorter lines, due to the addition of part of the D3 frequency band.

The upstream performance shown in Figure 8.6 shows virtually the same performance for profiles 12a and 17a, because they use the same upstream frequency bands, but they are both higher than the 8d profile on shorter lines, due to the addition of the U2 frequency band.

For both graphs, the 17a profile line stops at 1500 meters because the lower profiles are better suited for lines longer than 1500 meters.

The performance for VDSL2 profile 30a is provided below in Figure 8.7,[1] with both the downstream (the higher line) and upstream (the lower line) net bit rates on the same scale. This assumes that no upstream power back-off (UPBO) is applied. Once again, typical operating conditions are assumed (crosstalk from 12 VDSL2 profile 30a disturbers), and higher bit rates could be achieved for shorter lines

---

[1] Figure 8.7 was reproduced by permission of Vladimir Oksman of Infineon Technologies.

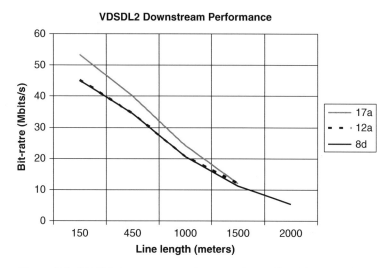

Figure 8.5 VDSL2 downstream performance for profiles 8d, 12a, and 17a

and when operating with less crosstalk. For example, profile 30a could achieve 240 Mbits/s downstream and 150 Mbits/s upstream on a near-zero length line.

Some deployment configurations, such as fiber-to-the-curb (FTTC), virtually eliminate crosstalk and, thus, the achievable bit rates would be substantially higher than shown in these graphs. A technique called vectoring, specified in ITU-T Recommendation G.993.5, cancels nearly all FEXT, approximately doubling the bit rate compared to VDSL2 operating in the presence of typical crosstalk. See Chapter 9 for more discussion of vectoring.

Profile 30a VDSL2 is the highest performance type of DSL, but it is only suitable for lines up to about one kilometer in length.

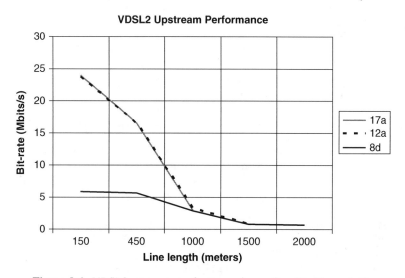

Figure 8.6 VDSL2 upstream performance for profiles 8d, 12a, and 17a

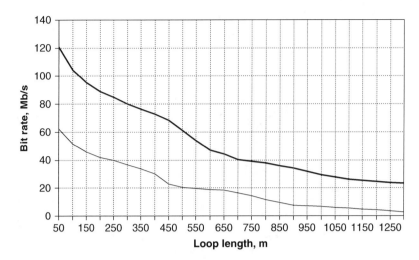

Figure 8.7 VDSL2 downstream and upstream performance for profile 30a. (Source: [Oksman 2009]. Reproduced with permission of Vladimir Oksman, Lantiq)

8.3 Basic Rate Interface ISDN

One could argue that repeatered T1 lines (symmetric 1.544 Mbs), E1 lines (symmetric 2.034 Mbits/s), and digital data service (DDS providing symmetric 56 and 64 kbits/s transport) were the first DSLs when they were introduced in 1962. However, these were primarily used for inter-office trunking and high-priced "special services" for large businesses. The T1, E1, and DDS lines were never called DSLs, and they lacked the critical ability to be provisioned quickly and at a sufficiently low price to support mass-market services.

The first DSL was the basic rate interface integrated services digital network (BRI-ISDN) access line. It was part of the internationally standard ISDN network that began trial service in 1985 to provide packet switched data, circuit switched data, and circuit switched digital voice service through an integrated, end-to-end digital, multi-service network. BRI transmitted a fixed bit rate, symmetric 160 kbits/s via a single telephone line from the customer to the local serving telephone office. Unlike ADSL and VDSL, BRI had no underlying POTS transmission. Instead, 64 kbits/s digitized voice was transported in a B channel.

BRI consisted of two 64 kbits/s circuit switched B channels, one 16 kbits/s packet switched D channel, and 16 kbits/s of overhead for framing, control, and management information. Later, when it became evident that 16 kbits/s was not enough capacity, the B channels were also used for packet switched data and, when combined with the D channel, a total symmetric data rate of 144 kbits/s was supported. BRI operated over lines up to 5.5 km (18 kft) in length without a mid-span repeater, and almost twice this distance with a repeater.

BRI was standardized in the CCITT (now called ITU-T) Recommendation G.961 in 1988 and the ANSI standard T1.601 in 1992. For most of the world, BRI employed the 2B1Q (2 binary bits encoded into 1 quaternary symbol) modulation technique with echo canceled transmission placing the transmitted signals in both directions in the frequency band from nearly zero to 80 kHz, with a PSD of a level of −38 dBm/Hz.

In addition to its use for ISDN, BRI was employed for packet-only services (called IDSL), and for digital transmission of two voice circuits over one pair of wires (DAML, digital added main line). Altogether, more than ten million BRIs were placed into service. Many lessons were learned from ISDN, and this led to subsequent success of high speed internet access using ADSL and, later, VDSL.

Section 2.3 of [1] provides more information about BRI-ISDN.

8.4 HDSL, HDSL2, and HDLS4

High bit rate digital subscriber line (HDSL) used much the same technology as basic rate interface ISDN (BRI) at a higher bit rate to support 1.544 Mbits/s and 2.048 Mbits/s fixed bit rate, symmetric transmission. The first HDSL service was provided in 1992. ANSI published the HDSL specification as TR28 in 1994, and the ITU-T produced the HDSL1 Recommendation (G.991.1) in 1998. Like BRI, HDSL used echo-canceled hybrid, 2B1Q modulation. The main lobe of the transmitted signals extended up to 196 kHz.

In North America, 1.544 Mb/s (DS1 rate) transport was widely used for trunks between central offices and for private-line services to businesses. HDSL transmitted a symmetric bit rate of 784 kbits/s via one pair of wires up to 3.6 km (12 kft), and two of these pairs were inverse multiplexed (today, we say "bonded") to carry the 1.544 Mbits/s payload plus overhead. The overhead consists of framing and the embedded operations channel (eoc).

In many other countries, the prevailing bit rate for high-speed transport was 2.048 Mbits/s (E1 rate), and this was conveyed by HDSL using three pairs of wires, each carrying one-third of the payload.

The deployed cost of a typical DS1-rate line using HDSL is about one-third the cost of the older repeated AMI-T1 technology. This is due to reduced provisioning labor for special circuit engineering, reduced occurrence of splicing repeaters into the line, and the elimination of most of the cost for repeaters and their waterproof housing. The large majority of HDSL lines have no mid-span repeaters, but a few HDSL lines have one or two repeaters to double or triple the reach.

In addition to being lower in cost to install, HDSL lines cost less to maintain and can be provisioned much more quickly than AMI-T1 lines. An additional benefit of HDSL in comparison to AMI-T1 is its greatly reduced crosstalk into other lines, which was later reduced further by the HDSL2 and HDSL4 versions of the technology.

In 2000, ANSI published the T1.418 HDSL2 standard, and this was rapidly adopted by the industry to further reduce the cost of DS1 bit rate transmission. Like the original HDSL, HDSL2 provides 1.544 Mbits/s fixed bit rate, symmetric transmission over lines up to 3.6 km. However, HDSL2 uses only one pair of wires, whereas the original HDSL technology used two pairs of wires. The doubling in transmission efficiency was accomplished by a more advanced transmission technique: trellis-coded pulse amplitude modulation (TC-PAM) with partially overlapped upstream and downstream transmission frequency bands. Upstream signals are transmitted from zero to 280 kHz, whereas downstream transmission is from zero to 400 kHz. The transmitted power for HDSL2 is 3 dB more than HDSL, but there are half as many transmitters, and also HDSL2 has much less out-of-band energy than HDSL.

In 2001, ANSI published the revised T1.418 standard to include the specification for HDSL4. HDSL4 uses the same TC-PAM technique as HDSL2, but it increased the maximum non-repeatered line length to 4.6 km. The 28% increase in line length was achieved using two pairs of wire, like the original HDSL technology. Of all the types of HDSL technology, HDSL4 produces the least crosstalk interference into other services (most notably ADSL), because it transmits at lower frequencies.

Several million HDSL, HDSL2, and HDSL4 lines are in service around the world. They are all optimized for operation at a fixed bit rate (either DS1 or E1 rate). Today, HDSL2 is deployed for DS1 transport for lines shorter than 3.6 km, and HDSL4 is used for lines longer than 3.6 km. HDSL4 can reach up to 9 km with a single mid-span repeater.

8.5 SHDSL

The SHDSL standard is officially called "single-pair high-speed digital subscriber line", but the technology is often thought of as "symmetric high bit rate DSL". The TC-PAM transmission technique used for HDSL2 is also used for SHDSL. Symmetric bit rates from 192 kbits/s to 5.7 Mbits/s are transmitted via one pair of wires, and higher bit rates can be achieved by bonding (i.e. combining) multiple pairs of wires to serve a customer. SHDSL is the technology of choice for symmetric bit rate

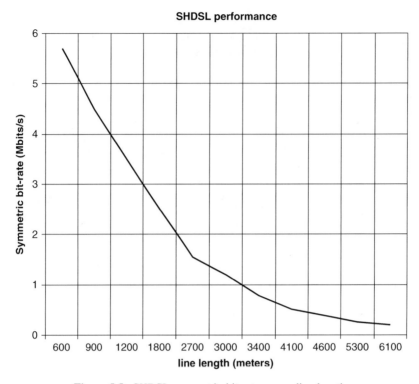

Figure 8.8 SHDSL symmetric bit rate versus line length

services, especially for lines with a length of one kilometer and longer, which corresponds to bit rates below 4 Mbits/s. VDSL2 can achieve higher symmetric bit rates for lines shorter than 1 km. Figure 8.8 shows the bit rate performance of SHDSL operating in the presence of crosstalk from other SHDSL lines. The SHDSL bit rates are the same for upstream and downstream.

In 2001, the ITU-T published the SDHSL standard as Recommendation G.991.2 and, later, Committee T1 published T1.422 for SHDSL in North America and ETSI published TS101524 (where it is called SDSL) for Europe.

SHDSL provides for only digital transmission with no POTS on the same line. Since POTS is not transmitted on the same line, the frequency band used for digital transmission extends to nearly zero Hertz; this aids performance on longer lines. For symmetric PSDs, the approximate frequency bands for transmission in both directions are as shown in Table 8.2.

Table 8.2 SHDSL frequency bands

SHDSL bit rate	Maximum transmitted frequency
192 kbits/s	70 kHz
1 Mbits/s	320 kHz
1.5 Mbits/s	470 kHz
2.3 Mbits/s	720 kHz
5.7 Mbits/s	1.3 MHz

G.992.1 includes specifications for bonding (combined transport) up to four pairs of wires to achieve a higher aggregate bit rate. See Section 3.7.4 about G.bond, which provides specifications for bonding up to 32 pairs of wires.

8.6   G.fast (FTTC DSL)

Recently, a new generation of DSL technology has been defined to transmit data between a customer premises and a network node up to 250 meters from the home. In many cases, the transceivers at the network end of the line would be contained within a wiring pedestal serving 16 or fewer lines. This new technology goes by many names, including fiber to the curb (FTTC), fiber to the distribution point (FTTdp), G.9701 and G.fast. The concept of G.fast is to provide the best of FTTH (fiber to the home) and ADSL, namely FTTH bit rates and customer self-installation like ADSL. It is estimated that G.fast will provide the following aggregate (downstream plus upstream) bit rates (without bridged taps):

- Up to 1 Gb/s at 200 feet.
- 500 Mb/s at 330 feet.
- 200 Mb/s at 660 feet.
- 150 Mb/s at 800 feet.

G.fast delivers bit rates comparable to what is practical with fiber or coax to the home. To facilitate simple installation of customer premises equipment, the transceiver at the customer end of the line would typically be located within the customer's residential gateway (RG) equipment. The G.fast signal would flow to the RG via the existing inside telephone wire and the existing outside "drop" telephone wire; thus, no new wire needs to be installed. Unlike VDSL2, which works marginally for customer self-installation, G.fast is specifically designed to work robustly with the challenging impedance and noise found in existing telephone wire inside the home.

To reduce the cost of powering the equipment in the pedestal at the distribution point, it has been proposed that pedestal equipment would be powered via DC current sourced by the RG (the residential gateway inside the customer premises) and fed via the drop wire from the RG to the pedestal. This "back fed" power would operate the G.fast transceiver and the common equipment in the pedestal. The RG could have a back-up battery to preserve service during a commercial power failure. It is anticipated that the transmitted signal would reside at frequencies from about 2 MHz to 105 MHz, but the upper and lower frequency bounds are configurable via MIB parameters. To avoid interference to and from FM radio transmissions (87.5 to 108 MHz in the USA), the G.fast signal may be configured to stop at 87 MHz. To avoid interference with legacy VDSL2 service in the same cable, the G.fast signal can be configured to start above 8.5 MHz.

The benefits of G.fast served from a small node at the distribution point include:

- support of hundreds of megabits per customer
- enabling simple customer self-installation of equipment inside the house, resulting in lower installation cost than FTTH
- elimination of the cost to feed network-based power to the distribution point serving node
- greater flexibility for adjusting the ratio of downstream/upstream bit rate (asymmetery), including operation with the same bit rate in both directions (symmetric bit rate)
- the distribution point serving node fits in a very small enclosure which can placed in a pedestal case, underground, on a pole, or attached to aerial cable. This avoids the difficulty of finding an acceptable place to put the much larger cabinet required for an ADSL or VDSL DSLAM

Legacy services such as POTS, ADSL, and VDSL could be provided to customers not taking the new G.fast-based services by having a metallic relay for each drop line in the distribution point serving node.

The relay could connect the drop wire to a wire pair in the distribution cable leading to legacy network serving equipment such as a DSLAM or voice switch. Of course, this would work only where the copper distribution cable and legacy network equipment is available.

The following aspects were decided for the draft G.fast standard:

- Time division duplexing (unlike ADSL and VDSL which use frequency division).
- DMT modulation using up to 2048 subcarriers (each 51.75 kHz) for a total bandwidth of 105 MHz (a later version of G.fast may go above 200 MHz).
- DMT symbol duration of 20 microseconds.
- 12 bits per DMT tone.
- maximum transmit signal power upstream and downstream: 4 dBm.
- the cyclic prefix may be between 0.6 and 4.98 microseconds.
- the windowing parameter (beta) is 64 or 128 samples.
- the time gap between upstream and downstream transmission is one DMT symbol.
- the TDD frames has 36 DMT symbols and a superframe has 8 TDD frames.
- each upstream and downstream superframe has one synch symbol.
- the degree of bit rate asymmetry is controlled by a MIB parameter which adjusts the amount of time allocated for upstream and downstream transmission.
- error control by trellis coding and Reed Solomon block coding plus physical layer retransmission based on ITU-T Recommendation G.998.4.
- startup negotiation using ITU-T Recommendation G.994.1 (G.hs).
- vectoring to cancel FEXT will be specified for G.fast.

The ITU-T expects to approve the G.fast standard early in 2014 as ITU-T G.9701 and G.fast chips are expected to be available by mid-2014.

Reference

1. Starr T, Cioffi JM, Silverman PJ. *Understanding Digital Subscriber Line Technology*. Prentice Hall; 1999.

9

Advanced DSL Techniques and Home Networking

9.1 Repeaters and Bonding

A repeater is a transmission device placed near the mid-point of a telephone line that receives an attenuated and distorted signal and then transmits a pristine digital copy at full power over the remaining portion of the line. A repeater placed at the mid-point of the line cuts the effective length of the line in half, thereby enabling a given bit rate to be sent approximately twice the distance. Two or more repeaters may be inserted into the line at approximately equidistant positions, thereby increasing the attainable line reach by a factor of three or more. For a fixed distance between the network and customer, repeaters can also be used to provide a higher data rate.

A repeater is essentially a pair of back-to-back receivers and transmitters. For each direction of transmission, the signal is received, the modulated symbols are detected, and then a new, full-power signal is transmitted to the remaining section of the line. A regenerator is similar to a repeater, but it does not detect the modulated symbol. A regenerator performs analog amplification and, possibly, filtering. It is a simpler device, but it provides less performance because the noise is amplified as well as the signal.

Repeaters are similar to bonding, because both techniques may be used to increase the total line length for a given service bit rate or to increase the service bit rate for a given line length. The price paid for bonding is the use of more wire pairs and more transceivers. The price paid for the use of repeaters is the addition of the repeaters (including the installation labor) and the equipment to power them.

Repeaters were widely used from 1960 to the mid-1990s for T1 and E1 carrier systems where the typical spacing between repeaters was slightly less than one kilometer. To reach the necessary distance, some T1-carrier lines had more than ten repeaters.

The use of repeaters diminished greatly as modern DSL and fiber technology became prevalent. The drawbacks of repeaters include poor reliability, added signal latency, and high cost. The cost of the repeater units is significant, but most of the total cost is due to labor, packaging, and power. Repeaters must be packaged in expensive waterproof housings that are spliced into the cable pairs by network technicians. Sometimes there may be no spare room in a manhole vault, and then an additional vault must be built. Repeaters are usually powered by a power supply at a central office that applies up to 190 volts DC to the wire pairs connected to the repeater. Another serious drawback of repeaters is the large amount of crosstalk they produce into other lines in the cable. As discussed in Chapter 7 (Section 7.4.7.1 Midpoint

Broadband Access: Wireline and Wireless – Alternatives for Internet Services, First Edition.
Steven Gorshe, Arvind Raghavan, Thomas Starr, and Stefano Galli.
© 2014 John Wiley & Sons, Ltd. Published 2014 by John Wiley & Sons, Ltd.

Crosstalk), crosstalk produced by a transmitter located near the mid-point of a line severely interferes with the transmission on other DSLs in the same cable.

While the use of repeaters has become relatively rare, the use of multi-pair bonding is growing, because bonding avoids most of the drawbacks of repeaters.

9.2   Dynamic Spectrum Management (DSM)

Section 10.1 discusses the ANSI T1.417 for spectrum management which specifies a static set of limits on the transmitted signal power and PSD (power spectral density) in order to limit the adverse effect of crosstalk into other DSLs. In 2012, ATIS published dynamic spectrum management (DSM) technical report issue 2 (ATIS-0900007). This report was created by ATIS COAST-NAI. Compared to the methods of spectrum management specified in T1.417, DSM uses more sophisticated methods to enable higher bit rates and longer line reach. Also, DSM techniques can help to reduce bit-error rates and improve service reliability.

DSM introduces the concept of a spectrum management center (SMC). The SMC is a central management system that has knowledge of most or all of the lines in the cable, and it is aware of the service attributes (e.g., minimum and maximum service bit rate) for each line. The SMC can maintain a long term history of the operating conditions for each line. Based on this knowledge, it can adjust parameters such as line bit rate, error correction coding for impulse noise protection (INP), transmitted signal power, and PSD independently for each line.

The ATIS DSM technical report defines four levels of DSM:

- Level 0 applies fixed limits to transmitted power and PSD, as specified in T1.417.
- Level 1 DSM minimizes excess margin, optimizes the transmission bit rate and optimizes the error correction coding. The level 1 DSM functions may be performed autonomously by each DSL modem, or the level 1 DSM functions may be controlled by a centralized management system on an individual line basis.
- Level 2 DSM includes level 1 techniques, and it also includes joint optimization controlled by a spectrum management center (SMC). The interaction between the transmitted signals on one line and the resulting affects to the reception on other lines is managed. Trade-offs are made between lines to benefit the lines with the greatest need for performance gain. The power spectra for shorter lines may be placed at higher frequencies in order to provide longer lines with less crosstalk at the lower frequencies.
- Level 3 DSM includes level 2 techniques and also vectoring. Vectoring is described in Section 9.3, and it requires a large amount of signal processing in the DSL modems.

For DSM levels 0 and 1, every line operates by itself with no knowledge or active coordination of what is happening on other lines in the cable. For DSM level 2, the general transmission parameters such as PSD are jointly coordinated for most or all of the lines in the cable. For DSM level 3, the degree of joint coordination is taken further by adjusting the transmitted and received signal on a symbol-by-symbol basis. The degree of teamwork increases with the DSM level.

See [1,2] for more information about DSM.

9.3   Vectored Transmission

Vectoring is a technique to reduce the effects of the crosstalk between DSLs operating in the same cable. It is sometimes called MIMO (multiple in multiple out) or coordinated transmission. When DSLs are operating on multiple pairs of wires in a cable, the pair-to-pair coupling of crosstalk results in each receiver having multiple inputs because it is receiving crosstalk signals from many other wire pairs. Also, each transmitter has multiple outputs since the transmitted signal crosstalks into many other DSLs.[1]

---

[1] The classical definition of MIMO is a physical system, such as a cable, with multiple input signals that produces multiple output signals.

DSL vectoring may consist of receiver-based noise cancellation techniques and transmitter-based precoding techniques. The receiver-based techniques attempt to discern the portion of the received signal that is noise by measuring the correlation of the received signal with other signal sources. This may consist of observing the correlation of the received signal with reference such as:

1. DSL signals received by other pairs of wires terminating on the same piece of equipment;
2. signals detected on a spare wire pair (not carrying any transmissions) in the same cable;
3. the signal detected on an antenna; or
4. the common-mode (longitudinal) signal on the same pair of wires.

The correlated signal is equalized by adjusting its phase and amplitude at each frequency. The equalized, correlated signal is subtracted from the received signal to cancel the received noise. Receiver-based noise cancellation provides noise cancellation to the extent that the noise is correlated with the reference. Provided that the reference contains the noise, receiver-based noise cancellation may be effective against crosstalk from DSLs in the same equipment (in-domain crosstalk), DSLs in other equipment (foreign crosstalk), or external noise such as RFI.

Transmitter-based vectoring uses a technique called precoding to alter the transmitted symbols in such a manner as to minimize the adverse affects of the far-end crosstalk (FEXT) into the victim line from other DSLs operating in the same cable. This is discussed in more depth later.

Vectoring depends on symbol-by-symbol knowledge of what is being sent on the other lines in the cable. As shown in Figure 9.1, the best case for vectoring is one-to-one where all the lines with broadband transmission in a cable terminate in the same vectoring group at both ends of the cable. A vectoring group is a set of DSL modems that jointly perform vectoring and, thus, are aware of each other's transmitted symbols. One-to-one vectoring applies for a trunking-type application where all the broadband pairs in a cable terminate on a "book-ended" pair of multi-port transmission units. For the case of one-to-one vectoring, both receiver-based and transmitter-based techniques may be employed at both ends of the line. In ideal cases, the performance of a one-to-one vectored system can be more than twice the performance of a non-vectored system.

Note that it does not matter if the other lines in the cable carry narrow-band signals such as POTS, but the vectoring performance gain may be reduced if the other lines have out-of-domain broadband signals.

A more common application of vectoring is one-to-many (see Figure 9.1) where a DSLAM serves single lines to many distributed modems at the customer end of the line. As a result, no coordination is possible between the customer-end modems. For this example, it is assumed that all of the ports in the DSLAM are within one vectoring group. For the one-to-many case, the downstream transmission may benefit from both DSLAM transmitter-based vectoring and CPE receiver-based vectoring (using signals from a spare pair, an antenna, or the common mode sensing). However, only DSLAM receiver-based vectoring is possible in the upstream direction because the CPE modems are not aware of each other's transmitted symbols.

The third case shown in Figure 9.1 is many-to-many. In this case, the DSLAM contains more than one independent vectoring group, or more than one DSLAM serves the lines in the cable. The customer-end modems are all independent of each other. Only receiver-based vectoring is effective for both upstream and downstream transmission. It might appear that transmitter-based vectoring might be useful in the downstream direction to reduce the crosstalk between the lines within a vectored group, but very little reliable performance gain would be achieved, because the performance gain would be limited by the extent of out-of-domain crosstalk from lines belonging to the other vectored group. Transmitter-based vectoring is effective only if no out-of-domain noise will be received, and this would not be likely if the lines in each vectored group are not carefully chosen to avoid out-of-domain coupling.

For transmitter-based vectoring, all of the lines in a vectored group transmit signals where the symbol and frame timing are synchronized. The frame contains SYNC symbols that carry a predetermined

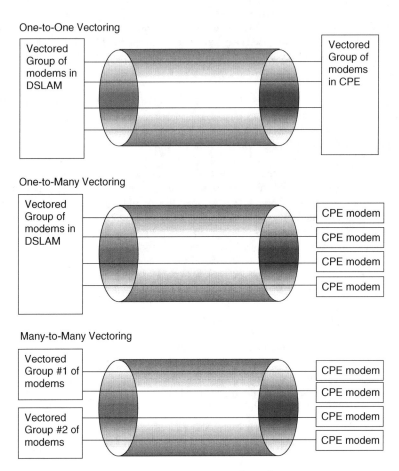

Figure 9.1 Vectoring configurations

pattern. The pattern in the SYNC symbols is orthogonal (i.e. different) for each line. Since the SYNC symbols carry a known pattern, the receiver can measure the error of each received SYNC symbol accurately. These received error samples are conveyed to the other end of the line via the vectoring back channel. Knowing the unique SYNC pattern sent by every line in the vectored group and the resulting far-end received error sample, the vectored transmitter can learn the far-end crosstalk (FEXT) coupling characteristics between its line and every other line in the vectored group. The matrix of FEXT coupling between all lines in a vectored group is updated continually in order to track changes in line conditions.

The degree of FEXT coupling is much greater between some pairs of lines than others. To reduce processing complexity, the transmitter-based vectoring function for a line performs FEXT compensation for crosstalk from the lines with the greatest FEXT coupling. With knowledge of the exact symbols that will be sent on the lines with the greatest FEXT into a victim line and the corresponding line-to-line coupling characteristics, the symbol to be transmitted is precoded (modified) to counteract the calculated FEXT interference at the victim receiver. For example, if the calculated impact of the composite FEXT from the worst disturbing lines would cause the next symbol on the victim line to be moved to the right in the received constellation-space, then the precoded victim symbol will be adjusted to the left of the constellation-space before it is transmitted, so that the symbol will be correctly received after the affects of

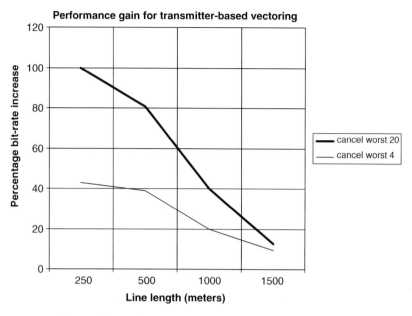

Figure 9.2   Performance gain for transmitter-based vectoring

FEXT noise. The effects of FEXT are cancelled by the transmitter sending a compensated symbol, and the nature of the compensation is recalculated for every symbol.

The performance gain provided by vectoring depends on the system correctly identifying the lines in the cable that will produce the greatest FEXT into the victim line. To do this, all the lines with potential crosstalk must be analyzed and then ranked in the order of the FEXT coupling. Due to the relative physical location of the wire pairs in the cable, a small number of the lines will cause most of the crosstalk into a given pair. Once the system identifies the worst crosstalkers, cancelling the crosstalk from as few as the six worst disturbers will provide a large performance gain. However, to get the best results, implementations of vectoring usually cancel crosstalk from all or nearly all of the disturbers. To put this a different way, the FEXT from only a few lines needs to be cancelled, but it is essential that these are the lines producing the greatest crosstalk into the victim line. Since FEXT has the greatest impact on shorter lines, transmitter-based precoding provides more benefit for shorter lines than longer lines.

Figure 9.2 shows the performance gain for transmitter-based cancellation of FEXT from the worst four and the worst 20 disturbers in a cable filled with VDSL2 lines with typical line conditions. For a line 250 meters long, canceling nearly all the FEXT disturbers would double the bit rate (100% gain) compared to a line with no vectoring, whereas cancelling the FEXT from only the four worst FEXT disturbers would provide a 43% bit rate gain. At 1500 meters, the performance gain from FEXT cancellation shrinks to about 10%.

The degree of performance gain from vectoring depends on the amount of noise in the cable. A large performance gain can be achieved in a cable with a high level of crosstalk. Very little vectoring gain would result in a cable with little crosstalk, but DSLs in the cable would be performing well anyway. Vectoring helps the most where the help is most needed.

For some implementations, it may not be practical to assure that all the broadband lines in a cable are connected to the modems in one vectored group. DSLAMs with more than 48 ports usually have the modems packaged on multiple line cards, and it is more complicated to provide the necessary full interconnection of symbol-by-symbol vectoring information between all of the modems in the DSLAM.

Some DSLAMs may not have adequate backplane bandwidth or vectoring engine capacity to cancel crosstalk across all the line cards in the DSLAM, resulting in each line card being a separate vectored group. If more than one DSLAM serves the lines in a cable, it would not be possible for the modems in the different DSLAMs to be in the same vectored group unless the DSLAMs had a special high-bandwidth interconnection for the vectoring information.

As of 2013, DSLAMs can support vector processing for up to 384 lines. Many service providers in Europe and North America conducted vectoring lab and field trials in 2012 and planned mass deployment in 2014 of vectored VDSL2 services with downstream bit rates up to 100 Mb/s and upstream speeds up to 10 Mb/s and, possibly, higher with bonded and vectored lines.

Telephone network operations and management also place practical limitations on transmitter-based vectoring. The operational support systems (OSSs) that perform the assignment of lines to equipment ports are, typically, not programmed to enforce the assignment of lines in specific binder groups or cables to specific line cards on a DSLAM. A next-generation line-assignment OSS could be developed to enforce the correct mapping of line card ports to cable wire pairs, but OSS development is prohibitively expensive and takes a long time. Furthermore, this process would be vulnerable to mistakes by installation and repair technicians.

A new DSLAM may be equipped initially with only one line card, and customers from all binder groups and cables would be served from this card. Later, another line card could be provisioned when more ports are needed. Again, new customer lines from all binder groups and cables will be served from the second card, and so on for additional line cards. As a result, each line card is likely to serve lines in different binder groups and cables.

Even if a careful, manual assignment of lines to line-card ports were to be performed when the DSLAM was initially installed, the service churn of customers discontinuing their service and others adding service would require an on-going, labor-intensive manual line assignment by highly skilled network technicians. This process is called "binder group management", and the resulting special service labor cost is prohibitive.

An automated metallic cross-connect switching matrix could be placed between the DSLAM modems and the cable to connect the pairs from a binder group automatically, so that these lines are always mapped to the same line card. Unfortunately, large metallic cross-connects with good broadband characteristics are so expensive that this approach is considered unattractive.

Since transmitter-based FEXT cancellation provides the greatest performance gain for lines shorter than 1 km, the simplest vectored solution is serving all of the broadband lines in a distribution cable from one DSLAM where all of the VDSL lines are connected to the same vectoring engine. In some cases, however, it may be necessary for more than one DSLAM to serve lines in the same cable. This may occur when a service provider adds a second DSLAM in a distribution area for more capacity, or if the national regulatory rules provide for multiple service providers to connect DSLAMs to the same cable. In these cases, the crosstalk between lines from different DSLAMs will not be cancelled, and the vectoring gains will be largely lost even if the number of uncoordinated lines is small. There have been proposals to develop vector processing that spans multiple DSLAMs. Cross-DSLAM vectoring is theoretically possible if the DSLAMs are located very close to each other, but it is not clear when cross-DSLAM vectoring might become practical.

There are methods that mitigate much of the effect of out-of-domain crosstalk between DSLAMs. Firstly, receiver-based techniques that look for correlated noise across multiple lines or common-mode signals can cancel a portion of the out-of-domain crosstalk and also foreign noise, such as AM radio ingress.

A more effective technique to mitigate cross-DSLAM crosstalk is to apply dynamic spectrum management (DSM) to shape the transmitted power spectral density (PSD) from the multiple DSLAMs so as to reduce the spectral overlap between the different DSLAMs. For example, in a case where one vectored DSLAM feeds lines into the same cable as a separate non-vectored VDSL DSLAM, the vectored DSLAM could place most of its energy at higher frequencies and the non-vectored DSLAM could place most of its energy at lower frequencies (where the crosstalk coupling is less).

With careful joint PSD optimization, it is possible for services from two DSLAMs without cross-DSLAM vectoring to co-exist with relatively small reduction in bit rate capacity (on the order of 10% reduction compared to vectoring without foreign noise). The ATIS-0900007 DSM Technical Report Issue 2 describes in detail the DSM techniques to mitigate uncancelled crosstalk with examples of the resulting bit-rates with and without DSM mitigation. Also, the Broadband Forum marketing document MD-257-issue2 provides an overview of vectoring and the impact of crosstalk from uncoordinated DSLAMs.

ITU-T Recommendation G.993.5 (G.vector) specifies the application of vectored transmission to VDSL2. See [1] for more information about vectoring.

9.4 Home Networking

Conveying data to the customer's home or business is of little value unless the data travels the last few meters to reach the computers, phones, televisions, and other networked devices within the premises. There are many standard and non-standard technologies to connect the customer premises equipment (CPE) to the DSL modem or residential gateway (RG). These include:

- Wireless WiFi LAN based on IEEE 802.11.
- Cat 5 and Cat 6 wired Ethernet LAN based on IEEE 802.3.
- Short wired cables with universal serial bus (USB) and FireWire (IEEE 1394).
- Inside telephone wire using HPNA^{TM} technology (per ITU-T Rec. G.9954).
- Inside coax wire using MoCA$^{®}$ or HPNA^{TM} technology.
- Inside power wire using Homeplug$^{®}$, universal powerline association (UPA), HD-PLC, and several other lower speed technologies. IEEE P1901 is working on powerline carrier technology standard.

The ITU-T G.hn Recommendations (G.9960 and G.9961) specify the physical layer and media access layer specifications for home networking via all forms of inside wiring (coax, powerline, telephone, and Cat 5). By facilitating home networking equipment that supports more than one type of wiring, connectivity to more places in homes will be possible without installing new inside wire. ITU-T Recommendation G.9963 specifies the multiple-in multiple-out (MIMO) enhancement for G.hn on inside power wires, which greatly improves the bit rate and home coverage. See section 13.4.1.3, [3] and www.homegridforum.org for more information about G.hn.

Femtocell wireless technology is another way to extend the benefits of broadband access throughout the home or business. Using licensed radio spectrum, the femtocell access point creates a low-power cellular telephone access point in the immediate vicinity of a home to provide voice, data, and video service to all types of cellular telephone devices. The femtocell access point uses broadband access such as DSL to link to the core network. Compared to traditional wide-area cellular service, a femtocell provides the benefits of greater service capacity and excellent radio signal quality. It is like living next to a traditional cellular base station and being the only customer within reach of the base station. More information about femtocells may be found at www.femtoforum.org.

References

1. Starr T, Sorbara M, Cioffi JM, Silverman PJ. *DSL Advances*. Prentice Hall; 2003.
2. ATIS Dynamic Spectrum Management (DSM) Technical Report, 0600007. ATIS; 2008.
3. Oksman V, Galli S. G.hn: The New ITU-T Home Networking Standard. *IEEE Communications Magazine*. 2009; **47**(10): 138–145.

Further Readings

1. Starr T, Cioffi JM, Silverman PJ. *Understanding Digital Subscriber Line Technology*. Prentice Hall; 1999.
2. Golden P, Dedieu H, Jacobsen K. *Fundamentals of DSL Technology*. Auerbach Publications; 2006.

10

DSL Standards

10.1 Spectrum Management – ANSI T1.417

ANSI standard T1.417 (now known as ATIS-0600417.2003) was originally published in 2001, specifying spectrum management guidelines for DSL transmission on telephone lines. This standard was developed by standards working group T1E1.4. Since that time, several national bodies have also established limits for the spectrum and power of transmitted DSL signals.

The signals transmitted by a DSL modem on one pair of wires can interfere with the transmission of another DSL using a different pair of wires in the same cable due to the crosstalk between the wire pairs. For the discussion of spectrum management, the signals from the *disturber* DSL modem are coupled from the *disturber* wire pair to the *victim* wire pair and then interfere with the reception at the *victim* DSL modem. As discussed in Section 7.4.7.1, the degree of crosstalk increases as the transmitted power and frequency of the disturbing transmitted increases. As is also described in that section, crosstalk is a two-way process, so each DSL modem is simultaneously a disturber and a victim with respect to the other line.

The purpose of the T1.417 DSL spectrum management standard is to set a limit on the interference produced by DSL modem crosstalk. T1.417 accomplishes this by limiting the allowable transmitted signal amplitude (in dBm/Hz) and maximum frequency as a function of the length of the line. T1.417 permits transmission at higher frequencies on shorter lines because the victim receivers are assumed to also be operating on shorter lines which have less attenuation. Thus, the victim receiver should be able to tolerate more interfering crosstalk noise than a victim DSL operating on a longer line.

Defining spectrum management limits for transmitted DSL power and maximum frequency is a tradeoff between the degree of protection for the potential victim DSLs versus the degree of performance restriction for the disturbing DSL. Lengthy debates were conducted in T1E1.4 about where to set the limits, but a limit had to be set to avoid an uncontrolled escalation in DSL crosstalk. In an environment of competitive DSL modem vendors and competitive service providers using unbundled lines, there was a tendency for DSL disturbers selfishly to transmit highly interfering signals to gain an increase in performance. Like people at a crowded party, there is a tendency to shout louder to overcome the noise but, if everyone shouts louder, then no one benefits. The development of T1.417 was motivated by the FCC mandated unbundling of telephone lines for competitive local exchange carriers (CLECs) to offer services from CLEC provided equipment. Coordinated engineering of all the transmission systems in cable is not feasible when the equipment is provided by multiple service providers.

Much of the debate during the development of T1.417 centered on the degree of protection for ADSL (used mostly by ILECs) versus the degree of performance restriction for SDSL (a 2B1Q symmetric DSL

Broadband Access: Wireline and Wireless – Alternatives for Internet Services, First Edition.
Steven Gorshe, Arvind Raghavan, Thomas Starr and Stefano Galli
© 2014 John Wiley & Sons, Ltd. Published 2014 by John Wiley & Sons, Ltd.

used mostly by CLECs). The FDM nature of ADSL permits it to achieve very high performance in a cable with only POTS and ADSL, but ADSL's performance is severely impacted by crosstalk for SDSL and other non-FDM type systems.

T1.417 primarily addresses spectrum management for DSL systems served from the same serving location (usually a central office), and it provides limited guidance regarding crosstalk between DSL systems originating at different locations (CO- and RT-fed DSL in the same cable).

The T1.417 limits on transmitted DSL signals were derived from limiting the performance degradation of a defined set of "basis system" types. The basis systems include analog voice service (POTS and an enhanced business service called P-phone), DDS per T1.410, BRI-ISDN per T1.601, HDSL per ITU-T G.991.1, various forms of ADSL, HDSL2 per T1.418, and a 2B1Q version of symmetric DSL at rates of 400 kbits/s, 1040 kbits/s, and 1568 kbits/s. The 2B1Q SDSL was included to address the interests of the CLECs using this technology. While SHDSL was not included as a basis system, the criteria protecting 2B1Q SDSL effectively protects SHDSL, too.

T1.417 defines the criteria to determine if a transmission system is spectrally compatible. A set of spectrum management classes are defined in terms of the power spectral density (PSD) limit for transmitted signal level as a function of frequency, total transmitted signal power, transverse balance, longitudinal (i.e., common mode) output voltage, and deployment guideline. The deployment guideline is the longest permitted line length in terms of equivalent working length (EWL). EWL is the length of a 26 AWG line with the same attenuation as the actual line without bridged taps. Thus, a 24 AWG line would have an EWL 1.33 times its physical length (with the length of any bridged taps ignored). The spectrum management classes defined in T1.417 are based on specific types of DSL technology, but any type of DSL qualifies if it complies with the PSD, power, and other criteria for that spectrum management class.

T1.417 also specifies the following spectrum management (SM) classes. Unless stated otherwise, all of the SM classes apply only for non-repeatered lines:

- SM class 1 (very low band symmetric) is based on BRI-ISDN with a PSD up to 115 kHz and total transmitted power less than or equal to 14 dBm. There is no EWL limit, but the line must not have loading coils.
- SM class 2 (low band symmetric) is based on SDSL up to 528 kbits/s with a PSD up to 238 kHz and total transmitted power less than or equal to 14 dBm. The EWL is limited to 3.5 km (11.5 kft) or less.
- SM class 3 (mid-band symmetric) is based on HDSL with a PSD up to 370 kHz and a total transmitted power less than or equal to 14 dBm. The EWL is limited to 2.74 km (9 kft) or less.
- SM class 4 is specifically defined to address HDSL2 (T1.418) with a downstream PSD up to 400 kHz and upstream PSD up to 300 kHz. The total transmitted power is limited to 17.3 dBm downstream and 17 dBm upstream. The EWL is limited to 3.2 km (10.5 kft) or less.
- SM class 5 (asymmetric) is based on ADSL1 (T1.413) with a downstream PSD up to 1.104 MHz and upstream PSD up to 138 kHz, with no overlap between the downstream and upstream bands. The total transmitted power is limited to 20.9 dBm downstream and 13 dBm upstream. There is no EWL limit, but the line must not have loading coils.
- SM class 6 (wide band asymmetric) is based on VDSL1 with a 998-type frequency plan extending up to 12 MHz. Note that SM class 6 is based on VDSL1, not VDSL2. The total transmitted power is limited to 14.5 dBm for both downstream and upstream. The EWL is limited to 4 km (13 kft) or less. While VDSL1 is qualified as spectrally compatible for line lengths up to 4 km, it will reliably operate on lines up to only 1.5 km.
- SM class 7 (very wide band symmetric) is based on SDSL transmitting between 1.168 Mbits/s and 1.568 Mbits/s with a PSD below 776 kHz. The total transmitted power is limited to 14 dBm. The EWL is limited to 2 km (6.5 fkt).
- SM class 8 (wide band symmetric) is based on SDSL transmitting between 784 kbits/s and 1.168 Mbits/s with a PSD below 584 kHz. The total transmitted power is limited to 14 dBm. The EWL is limited to 2.29 km (7.5 kft).

- SM class 9 (overlapping asymmetric) is based on a version of ADSL1 (T1.413) where the upstream band overlaps the downstream band. The downstream PSD may extend from 25 kHz to 1.104 MHz, and the upstream may extend from 25 kHz to 138 kHz. The total transmitted power is limited to 20.9 dBm downstream and 13 dBm upstream. There is no EWL limit, but the line must not have loading coils.

Technology-specific spectral compatibility guidelines are provided for SDSL using 2B1Q. SDSL is spectrally compatible for all non-loaded lines for SDSL operating at 300 kbits/s and less. SDSL operating at 2.32 Mbits/s is spectrally compatible for line lengths up to 1.52 km (5 kft).

Technology-specific spectral compatibility guidelines are also provided for SHDSL (ITU-T Rec. G.991.2). SHDSL is spectrally compatible for all non-loaded lines for SHDSL operating at 592 kbits/s and less. SHDSL operating at 2.32 Mbits/s is spectrally compatible for line lengths up to 2.6 km (8.55 kft). Since SHDSL is more spectrally efficient than SDSL using 2B1Q modulation, SHDSL is spectrally compatible at higher bit rates and on longer lines. Repeatered SHDSL is spectrally compatible for bit rates of 634 kbits/s and less.

Technology-specific guidelines are also provided for HDSL4 technology, with it being spectrally compatible for all non-loaded lines when operating without repeaters. For HDSL4 lines with one repeater, specific spectral compatibility guidelines are provided for lines with the repeater up to 3.35 km (11 kft) from the central office.

Other than the technology-specific sections mentioned above, T1.417 does not place any restrictions on the modulation method, coding technique, transport protocol, or other aspects of the DSL transmission technique. The spectral compatibility criteria pertain only to transmitted PSD, total power, balance, output voltage, and EWL.

For systems that do not fit into the SM classes and technology-specific sections, T1.417 specifies an analytical method (called "Method B") by which the spectral compatibility limits for any well-defined transmitted signal may be determined.

To protect against the potentially harmful effect on fluctuation crosstalk, T1.417 specifies a limit on the short-term stationary behavior of transmitted signals. The signal power may not vary by more than 8 dB between adjacent 1.25 ms intervals.

For more information about spectrum management, see [1,2].

10.2   G.hs – ITU-T Rec. G.994.1

There are many types of DSL technology, and each of them has many alternative operating modes and options. To serve a larger potential market, many DSL modems are capable of operating in a large number of modes. For example, most VDSL2 modems are also able to operate as ADSL2 and ADSL2plus modems. When DSL modems are powered up and connected to a line, they need to determine which mode and options will apply. A multi-mode capable modem might choose to operate as a VDSL2 modem on a short line, but it would operate as an ADSL2plus modem on a longer line if the modem at the other end of the line were also capable of that type of operation. While operating as an ADSL2plus modem, it might choose the Annex A mode when operating in the United States, Annex B for Germany, and Annex C for Japan. If greater protection against impulse noise protection were needed, the ADSL2plus modem would need to negotiate with the modem on the other end of the line to see whether it supported the optional extended INP settings.

The success of DSL technology has led to a vast number of alternative modes and options. ITU-T Rec. G.994.1 (generally known as G.hs for handshake) specifies a set of parameters describing all standard DSL modes and options. G.hs also specifies the messages to convey the parameters and a modulation method to convey these messages. G.hs must surely be the longest telecommunications standard ever written. The number of supported parameters is huge and, as a result, G.hs is more than 850 pages long. It provides the protocol, messages, and parameters by which the two DSL modems connected to a line determine each other's capabilities and then negotiate a common set of operating parameters.

When the pair of DSL modems start up, G.hs is the first phase of communication. Once the two modems have selected a common set of operating parameters, they both transition from G.hs operation to the selected operating mode, such as operating as ADSL2plus, Annex A, with packet transport mode. G.hs is used for the initial handshake between ADSL2, ADSL2plus, VDSL2, and SHDSL modems, but it is not used for older versions of DSL such as BRI-ISDN, HDSL, ADSL1, and VDSL1.

To assure robust operation for all types of line conditions, G.hs transmits its messages at an effective bit rate of approximately 500 bits/s, using binary differential phase shift keying (DPSK) modulation, with frequency diversity provided by sending the same information simultaneously in multiple tones. A set of 4 kHz tones is used for G.hs by SHDSL modems, and a set of 4.3125 kHz tones is used for G.hs by ADSL2/plus and VDSL2 modems. The G.hs message frames have a 16-bit CRC checksum used by the receiver to detect transmission errors. An ACK message acknowledges the correct reception of a G.hs message from the other end, and a NAK message indicates the receipt of a corrupted or unsupported message. Both duplex transmission (messages may be sent simultaneously in both directions) and half-duplex transmission (messages are sent in only one direction at a time) modes are supported.

Either end can start the G.hs exchange. A typical startup message would consist of the customer-end modem sending a CLR (capability list request) message providing the Capability List for the customer-end modem and requesting the other end to respond with its capability list. The capability list includes parameters for every mode and option that the modem is capable of supporting. The network-end modem would respond with a CL message with its Capability List of parameters. Note that the CL parameters offered by the network-end modem may depend upon what was offered by the customer-end modem in its CLR message. The customer-end modem would send an ACK (acknowledge) message to acknowledge receipt and then send a MS (mode select) message listing the exact set of parameters it has selected based upon its knowledge of what both it and the network-end modem are jointly capable of performing. Upon receipt of the MS message, the network-end modem would respond with an ACK message to acknowledge that it has received the MS message and is prepared to enter the specified mode of operation. Both modems will then transition to the common operating mode at the same time.

G.hs defines a generic data structure for the parameters. This data structure permits the parameters to have virtually any length. The parameters are structured in an expandable tree structure with two first-level parameter types: NPar(1) parameters have no second-level sub-parameters (subtending branches), while SPar(1) parameters have associated second-level sub-parameters. This structure continues with second-level parameters NPar(2) and Spar(2) as well as the third-level parameters. This data structure permits the addition of more parameters and sub-parameters in later versions of G.hs. Furthermore, if a newer version modem offers new parameters to an older version modem that does not know the meaning of the new parameters, the older modem can still understand all of the remaining parameters that are within its capabilities.

The protocol also conveys the version of G.hs being used so, if a modem with a newer version of G.hs communicates with a modem with an older version of G.hs, the newer modem can revert to the older type of G.hs. G.hs messages contain a non-standard information field to convey vendor proprietary information that is not specified within G.hs.

See [1] for more information about G.hs.

10.3   PLOAM – ITU-T Rec. G.997.1

ITU-T Recommendation G.997.1 is usually called PLOAM, but its official title is *Physical Layer Management forDigital Subscriber Line(DSL) Transceivers*. PLOAM defines *Simple Network Management Protocol, Management Information Base* (SNMP MIB) elements for the management of ADSL1, ADSL2, ADSL2plus, and VDSL2 modems. SNMP is specified in [3].

While the ADSL or VDSL is in its normal operating mode (also known as "showtime"), PLOAM communicates management information between a network management system (NMS) and the DSL modems at both the network end (xTU-C) and the customer end (xTU-R) of the line. This information is conveyed between the xTU-C and xTU-R via indicator bits and an embedded operations channel (EOC). The indicator bits convey time-critical information via dedicated bits in the DSL frame. The EOC conveys MIB elements via a message-based protocol. The NMS uses the MIB elements defined in PLOAM to perform DSL configuration, fault, performance, and diagnostic management.

Management information is conveyed both to and from the NMS, which sends management information to change the operating parameters of the DSL modems and to command the modem to perform a specified action. The DSL modem sends autonomous information (such as failure alarms) and requested information such as diagnostic results and performance monitoring counters to the NMS.

PLOAM can retrieve performance monitoring information from the xTU-C and xTU-R. This includes the number of code violations (CV); error seconds (ES); severely error seconds (SES, a second with 18 or more CRC-8 errors); loss of signal seconds (LOSS); unavailable seconds (UAS, having ten contiguous SES); forward error control seconds (FECS); impulse noise events; modem initializations (restart); and incorrect CRC-8. For each of these performance monitoring parameters, separate 16-bit history registers are maintained for the current and previous 24-hour period, the current 15-minute period, and the previous sixteen 15-minute periods.

The inventory information reported by the DSL modem includes: vendor ID; version number; serial number; self-test result; and equipment capabilities. The capabilities may be derived from G.hs information.

The line diagnostic information reported by the DSL modem includes: line attenuation; signal-to-noise ratio margin (SNR) per subcarrier; attainable data rate; actual PSD; actual transmitted power; H(f) channel characteristic per subcarrier; quiet line noise (QLN) per subcarrier; bits loaded per subcarrier; gain setting per subcarrier; actual data rate; and actual framer overhead settings.

The NMS control of the DSL modem configuration includes: adjustment of the transmitted signal power; transmitted PSD shape; RFI band notching; upstream power back off control parameters; upstream/downstream transition frequency; impulse noise protection; interleaving delay; minimum/maximum noise margin; and minimum/maximum bit rate.

10.4   G.bond – ITU-T Recs. G.998.1, G.998.2, and G.998.3

In the context of DSL technology, "bonding" refers to the use of two or more DSLs to convey a service to a customer. The bonded wire pairs have the same endpoints at the network end and customer end, and each bonded wire pair conveys a unique DSL signal which is carrying a portion of the total service payload. The bonded wire pairs are not electrically connected to each other,[1] but the traffic carried by each bonded DSL is inverse multiplexed to form an aggregate transport for the total payload. As a result, each pair of wires in a bonded DSL system carries a fraction of the total bit rate.

Compared to a single line DSL, bonding two DSLs can convey twice the bit rate over the same distance, and bonding n DSLs can convey n times the bit rate. At the cost of using additional pairs of wires and additional DSL modems, higher aggregate bit rates can be achieved.

Bonding can also be used to deliver the same bit rate over longer lines. For flexible bit rate DSL technologies, the achievable line length increases as the bit rate decreases. For a given aggregate service bit rate, the bit rate per line will be only half as much if the service is conveyed by bonding two DSLs instead of carrying the service via single-line DSL. As a rule of thumb, for the same aggregate bit rate, two-line bonding can reach approximately 1.6 times further than a single-line DSL, and bonding four DSLs will reach approximately 2.2 times further than a single-line DSL.

While the ITU-T standards provide specifications for bonding up to 32 DSLs, the most common applications bond 12 or fewer lines. Although the earlier inverse multiplexed ATM (IMA)

---

[1] Physically connecting two pairs of wires in parallel would reduce the dc resistance of the path, but it would impair DSL transmission due to the resulting imbalance and increased capacitance. Thus, the wire pairs should be electrically separate.

specification required every line in a bonded group to transmit exactly the same bit rate, G.bond (ITU-T Recs. G.998.1-3) allows for each wire pair in the bonded group to transmit at a different bit rate. Even though wire pairs of the same length and in the same cable will have approximately the same transmission capacity, the actual maximum bit rate will vary slightly between the lines. More aggregate throughput can be achieved by allowing each line to run at its maximum bit rate instead of forcing all lines in a bonded group to transmit at the bit rate of the lowest speed line in the bonded group.

G.bond consists of a suite of three ITU-T Recommendations: G.998.1 specifies DSL bonding for ATM transport; G.998.2 specifies DSL bonding for Ethernet-based transport; and G.998.3 specifies DSL bonding for synchronous transport mode (STM, also known as time division inverse multiplexing (TDIM)). Systems designed to one of these specifications are optimized to transport the specified type of traffic, but they are also capable of transporting the other types of traffic with less efficiency. All three G. bond Recommendations apply for bonding ADSL, VDSL, and SHDSL lines.

G.998.1 is nearly identical to the ATIS T1.427.01 standard for ATM bonding. This method for ATM bonding has very little overhead, because the bonding sequence identification (SID) is accomplished using bits within the existing ATM cell 5-byte header. The small overhead results from the bonding status and control messages.

Ethernet-based bonding is specified in ITU-T G.998.2, which is nearly identical to the ATIS T1.427.02 standard, and these were both derived from the earlier IEEE 802.3ah (Ethernet in the first mile) standard. Of the three ITU-T bonding recommendations, G.998.2 is most efficient for transporting IP packet type traffic. The bonding engine breaks Ethernet frames into fragments and adds a two-octet fragment ID header to each fragment. The fragments are then distributed among the bonded pairs, and then the fragments are reassembled into whole frames at the other end of the line. The resulting bonding overhead is small – typically less than 0.5%.

Each line within the bonded group may transmit at a different bit rate, and lines may be seamlessly added or removed from the bonded group during operation. In the event that one line fails during operation, the remaining lines within the bonded group will increase their transmission rates as needed up to their maximum capacity to carry the aggregate bit rate. If the remaining lines have adequate spare capacity, this would provide immediate and automatic protection against a line failure. Of course, if the remaining lines were unable to support the aggregate bit rate, then the Ethernet frame rate would be reduced to the available transport capacity.

G.998.3 is nearly identical to the ATIS T1.427.03 standard for STM bonding. The specification provides efficient bonded transport with seamless in-service addition and removal of lines to the group of bonded lines. The bonding method is based on 12 ms super-frames which are comprised of six frames, and each frame contains sixteen bonding sub-blocks. The 125 ms bonding sub-blocks are dispatched in a cyclical manner across the available lines in the bonding groups.

The Broadband Forum's TR-159 specifies the management of network elements for bonded DSL based on ITU-T G.998.1, G.998.2, and G.998.3.

10.5   G.test – ITU-T Rec. G.996.1

ITU-T Recommendation G.996.1 provides a common, defined set of twisted-pair line characteristics, in-home wiring models, and noise models to be used for performance testing of DSL modems. The configuration and calibration of testing equipment in a laboratory is also discussed. The noise models include crosstalk (NEXT and FEXT), impulse noise, and radio frequency interference (RFI) ingress. Testing of digital transmission and analog POTS is addressed. G.996.1 does not specify pass/fail performance requirements, which are addressed by the Broadband Forum testing specifications.

10.6   G.lt – ITU-T Rec. G.996.2

G.lt specifies *line testing* functions built into DSL modems at both the network and customer ends of the line. In particular, G.lt specifies single-ended line testing (SELT) functions to supplement the dual-ended line testing (DELT) functions already included in the ADSL, VDSL, and SHDSL standards. DELT functions require communication between two functioning DSL modems to perform diagnostic functions including monitoring the bit-error rate, SNR margin, signal attenuation, quiet line noise (QLN), and the per-tone bit loading.

The SELT type testing can perform diagnostic testing of a wire pair with or without communication to a DSL modem on the other end of the line. SELT measurements can be performed if there is no modem at the other end of the line, if the line is open, or if the line is shorted. The SELT diagnostic functions are equivalent to having a line test-set built into the DSL modem.

The network maintenance support system can directly access the SELT test results from the network-end modem. If a customer calls their service provider to report trouble with their DSL, the customer support center could ask the customer to run a diagnostic program in their computer, and then the test result could be read by the customer over the phone to the technician at the customer support center. For example, the SELT tests might identify a missing in-line filter which the customer could then be instructed to fix the problem. G.lt specifies the measurement and reporting of:

- uncalibrated echo response versus frequency; this is the voltage reflected back to the modem after a certain waveform is applied to the line;
- quiet line noise versus frequency;
- missing in-line filter;
- estimated line length;
- estimated line attenuation;
- line open or shorted;
- line topology, including the distance to each bridged tap and the tap length;
- estimated bit-rate capacity for upstream and downstream.

Work has begun to develop specifications for metallic line testing (MELT) for a future version of G.lt. MELT tests would include measurements of resistance, capacitance, DC voltage, and AC voltage as measured between the tip and ring wires, tip-to-ground, and ring-to-ground. The MELT measurement functions would be performed by a special measurement function beyond what is normally included in a DSL modem.

10.7   Broadband Forum DSL Testing Specifications

The ITU-T, ATIS, and ETSI DSL standards specify the signals transmitted across the telephone line and define functional requirements for the DSL modems, but these standards do not provide comprehensive equipment testing specifications to verify whether a DSL modem fully interoperates with another DSL modem, possibly made by a different vendor, and to verify that the modem provides an acceptable degree of performance. Performance is defined as the bit rate and error rate achieved for a variety of line lengths and noise conditions. The Broadband Forum (BBF) has produced a series of technical reports (TRs) that specify the configuration and calibration of the testing laboratory's equipment, a complete specification of the series of tests to be performed, the pass/fail criteria for each test, and the format for reporting the results of the testing.

The Broadband Forum bases its interoperability activities on the DSL standards published by the ITU-T, ATIS, and ETSI. The BBF interoperability program consists of three activities: plugfests, testing specifications, and qualification of testing laboratories. The BBF organizes and develops tests plans for plugfests, which are private events where vendors meet to experiment by interconnecting prototypes from different vendors to learn what functions work correctly and which do not. The plugfests are conducted at independent testing laboratories, such as the University of New Hampshire's IOL.

The BBF's website (www.broadband-forum.org) lists a set of recognized DSL interoperability testing laboratories meeting the BBF interoperability program qualifications. These testing labs agree to test DSL equipment according to the BBF testing specifications.

The BBF's Testing and Interoperability working group develops detailed technical reports that fully specify the required set of interoperability, performance, and functional verification tests to be performed, how the testing lab must perform the tests, and criteria for passing each test. The following BBF TRs provide the testing specifications for each type of DSL Technology:

- BBF TR-60 – SHDSL (ITU-T G.991.2).
- BBF TR-67 – ADSL1 (ITU-T G.992.1).
- BBF TR100 – ADSL2 and ADSL2plus (ITU-T G.992.3&5).
- BBF TR114 and 115 – VDSL2 (ITU-T G.993.2).

The tests specified in these TRs include:

- upstream and downstream bit rate for various length lines and various noise conditions;
- upstream and downstream bit rate for lines with bridged taps;
- packet throughput;
- errors caused by various types REIN and SHINE impulse noise;
- errors during 8-hour period (this is called a "stress test");
- CRC error reporting accuracy;
- SNR margin reporting accuracy;
- DMT bit-swap performance.

Downloadable copies of all BBF TRs are freely available at www.broadband-forum.org.

10.8   Broadband Forum TR-069 – Remote Management of CPE

The Broadband Forum's TR-069 specification has been widely adopted for wide area network (WAN) remote management of customer premises equipment (CPE) for customers served by DSL and, to a growing extent, customers served by fiber-to-the-home, wireless data, and other types of broadband access. TR-069 defines the protocol for communication of management information between the CPE and the service provider's auto-configuration server (ACS). The ACS may have a "northbound interface" to pass along some of the TR-069 management objects to the service provider's service configuration management system. TR-069 with several associated BBF TRs, also specifies the elements of the object model for the management of various types of CPE, including residential gateways (RGs), set-top boxes (STBs), and voice-over-IP (VoIP) devices.

TR-069 enables the automatic detection of connected CPE, automatic configuration of the CPE operating characteristics, download of CPE firmware (and firmware updates) from the ACS, inspection of CPE settings and capabilities by a remote service management system, control of CPE settings by a remote service management system, retrieval of performance monitoring history and fault indications, and initiation of CPE diagnostic tests with the results reported to a remote service management system. As a result, CPE may be installed, removed, or added, and many types of CPE and home networking troubles can be quickly corrected without sending a technician to the customer's premises and without mailing a disk to the customer. TR-069 management information flows over the DSL, fiber, or wireless WAN link to the customer's residential gateway (RG) or network terminal (NT), and then continues through the customer's local area network (LAN) to networked devices in the premises.

The protocol consists of *remote procedure calls* (RPC) over an *Extensible Mark-Up Language* (XML is specified by W3C) syntax using the *Simple Objected Access Protocol* (SOAP is specified by W3C) with the *Hypertext Transfer Protocol* (HTTP – IETF RFC 2616). Below this, standard *Internet Transport Layer Security* is provided by the *Secure Socket Layer* (SSL/TLS – IETF RFC 2246) operating above the standard transmission control protocol for the internet protocol (TCP/IP – IETF RFC 1122). The security protocol assures the fidelity and confidentiality of the communications between the ACS and the CPE, and this helps to prevent the customer from gaining unauthorized access to services.

Other BBF TRs related to TR-069 include:

- TR-064 CPE LAN side management;
- TR-098 internet gateway device data model;
- TR-106 CPE object model for LAN side devices;
- TRs-104 and 110 VoIP device provisioning and configuration;
- TR-111 Remote management of home devices;
- TR-135 Data model set-top boxes (STBs);
- TR-140 Data model for storage enabled devices;
- TR-142 Management of passive optical network (PON) devices;
- TR-157 Components objects for CWMP;
- TR-181 Updated Device Model;
- TR-196 Femto access point data model;
- TR-232 Bulk data collection.

References

1. Starr T, Sorbara M, Cioffi JM, Silverman PJ. *DSL Advances*. Prentice Hall: 2003.
2. ANSIT1.417-2003. Spectrum Management for Loop Transmission Systems. ANSI; 2003.
3. IETFRFC 1157. A Simple Network Management Protocol (SNMP). IETF; 1990.

11

The DOCSIS (Data-Over-Cable Service Interface Specification) Protocol

11.1 General Introduction

Cable television companies, commonly known as Multiple System Operators (MSOs), have traditionally specialized in delivering video signals to subscribers. In order to compete with telephone network operators for triple-play services, MSOs have upgraded their networks to support broadband data and voice service. A key element of this upgrade is the deployment of an IP-based core data network. This network enables high-speed data access, which has become the "killer" service for the MSOs. The customer applications making use of their high-speed IP networks include online gaming, the acquisition and sharing of music, instant messaging, and video delivery. IPTV (Internet Protocol Television) on-demand video delivery is a service pioneered by the MSOs. Voice over IP (VoIP) has also become an important MSO application. The Data-Over-Cable Service Interface Specification (DOCSIS) is the heart of providing IP-based services over the access network to the subscribers. This chapter focuses on DOCSIS, especially its physical [1] and data link layers [2], and briefly mentions other related companion protocols.

11.2 Introduction to MSO Networks

From its inception, the community antenna television (CATV) network, also commonly known as the cable television network, was designed for broadcast service delivery. The network was constructed with a head end office feeding the broadcast TV signals over a coaxial cable distribution network to the individual subscribers.[1] The coaxial cables were deployed in a physical bus topology with coaxial drop connections to individual subscribers implemented as taps on the coaxial feeder cable.

These coaxial cables had the potential for much higher bandwidth than the twisted pair wires of the telephone company. The early cable networks, however, were built for unidirectional transmission (downstream) of frequency division multiplexed analog television signals. As it became apparent that high-speed Internet access was becoming an increasingly important residential service, cable companies

---

[1] In smaller communities, the local distribution hub used a satellite receiver to access the TV signals from the head end over a satellite network, hence the name "community antenna" TV.

Broadband Access: Wireline and Wireless – Alternatives for Internet Services, First Edition.
Steven Gorshe, Arvind Raghavan, Thomas Starr, and Stefano Galli.
© 2014 John Wiley & Sons, Ltd. Published 2014 by John Wiley & Sons, Ltd.

began upgrading their networks to support bidirectional residential digital data connectivity. As cable television companies began supporting voice and data services over their networks, they became commonly known as Multiple System Operators (MSOs).

This bidirectional digital service capability required a major upgrade to the cable network.[2] The original cable network used amplifiers to boost the downstream analog signals, and bidirectional data transmission required retrofitting these amplifier nodes to include bidirectional digital repeaters to support upstream data transmission. The increasing data bandwidth demands and the decreasing cost of fiber optic transmission motivated a network that used fiber to connect from the head end office to optical nodes located closer to the subscribers, with coaxial cables continuing to be used for the connections to the subscribers. These networks, known as hybrid fiber/coax (HFC) networks, allowed increased data service bandwidth by reducing the number of subscribers sharing each coaxial feeder.

While up to 2000 subscribers may be connected to a coaxial drop, it has become common with HFC to have 500 or fewer subscribers connected to the coaxial cable from an optical node. In order to support increased data rates per subscriber, cable television companies are moving to 64 subscribers per coax cable. MSOs have also begun using other techniques to increase the data capacity of their HFC networks, such as reclaiming additional analog channels for data transport and using switched multicast rather than broadcast service delivery methods.

As the MSOs implemented wide deployment of data connectivity, the Voice over Internet Protocol (VoIP) provided a natural mechanism to enable them to provide voice services. This combination of voice, data, and video services is known as "triple-play". Since the feeder coax cable is a medium shared by all the subscribers tapping into that cable, a medium access control (MAC) protocol is required to enable subscribers to transmit upstream.

The protocols that enabled subscribers to have bidirectional data transmission over the cable network, including the MAC, were developed by CableLabs.[3] The core protocols are defined in the Data-Over-Cable Service Interface Specification (DOCSIS). This chapter gives a mid-level tutorial of DOCSIS, including DOCSIS 3.0.

Note that each MSO typically has somewhat different network deployment models. Consequently, there has been considerable variety in network and specific physical layer DOCSIS implementations, but the breadth of these physical layer implementations is beyond the scope of this book. The interested reader can learn more in [3]. As the DOCSIS standards have evolved, equipment compliance certification has resulted in interoperability and greater commonality of deployment models.

11.3   Background on Hybrid Fiber Coax (HFC) Networks

The HFC network deployed by MSOs is illustrated in Figure 11.1, although, as previously noted, there is considerable variety in the implementation details used by different MSOs. The primary hubs are typically the locations of the video head ends and the connections to the Internet Service Providers (ISPs). The secondary hubs often consist of telephony switching equipment, servers for locally cached video, interactive data service interfaces, and the CMTS (Cable Modem Terminating System) equipment. The core network that interconnects the hub nodes is typically built with fiber rings and spurs. Similar to the telephone networks, the rings provide the diverse routing to allow automatic protection against network and node faults.

The Fiber Nodes (FNs) subtend from the secondary hubs as spurs in a star configuration. The primary function of the FN is to convert the signals from the optical domain of the fiber to the electrical domain of the coax distribution network cable that connects to the Cable Modem (CM) at each home. The typical FN has been designed as a sealed boxed for remote aerial or pole mounting with a direct network power feed. The FN also supplies AC electrical power (50 Hz or 60 Hz) over the coax cable, which is used to power

---

[2] Note that at the same time, bandwidth of the systems was typically upgraded to 750 or 860 MHz.

[3] CableLabs is a research and development consortium of North and South American cable companies.

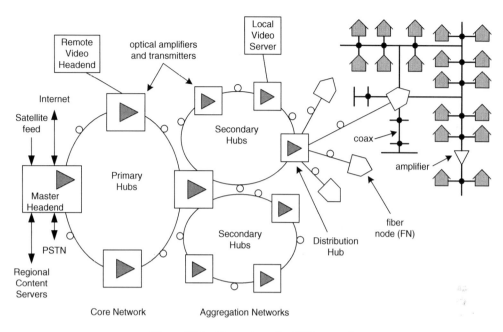

Figure 11.1 Illustration of a HFC network

amplifiers to boost the signals for transmission over longer coaxial lengths. Cable companies typically use no more than five amplifiers on a coaxial run.

The distance between amplifiers can vary, depending on factors that include the amplifier technology, the type of coaxial cable, the number of taps, and the maximum frequency of the RF signal. However, a spacing of 0.33 miles (0.53 km) between amplifiers is typical with current technology. The power and maintenance requirement of the amplifiers, and the decreasing cost of fiber optic technology, is leading MSOs to run the fiber closer to the subscriber. The "fiber deep" architecture places the FN close enough to the subscribers that few or no line amplifiers are required on the coax segments. In these applications, the subscribers are typically within one mile or 1.5 km of the FN, which requires three amplifiers, or within 0.125 miles (0.2 km), with no line amplifiers on the coaxial cable.

Note that while DOCSIS supports a maximum distance of 100 miles between the CMTS, which typically would be located at a distribution hub or head end, and a CM at the customer's premise, an overall distance of 15 miles is more typical.

The head end modulates a radio frequency (RF) carrier signal with the video signal associated with that RF channel slot. These RF signals are then combined through frequency division multiplexing (FDM). In North America, each RF channel occupies a 6 MHz band, while 8 MHz bands are typically used in Europe.[4] Total frequency ranges of up to 750 MHz and 860 MHz have been common since they were introduced in the mid-1990s, with a top frequency of 1128 MHz being introduced in 2008. An 860 MHz system provides up to 120 channels with 6 MHz channel spacing, or 100 channels with 8 MHz spacing, and a 1128 MHz system provides 179 channels with 6 MHz. Control information, including information

---

[4] Note that originally, the frequency plans were different for different manufacturers. The National Cable Television Association (NCTA) developed the first standard plan. The EIA/CEA-542 Rev. A is the current standard plan. The Rev. A version added support for digital channels.

about channel identification and programming, can be sent in-band with each channel or out of band on a separate RF channel. The combined FDM signal modulates a linear laser for transmission over the optical network. The frequency spectrum and other physical layers topics are covered in Section 11.7.

While video signals were originally analog signals, it is now common to also support the transmission of digitally encoded video signals. These digital video streams typically use a compression protocol such as MPEG2 or MPEG4, and are modulated into the RF channel with quadrature amplitude modulation (QAM). MPEG compression allows carrying typically up to five standard definition quality digital video channels or two high definition video channels within a single 6 MHz channel slot. Note that additional video channels can be carried at the expense of lower quality due to more extensive MPEG compression.

Premises distribution networks (PDNs) carry the data and video information within the subscriber's premises. The PDN can use Ethernet, wireless LAN, coaxial cable, telephone wiring, or power lines.

Devices called telephone adapters are used to convert the telephone signals to a format such as VoIP that can be carried over the cable network. Such devices are also referred to as voice gateways. A device capable of converting multiple types of information for transport over the cable network is called a multimedia terminal adapter (MTA) or an integrated access device (IAD). The Cable Modem Terminating System (CMTS) discussed below communicates with the MTA using a portion of a data channel.

11.4   Introduction to DOCSIS

As noted above, the DOCSIS protocol and its related protocols were developed by CableLabs, which is an organization owned and funded by the various MSOs. The DOCSIS specification is very detailed, covering the aspects of data services over MSO HFC networks from Layer 1–7. More specifically, DOCSIS specifies the physical and MAC layer protocols, and an associated provisioning and management framework.

At the physical layer, DOCSIS specifies how the downstream and upstream data is carried over the RF channel slots, including the line coding and modulation formats. Since the coaxial cable is a shared medium, DOCSIS specifies the MAC protocol by which the subscribers can access the medium in the upstream direction. At the service level, DOCSIS defines different types of service flows based on the required service level agreement (SLA) performance that must be met. DOCSIS also specifies the encryption protocol that prevents subscribers from accessing the data associated with other subscribers on the shared medium. The DOCSIS cable modem authentication protocols allow the subscriber to obtain the services to which they are entitled, but not services for which they have not subscribed. The authentication protocol also allows a subscriber to move a cable modem to different locations within the MSOs network and still obtain service.

11.5   DOCSIS Network Elements

The elements of the DOCSIS network are shown in Figures 11.2 and 11.3. The Cable Modem Terminating System (CMTS) is the host of the DOCSIS protocol and the Cable Modem (CM) is the customer premises terminal that interconnects the subscriber's equipment with the DOCSIS network. The CMTS and CM are interconnected with the HFC network. The other two network elements in the HFC are the Fiber Node (FN) and the RF combiner shelf, whose functions include the optical to electrical (O/E) and electrical to optical (E/O) conversions. The RF combiner shelf interconnects with the CMTS over coax cables within the hub office. The functions of each of these systems are described in the remainder of this section. The details of some of these functions are discussed further in the sections that describe the various DOCSIS functions.

The remainder of this chapter covers aspects of DOCSIS in a network layer order, beginning with Layer 1.

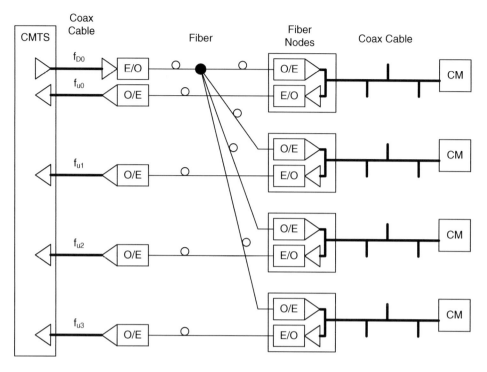

Figure 11.2 Illustration of a DOCSIS network

11.5.1 CMTS (Cable Modem Terminating System)

The CMTS is the DOCSIS host, originating the downstream DOCSIS protocol for the CMs and terminating the upstream DOCSIS signals from the CMs.

As illustrated in Figure 11.3, the CMTS consists of an interface to the MSO data network, line cards that provide the DOCSIS interface to the HFC, a packet switch, and interfaces to the different operations and

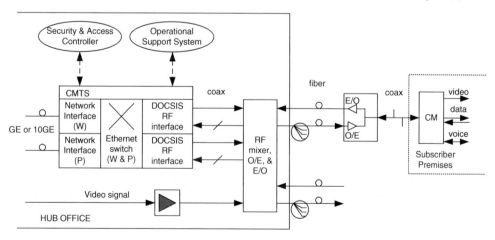

Figure 11.3 DOCSIS access network illustration

management systems. The MSO data network is an IP network that uses Ethernet for its lower layers. The network interfaces are typically Gigabit/s Ethernet (GE) today, but are expected to migrate to 10 Gigabit/s Ethernet (10GE). Due to the number of subscribers that would be affected by a network interface failure, these units are protected.

The CMTS uses an Ethernet switch to route the data packets between the network interface and the appropriate HFC connection. These switch units also typically use redundancy to protect against unit failures.

The CMTS line units provide the data portion of the interface to the HFC network, sending and receiving RF signals over a coax interface to the RF combiner shelf. In the downstream direction, the same RF signal is typically broadcast to multiple FNs. In the upstream direction, where HFC bandwidth is more limited, it is more common for each FN to have its own fiber and RF connection to the CMTS, either over a separate fiber or a separate set of wavelengths on a shared fiber (see Figure 11.2). The line units are typically protected in a 1: n manner using an external relay switch shelf to make the RF coax connections to the n working line units or to a protection line unit. The external relay switch shelf is controlled by the CMTS.

The CMTS is controlled by the DOCSIS back office, which provisions the CMTS so that it can provide the appropriate service level agreement (SLA) for each service flow for each CM. See Section 11.9.4.1 for the discussion of service flows.

The CMTS works with the security access controller to determine which, if any, service a CM is authorized to receive from that MSO. It is this DOCSIS feature that allows a subscriber to move the CM to any location served by that MSO (i.e., any home within the domain of that DOCSIS management system) and receive the appropriate services. This feature stands in contrast to DSL or PON customer premises equipment, which is tied to a location rather than to a subscriber.

A later evolution of the CMTS introduced in the DOCSIS 3.0 is the modular CMTS (M-CMTS) [4]. The M-CMTS breaks the CMTS functions into separate modular pieces of equipment, each located in its own shelf. These functions include downstream RF interfaces, upstream RF interfaces, traffic shaping, routing, and timing/synchronization. For example, separating the upstream and downstream portions of the CMTS allows each to be scaled separately based on the demand patterns rather than requiring upgrading an entire CMTS. This separation is especially attractive with the far greater DOCSIS 3.0 downstream bandwidth capacity from channel bonding, although it can also be used with DOCSIS 2.

Separating the downstream signal processing into a separate module allows implementations with fewer edge QAM (EQAM) devices for the faster data signals, thus providing a substantial equipment cost reduction. Having a separate EQAM module allows sharing the HFC bandwidth between the MPEG-TS Video services and DOCSIS. Similarly, the ratio of RF interface units to router or traffic management processing capacity can be matched to the needs of the application. The DOCSIS Timing Interface Server (DTIS) is the name given in DOCSIS 3.0 to the module that provides accurate clock and DOCSIS timestamp synchronization between the separate upstream and downstream modules. The M-CMTS approach also allows different equipment vendors to supply different modular components.

11.5.2 CM (Cable Modem)

The CM provides the subscriber-side termination for the DOCSIS protocol. In the downstream direction, it demodulates the RF signal, identifies the packets intended for that CM, and converts the packets into the appropriate subscriber interface signal. For data traffic, this subscriber interface may be Ethernet or a home network protocol such as the MOCA (Multimedia over Coax Alliance) standard that uses the in-home coax television wiring as the physical layer of a LAN for data distribution. As previously noted, voice is carried over DOCSIS networks as Voice over Internet Protocol (VoIP). The CM provides the VoIP processing and a normal telephone interface to the subscriber. The video signal is typically passed through the CM to a set-top box (STB). In the upstream direction, the CM participates in the DOCSIS MAC protocol to obtain upstream bandwidth, performs the packet multiplexing, and transmits its upstream data at the appropriate time by modulating it onto the appropriate RF channel.

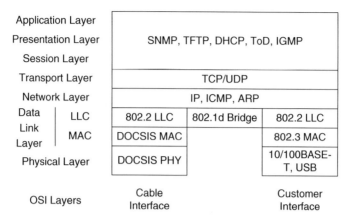

Figure 11.4 DOCSIS protocol layer stack

The protocol stack processed by the CM is shown in Figure 11.4. The CM bridges between the DOCSIS protocol, which it terminates/originates on the network side, and the Ethernet/LAN interface it provides to the subscriber. The higher layers of the protocol stack provide the functionality needed for OAM&P and maintaining SLAs for customer data streams.

Note that in contrast to DSL terminals and PON ONUs, DOCSIS can potentially connect to multiple CMs within a single home. In other words, there is no DOCSIS equivalent of the gateway functions performed by DSL or PON customer premises equipment.

11.5.3   FN (Fiber Node)

The FN, which is also described in Section 11.3, is a relatively simple node with only very basic management functionality. In the downstream direction, it performs the O/E function and transmits the downstream signal over the coax network. It also provides the power feed for any amplifiers used on the coax network. The coax network is bidirectional. In the upstream direction, the FN extracts the RF signals from the subscribers and performs the E/O function to transmit them toward the CMTS.

Simplicity, size and low power consumption have been key considerations.

11.5.4   RF Combiner Shelf

In the downstream direction, the RF combiner shelf takes the RF signals from the CMTS and the video signal feed, performs the RF mixing, and performs the electrical to optical (E/O) conversion to send the merged signal over the HFC network. In the upstream direction, the RF combiner shelf performs the optical to electrical (O/E) conversion and sends the RF signals to the CMTS.

11.6   Brief History of the DOCSIS Protocol Evolution

DOCSIS is a family of protocols that have added functionality with each new release. The added functionalities of the major releases are summarized here.

One of the key aspects of each DOCSIS release is that a CMTS or CM that supports the new release is backward compatible with a counterpart that only supports an earlier release. For example, a CMTS that supports DOCSIS 2.0 can communicate seamlessly with a CM that only supports DOCSIS 1.0, and vice versa.

11.6.1   DOCSIS 1.0

DOCSIS 1.0 was the initial release. Its primary purpose was to provide broadband internet access, and this release specified the minimal features required set to support that service. In support of that service, it defined a standard provisioning protocol based on Dynamic Host Configuration Protocol (DHCP) and Trivial File Transfer Protocol (TFTP). It also defined a network management protocol built on Simple Network Management Protocol (SNMP). Address and port filtering were also included in DOCSIS 1.0.

Since coax is a shared medium, it was important to provide link-layer encryption from the beginning in DOCSIS 1.0. Rate limiting was also an important feature of DOCSIS 1.0 for the shared coax medium.

11.6.2   DOCSIS 1.1

The DOCSIS 1.1 release fleshed out the DOCSIS with several enhanced capabilities.

Support for guaranteed Quality of Service (QoS) was a key addition. These guarantees included bandwidth, latency, and jitter performance. Support for dynamic service creation and deletion was also added.

In order to increase the network performance, the MAC layer was enhanced for more bandwidth efficiency. One such enhancement was support for the concatenation of data frames, so that multiple data packets/frames could be sent within a single upstream bandwidth grant rather than requiring a separate grant for each packet. Another enhancement was support for fragmenting long data packets in order to minimize their impact on the latency and jitter of other packet flows. Payload header suppression was also added to increase the bandwidth efficiency for services that have large per-packet headers that differ little from packet to packet.

The management protocol was upgraded to use SNMPv3. This upgrade added a view-based access control model (VACM) for the SNMP management. It also supported the additional privacy and authentication capabilities discussed below. In addition, it uses MIB kickstart values that allow the CMTS to determine the CMs public encryption key and to begin the exchange to set up the authentication and privacy keys that it will use with that CM.

The Dynamic Channel Change capability allows the CMTS to change which downstream or upstream RF channel the CM uses. IP Multicast support was added in order to better support the delivery of video services.

A key aspect of DOCSIS 1.1 is increased security through a stronger encryption method and a CM authentication protocol. Secure software download to the CMs was also added. The new MIBs included support for QoS, BPI+, and IGMP.

Other features of DOCSIS 1.1 were subscriber management, account management, and fault management capabilities.

11.6.3   DOCSIS 2.0

DOCSIS 2.0 preserved everything from DOCSIS 1.1, and added new physical and MAC layer capabilities.

The upstream channels of DOCSIS 2.0 support two different modes of operation. The first is Advanced TDMA (A-TDMA), which is a direct extension of the DOCSIS 1.0 and 1.1 concepts. The other mode is a new synchronous CDMA (S-CDMA) technology. S-CDMA shares the RF channel by having subscribers transmit continuously into the shared bandwidth using code division spread spectrum techniques, rather than using the time sharing methods of TDMA. A-TCMA and S-CDMA allowed the use of portions of the spectrum that were previously unusable.

A MAC enhancement introduced in DOCSIS 2.0 is the concept of logical channels for the upstream direction. A logical channel is a MAP (upstream bandwidth assignment map) entity that receives its bandwidth allocations through a MAP message associated with its unique channel ID. It is completely described by its MAP messages and UCD (Upstream Channel Descriptor). Multiple logical channels can be supported over the same physical channel.

DOCSIS 2.0 also provided for autonomous load balancing. During the time in which a CM is registered, the CMTS uses this feature to control the dynamic channel changes in the upstream and downstream direction for that CM. The CMTS controls the channel change through management messages (i.e., the Dynamic Channel Change (DCC) messages).

11.6.4   DOCSIS 3.0

One of the key features of DOCSIS 3.0 is the ability to bond individual physical channels together to create higher bandwidth logical channels. Channel bonding is supported in both the upstream and downstream directions. This mechanism allows much higher data rates (e.g., over 100 Mbit/s downstream), yet is also interoperable with legacy CMs.

DOCSIS 3.0 adds bonded channel group support to the autonomous load balancing feature. CMs supporting only a single channel in both the upstream and downstream directions are called "single-channel" CMs, which are handled in the same manner as with DOCSIS 2.0. The CMTS controls the load balancing of channels used by "multiple-channel" CMs through the Dynamic Bonding Change (DBC) messages. Load balancing for single-channel CMs requires moving the CM to a different upstream and/or downstream channel. In contrast, for multiple-channel CMs, the CMTS moves the service flow to a different set of upstream or downstream channels.

Another key feature of DOCSIS 3.0 is support for IPv6. IPv6 provides a greatly expanded address space and allows improved operation capabilities. IPv6 also makes support for IP Set-top Boxes more practical.

IP multicast is another important feature of DOCSIS 3.0. The driving application is IPTV-type services. Using IP multicast allows using standard protocols to manage the IP video services in a switched video manner. Bandwidth efficiency is gained by only delivering video programs to the subscriber when a viewer is present. IP multicast also allows QoS provisioning, so that the video quality is not impacted by network congestion.

Another DOCSIS 3.0 feature is support for stronger traffic encryption. DOCSIS 3.0 supports 128-bit AES instead of the 56-bit DES encryption used by previous generations of DOCSIS. It uses CBC mode (cipher block chaining) where the residual of the encryption of a block is the "seed" for the encryption of the next block.

11.6.5   Regional History and Considerations

Several competing regional standards emerged in the 1990s to address two-way transmission over HFC networks, and the DOCSIS protocol subsequently added region-specific variations. These are briefly summarized here.

The primary early cable standards work included:

- IEEE 802.14. Launched in 1994, this project was to develop a MAC for both ATM and IP traffic over HFC networks. It failed to achieve market success primarily due to the timing of its completion.
- DAVIC/DVB. In 1994, the Digital Audio Visual Council (DAVIC) was formed in Switzerland to promote digital audio-visual services and applications. One of the standards it produced, Digital Video Broadcast Return Channel for Cable (DVB-RCC) achieved initial success. Later, the European MSOs moved to EuroDOCSIS in order to take advantage of the DOCSIS economies of scale.

"EuroDOCSIS" was developed by CableLabs to address the European regional distinctives. The main difference in Europe was the use of 8 MHz channel bandwidth (conforming to the PAL standards) rather than the 6 MHz used in North America. The wider channel bandwidth allows for higher downstream bandwidth per user.

Most MSOs in Japan use the North America version of DOCSIS, although some use a variation with a different upstream channel clock. Specifically, the master clock for this version is based on 9.216 MHz rather than the 10.24 MHz used by DOCSIS and EuroDOCSIS. Using this master clock produces

Table 11.1 Correlation of DOCSIS and ITU-T Standards

DOCSIS version	ITU-T Recommendation
DOCSIS 1.0	ITU-T J.112 Annex B (1998)
DOCSIS 1.1	ITU-T J.112 Annex B (2001)
DOCSIS 2.0	ITU-T J.122 Annex B
DOCSIS 3.0	ITU-T J.222, J222.0, J.222.1, J222.2 and J222.3

Notes:

1. ITU-T J.112 Annex C covers the DOCSIS 1.1 variant used in Japan.
2. ITU-T J122 Annex F covers EuroDOCSIS and Annex J covers the variant used in Japan.

Table 11.2 DOCSIS downstream physical layer parameters

Region	North America		Europe	
Standard	ITU-T J.83-B		ETSI EN 300 429 (DVB)	
Modulation	64QAM	256QAM	64QAM	256QAM
Channel spacing	6 MHz	6 MHz	8 MHz	8 MHz
Symbol rate (MBaud)	5.057	5.361	6.952	6.952
Raw data rate (Mbit/s)	30.34	42.88	41.71	55.62
TCM rate	14/15	19/20	N/A	N/A
Reed-Solomon FEC	(128,122)	(128,122)	(204,188)	(204,188)
Post-FEC data rate (Mbit/s)	26.97	38.80	38.44	51.25

upstream channel widths that are a power-of-two division of 6 MHz rather than the 6.4 MHz of DOCSIS and EuroDOCSIS.

The different versions of DOCSIS have been ratified by the ITU-T as international standards [5–10]. The correlation between DOCSIS standards and their ITU-T standards are shown in Table 11.1. Some of the specific regional differences are covered in the next section, and they are summarized in Table 11.2.

11.7 DOCSIS Physical Layer

A typical MSO cable frequency spectrum is illustrated in Figure 11.5. The analog channel video spectrum typically goes down to around 54 MHz. Below that frequency, noise impairments are too large for quality analog video transmission. However, the 5–42 MHz band is robust enough for digital modulation formats and is the spectrum region chosen for DOCSIS upstream transmission. An enhancement to the original analog video spectrum has been the addition of new higher frequency channels to support digital signals. The digital signals carried over these channels include digital-encoded video and the DOCSIS downstream channels.

11.7.1 DOCSIS Downstream Transmission

The parameters of the DOCSIS downstream signals are summarized in Table 11.2. The North American MSOs follow the ITU-T J.83-B standards, while the European MSOs follow the ETSI EN 300 429 Digital Video Broadcast (DVB) standards. As noted above, one of the main differences between North American and European cable systems is that a channel width of 6 MHz is used in North America, while 8 MHz channels are used in Europe. This difference obviously affects the data rate that is feasible per channel. Both regions use 256QAM modulation for channels with adequate signal to noise ratios, and drop back to 64QAM on channels that can not support 256QAM.

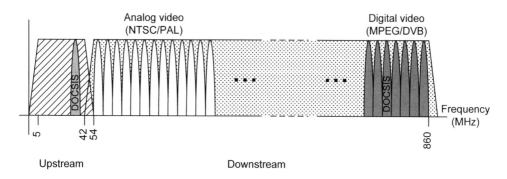

Figure 11.5 DOCSIS spectrum use

Both regions also use shortened Reed-Solomon block forward error correcting codes (FEC), although each region has chosen a different specific FEC. The error correction provided by the FEC increases the effective signal-to-noise ratio of the channel, thus allowing an increase in data rates beyond what the additional FEC overhead consumes. The European RS(204,188) provides more error correction capability than the North American RS(128,122), at the expense of added decoder complexity. The RS(204,188) corrects up to 8 symbol errors, while the RS(128,122) corrects up to 3 symbol errors. The North American specification also supports trellis code modulation (TCM).

Depending on the quality of the channel, DOCSIS thus supports per-channel data delivery rates of roughly 27–39 Mbit/s in North America and 38–51 Mbit/s in Europe.

DOCSIS defines a downstream transmission convergence (TC) sublayer that provides a common interface between the MAC and physical medium dependent (PMD) sublayer. This TC sublayer allows multiplexing and demultiplexing MPEG video and DOCSIS data over the PMD sublayer. The downstream TC signal is a continuous stream of 188-byte long MPEG packets. As illustrated in Figure 11.6, the MPEG frame begins with a 4-byte header that identifies whether the frame contains an MPEG video payload or a DOCSIS MAC payload. The header conforms to the MPEG standard, but the inclusion of the adaptation_field is not allowed in these MPEG packets.

When the MPEG packet carries DOCSIS data, the DOCSIS payload occupies the MPEG packet payload area. A DOCSIS MAC frame can begin anywhere within the MPEG packet. Each MPEG packet can carry a single DOCSIS MAC frame with stuffing to fill the remainder of the MPEG packet payload area or, alternatively, it can contain multiple DOCSIS MAC frames that are either concatenated together or separated by optional stuff bytes. It is also possible for a DOCSIS MAC frame length to exceed the length of the MPEG packet payload.

These different alignments of the DOCSIS MAC frame to MPEG packet are handled with the pointer_field. If the MPEG header's payload_unit_start_indicator (PUSI) is set, the pointer_field occupies the first byte after the MPEG header. The pointer_field tells the CM how many of the subsequent packet bytes it must skip before the beginning of the DOCSIS MAC frame. It also tells the CM whether any stuff

MPEG TC Header (4 bytes)	Pointer_field (1/0 bytes)	DOCSIS Payload (183/184 bytes)

Figure 11.6 MPEG packet format for DOCSIS use

Table 11.3 DOCSIS upstream physical layer parameters

Version	DOCSIS 1.x	DOCSIS 2.0
Format	Bursted F/TDMA	Bursted F/TDMA, F/S-CDMA
Modulation	QPSK, 16QAM	QPSK, 8QAM, 16QAM, 32QAM, 64QAM, 128QAM
Channel width (MHz)	0.2, 0.4, 0.8, 1.6, 3.2	0.2, 0.4, 0.8, 1.6, 3.2, 6.4
Symbol rate (Mbaud)	0.16, 0.32, 0.64, 1.28, 2.56	0.16, 0.32, 0.64, 1.28, 2.56, 5.12
Raw data rate (Mbit/s)	0.32–10.24	0.32–35.84
Pre-equalization	8-tap FIR (optional in 1.0)	24-tap FIR
Trellis coded modulation rate	N/A	$n/n+1$ (optional)
Reed-Solomon FEC (See note 1)	$T=0$–10; $k=16$–253	$T=0$–16; $k=16$–253
Post-FEC data rate (Mbit/s)	0.14–10.24	0.11–30.72

Note 1 – For the FEC, $T=$ the number of symbol errors the code is capable of correcting, and $k=$ the number of data symbols per FEC block.

bytes follow the DOCSIS frame. DOCSIS uses all-ones stuff bytes to fill any gaps between contiguous DOCSIS MAC frames.

11.7.2 DOCSIS Upstream Transmission

The DOCSIS upstream transmission parameters are summarized in Table 11.3, which illustrates the enhancements introduced by DOCSIS 2.0. As illustrated in Figure 11.5 above, the upstream channels are located in the frequency band of roughly 5–42 MHz. The upstream transmitter is illustrated in Figure 11.7, which highlights which blocks are unique to the A-TDMA and S-CDMA modes described below.

The upstream signal is scrambled in order to achieve nominally equal probabilities for all of upstream signal constellation points. The scrambler uses the $g(x)=x^{15}+x^{14}+1$ scrambler polynomial, and it is reset to a seed value at the beginning of each upstream burst. The seed value is calculated in response to the Upstream Channel Descriptor (UCD) message from the CMTS.

Since the upstream channels occupy a relatively noise part of the cable spectrum, they are specified to support a wider range of modulation options. The most efficient modulation option that can be supported by the channel is chosen.

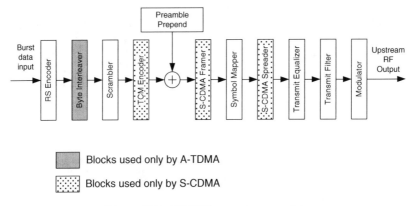

Figure 11.7 DOCSIS upstream transmitter

Adaptive pre-equalization filters are specified to accommodate the linear distortion of the coax cables in this frequency range. There are between 1–4 filter taps per symbol. The CMTS determines the upstream channel characteristics and sends pre-equalizer coefficients to the CM in the ranging response (RNG-RSP) message, which is discussed in Section 11.8.2. A default pre-equalizer coefficient setting is used prior to the CM receiving its initial ranging response, or whenever it changes its upstream channel or symbol rate.

A range of Reed-Solomon FEC codes are specified for the upstream signal. The code block size and ratio of data to error check bits is determined by the characteristics and impairments encountered on the channel. DOCSIS 2.0 also introduces the option of trellis coded modulation (TCM) for additional error performance improvement. An $n/n+1$ trellis code doubles the number of bits encoded into each transmitted symbol, but achieves an effective increase in signal-to-noise ratio by examining multiple sequential symbols to determine the most likely sequence of symbols produced by the trellis encoder. The resulting range of per-channel upstream data rates, not including the FEC overhead, ranges from 110 Kbit/s to 30.72 Mbit/s.

As noted in 11.6.3, DOCSIS 2.0 introduced a CDMA transmission mode in addition to the TDMA transmission mode used by DOCSIS 1.x. The characteristics of both modes are described in the following sections.

The CM can use either a fixed length for all upstream FEC codewords, or it can use a shortened codeword for the last codeword of the burst. As can be seen from the k parameter in Table 11.3, the minimum number of information bytes in the codeword is 16, and a full-length code word has 253 information bytes. If a shortened last codeword is used, the last codeword will still contain the same number of total bits as a full-length codeword, but it will pad with zeros between the end of the information bytes and the FEC check byte location.

11.7.2.1   TDMA Upstream Transmission Mode

The TDMA mode is similar to that used by most PON networks. While the downstream signal is broadcast to all CMs, the CMs time-share the upstream bandwidth by being granted transmission times. Only one CM transmits at a time on a given channel, and it uses the channel's entire bandwidth during that transmission.

The upstream transmission burst from a CM comprises a preamble, the data burst, and a guard time. The preamble allows the CMTS to adapt its equalizer receiver gain setting to the signal level of the CM's burst, and to acquire synchronization for clock and data recovery in order to receive the data portion of the burst. The preamble can be up to 1024 bits long. As with PON systems, the guard time is allocated to prevent the upstream transmissions from different CMs from overlapping at the CMTS, due to differences in the propagation paths between the CMTS and the different CMs beyond what is resolved by the ranging protocol.[5]

The primary advantage to TDMA is that it provides better immunity to narrowband interferers (i.e., noise sources with a narrow frequency range).

The Advanced TDMA (A-TDMA) of DOCSIS 2.0 uses a maximum channel width of 6.4 MHz, and it adds support for 8QAM, 32QAM and 64QAM modulation. In addition, it specifies a much-improved 24-tap pre-equalizer and enhanced Reed-Solomon error correction with optional interleaving. The result is much higher throughput and robustness for upstream data transmission. The robustness against worst-case cable plant impairments allows the use of previously unused portions of the upstream spectrum.

11.7.2.2   Synchronous CDMA (S-CDMA) Upstream Transmission Mode

S-CDMA is a direct sequence spread spectrum technique that allows multiple CMs to transmit simultaneously over the same channel. The spread spectrum transmission is combined with TDMA

---

[5] The guard time between bursts can be specified in the range between 5 and 255 symbol periods.

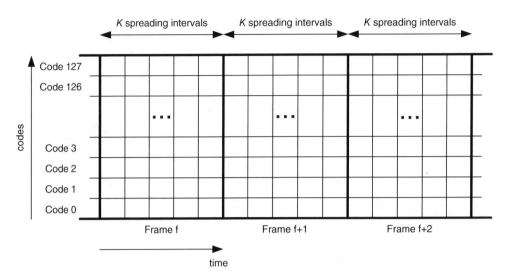

Figure 11.8 S-CDMA framing structure

such that CMs can have longer, overlapping burst transmission times. In other words, S-CDMA preserves most of the features of A-TDMA but adds CDMA spreading with optional trellis coded modulation. By allowing longer continuous transmission instead of relatively short burst transmission, S-CDMA avoids the bandwidth wasted by the burst preamble and guard times of multiple shorter bursts. The result is a slightly more efficient upstream transmission. CDMA also provides better immunity to burst noise.

S-CDMA uses a set of 128 orthogonal codes with 128QAM and an optional trellis code modulation. Specifically, the S-CDMA frame, which is illustrated in Figure 11.8, is configurable with up to 128 codes by K spreading intervals per frame. As illustrated in Figure 11.9, the mini-slots of the S-CDMA frame are determined by a combination of the frame number and the CDMA spreading code, with mini-slots mapped onto a group of consecutive codes. A spreading interval is the number of timeticks required to transmit a CDMA symbol. The transmitted waveforms of the different CMs are synchronized to within 1 ns, which maintains their orthogonality.

The physical layer maps the mini-slots to frames.

A CM achieves synchronization through the combination of it recovering the 10.24 MHz clock from the downstream signal sent by the CMTS, and of the ranging protocol that determines the CM distance from the CMTS. The frame format for the S-CDMA transmission is discussed in 11.9.1.2, and the ranging protocol is described in 11.8.2.

11.7.2.3 DOCSIS Channel Bonding

As noted above in Section 11.6.4, DOCSIS 3.0 introduces a mechanism to bond multiple RF channels together between the CMTS and a CM. These RF channels are the same ones used by previous versions of DOCSIS. A set of bonded channels is referred to as a Transmit Channel Set and/or Receive Channel Set. Packets associated with the same service flow are distributed over this set of channels. This mechanism allows much higher data rates, yet is also interoperable with legacy CMs. For North American systems, the individual 6 MHz downstream physical channels support 40 Mbit/s and the individual 6.4 MHz upstream channels support 30 Mbit/s. DOCSIS 3.0 supports at least 160 Mbit/s downstream and 120 Mbit/s upstream, in each case bonding four physical channels together. In addition to higher channel rates,

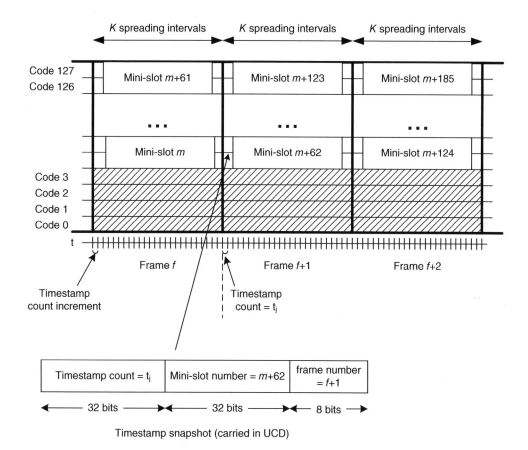

Figure 11.9 S-CDMA mini-slot and timestamp snapshot illustration

dynamic distribution of packets across the different channels in a downstream bonded group maximizes the achievable statistical multiplexing gain.[6]

Latency differences between the channels can result in the packets sent on different channels arriving out of sequence. A sequence number added to each packet enables the receiver to restore the packet stream properly before forwarding it. The sequence numbers also enable the detection of lost packets.

A channel within a Transmit or Receive Channel Set can potentially become unusable, for example if a CM loses an upstream and/or downstream channel during normal operation. This situation can also occur if the CM cannot acquire that channel during registration and/or DBC. This mode of CM operation is referred to as "partial service" mode in the respective upstream or downstream channel. The CM communicates this situation to the CMTS by either not giving a response (e.g., the REG-ACK or DBC-RSP) when it is unable to acquire that channel, or through the CM-STATUS message in the event of a channel being lost during normal operation.

For the upstream channel case of partial service, the CM does not transmit anything on the unusable upstream channel except that it will respond to unicast ranging opportunities for that channel in an attempt

---

[6] Dynamic channel assignment and bonding works best with bursty data that could benefit from statistical multiplexing. Services such as IP video streams work most efficiently with a relatively constant bandwidth assignment.

to re-establish communications on it. When the CMTS becomes aware of the situation, it will provide unicast ranging transmission opportunities for the CM for that upstream channel.

For downstream partial service, a CM will continue to attempt to acquire a lost non-primary channel. If a primary downstream channel is lost, the CM will re-initialize its MAC. If this occurs during normal operation, it quits transmitting data on all upstream channels while it attempts to re-acquire the primary downstream channel. The CMTS does not send unicast packets over an unusable downstream channel.

Partial service can be resolved either by the re-acquisition of the channels, or by using DBC to switch to a different channel set.

Downstream Channel Bonding

A group of downstream 6 Mhz (or 8 MHz) RF channels that are bonded together for carrying a service flow is called a Downstream Bonding Group (DBG). Each CM is tuned to receive its data on one more DBGs. Note that individual channels can be used simultaneously as part of multiple DBGs (i.e., DBGs can overlap) or as both part of a DBG and as a single channel to serve legacy CMs that only support reception over a single channel at a time. For the latter case, QoS can be maintained by dynamic balancing of the legacy CMs across the DBG.

At a given point in time, a service flow can be scheduled by the CMTS over one or more channels (i.e., can be assigned to this channel set). This set of channels, referred to as the Downstream Channel Set (DCS), consists of either a single downstream channel or a DBG. A 16-bit DCS ID is assigned by the CMTS to each DCS. A "bonded" service flow is one assigned to a DBG, while a "non-bonded" service flow is one assigned to a single channel.

The CMTS uses a separate sequence number space for each such CM, and also for each set of CMs in a multicast session that receives sequenced frames. A CM using a downstream DBG performs resequencing for only those received frames that it forwards to CPE. Specifically, a 20-bit Downstream Service ID (DSID) in the Downstream Service Extended Header is used to identify the CM or CM set to which a downstream sequences of packets is destined. The CM can use the DSID to filter the packets that need to be resequenced for forwarding to CPE. See 11.9.4.5 for more discussion of the use of the DSID.

The DSID value and the individual packet streams are independent in the sense that the CMTS may assign one or more DSID values to the same service flow. An example application for using multiple DSID values for the same group service flow is to distinguish IP multicast sessions.

Each CM is configured by the CMTS to receive downstream data on a set of channels called its Receive Channel Set. The CMTS can assign these such that overlap is minimized for CMs connected to the same fiber node. Dynamic load balancing can then optimize bandwidth usage between the CMTS and the fiber node, thus increasing the overall throughput to the set of CMs.

Upstream Channel Bonding

The CMTS uses the grant process to control upstream bonding. When the CMTS receives an upstream bandwidth request from a CM for a particular service flow, it decides whether to use one or more of the channels associated with that service flow in the grant for that request. In order to allow real-time load balancing for better statistical multiplexing gain, the allocation of bandwidth across the individual upstream channels is also the responsibility of the CMTS. The CMTS sends the grants over multiple channels, with each grant containing a transmit time and a particular size. The CM divides its upstream transmission accordingly, using sequence numbers so that the CMTS can reconstruct the stream. This is especially important in the upstream direction, since the grants on the different upstream channels can be scheduled for different times relative to each other.

11.8   Synchronization and Ranging

The DOCSIS CMs are synchronized to the CMTS in order to aid the CMTS recovery of upstream data bursts and maximize the upstream bandwidth utilization. The upstream symbol transmission timing of the

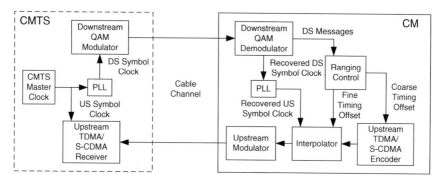

Figure 11.10 DOCSIS synchronization block diagram

CMs is coordinated by the CMTS, based on a combination of their synchronization and knowledge of the CMs distance (range) from the CMTS.

11.8.1 Synchronization

The basic blocks of the DOCSIS CMTS CM synchronization are illustrated in Figure 11.10. The CMTS is the master clock source for the network, with the CMs deriving their clock and symbol timing from the downstream DOCSIS signal. The downstream demodulation and synchronization follow ITU-T Recommendation J.83, Annexes A, B, and C [11]. The CMTS 10.24 MHz master clock and its symbol rate are locked together. The relationship between the clock and symbol period is defined as the ratio of M and N, with M and N being 16-bit integers specified by TLV parameters in the UCD message.

Each CM also maintains a local timestamp counter that is initialized by the SYNC message from the CMTS. The local timestamp is used by the ranging mechanism to provide timing offset information. The DOCSIS timestamp counter mechanism for ranging is similar to the one used for EPON, described in Chapter 3.

Specifically, the timestamp is a 32-bit counter run by a clock with a frequency of 10.24 MHz ± 5 ppm that is externally supplied to both the upstream PHY and MAC. The CMTS MAC layer maintains the master timeslot counter, while the master mini-slot and master frame counters reside in the PHY layer. The timestamp at the CMTS represents the counter value at a fixed time relative to the instant at which a Time Synchronization MAC Management message's first byte is transferred to the Downstream Physical Media Dependent Sublayer from the Downstream Transmission Convergence Sublayer (the offset time can be zero or a fixed value). All of the PHYs are synchronized to the MAC timestamp value by means of a frame synchronization pulse. At every frame boundary, the PHY captures a timestamp snapshot that will be available for transmission in a UCD.

Either the CMTS or CM can own the Start Reference Point and End Reference Point. Whichever one owns the Start Reference Point maintains this field as a copy of its local timestamp. If the CM owns the Start Reference Point, the CMTS places all zeros into the Timestamp Start field and, if the CMTS owns it, the CM copies the value from the identical field of the DPV-REQ (DOCSIS Path Verify Request) message into the Timestamp Start field. The Timestamp Start then corresponds to the timestamp value when the sender injects a DPV packet into the data stream at the DPV reference point. The Timestamp End field is initialized by the CMTS to all zeros. The CM copies into the Timestamp End field either its local timestamp, if it owns the End Reference Point, or otherwise the identical value from the DPV-REQ message.

The CMTS establishes a global timing reference for TDMA channels by transmitting the SYNC messages downstream at a nominal frequency. Since the SYNC message contains the timestamp, the CM knows exactly when the CMTS sent the message. The CM compares the received timestamp to the time at which it was actually received, and updates its local clock references accordingly. The CMTS establishes an additional global timing reference for S-CDMA channels by sending both the SYNC and UCD MAC messages downstream at a nominal frequency.

In general, the CM achieves MAC synchronization after it receives at least two SYNC messages within 200 ms (the maximum SYNC interval) and has verified that the received clock accuracy falls within the specified limits. In the "Locked" mode, the CM synchronizes (locks) to the received downstream symbol clock in order to derive its upstream transmission clock. In the "Not Locked" mode, the CM does not use the received downstream symbol clock but, instead, derives its upstream timebase from the received SYNC messages.

In topologies where a CM uses multiple downstream and upstream channels, the timestamp information can be sent on either a single downstream channel or on multiple channels. If the downstream channel timestamps are unsynchronized, then each upstream channel must be tied to a downstream channel for its timestamp information, and each downstream channel should only send timing messages for the upstream channels associated with that downstream channel. If all the downstream channels are synchronized (i.e., the timestamps are derived from a common clock and timebase), then timing messages for all upstream channels are sent on all downstream channels, and an upstream channel can use any downstream channel for its timestamp information.

With DOCSIS 3.0, the CMTS uses each CM's Primary Downstream Channel for communicating the global timing reference to it. The CM uses this timing reference as the timebase for upstream burst timing on all the upstream channels it uses.

11.8.2 Ranging

The CMTS determines the distance (range) between itself and each CM, which it takes into account in the upstream burst transmission grant timing that it sends to each CM. As discussed in this section, TDMA and S-CDMA have different ranging requirements. The ranging protocol supports a maximum end-to-end (one way) delay between the CMTS and CM of 800 µs, which corresponds to roughly 100 miles (161 km).

There are two steps to the ranging process. The first is an initial maintenance step that provides the coarse timing alignment. The second step is a periodic station maintenance that provides the fine timing alignment. The reason for performing the ranging, power, and frequency checks on a periodic basis is to accommodate the physical variations of the HFC plant over the course of the day. For example, the HFC plant can be affected by temperature variations and cable stretch.

In the first stage of ranging, the CM achieves timing synchronization with the downstream symbol stream and analyzes a received Upstream Channel Descriptor MAC (UCD-MAC) message to determine the upstream channel characteristics. Next, the CM scans the Bandwidth Allocation MAP message for a Broadcast Initial Maintenance Region, which the CMTS schedules such that it is large enough to accommodate the worst-case round trip delay. The CMTS schedules the Broadcast Initial Maintenance transmit opportunities to align with, and span an integer number of, S-CDMA frames.

The CM then transmits the appropriate ranging message in a Broadcast Initial Maintenance region. It sends the Bonded Initial Ranging Request (B-INIT-RNG-REQ) if it detects a MAC Domain Descriptor on its candidate Primary Downstream Channel and is performing its initial ranging after power-up or reinitialization on the first upstream channel. It sends an Initial Ranging Request (INIT-RNG-REQ) if the UCD indicates that the upstream channel supports DOCSIS 2.0 (Type 3 upstream channel) or DOCSIS 3.0 (Type 4 upstream channel), and the CM is not performing its initial ranging after power-up or reinitialization on the first upstream channel, and the MAC Domain Descriptor is not present. The INIT-RNG-REQ is used when a CM is initializing on a secondary channel (i.e., a channel on which the CM attempts to range after it has received a ranging response on a different channel, unless that ranging response contained an Upstream Channel ID Override. The CM sends the Ranging Request (RNG-REQ) otherwise.

The CM uses its internal delay as the initial value for the timing offset (i.e., the amount of delay that would occur if the CM is next to the CMTS with no cable plant delay). It uses the Ranging Backoff Start parameter as the starting point for the back-off window of its initial ranging contention transmission, and the Ranging Backoff End as the final window value. The parameter value (0–15) represents the power of 2 value expressing the window size. See Section 11.9.2.2 for a detailed description of the contention and backoff protocols.

The CMTS responds to a CM's ranging request message with a Ranging Response (RNG-RSP) message addressed to that CM. If the CM has not retained a previous SID, the Ranging Response assigns either a temporary SID or a Ranging SID to the CM until the completion of the registration process. The adjustments for RF power level[7] and frequency, as well as the timing offset corrections are also sent within the Ranging Response message. The result of the ranging adjustment is that, from the CMTS receiver's perspective each CM appears to be located adjacent to the CMTS.

In summary, the first stage of the ranging mechanism begins with the CM sending a ranging request message, and the CMTS responds with a ranging response message that contains the Timing Adjust, in addition to power and frequency adjustment information and pre-equalizer parameters.

The second step begins with the CM waiting for a Unicast Initial Maintenance region assigned to the SID it is initially using. This SID can be the temporary SID or Ranging SID assigned during the first stage of the ranging process, or a SID assigned previously if the ranging is occurring due to an Upstream Change Request (UCC), Dynamic Change Request (DCC) or Upstream Channel Descriptor (UCD) change. After this, the station range is maintained through the CM responding to Station Maintenance messages sent to it. In each of these cases, the CM responds by using its current SID to send a RNG-REQ message that includes power level and timing offset corrections if they are required.

The CMTS responds to these RNG-REQ messages by sending any additional required fine tuning in a RNG-RSP. The CM and CMTS continue to exchange RNG-REQ and RNG-RSP messages until either the CMTS sends a Ranging Successful indication in the response, or the CMTS aborts ranging. After completion of the ranging process, the CM participates in the normal upstream traffic flow.

For TDMA transmission, the goal is to prevent the upstream bursts from different CMs overlapping at the CMTS receiver, while minimizing the guard time between the bursts in order to maximize the upstream bandwidth utilization. These goals are the same as for PON and wireless systems using TDMA. Consequently, upstream TDMA transmissions require only the relatively coarse synchronization alignment of ±800 ns.

In the case of S-CMDA upstream transmissions, the synchronization requirements are much stricter. The S-CDMA ranging is also critical to synchronize the arrival times of the transmitted upstream CDMA symbols. The synchronization of the CDMA arrival times at the CMTS receiver is the key to maintaining the relative orthogonality of the CDMA codes used by the different CMs so that the CMTS can distinguish the symbols from each CM when they overlap in time. Specifically, the S-CDMA symbols must be aligned within ±1 ns at the CMTS in order for the codes from the CMs to be orthogonal.

As noted above, the CM timing synchronization, transmit frequency, and transmit power are also established during the ranging process. At the end of the ranging procedure, the transmit frequency error is less than 10 Hz, and the transmit power level error is less than 1 dB.

The interval between the periodic maintenance opportunities must be no more than "T4" seconds, although it is common for the CMTS to provide the opportunities much more often. T4 is typically 30 seconds, although a 10× multiplier option is allowed to give a T4 = 300 seconds. The CM declares a timeout error if it fails to detect a ranging opportunity within the T4 seconds.

---

[7] Note that DOCSIS 3.0 adds a Dynamic Range Window that allows the CMTS to specify a 12dB range for the transmit power used by a CM for channels in its Transmit Channel Set. This information can be communicated in the RNG-RSP message, the REG-RSP-MP or the dynamic bonding change request (DBC-REQ) message.

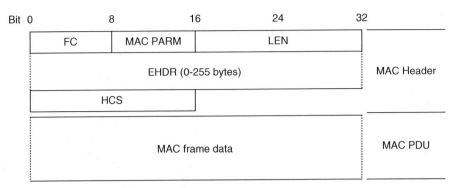

Figure 11.11 DOCSIS generic MAC frame format

11.9 DOCSIS MAC Sub-Layer

The CMTS uses the DOCSIS MAC sub-layer to control access to the physical layer channel in both the downstream and upstream directions. The upstream and downstream channels are separate physical channels (e.g., separate frequencies), with all communications going between the CMTS and CMs. In other words, there is no direct CM-to-CM peer-to-peer communication.

As illustrated in Figure 11.11, the DOCSIS MAC frames include a MAC header and an optional payload field to carry a variable length protocol data unit (PDU). The MAC header fields are defined as follows.

FC is the Frame Control that defines the PDU type and format. The four PDU types supported by DOCSIS include:

- variable length packets (18–1518 bytes);
- ATM cells ($n \times 53$ bytes);
- a MAC-specific header with no PDU; and
- a reserved PDU type for future use.

The five categories of MAC-specific headers are:

- the timing header;
- the MAC management header;
- the request frame;
- the fragmentation header; and
- the concatenation header.

The Timing header is used with a downstream packet data PDU to communicate the Global Timing Reference to the CM. As explained in 11.8.1 above, the CMTS sends a timestamp in each SYNC message to indicate the exact time it sent the message, and the CM adjusts its local clock to the timestamp.

The MAC_PARM field is used for MAC control. Its specific use depends on the frame type specified by the FC field. The MAC_PARM field indicates:

- The number of concatenated MAC frames for a concatenated MAC header;
- The amount of requested bandwidth for a REQ message; and
- An extended header length field if the EHDR_ON = 1.

The LEN field typically indicates the length of the MAC frame. The exception is the REQ message, in which the LEN field carries the CM SID.

The extended header field (EHDR) is an optional extension to the MAC frame format. It is used to support such additional functions as frame fragmentation and data link security.

The HCS is a CRC-16 header check sequence field to detect errors in the MAC header. It uses the CRC-CCITT generator polynomial $g(x) = x^{16} + x^{12} + x^5 + 1$.

It is common for MAC messages to use Type-Length-Value (TLV) fields in which the first byte of the field indicates the type of element specified by the TLV, the second byte indicates the length of the TLV, and the remaining byte(s) carry the information being communicated by the TLV. The TLV format, commonly used in IP, provides a simple mechanism for adding the communication of new parameters. A CM or CMTS that does not understand a TLV ignores it.

11.9.1 Downstream MAC

As discussed in Section 11.7.1, the DOCSIS MAC frames are sent downstream encapsulated into a stream of MPEG packets. The CMTS schedules its own downstream transmissions. QoS control is provided by traffic priority, reserved data rates and token bucket rate limiting for policing and shaping.

11.9.1.1 Downstream Multicast

Prior to DOCSIS 3.0, DOCSIS provided limited support for multicast transmissions. Downstream multicast capabilities were enhanced with DOCSIS 3.0 in order to support multimedia services such as Internet Protocol Television (IPTV) that are based on IP Multicast. The new features were added such that backward compatibility was maintained with DOCSIS 2.0 multicast mode. Specifically, the new features added for DOCSIS 3.0 are:

- Source Specific Multicast (SSM) traffic forwarding for IGMP v3 [12] and MLDv2 [13] CPE devices.
- Bonded multicast traffic support.
- QoS provisioning support for multicast traffic.
- IPv6 multicast traffic support.
- Tracking CPEs explicitly when they are part of a multicast group at the CMTS in order better to support functions including usage tracking, billing, and load balancing.

Rather than using the IGMP (Internet Group Management Protocol) snooping that was required with DOCSIS 2.0, DOCSIS 3.0 uses simplified mechanisms at the CM. IGMP/MLD client messages arriving at the CM are forwarded transparently to the CMTS. The CMTS Layer-2 control mechanism establishes the forwarding of downstream multicast packets to specific CM interfaces by labeling all multicast packets with a Downstream Service Identifier (DSID). (See Section 11.9.4.5 for additional discussion of the DSID usage). The CMTS communicates the DSID and its associated group forwarding attributes to the CM, and uses the DSID to identify the set of CMs that are intended to receive the multicast packets with that DSID. The CM uses the DSID to identify the CPE-side interfaces to which the packet should be forwarded. This DSID filtering and forwarding mechanism is based on the one defined for IPv6 multicast traffic, which includes Neighbor Discovery and Router Solicitation. Note that the CMTS has the option of encrypting multicast packets on a per-session basis, through a Security Association that it communicated to a CM.

The DOCSIS Service Flow and Classifier QoS constructs are extended to support QoS for multicast traffic. Group Classifiers are used by the CMTS to classify which multicast packets are destined to a particular group of CMs on a network. The Group Service Flows can be statically provisioned by the cable operator.

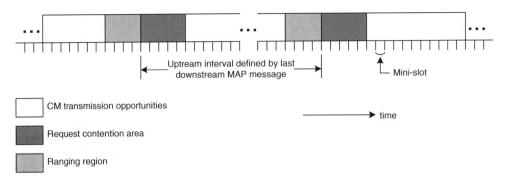

Figure 11.12 DOCSIS upstream frame format illustration

11.9.2   Upstream MAC

11.9.2.1   Upstream Signal Format

As illustrated in Figure 11.12, the upstream channel is divided in time into mini-slots that are numbered relative to a CMTS master clock. Any mini-slot in which a CM is allowed to start transmitting constitutes an upstream transmit opportunity. The CMTS synchronizes the mini-slot numbering at the CMs by sending SYNC (synchronization) packets. For TDMA mode, the mini-slots are $(2^n)(6.25\ \mu s)$ in duration, where $n = 1, \ldots, 7$ and $6.25\ \mu s$ is the inverse of the lowest symbol rate (160 KHz). The maximum length of an upstream burst is 255 mini-slots.

While the TDMA upstream transmissions are separated only in time, the S-CDMA upstream transmissions are separated by both time and CDMA spreading code. The mini-slot structure with S-CDMA is described in Section 11.7.2.2. The physical layer maps the mini-slots to frames.

The upstream DOCSIS MAC frames are preceded by a physical medium-dependent (PMD) header that indicates the mini-slot boundary to the MAC layer. MAC frame transmission must begin at a mini-slot boundary.

11.9.2.2   Upstream Bandwidth Requests

The CMTS grants upstream transmission times to the CMs a through "MAP" messages that it transmits downstream. The CMTS transmits one MAP message per upstream channel during each MAP interval. The MAP interval is provisionable, with 2–5 ms being typical. See Section 11.9.2.3 for more on the MAP messages and bandwidth grants, and the bandwidth request and grant example of Figure 11.13.

The access control is reservation based. One of the reservation mechanisms pre-schedules reservations for upstream transmission at periodical intervals. The other reservation mechanism is for the CMs to send their reservation requests in contention slots. This latter scheme is commonly known as "Slotted-ALOHA".

With contention-based reservations, a CM with data to send waits for a request contention interval, which is specified in the downstream MAP messages (see Figure 11.12). The CM then selects a sub-interval at random in which to transmit a six-byte REQ message. Specifically, the CM chooses a random number offset value from the initial backoff window starting time specified by the Data Backoff Start value in the most recently received MAP message (See Figures 11.13 and 11.14). It sends the request message so that it will arrive at the CMTS at the time determined by the random offset number. The REQ message uses a Service ID (SID) to identify the CM and specifies the number of mini-slots the CM needs (see Sections 11.9.2.3 and 11.9.4.1 for the description of the SID and its uses). If it receives no response from the CMTS to this request, it assumes that it was lost (e.g., due to a collision with another CM transmitting in the same sub-interval) and it enters a contention resolution process for resending the request.

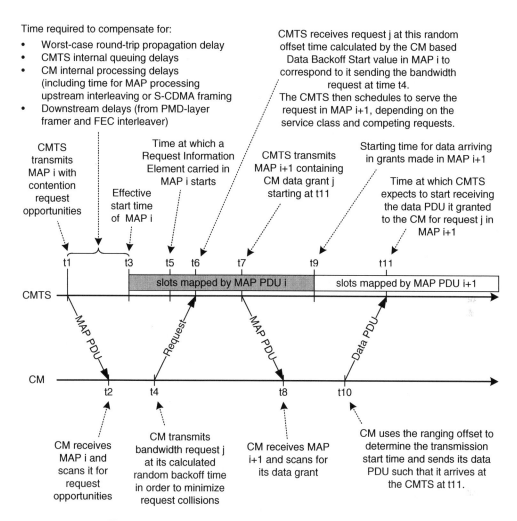

Time required to compensate for:
- Worst-case round-trip propagation delay
- CMTS internal queuing delays
- CM internal processing delays (including time for MAP processing upstream interleaving or S-CDMA framing
- Downstream delays (from PMD-layer framer and FEC interleaver)

CMTS receives request j at this random offset time calculated by the CM based Data Backoff Start value in MAP i to correspond to it sending the bandwidth request at time t4.
The CMTS then schedules to serve the request in MAP i+1, depending on the service class and competing requests.

CMTS transmits MAP i with contention request opportunities

Time at which a Request Information Element carried in MAP i starts

Effective start time of MAP i

CMTS transmits MAP i+1 containing CM data grant j starting at t11

Starting time for data arriving in grants made in MAP i+1

Time at which CMTS expects to start receiving the data PDU it granted to the CM for request j in MAP i+1

t1   t3   t5 t6   t7   t9   t11

slots mapped by MAP PDU i   slots mapped by MAP PDU i+1

CMTS

MAP PDU   Request   MAP PDU   Data PDU

CM

t2   t4   t8   t10

CM receives MAP i and scans it for request opportunities

CM transmits bandwidth request j at its calculated random backoff time in order to minimize request collisions

CM receives MAP i+1 and scans for its data grant

CM uses the ranging offset to determine the transmission start time and sends its data PDU such that it arrives at the CMTS at t11.

Figure 11.13 DOCSIS bandwidth request and grant illustration

The contention resolution process uses a *p*-persistency with truncated binary exponential backoff algorithm. The CM chooses a random number within an initial range (the initial backoff window) and waits that number of contention intervals before re-sending its request. If there is no response to that request, the CM chooses another random number from a window of numbers with a range that is twice that of the initial window. This process repeats, with the window range doubling each time until there is either a successful response to the request or 16 re-tries have been attempted. The CMTS controls the initial backoff window and maximum backoff window values. The window range is determined by the following formula, where the window is from 0 to the range value, and k is the number of collisions:

$$\text{Range} = 2^{\max\ [(\text{initial window}+k-1),(\text{maximum window})]} - 1$$

Note that a CM can only have one pending request per SID at a time.

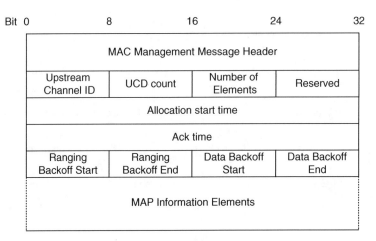

Bit 0   8   16   24   32

MAC Management Message Header			
Upstream Channel ID	UCD count	Number of Elements	Reserved
Allocation start time			
Ack time			
Ranging Backoff Start	Ranging Backoff End	Data Backoff Start	Data Backoff End
MAP Information Elements			

Figure 11.14 DOCSIS MAP message format

It is possible for MAP messages to be lost due to transmission errors, but the CM can recover from this situation as follows. The CM will continue to receive MAP messages for other upstream channels. It can see that the CMTS has no outstanding requests for the SID Cluster associated with the lost MAP when it has no pending grant for that SID Cluster and these other MAP messages arrive with an acknowledgment time for all channels that is past the time of the request associated with the lost MAP. At this point, the CM sends a new request for that Service Flow to send the data associated with the prior request and any additional data that has subsequently arrived.

11.9.2.3 Upstream Bandwidth Allocation

The CMTS queues and prioritizes the different REQ request messages and reserves future upstream mini-slots for the associated CMs to send their data. The scheduling algorithm takes into account parameters including the requested class of service and competing requests from other CMs for the same upstream channel. Quality of Service (QoS) and other MAC parameters are discussed further in Sections 11.9.4.1 and 11.10 on CM provisioning.

The MAP message from the CMTS tells a CM which mini-slots it can use to transmit data in the upstream frame of a given channel. The MAP message also communicates which mini-slots are available for contention based upstream request transmissions (as discussed above), and which mini-slots are available for new CMs to announce themselves to the CMTS (see Figure 11.12). The MAP message mini-slot allocations times are specified such that the upstream data arrives at the CMTS at the appropriate time in the upstream frame. This requires the CMTS to take into account the round trip propagation delay between it and the CM, and the associated processing time required by the CM. See Section 11.8.2 for the description of the ranging protocol and its round trip delay time adjustment. The request and grant message flow and processing are illustrated in Figure 11.13.

If the CMTS cannot service a CM bandwidth request in the next MAP message, it must send a zero-length grant to that CM as an acknowledgment that it has received the request from the CM. When a single downstream channel is associated with multiple upstream channels, it must transmit a separate allocation MAP per upstream channel.

The MAP message is a type of MAC management message (see Section 11.9.3). The message portion of the MAP message is shown in Figure 11.14. The UCD count indicates the number of times a change has been made to the UCD parameters. The fields associated with the Ranging Backoff are described in 11.8.2, and the fields associated with Data Backoff are described in 11.9.2.2.

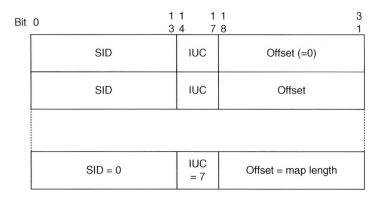

Figure 11.15 DOCSIS MAP message upstream bandwidth allocation IEs

The 32-bit mini-slot time count associated with the MAP goes from zero to $2^{26M}- 1$. The CMTS matches LSBs 0 to 25–M of the MAP time value to MSBs $6 + M$ to 31 of the SYNC timestamp counter, where M is the number of clock ticks per minislot. As a result, the start time for mini-slot N begins at a timestamp reference $(N)(2^M)(64)$. The 26-M LSBs of the Allocation start and Ack times are the MAP start and Ack times for TDMA channels.

For S-CDMA channels, the situation is somewhat different since the mini-slot size depends on the modulation rate, the number of codes per mini-slot and the spreading intervals per frame. The S-CDMA frame length is not necessarily a 2^k multiple of the 10.24 MHz reference clock. Consequently, the 32-bit timestamp counter role over point would not coincide with the S-CDMA frame boundary. The location of the S-CDMA frame boundary relative to the timestamp counter must be identified periodically for each upstream channel through a timestamp snapshot. The CMTS maintains both frame and mini-slot counts. It samples these counts and the timestamp count periodically on a frame boundary, and transmits the values as a timestamp snapshot in the UCD message, as illustrated in Figure 11.9.

The bandwidth allocation information elements (IE) of the MAP message are illustrated in Figure 11.15. The Service Identifier (SID) is a 14-bit field that identifies the Service Flow for which the grant is being made. See Section 11.9.4.1 for more on service flows. The Interval Usage Code (IUC) identifies the type of allocation MAP IE. The offset is the starting time offset. The time unit of the offset value is in terms of mini-slots relative to the beginning of the first mini-slot allocated to the CM (the Allocation start time). Hence, the first IE has an offset of 0. The end of an allocation is implied as being one less than the offset time specified in the next MAP IE. Consequently, the string of MAP IEs must end with a NULL MAP IE (SID = 0, IUC = 7). Note that no SID can be granted more than 255 mini-slots.

There are five 14-bit SID types defined for the MAP messages. As shown in Table 11.4, the SID types indicate whether this IE is intended for all CMs, multi-cast to an administrative set of CMs, or unicast to a specific CM or Service Flow within that CM.

Table 11.4 DOCSIS MAP IE SID types

MAP SID value	SID type
0x3FFF	Broadcast (for all stations)
0x300 – 0x3FFE	Multicast (with an administrative assignment)
0x200 – 0x3DFF	Expanded Unicast (designates a specific CM or service within that CM)
0x0001 – 0x1FFF	Unicast (designates a specific CM or service within that CM)
0x0000	Null address

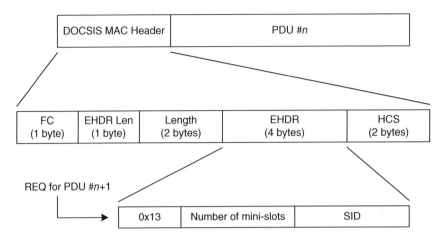

Figure 11.16 REQ Piggybacking illustration

11.9.2.4 Techniques for Improved Efficiency

The upstream MAC latency is affected by the frequency at which the MAP messages are sent (the MAP Rate), how far into the future the CMTS does the scheduling (the MAP Advance), and the likelihood of collisions during the contention requests. One way to reduce the latency is to have the REQ messages "piggyback" onto the upstream data transmissions. The other method is to concatenate multiple packets into the same upstream transmission.

REQ piggybacking works by including the REQ for the next upstream packet in the DOCSIS MAC header of the current upstream packet. This is illustrated in Figure 11.16. Piggybacking avoids the latency associated with waiting for a collision interval and the latency of REQ collisions.

With concatenation, the CM combines multiple smaller packets into a single larger upstream transmission frame. The CM makes its REQ for the number of mini-slots required for this concatenated frame. The concatenated frame begins with a Concatenation Header, followed by the individual packets. The CMTS uses the Concatenation Header to deconstruct the frame into the individual packets for forwarding. Concatenation reduces latency by avoiding multiple REQ and Grant cycles, with their potential associated collisions. It also minimizes overhead by using a single DOCSIS header for the multiple packets.

11.9.3 MAC Management Messages

The MAC management message format is illustrated in Figure 11.17. The format of the messages follows ISO/IEC-8802-2 [14]. The Destination Address (DA) of the MAC frame can be unicast to a specific CM or a multicast address. The Source Address (SA) indicates the CM or CMTS sending the message.

Within the MAC management header, the DSAP and SSAP are the Destination and Source Service Access Point addresses, respectively. The DSAP is always set to 00, and the SSAP is also set to 00 except when the frame is a RNG-REQ, INIT-RNG-REQ, or B-INIT-RNG-REQ. The control field is set to 03 to indicate an unnumbered information frame. The type field indicates the type of message, and the version indicates the DOCSIS versions that are compatible with this type of message. Version number 1 messages are compatible with all DOCSIS CM and CMTS systems. Version number 2 messages are compatible with equipment supporting DOCSIS 1.1, 2.0, and 3.0 (i.e., equipment beyond DOCSIS 1.0). Equipment supporting DOCSIS 2.0 and 3.0 is compatible with version number 3 messages, and only DOCSIS 3.0 equipment is compatible with version number 4 messages.

The important types of MAC management messages are summarized in Table 11.5.

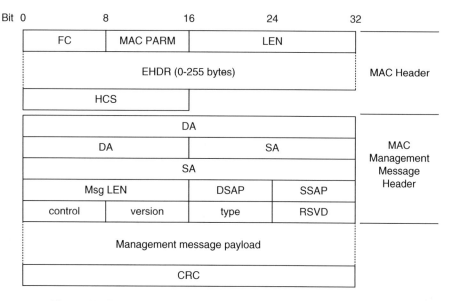

Figure 11.17 DOCSIS MAC header and management message format

11.9.4 MAC Parameters

The MAC parameters and components include:

- service flows;
- classifiers;
- upstream scheduling types;
- QoS parameters and parameter sets;
- payload header suppression;
- dynamic services;
- fragmentation.

11.9.4.1 Service Flows and Quality of Service (QoS)

DOCSIS Service Flows provide a MAC layer mechanism to support different QoS for transporting different traffic types between upper layer entities over the RF channels. In other words, the QoS is defined and managed on a per Service Flow basis. Packets are classified and mapped into Service Flows, with scheduling performed for each Service Flow such that its QoS parameters will be satisfied. The Service Flow is specified on a unidirectional basis and includes shaping, policies and priorities, and QoS parameters such as jitter, end-to-end delay and throughput for the traffic defined for that flow. Each CM requires at least two Service Flows, namely the Primary Upstream and Primary Downstream Service Flows. Service Flows are allowed to exist without being activated for carrying traffic.

The 32-bit Service Flow ID (SFID) is assigned by the CMTS. The CMTS also assigns a 14-bit service ID (SID) in order to implement the QoS for that Service Flow. Each SID is associated with a specific CM, and the CM uses its SID when requesting bandwidth. A basic CM will only support a single SID in each direction. CMs supporting multiple SIDs use them to support multiple service classes.

Table 11.5 DOCSIS MAC Management messages.

Message type	Type value	Summary description
SYNC	1	The SYNC message carries a timestamp that is value of the CMTS 32-bit counter that is locked to its 10.24 MHz master clock
UCD	2, 29, 35	Upstream channel descriptor – the CMTS transmits it periodically to define the upstream channel characteristics. It includes the upstream channel ID, the number of upstream mini-slot channels, the ID of the downstream channel carrying this message, a counter that is incremented whenever any of the channel descriptors is changed, and the various channel TLV parameters. Message type values 2, 29, or 35 are used for equipment supporting version 1, 3, or 4 messages, respectively.
MAP	3	Upstream bandwidth allocation MAP – used by the CMTS to communicate the upstream transmission start time, the ACK time, and ranging and data backoff information to the CMs.
RNG-REQ	4	Ranging request – sent by the CM at initialization, and periodically as requested by the CMTS, to determine the round trip delays and for adjusting the CM output power.
RNG-RSP	5	Ranging response – sent by the CMTS to assign or confirm the CM's SID. It also communicates various information parameters including those for adjusting the CM's timing (Ranging Offset), and transmitter power, frequency and equalization.
REG-REQ	6	Registration request – Sent by the CM to communicate its configuration file settings, all its modem capabilities, and initialization SID after it receives a CM parameter file.
REG-RSP	7	Registration response – sent by the CMTS to communicate the CM SID and various parameters including those associated with service flows, modem capabilities, and vendor-specific information.
UCC-REQ	8	Upstream Channel Change request – sent by the CMTS
UCC-RSP	9	Upstream Channel Change response – sent by the CM to indicate that is complying with the UCC-REQ.
BPKM-REQ	12	Privacy key management request (for DOCSIS SECv3.0)
BPKM-RSP	13	Privacy key management response (for DOCSIS SECv3.0)
REG-ACK	14	Registration acknowledgment – sent by the CM to acknowledge and confirm the QoS parameters it received in the REQ-RSP.
DSA-REQ	15	Dynamic service addition request – sent by either the CMTS or a CM to dynamically create a new Service Flow
DSA-RSP	16	Dynamic service addition response
DSA-ACK	17	Dynamic service addition acknowledgment
DSC-REQ	18	Dynamic service change request – sent by either the CMTS or a CM to dynamically change the parameters of an existing Service Flow
DSC-RSP	19	Dynamic service change response
DSC-ACK	20	Dynamic service change acknowledgment
DSD-REQ	21	Dynamic service deletion request – sent by either the CMTS or a CM to dynamically remove an existing Service Flow
DSD-RSP	22	Dynamic service deletion response
DCC-REQ	23	Dynamic channel change request – sent by the CMTS to cause a CM to either use a different upstream channel, receive its data on a different downstream channel, or both
DCC-RSP	24	Dynamic channel change response
DCC-ACK	25	Dynamic channel change acknowledgment
DCI-REQ	26	Device class identification request – a CM supporting this capability sends this message immediately after the CMTS indicates its completion of the ranging process, and waits to continue its initialization until it has received a DCI-RSP from the CMTS.
DCI-RSP	27	Device class identification response

Table 11.5 (*continued*)

Message type	Type value	Summary description
UP-DIS	28	Upstream transmit disable
INIT-RNG-REQ	30	Initial ranging request – sent by a CM that has yet to be registered
TST-REQ	31	Test request
DCD	32	Downstream channel descriptor
MDD	33	MAC domain descriptor – transmitted by the CMTS periodically on all downstream channels in the MAC domain, with a different MDD used for each channel
B-INIT-RNG-REQ	34	Bonded initial ranging request
DBC-REQ	36	Dynamic bonding change request – sent by the CMTS in order to change the channel bonding parameters for the downstream and/or upstream channels, or to change the downstream multicast parameters
DBC-RSP	37	Dynamic bonding change response
DBC-ACK	38	Dynamic bonding change acknowledgment
DPV-REQ	39	DOCSIS path verify request – used to determine the latency within the DOCSIS system. It may be sent by the CMTS to either a CM unicast MAC address or a multicast MAC address
DPV-RSP	40	DOCSIS path verify response
CM-STATUS	41	CM status report – sent by the CM on any available channel in response to certain types of events (e.g., MDD timeout or recovery, QAM/FEC lock failure or recovery, various timeouts, and whether the CM has begun to operate on battery or returned to AC power)
CM-CTRL-REQ	42	CM control – used by the CMTS to enforce certain actions (e.g., upstream muting, CM re-initialization, setting the upstream or downstream channel ID, or disabling forwarding)
CM-CTRL-RSP	43	CM control response
REG-REQ-MP	44	Multipart registration request
REG-RSP-MP	44	Multipart registration response
	46–255	Reserved for future use

Service Flows can be grouped together to form a Service Class. Upper layer entities and external applications can then use Service Classes as a globally consistent mechanism for requesting a Service Flow with the QoS parameters it desires.

A set of QoS parameters is provisioned through negotiation between the CM and CMTS when a Service Flow is created in the process of registering a CM. Dynamically created Service Flows begin without a provisioned set of QoS parameters. The CMTS and CM reserve resources for the admitted QoS parameter set, which is the set of Service Flow Encodings describing the QoS attributes of a Service Flow or Service Class.

DOCSIS 3.0 requires that the CMTS must be configurable to overwrite the packet's DiffServ Field setting. It also requires that the downstream packet queuing may be prioritized at a CM's Cable Modem to Customer Premises Equipment Interface (CMCI) using the Traffic Priority. In addition, packet queuing at the CMTS may be based on the DiffServ field in either the upstream or downstream direction. Also, a CM may reclassify downstream packets in order to provide enhanced service onto the subscriber-side network.

The CM can either make its bandwidth requests to the CMTS on a per-Service Flow basis, or based on all of the data in its queues for upstream transmission. The latter approach can be more efficient but, since the CMTS lacks knowledge of the packet boundaries, its upstream bandwidth grants can cross packet boundaries. Consequently, it needs to use the Continuous Concatenation and Fragmentation techniques described in Section 11.9.4.7.

The bandwidth request from the CM must take into account the total amount of bandwidth it needs for the upstream transmission, including the MAC-layer overhead associated with each packet. Since the CM cannot know how many segments the CMTS will use when it fragments the grant, it does not include any estimate of segment header bandwidth in its request. It is the responsibility of the CMTS to add in the bandwidth required for the segment headers.

When the CM sends a bandwidth request for a given Service Flow, it sends the request on any upstream channel that can be used by that Service Flow. The CMTS can then grant the bandwidth on any of the upstream channels, or combinations of channels, associated with that Service Flow.

DOCSIS provides further flexibility by allowing the grouping of all SIDs using an upstream channel within an upstream bonding group to be treated as a group of SIDs, and to make bandwidth requests/grants using a single SID for that upstream channel. This construct is called a SID Cluster. The SID Cluster is assigned to a specific CM service flow, and when the CM requests bandwidth for that service flow using the SID Cluster, it must send the request on the associated channel.

When channel bonding is used, the CM must keep track of the mini-slot state of all the channels in that bonded group. Specifically, it must remember the mini-slot count on all the channels of that bonding group at the precise time that it sends its request on one of the channels. The next later mini-slot count value is used if the mini-slot for the other channel(s) does not begin at exactly the same time as the one of the requesting channel.

11.9.4.2 Classifiers

Classifiers are used to map traffic into the defined QoS Service Flows. Packet header parameters (from layers 2, 3 or 4) are used to identify the traffic types. For example, the packet header information can be the MAC or IP addresses, TCP/UDP port numbers, or TOS/DSCP. More than one classifier can point to one Service Flow. The order in which different classifiers are searched and applied is determined by the priority field associated with each classifier type. The incoming packets are checked against the classifier list, with the packet sent to the appropriate Service Flow when a match is found. The packet is sent to the Primary Service Flow by default when no explicit match is found. Downstream classifiers are applied by the CMTS. The upstream classifiers are applied at the CM and, potentially, at the CMTS, for the purpose of policing the upstream packet classification.

11.9.4.3 Upstream Schedule Types

DOCSIS uses four types of upstream scheduling: Best Effort, Non-Real-Time Polling Services, Real-Time Polling Services, and Unsolicited Grant Services:

- Best Effort (BE) services use a simple bandwidth request from the CMs (using the contention slots) and bandwidth grants from the CMTS. The CMTS makes the grants according the available upstream bandwidth and priorities of other service types. The Best Effort traffic is allocated bandwidth on a first-come-first-serve basis, with no prioritization. The applications that use Best Effort scheduling include standard IP data services such as http and FTP. Piggybacked requests are allowed with BE.
- Both of the Polling Services provide the CM a direct, deterministic method to make bandwidth requests without the latency uncertainties of having to use the contention request mechanism.
  - Non-Real-Time Polling Services (nrt-PS) rely on the CMTS polling the CMs to determine their bandwidth needs on a nominally periodic basis. The period is not guaranteed, however, and could be random. Periods of one second or less are typical. The application for nrt-PS is premium data services.
  - With Real-Time Polling Services (rt-PS), the CMTS polls the CMs with guaranteed regularity. The primary applications are those such as videoconferencing where latency and jitter guarantees are required.
- With Unsolicited Grant Services (UGS), the CMTS provides bandwidth grants to the CMs without receiving a request from the CM or having polled it in advance. There are two types of UGS:

- ○ The first is a simple unsolicited bandwidth grant that the CM is guaranteed to receive on a regular basis. A typical application for this type of Unsolicited Grant is voice services. Once the voice connection is established, the CMTS knows that information can continue to arrive at the CM and can schedule the grants in advance.
- ○ The second type of UGS is UGS with Activity Detection (UGS-AD). The activity detection serves as the trigger to switch between UGS and rt-PS. UGS is used during periods when activity is detected, and rt-PS is used during periods when no activity is detected (i.e., "silent" periods). The UGS mode allows the CM to send voice information when the talker is speaking without the bandwidth penalty associated with poling. When there is no activity, poling increases the bandwidth efficiency by not allocating upstream bandwidth when it is not needed. In other words, UGS-AD allows more efficient transport of voice services. Note that piggybacking is not allowed with UGS.

11.9.4.4   QoS Parameters

The DOCSIS QoS mechanism provides management tools to guarantee or limit the subscriber data rates, and to provide latency guarantees. There are conceptually three types of QoS parameter sets. The provisioned QoS parameter set is defined and provisioned into the CM configuration file when a service flow is created. From this set, the CMTS and potentially the CM establish the admitted QoS parameter set for which the CMTS and CM reserve resources, including upstream bandwidth and, possibly, resources such as memory. The parameter set and associated resources that are actually committed to an active service flow are the active QoS parameter set.

The CMTS uses a logical function called the authorization module to determine whether a change to the classifiers or QoS of a service flow should be allowed or denied. With a static authorization model, admission and activation requests are allowed as long as the resulting active QoS is a subset of the admitted QoS parameter set which, in turn, is a subset of the provisioned QoS parameter set. With a dynamic authorization model, a policy server is used to determine in advance which CM requests will be permitted. If a request is made that does not match what was signaled in advance from the policy server, the request is either denied or is forwarded to the policy server for a real-time query. The dynamic model also allows parameter changes to existing service flows through a dynamic service change request (DSC-REQ). In summary, the authorization module is responsible for defining the range of acceptable admitted and active QoS parameters in order to make the decisions to allow or deny new requests.

11.9.4.5   Payload Header Suppression (PHS)

Some types of traffic have relatively static PDU headers for each packet, with the header information being large relative to the PDU data. For example, VoIP packets have a total of 54 bytes of header (14 for the Ethernet header, 20 for the IP header, 8 for the UDP header and 12 for the RTP header) and only 80 bytes of voice data. Consequently, it is very bandwidth-inefficient to send all the header information with each packet. PHS, which can be used for both upstream and downstream traffic, allows sending the repetitive header information once and caching it at the receiving end. The basic concepts of PHS are described in this section.

PHS is used to remove all or part of the client headers following the DOCSIS MAC frame's Extended Header field. Note that the bandwidth requests and grants take the suppression into account. All of the packets in a Service Flow to which PHS is applied will have a fixed length compressed header. When a client packet arrives at the DOCSIS MAC, the Classifier maps the packet into a Service Flow and, if that packet is eligible for PHS, it is also uniquely mapped to the appropriate PHS Rule (PHSR). The PHSR is established ahead of time between the CM and CMTS through a set of TLVs, and it provides the details regarding which header bytes are being suppressed by the sender for subsequent regeneration by the receiver. The PHS overhead contains an index value that identifies the PHSR being used for that packet. Multiple PHSRs can be used for each Service Flow. Note that the CM and CMTS delete the PHSR associated with a Service Flow when either that Service Flow or the Classifier is deleted.

The sender uses the PHS Field (PHSF) to communicate to the receiver a snapshot of the uncompressed client packet header that allows the receiver to reconstruct the packet header information after receiving the packet with a suppressed header.

PHS also supports suppressing a consistent subset of the client packet header bytes if some of those byte contain values that can change from packet to packet (e.g., sequence numbers or checksums). The PHSF includes all the suppressed and unsuppressed bytes, and a PHS Mask (PHSM) field is used to identify which bytes of the PHSF are suppressed and which are unsuppressed. The PHSM contains a bit field in which each bit of the bit field is associated with a byte of the client packet header. A 0 in a PHSM bit indicates that the associated header byte is not suppressed, while a 1 indicates that it is suppressed.

PHS supports the option of verifying whether on not individual client bytes should be suppressed. Specifically, the PHS Verify (PHSV) flag tells the sender whether or not to verify the client packet header bytes prior to performing suppression. Specifically, it verifies whether each client packet header byte that the PHSM indicates is to be suppressed contains the same information as the corresponding PHSF bytes. If the PHSV is not present, the default is to perform the verification. When the verification fails, the suppression operation will not be used for that packet.

There are two types of PHS, one pertaining to unicast traffic, and the other pertaining to downstream multicast traffic. These types can be summarized as follows:

- PHSI-indexed PHS, which can be used for unicast traffic in either direction, uses an 8-bit PHS Index (PHSI) assigned by the CMTS to uniquely identify the PHSR being used for that packet. A unique PHSI is used per CM in the downstream direction and a unique PHSI is used per Service Flow in the upstream direction. The respective PHSI values in the two directions are independent of each other. PHSI-indexed PHS was introduced with DOCSIS 1.1. It was designed for UGS (Unsolicited Grant Service), although it can be used for any Service Flow type.
- Downstream Service Identifier (DSID)-indexed PHS was introduced with DOCSIS 3.0 for use with DSID-labeled downstream multicast traffic. The 20-bit DSID is carried in the downstream Extended Header and is used with DSID-indexed PHS to uniquely identify the PHSR for that packet. The CMTS Classifier maps the packet into a Group Service Flow. However, this Group Service Flow is not communicated to the CMs. The CMTS and CM use the DSID carried in the Downstream Service Extended Header to identify the PHSR pertaining to that packet. One PHSR is supported per multicast DSID, and each CM must be capable of supporting at least 16 DSID-indexed PHSRs.

11.9.4.6 Dynamic Services

Dynamic Services allow a Service Flow to be created or deleted on demand rather than being permanently configured for the CM. It also allows the QoS parameters of the Service Flow to be changed on the fly. Either the CM or the CMTS can initiate a Dynamic Service. When a CM wants to add or change a Dynamic Service, it sends the appropriate request to the CMTS. The CMTS then responds to establish or modify the service. The CM completes the operation by acknowledging the CMTS response. The deletion of a Dynamic Service by the CM works the same way, except that there is no need for the CM to acknowledge the CMTS response.

11.9.4.7 Fragmentation

Large packets can create scheduling difficulties for real-time services. In order to avoid these problems, the CMTS can force a CM to fragment a large packet into smaller pieces. A fragmentation header (and a fragment CRC) is attached to each fragment so that the CMTS can reassemble the original packet. The CMTS takes into account the additional fragment overhead when determining these grants. Note that Concatenation frames can also be fragmented. There are two fragmentation modes.

The first of the fragmentation modes is called the multiple grant mode. As the name implies, in this mode multiple outstanding partial grants are allowed for a SID. These grants can occur either within the same MAP message or across consecutive MAP messages. When the grants span multiple MAP messages, the CMTS sends a zero length grant called a grant pending message in that MAP message and all subsequent MAP messages until the CMTS can grant the remaining bandwidth for that request.

The second fragmentation mode is called the piggyback mode. In this mode, the CMTS does not maintain fragmentation state information. Only a single outstanding partial grant is allowed. The CM forms a packet fragment to fill the current grant, and it calculates how much of the packet remains to be sent. The CM uses the piggyback field of the fragmentation overhead to request bandwidth for this remainder, including the required physical layer and fragmentation overhead bandwidth. The CMTS uses the fragment header check sequence to determine when a payload fragment is ready for re-assembly into the MAC frame.

The piggyback mode is generally simpler to implement, since it does not require the extensive fragment state information to be kept by the CMTS.

The Multiple Transmit Channel Mode available in DOCSIS 3.0 allows an additional concatenation method. Here, it is possible to use a continuous concatenation and fragmentation (CCF) mode, in which each service flow is treated as a separate continuous stream of data regardless of the channel on which it is transmitted.

CCF uses a segment header at the beginning of each segment that contains a pointer to the beginning of the first MAC header that is contained in that segment. This pointer allows the MAC frame boundaries to be recovered in the event that a segment is lost. The segment header also contains a sequence number that is used by the CMTS to align the different segments from this service flow. This is necessary due to the overlapping segments transmitted on different channels and variations in channel propagation delays. The CM increments the sequence number of the segment headers in the order in which it fills the segments for that service flow. Note that there is no concatenation or fragmentation header inserted with the data in CCF. Since this is a continuous stream, the CM completely fills each segment and must pad the end of a segment if it does not have enough data to fill it. In the downstream direction, the CCF grants are packed together with the data in the downstream stream.

11.10   CM Provisioning

The CM initialization and provisioning process is illustrated in Figure 11.18. The first thing a CM does when it connects to the network is acquire a downstream channel. The process includes scanning the digital channels, acquiring lock for the QAM, FEC and MPEG, and identifying the DOCSIS Program ID (PID). The next step is the upstream channel acquisition. The upstream channel is selected based on Upstream Channel Descriptor messages from the CMTS. Note that the CMTS can subsequently move the CM to a different upstream or downstream channel.

The CM then goes through the initial ranging process, which includes finding the Initial Maintenance interval in the MAP messages. At this point, the CM sends the RNG-REQ and receives the RNG-RSP messages, as described in detail in Section 11.8.2.

After the initial ranging process, the CM goes through a DHCP set-up process in order to establish IP communications. The CM includes its modem capability information in its DHCP discovery message. After the DHCP process is completed, the CM requests and receives the TFTP (Trivial File Transfer Protocol) from the cable operator, which it uses to obtain additional provisioning information, including the Download Configuration File. The Download Configuration (Config.) file defines the QoS parameters, filters, and privacy (encryption) parameters for the CM. As explained above, the QoS parameters include rate limiting and bandwidth guarantees. The filters include such things as blocking Windows file sharing (NetBIOS, SMB, CIFS).

The CM then requests and receives its time of day (ToD) provisioning and completes the registration process handshake. The registration request message from the CM (REG-REQ) communicates the QoS parameters for which the CM is configured, and repeats sending its modem capability information. The CMTS can turn off any CM capabilities that it cannot support. The final configuration stage, not shown in Figure 11.18, is the CM authentication and encryption key exchange with the CMTS.

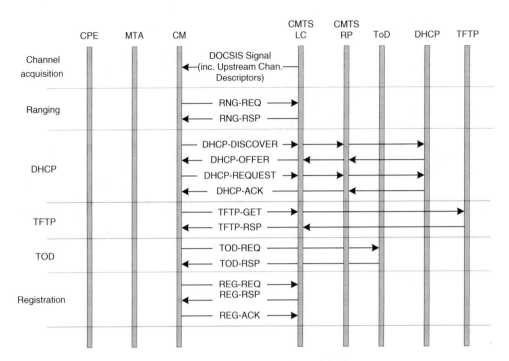

Figure 11.18 CM initialization and provisioning steps

11.11   Security

There are two factors that necessitate a more extensive security system for DOCSIS than for the FTTH protocols. The first is that the coaxial cable is a shared medium in which both the downstream and upstream physical layer signals are visible to all CMs sharing the same coaxial cable domain. FTTH systems can exploit the fact that only the downstream signal is visible to all ONUs. The second consideration is that while the ONU of a FTTH network is typically owned or provided by the network operator, the DOCSIS service model allows customers to provide their own CMs. Thus, an authentication mechanism is needed in order to prevent theft of service. While an extensive discussion of the DOCSIS 3.0 security protocols [15] is beyond the scope of this book, some of the basic concepts and capabilities will be discussed in this section.

The DOCSIS security features are controlled by the CMTS. It enforces encryption of the traffic flows in order to defend against unauthorized access to the data streams, performs an authentication of the CMs, and is the server for the distribution of encryption keys to the CMs.

DOCSIS 3.0 uses DOCSIS Baseline Privacy Plus (BPI+) architecture, which is an enhancement to the original DOCSIS Baseline Privacy scheme. BPI+ deals with authentication of the CMs, exchange of keys, and encrypted traffic session establishment between the CM and CMTS.

The security protocol begins with the CM provisioning. A new CM on the network is not initially trusted, but the level of trust for a CM increases as it successfully completes the different provisioning steps. The first step is device authentication. Additional steps include validation of the registration request and service authorization. The CM has achieved trusted status for carrying subscriber data services after it has been completely provisioned with its DOCSIS 3.0 security features enabled.

The authentication process proceeds as follows. First, after completion of ranging, but before the DHCP exchange, the Early Authentication and Encryption (EAE) process authenticates the new CM, using the

BPI+ authorization exchanges (if EAE is enabled at that CM). The CM must provide its unique X.509 digital certificate to the CMTS when it makes its authorization request. Next, EAE handles the Traffic Encryption Key (TEK) exchanges for the Primary SAID (Security Association Identifier) of the CM. Then, during CM initialization, EAE encrypts the provisioning traffic and REG-REQ-MP message. Note that the authentication protocol uses a three-level certificate hierarchy, with the Root Certification Authority (CA) belonging to CableLabs. A Manufacturer's CA at the second level is signed by the Root CA, and the Manufacturer's CA signs the individual CM certificate for CMs made by that manufacturer. This process ensures that only legitimate CMs receive encryption keys.

BPI+ is comprised of two protocols. The first protocol encapsulates and uses 128-bit AES encryption on the packet data within DOCSIS MAC frames that are transported across the network. It includes a cryptographic suite set that pairs algorithms for data encryption and authentication, and it defines the rules for applying these algorithms. Note that no encryption is applied to the DOCSIS MAC frame header. Also, the only management message that is encrypted is the REG-REQ-MP when EAE is enabled.

The second of the BPI+ protocols is the Baseline Privacy Key Management protocol, which enables the CMTS to securely distribute keying data to the CMs and to synchronize the CMTS and CM keying data. It also allows the CMTS to grant conditional network service access.

DOCSIS provides for the secure provisioning of CMs in order to both protect against service theft and also attacks against the CM or network. The secure provisioning processes cover DHCP, ToD and TFTP at the IP layer, and MAC layer registration.

DHCP packets between the CM and CMTS are encrypted when EAE is enabled for the CM. DOCSIS 3.0 supports IPv6, with IPv6 provisioning at the CMs using the lightweight DHCPv6 authentication protocol. Since the lightweight DHCPv6 authentication protocol uses DHCPv6 messages to distribute Reconfiguration keys to the CMs, it is critical to encrypt the DHCPv6 messages. EAE provides the means to protect the Reconfiguration key value at a CM that is provisioned to accept DHCPv6 Reconfiguration messages.

Since the TFTP protocol is used for file downloads in the provisioning process, it is especially important to encrypt the TFTP packets in order to keep the CM configuration information confidential. Also, encrypting the TFTP packets ensures that the level of service at a CM will be the same as that described in its configuration file. Enabling EAE for the CM protects sensitive configuration information. The CMTS provides both a TFTP server and client, acting as the TFTP server for CMs to download configuration files, and acting as the client for downloading these configuration flies from the provisioning system's TFTP server. The CMTS is referred to as a TFTP Proxy when it serves both roles at the same time. While TFTP Proxy support is the default, the CMTS must be capable of disabling it. The CMTS uses this proxy function to prevent disclosure of the configuration file server's IP address.

A CMTS acting as a TFTP Proxy can learn the name of a CM's configuration file through the DHCP configurations that an authorized DHCP server offers to that CM. The configuration file that a CM is authorized to download is identified by this file name. Hence, knowing this file name allows the CMTS to perform Configuration File Name Authorization, in which it discards any CM TFTP requests for which the learned configuration file name differs from the name of the one requested by a CM.

Since a CMTS enabled as a TFTP Proxy downloads the configuration files on the CM's behalf, it also can (and must be able to) perform Configuration File Learning in which it learns about the CM's configuration files. Knowing the contents of this file allows the CMTS to enforce consistency between a CM's Registration and its downloaded configuration file. The CMTS responds with an Authentication Failure if there is a mismatch.

DOCSIS also provides a secure mechanism for downloading software to a CM. The software download protocol includes authentication and integrity checking features that protect the CM from attacks by hackers.

An additional security feature with DOCSIS 3.0 is Certification Revocation. This feature allows for identifying CMs with revoked manufacturer certificates. The Certification Revocation List is maintained and distributed by a DOCSIS Root Certificate Authority.

11.12   Introduction to Companion Protocols

11.12.1   The PacketCable™ Protocol

The initial version of PacketCable was used for providing telephone service over the cable network, but the protocol has evolved to support advanced multimedia services. The PacketCable protocol runs over DOCSIS and uses standard IP packet data communications within a managed IP network structure. Specifically, PacketCable is an enhancement to the IP Session Initiation Protocol (SIP) that creates an IP multimedia subsystem. The resulting system is able to assign and manage the different QoS requirements for voice, video, and data services. PacketCable 3.0 was made to work with the DOCSIS 3.0 channel bonding and IPv6 capabilities.

The key motivation for PacketCable was to have a specification that would allow MSOs to construct their networks with equipment from multiple vendors, and to have all the equipment interoperate to provide the desired services.

11.12.2   The OpenCable™ Protocol

The OpenCable protocols and specifications have been developed by CableLabs for next generation digital consumer equipment. They include a reference design with the standard interfaces to the consumer products, head end, and core network. The objective for this project is to create a competitive environment for consumer equipment without any single company dominating.

11.12.3   PacketCable Multimedia (PCMM)

In order to provide flexible services to the cable network, DOCSIS uses PCMM. As illustrated in Figure 11.19, the PCMM framework consists of an applications manager and a policy server that interact with customer application and CMTS software. The applications manager interacts with the customer application in an application-specific manner (e.g., through a web page for video selection), and translates the application specific information into a format that can be understood by the policy manager. The policy server verifies whether the customer is authorized to use that service or access the requested item (e.g., a movie). The policy server interacts with the CMTS to enable reliable delivery of the service by dynamically creating the proper QoS profile.

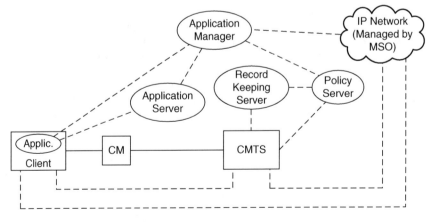

Figure 11.19 PCMM framework

11.13   Conclusions

The MSO infrastructure is widely deployed in much of the world. Due to its use of a combination of fiber for the access distribution plant and coaxial cable to the home, it has many advantages over copper telephone line connections for providing broadband data services to home users. The channel bonding capabilities introduced with DOCSIS 3.0 allow it to compete with at least some FTTH deployments for downstream data traffic rates.

However, the nature of the legacy video spectrum assignments leaves relatively little bandwidth available for upstream data transmission. It is also more limited in the maximum potential overall throughput that it can achieve in both the downstream and upstream directions, relative to an FTTH protocol such as those described in Chapters 3 and 4. The telephone carriers have used FTTH to leapfrog the bandwidth capabilities of the MSOs. The MSOs will ultimately need to address this with an all-fiber approach like the one described in Chapter 6.

References

1. CM-SP-PHYv3.0-I03-070223 Data-Over-Cable Service Interface Specifications 3.0 – Physical Layer Specification; 2013.
2. CM-SP-MULPIv3.0-I03-070223 Data-Over-Cable Service Interface Specifications 3.0 – MAC and Upper Layer Protocols Interface Specification; 2013.
3. Ciciora W, Farmer J, Large D, Adams M. *Modern Cable Television Technology*, 2nd edn. San Francisco: Morgan Kaufmann; 2003.
4. Chapman J. Next Generation CMTS – An Architectural Discussion. Proc. of the Emerging Tech. conference; 2008.
5. ITU-T J.112. Transmission systems for interactive cable television services; 2008.
6. ITU-T J.122. Second-generation transmission systems for interactive cable television services – IP cable modems; 2007.
7. ITU-T J222.0. Third-generation transmission systems for interactive cable television services – IP cable modems: Overvie ; 2007.
8. ITU-T J.222.1. Third-generation transmission systems for interactive cable television services – IP cable modems: Physical layer specification; 2007.
9. ITU-T J222.2. Third-generation transmission systems for interactive cable television services – IP cable modems: MAC and Upper Layer protocols; 2007.
10. ITU-T J222.3. Third-generation transmission systems for interactive cable television services – IP cable modems: Security services; 2007.
11. ITU-T J.83. Digital multi-programme systems for television, sound and data services for cable distribution; 2007.
12. IETF RFC 3376. Internet Group Management Protocol; 2002.
13. IETF RFC3 810. Multicast Listener Discovery; 2004.
14. ISO/IEC 8802-2. Information Technology – Telecommunications and information exchange between systems – Local and metropolitan area networks – Specific requirements – Part 2: Logical link control; 1998.
15. CM-SP-SECv3.0-I03-070223 Data-Over-Cable Service Interface Specifications 3.0 – Security Specification; 2007.

Further Readings

1. National Cable Television Association (NCTA), EIA/CEA-542 Rev. C, cable television channel identification plan; 2009.
2. ETSI EN 300 429, Digital Video Broadcasting (DVB); Framing structure, channel coding and modulation for cable systems; 1998.
3. Data-Over-Cable Service Interface Specifications 3.0 – Operations Support System Interface Specification; 2013.
4. Ovadia S. *Broadband Cable TV Access Networks*, Upper Saddle River, NJ: Prentice Hall; 2001.
5. CM-SP-PHYv3.0-I03-070223 Data-Over-Cable Service Interface Specifications 3.0 – Physical Layer Specification; 2007.
6. Young C. CATV Hybrid Amplifier Modules: Past, Present and Future, RFMD white paper WP090529; 2012.

12

Broadband in Gas Line (BIG)

12.1 Introduction to BIG

Broadband in gas line (BIG) has been proposed as an alternative broadband access technology using natural gas pipelines as a microwave waveguide from a network access node to customer's homes and businesses. To a large degree, the BIG access architecture would be similar to coaxial cable systems. Both BIG and coaxial cables have a hundreds of homes and businesses sharing access to a common transmission medium connecting to a network access node. In the United States, natural gas lines serve 63 million residential customers and more than five million commercial enterprises [1]. There is no known deployed service using BIG technology, and it is not clear if any field trials have been conducted.

12.2 Proposed Technology

The leading proponent of BIG technology is the StartUp company Nethercomm Corp. In theory, natural gas pipelines would provide an existing low-loss, low-noise, multi-drop radio waveguide to customers. Ultra-wideband radio technology has been proposed for BIG. The claimed media bit-rate capacity is approximately five to ten Gbps (probably combined upstream plus downstream) over a distance of ten kilometers or more. Like coaxial cable, this media bit-rate capacity would be shared by all customers connected to a distribution line. With a typical number of customers per line, this would yield approximately 40 Mbit/s per customer. There is no known independent verification of the achievable bit rate for various field conditions.

The ultra-wideband modem at the customer's home would attach to the gas line near the gas meter, and the modem would be powered via commercial power from the home. Thus, installation of the customer's BIG modem would require power wiring routed to an available electrical outlet in the home. The power wire connection could also be used for in-home broadband distribution using power line LAN technology. In-home broadband distribution via coaxial cable or phone wires would require the connection of these wires to the BIG modem.

12.3 Potential Drawbacks for BIG

Since active electronics and microwave energy would be placed in the flammable natural gas, the BIG system must be designed so as not to cause fires. Metal pipelines should serve as effective microwave waveguides, but microwave containment could be lost if there were any plastic pipe in the path, even for a short distance.

Broadband Access: Wireline and Wireless – Alternatives for Internet Services, First Edition.
Steven Gorshe, Arvind Raghavan, Thomas Starr, and Stefano Galli.
© 2014 John Wiley & Sons, Ltd. Published 2014 by John Wiley & Sons, Ltd.

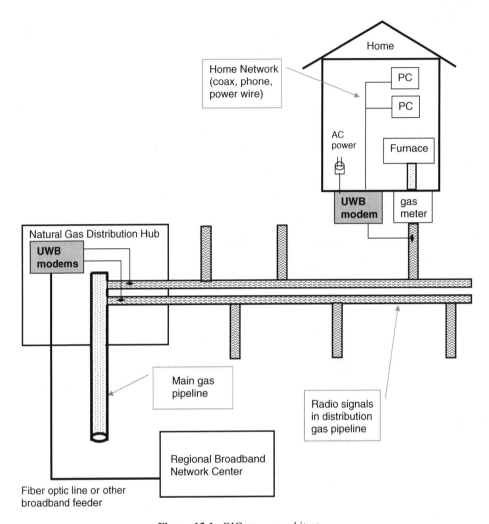

Figure 12.1 BIG access architecture

BIG shares the same marketing drawback as BPL (broadband power line). In nearly all areas it would be late to market because DSL, coaxial cable, fiber, and wireless broadband have already captured the customers. There are very few potential customers with gas line connections who do not already have broadband service available by other means. In order to compel customers to switch from their existing broadband service to a new service, the new service would have to be either much lower cost or a much better service. Accomplishing either of these is doubtful while making an attractive profit. Like other fixed broadband technologies, BIG lacks the mobility provided by wireless service.

Like fiber to the home (FTTH), BIG lacks native distribution within the home. The natural gas pipeline within the home would not used for the in-home distribution. Additional cost would be needed for in-home networking, possibly via inside coax, phone, or power wiring. The BIG modem by the gas meter could use IEEE 802.11 (WiFi) with a directional antenna to distribute wireless signals throughout the house. However, given the typical location for the gas meter, the WiFi signal quality could be poor for some parts of the house.

12.4   Broadband Sewage Line

The possibility of using sewage pipes as a microwave waveguide has also been discussed, but it is not known if its implementation is practical. What is certain is that sewage pipes, utility conduits, tunnels, bridges, and elevated railways are often used as convenient routes for new fiber optic cables. Consideration has also been given to routing fiber optic cables within natural gas pipelines.

Reference

1. American Gas Association http://www.aga.org/Kc/aboutnaturalgas/; 2006.

13

Power Line Communications

13.1 Introduction

Power Line Communications (PLC) is an old idea that dates back to the early 1900s, when the first patents were filed in this area [1]. Since then, utility companies around the world have been using this technology for remote metering and load control [2,3], using at first single carrier narrowband solutions operating in the Audio/Low Frequency bands that achieved data rates ranging from few bps to a few kbit/s. As technology matured and the application space widened, broadband PLC systems operating in the high frequency band (2–30 MHz) and achieving data rates up to a 200 Mbit/s started to appear in the market. Today broadband PLC devices use frequencies up to 100 MHz and can achieve data rates up to half a Gbit/s.

PLC can be used to provide broadband internet access to residential customers. Access to the internet is today becoming as indispensable as access to electrical power and, since many devices that access the internet are normally plugged into an electrical outlet, the unification of these two networks seems a compelling option. Additionally, PLC also allows reusing existing power line cables within a building to provide a broadband LAN within the home or office. The major advantage offered by PLC home networks is the availability of an existing infrastructure of wires and wall outlets, so that new cable installation is averted.

Besides the traditional access and LAN applications, PLC also has other interesting applications. Today in the construction of vehicles, ranging from automobiles to ships, from aircraft to space vehicles, separate cabling is used to establish the underlying physical layer of a local command and control network. As LAN technology and associated networking protocols continue to advance, local command and control networks are evolving towards broadband local networks, supporting a proliferation of sophisticated devices and software-based applications. Without exception, vehicles have a power distribution system based on metallic conductors of some type and this power distribution network may well perform a double duty – as an infrastructure supporting both power delivery, and also as broadband digital connectivity.

Finally, with the new impetus given by various administrations in the world to the modernization of their national power grids (Smart Grid), PLC will certainly have an important role in Smart Grid applications as power lines become the information highway dedicated to the management of energy transmission and distribution.

Broadband Access: Wireline and Wireless – Alternatives for Internet Services, First Edition.
Steven Gorshe, Arvind Raghavan, Thomas Starr, and Stefano Galli.
© 2014 John Wiley & Sons, Ltd. Published 2014 by John Wiley & Sons, Ltd.

There are several PLC technologies available today which can basically be grouped in three classes of PLC technologies [4]:[1]

- **Ultra Narrow Band (UNB)**: technologies operating at very low data rate ($\sim$100 bps) in the ultra low frequency (0.3–3 kHz) band or in the upper part of the super low frequency (30–300 Hz) band. Examples of UNB-PLC are Ripple Carrier Signaling (RCS), which operates in the 125–2000 Hz and is able to convey several bps using simple amplitude shift keying modulation; the AMR Turtle System, which conveys data at extremely low speed ($\sim$0.001 bps); the Two-Way Automatic Communications System (TWACS), which can carry data at a maximum data rate of two bits per mains frequency cycle, that is, 100 bps in Europe and 120 bps in North America. UNB-PLC have a very large operational range (150 km or more). Although the data rate per link is low, deployed systems use various forms of parallelization and efficient addressing that support good scalability capabilities.
- **Narrowband (NB)**: Technologies operating in the VLF/LF/MF bands (3–500 kHz), which include the European CENELEC (Comité Européen de Normalisation Électrotechnique) bands (3–148.5 kHz), the US FCC (Federal Communications Commission) band (9–490 kHz), the Japanese ARIB (Association of Radio Industries and Businesses) band (10–450 kHz), and the Chinese band (3–500 kHz). We can identify two sub-categories in NB-PLC:
  - *Low Data Rate (LDR)*: Single carrier technologies capable of data rates of few kbit/s. Typical examples of LDR NB-PLC technologies are devices conforming to the following recommendations: ISO/IEC 14908-3 (LonWorks), ISO/IEC 14543-3-5 (KNX), CEA-600.31 (CEBus), IEC 61334-5-2, and IEC 61334-5-1 (FSK and Spread-FSK). Non-SDO based examples are Insteon, X10, HomePlug C&C, SITRED, Ariane Controls, and BacNet.
  - *High Data Rate (HDR)*: Multicarrier technologies capable of data rates ranging between tens of kbit/s and up to 500 kbit/s. Typical examples of existing HDR NB-PLC technologies are ITU-T Recommendations G.9902 (G.hnem), G.9903 (G3-PLC), and G.9904 (PRIME). Additional non-SDO examples are PRIME and G3-PLC, which started as open specifications developed in industry alliances, but now have been approved as ITU-T standards.
- **Broadband (BB)**: Technologies operating in the HF/VHF bands (1.8–250 MHz) and having a PHY rate ranging from several Mbit/s to several hundred Mbit/s. Typical examples of BB-PLC technologies are devices conforming to the TIA-1113 (HomePlug 1.0), IEEE 1901, ITU-T G.hn (G.9960/G.9961) recommendations. Non-SDO based examples are HomePlug AV 2.0, HomePlug Green PHY, UPA Powermax, and Gigle MediaXtreme.

This chapter will focus mostly on BB-PLC applications, with particular attention to the access and in-home environments. Nevertheless, other applications related to Smart Grid will also be addressed as the role of PLC for Smart Grid is becoming increasingly important [4]. For an in-depth review of all PLC applications, see [2–4,24].

13.2   The Early Years

The first PLC application that was put in place by power utilities involved voice and data communications over High voltage (HV) lines which typically bear voltages above 100 kV and span very large geographical distances. HV lines have been used as a communications medium for voice since the 1920s (power carrier systems) [1,4]. At that time, telephone coverage was very poor, so engineers operating power plants and transformer stations needed an alternative way to communicate for operations management with colleagues stationed tens or hundreds of kilometers away. HV lines are good waveguides, and a transmit power of 10 W was sufficient to span distances up to 500 km.

Initially, voice communication over HV lines was analog and based on Amplitude Modulation, typically single-sideband. A reasonable band of some 100 kHz is available on HV lines in the Low

---

[1] BB-PLC technologies devoted to internet-access applications have also been referred to as Broadband over Power Lines or BPL, whereas LDR NB-PLC technologies have been referred to as Distribution Line Carrier or Power Line Carrier.

Frequency (LF) band (30–300 kHz) and sometimes[2] up to 500 kHz for transmitting over distances of few hundred kilometers, and Frequency Division Multiplexing (FDM) was used to accommodate multiple voice channels. Also digital communications techniques started to be introduced for telemetering and telecontrol, although very low data rates in the order of few hundred bps were achieved.

Another important driver for the original interest of utilities in PLCs was load management, that is, the capability of switching on/off appliances responsible for high energy consumption as water heaters (Ripple Carrier Signaling or RCS [2]). Utilities have used RCS since the 1930s to control peak events at demand side by issuing control signals to switch off heavy duty appliances in order to balance generation and demand.[3] This application was unidirectional and was carried over the Medium Voltage (MV) and Low Voltage (LV) levels, which carry typically 1–100 kV and less than 1 kV, respectively. RCS used the low part of the audio band (125 Hz to 2 kHz) and simple Amplitude Shift Keying (ASK) modulation, conveying several bps.

The reason why RCS was unidirectional is that it required a high transmit power in the order of tens of kW, and, exceptionally even 1 MW. Although the MV and LV part of the grid offer very good propagation characteristics (low loss, low interference) to signals in the audio band, the impedance of a heavily loaded network is very low in the Ultra Low Frequency (ULF) band so that transmit power depends on the load of the network. On the other hand, signals in the audio band propagate easily through the MV/LV transformer without any need for special (and costly) coupling equipment for by-passing the transformer. Distances of tens of kilometers can be reached with RCS so that, generally, there are few injection points located on the MV section. RCS has been quite successful, especially, in Europe, and its use has been extended to include other applications such as day/night tariff switching, street light control, and control of the equipment on the power grid.

Other examples of UNB-PLC are the AMR Turtle System which conveys data at extremely low speed (~0.001 bps) and the Two-Way Automatic Communications System (TWACS) that can carry data at a maximum data rate of two bits per mains frequency cycle (i.e., 100 bps in Europe and 120 bps in North America). UNB-PLC have a very large operational range (150 km or more). Although the data rate per link is low, deployed systems use various forms of parallelization and efficient addressing that support good scalability capabilities. Despite the fact that these UNB solutions are proprietary, they are very mature technologies, they have been in the field for at least two decades and have been deployed by hundreds of utilities.

As years passed, sophistication increased and the devices used for load management in MV/LV distribution networks started using more efficient modulation methods thus allowing higher data transfers. The transmit power requirements decreased so that bi-directional data transfers could also be supported. The decrease in transmit power was reached by increasing the frequency of the carrier signal which also implied that it was not always possible to go through the MV/LV transformer.

13.3 Narrowband PLC*

Recognizing the increasing desire for the use of higher (above audio) frequencies, the European Committee for Electrotechnical Standardization (Comité Européen de Normalisation Electrotechnique, CENELEC) issued in 1992 standard EN 50065 (revised in 2011). The CENELEC EN 50065 standard allows communication over LV distribution power lines in the frequency range from 3 kHz up to 148.5 kHz[4] and addresses four frequency bands:

- the A-band (3 kHz–95 kHz), reserved for power utilities only;

---

[2] The band between 300 kHz and 500 kHz may be used when the risk of frosting is low since this frequency range is not very tolerant to frost and attenuation would grow very quickly with distance [2].

[3] Interestingly, this old method of load control is today again at the center of discussion in the context of Smart Grid – see Section 13.9 for more details. However, RCS is a unidirectional technique while today power utilities demand a two-way communications link in order to receive an acknowledgement that the issued command has been carried out.

[4] Prior to publication of EN 50065, several European countries allowed free usage of this spectrum, although a maximum transmit power was specified. Compliance to EN 50065 is now mandatory.

*Source: [4]. Reproduced with permission of IEEE.

- the B-band (95 kHz–125 kHz), that can be used by all applications without any specific access protocol'
- the C-band (125 kHz–140 kHz), reserved for home network systems with a *mandatory* access protocol CSMA;
- the D-band (140 kHz–148.5 kHz), which is specified for alarm- and security systems without any access protocol.

Note that CENELEC specifications regulate only spectrum usage and do not mandate any modulation or coding schemes.

EN 50065 mandates a CSMA/CA mechanism in the C-band and stations that wish to transmit must use the 132.5 kHz frequency to inform that the channel is in use. This mandatory protocol defines a maximum channel holding period (1 sec), a minimum pause between consecutive transmissions from the same sender (125 ms), and a minimum time for declaring the channel is idle (85 ms). Note that CENELEC specifications regulate only spectrum usage and the CSMA/CA protocol but do not mandate any modulation or coding schemes. Interestingly, the coexistence mechanism defined in EN 50065 is PHY/MAC-agnostic and several NB-PLC standards were developed after EN 50065 was ratified [4].

In other countries, regulations are different. For example, in the USA, the use of the 10 kHz to 490 kHz band is allowed by the FCC. In Japan, ARIB has allowed the use of the 10 kHz to 450 kHz band. Neither FCC nor ARIB assign bands to utilities for exclusive use.

The use of these bands has been very successful, and power utilities across the world – and especially in Europe – have made wide use of the CENELC standard for meter reading and other low data rate applications. Most PLC technologies used in the CENELEC bands are single carrier and provide at most data rates of few tens of kbit/s.

Today, there is also a growing interest in HDR NB-PLC solutions operating in the CENELEC/FCC/ARIB bands which are able to provide higher data rates than LDR NB-PLC. For example, the Powerline Related Intelligent Metering Evolution (PRIME) initiative has gained industry support in Europe and has specified an HDR NB-PLC solution based on Orthogonal Frequency Division Multiplexing (OFDM), operating in the CENELEC-A band, and capable of PHY data rates up to 125 kbit/s. A similar initiative, the G3-PLC Alliance, was also recently launched for the development of G3-PLC which is an OFDM-based HDR NB-PLC specification that supports IPv6 and can operate in the 10–490 kHz band. Both G3-PLC and PRIME are now international standards and are specified in the ITU-T G.9903 and G.9904 Recommendations, respectively.

13.3.1   Overview of NB-PLC Standards

One of the first LDR NB-PLC standards ratified was the ANSI/EIA 709.1 standard, also known as LonWorks. Issued by ANSI in 1999, it became an international standard in 2008 (ISO/IEC 14908-1). This seven-layer OSI protocol provides a set of services that allow the application program in a device to send and receive messages to from other devices in the network without needing to know the topology of the network or the functions of the other devices. LonWorks transceivers are designed to operate in one of two frequency ranges depending on the end application. When configured for use in electric utility applications, the CENELEC A-band is used, whereas in-home/commercial/industrial applications use the C-band. Achievable data rates are in the order of few kbit/s.

The most widespread PLC technologies deployed today are based on Frequency Shift Keying (FSK) or Spread-FSK as specified in the IEC 61334-5-2 and IEC 61334-5-1 standards, respectively. The availability of standards for these technologies goes from recommendations that specify the stack of communications protocols from the physical up to the application layer (IEC 62056-53 for COSEM) thus facilitating the development of interoperable solutions. Such AMR/AMI solutions are now provided by a number of companies and are widely and successfully implemented by utilities. Spread-FSK based NB-PLC devices are being currently deployed in Europe.

An effort was made in the ITU-T to standardize both PRIME and G3-PLC technologies and this was successfully accomplished in the ITU-T Study Group 15 Question 15 (Q15/15). Both of these specifications have recently been approved as international standards by the ITU-T and are open specifications available online. The following ITU-T recommendations have received final approval in late 2012:

- G.9901 "Narrowband OFDM Power Line Communication Transceivers–Power Spectral Density (PSD) Specification". This recommendation contains all the material with regulatory relevance and implications, such as OFDM control parameters that determine spectral content, PSD mask requirements and the set of tools to support reduction of the transmit PSD.
- G.9902 (G.hnem) "Narrowband OFDM Power Line Communication Transceivers for ITU-T G.hnem Networks". This recommendation contains the PHY and the DLL specifications for G.hnem NB-PLC transceivers.
- G.9903 (G3-PLC) "Narrowband OFDM Power Line Communication Transceivers for G3-PLC Networks". This recommendation contains PHY and the DLL specifications for G3-PLC NB-PLC transceivers.
- G.9904 (PRIME) "Narrowband OFDM Power Line Communication Transceivers for PRIME Networks". This recommendation contains the PHY and the DLL specifications for PRIME v. 1.3.6 NB-PLC transceivers.

Another SDO effort in the area of HDR NB-PLC is IEEE 1901.2, which was published in December 2013 and has many similarities with G3-PLC (ITU-T Recommendation G.9903).

13.4 Broadband PLC*

As NB-PLC started to be progressively successful, BB-PLC started to appear as well - initially for internet access applications and then for LAN applications [4]. The first wave of interest into the use of BB-PLC for internet access started in Europe. In 1997, Nortel announced a partnership with UK power utility United Utilities under the name of NorWeb for the development of a technology to provide access service to residential customers via PLC. Limited trials of broadband internet access through PLs were conducted in Manchester and NorWeb prototypes were able to deliver data at rates around 1 Mbit/s. However, higher than anticipated costs and growing EMC issues caused the early termination of the project in 1999. Other projects in Europe led by Siemens and Ascom encountered a similar fate. On the other hand, a multi-year project funded by the European Community (The Open PLC European Research Alliance, OPERA) led most of the recent research efforts in the field of BB-PLC for internet access.

Given the initial disappointing results in using PLC for internet access applications, industry interest started shifting towards in-home applications in early 2000. In the last decade, several industry alliances have been formed with a charter to set technology specification mostly for in-home PLC, for example, the HomePlug Powerline Alliance (HomePlug), Universal Powerline Association (UPA), High Definition Power Line Communication (HD-PLC) Alliance, and The HomeGrid Forum. Products allowing PHY data rates of 14 Mbit/s (HomePlug 1.0), then 85 Mbit/s (HomePlug Turbo), and then 200 Mbit/s (HomePlug AV, HD-PLC, UPA) have been progressively available on the market over the past several years. However, none of these technologies are interoperable with each other.

BB-PLC for internet access started to regain some traction, as failures sometimes have short memory and the Norweb issues became forgotten. The rapid advances in signal processing and the availability of in-home PLC products probably inspired some optimism. Thus, in the past decade, several field trials and some commercial deployments of access PLC have started to appear worldwide.

In Europe, especially in Germany, large deployments involving some thousand customers were performed in Mannheim and Ellwangen. In Asia, some field trials are already in an advanced phase. In Korea, a PLC Forum was founded in 2000 for standardization for application of PLC systems, such as internet access, home automation and digital appliances.

*Source: [4]. Reproduced with permission of IEEE.

In the United States, power utilities partnered with vendors to conduct field trials (e.g., Southern Company – Ambient Corporation; Main.Net – ConEd and PowerTrust). Two of the largest deployments have involved Current Communications. The largest operational deployment to date began in 2004 as a joint venture between Current Communications and Cinergy Corporation (which provides electricity to the greater Cincinnati, Ohio area) and parts of Northern Kentucky. In December of 2006, Current Communications announced another large-scale joint venture with TXU Corporation of Dallas to offer BPL service to two million homes and businesses in the Dallas-Fort Worth area. Thanks to federal subsidies, BB-PLC is also being deployed in US rural communities where other means of broadband access are not present or are too expensive to deploy.

More recently, Scottish Power started trials on access PLC for around 1000 homes in Liverpool, UK, in 2011.

To further confirm this renewed interest in BB-PLC, two new BB-PLC standards have been ratified: IEEE P1901 and ITU-T G.hn.

BB-PLC is also being tested today over HV lines. The feasibility of sending high data rate PLC signals over HV lines has been reported recently by the US Department of Energy, American Electric Power, and Amperion, who jointly tested successfully a BB-PLC link over a 69 kV and 8 km long line with no repeaters. Data rates of 10 Mbit/s with latency of about 5 ms were reported while complying with FCC emission limits. The trial employs multiple 5 MHz bands in the range of 2–30 MHz using DS2 (UPA) chips to communicate over two contiguous 69 KV lines. Next steps for this project are raising the applicable voltage to 138 kV and also extending repeater spacing. Other international activities involving PLC over HV lines can be found in Italy, Spain, and Korea.

13.4.1   Overview of BB-PLC Standards

In the next few sections, we will give an overview on the latest standardization developments that occurred in BB-PLC. More details about the IEEE 1901 Broadband over Power Lines Access Standards will be given in Section 13.8.

13.4.1.1   The TIA-1113 Standard (HomePlug 1.0)

The world's first BB-PLC ANSI standard to be approved is the TIA-1113. The standard is largely based on the HomePlug 1.0 specifications and defines a 14 Mbit/s PHY based on OFDM. Carriers are modulated with either BPSK or QPSK depending on the channel quality and operational functionality. The Media Access Control (MAC) for HomePlug 1.0 is based on a CSMA/CA scheme that features an adaptive window size management mechanism in conjunction with four levels of priority. Products based on the TIA-1113/HomePlug 1.0 specifications have experienced a good success in the in-home and industrial markets. Despite its design targets in home or industrial applications, HomePlug 1.0 has often been used in access trials.

13.4.1.2   The IEEE 1901 Broadband over Power Lines Standard

The IEEE 1901 Working Group was established in 2005 to unify PL technologies with the goal of developing a standard for high-speed (>100 Mbit/s) communication devices using frequencies below 100 MHz and addressing both HAN and access applications [5,6]. The standard was approved in September 2010 and it defines two BB-PLC technologies [7]: an FFT-OFDM based PHY/MAC and a Wavelet-OFDM based PHY/MAC. Another key component of the IEEE 1901 standard is the presence of a mandatory coexistence mechanism called the Inter-System Protocol (ISP) that allows 1901-based PLC devices to share the medium fairly regardless of their PHY differences. Furthermore, the ISP also allows IEEE 1901 devices to coexist with devices based on the ITU-T G.hn standard. The ISP is a new protocol that is unique to the PL environment [25].

The FFT-OFDM IEEE 1901 PHY/MAC specification facilitates backward compatibility with devices based on the HomePlug AV specification. Similarly, the Wavelet-OFDM IEEE 1901 PHY/MAC specification facilitates backwards compatibility with devices based on the HD-PLC specifications of the HD-PLC Alliance led by Panasonic. The multi-PHY/MAC nature of the IEEE 1901 standard does not descend from a technical necessity, but is simply the consequence of a compromise caused by the lack of industry alignment behind a single technology. On the other hand, we can consider the multi-PHY/MAC nature of the IEEE 1901 standard as the first step towards that further consolidation of PLC technologies that will inevitably happen in the future.

Devices conforming to the standard must be capable of at least 100 Mbit/s and must include ISP in their implementation. Mandatory features allow IEEE 1901 devices achieving ~200 Mbit/s PHY data rates, while the use of optional bandwidth extending above 30 MHz allows somewhat higher data rates to be achieved. However, data rate improvements due to the use of higher frequencies are often marginal and are characterized by short range, due to the higher attenuation of the medium and the presence of TV broadcast channels above 80 MHz.

Since the IEEE 1901 standard is the only standard for the use of PLC as an access technology, more details will be given in Section 13.8.

13.4.1.3   The ITU-T G.hn Home Networking Standard

The ITU-T started the G.hn project in 2006 with a goal of developing a worldwide recommendation for a unified HAN transceiver capable of operating over all types of in-home wiring: phone lines, PLs, coax and Cat 5 cables, and bit rates up to 1 Gbit/s. The PHY of G.hn was ratified by the ITU-T in October 2009 as Recommendation G.9960, while the Data Link Layer was ratified in June 2010 as Recommendation G.9961. The technology targets residential houses and public places, such as small/home offices, multiple dwelling units or hotels; it does not address PLC access applications as IEEE 1901 does. Compliance to ITU-T Recommendations G.9960/G.9961 does not require the support for coexistence. Thus the support of ISP is optional for G.hn-compliant transceivers.

Past approaches emphasized transceiver optimization for a single medium only, that is, either for PLs (HD-PLC Alliance, HomePlug, UPA), or phone lines (HomePNA), or coax cables (MoCA) only. The approach chosen for G.hn was to design a single transceiver optimized for multiple media, that is, for power, phone and coax cables. Thus, G.hn transceivers are parameterized so that relevant parameters can be set depending on the wiring type. A parameterized approach allows, to some extent, optimization on a per media basis to address channel characteristics of different media without necessarily sacrificing modularity and flexibility. G.hn also defines an interoperable low complexity profile for those applications that do not require a full implementation of G.hn.

The G.hn WG engaged in a year-long debate about the selection of the advanced coding scheme. The two competing proposals were based on a Quasi-Cyclic Low Density Parity Check (LDPC) code and a Duo-Binary Convolutional Turbo Code. The Turbo Code proposed in G.hn was meant to be an improvement over the one specified in the IEEE 1901 FFT-OFDM PHY/HomePlug AV as it allowed a higher level of parallelism and better coding gain. Following the comparative framework proposed in [8], the G.hn Working Group selected the LDPC code as the only mandatory FEC.

More recently, support for MIMO and neighbor network interference rejection have been added to Ghn and standardized in the ITU-T. Additional information on G.hn can be found in Section 9.4.

13.4.1.4   PLC Co-Existence

PL cables connect LV transformers to a set of individual homes or set of multiple dwelling units without isolation [4]. Signals generated within the premises interfere with each other and also with signals generated outside the premises. As the interference increases, both from indoors and outdoors sources, PLC stations will experience a decrease in data rate as packet collisions increase, or even suffer complete service interruption. Hence, PL cables are a shared medium (like coax and wireless) and do not provide links dedicated exclusively to a particular subscriber. As a consequence, the PLC channel is interference-limited, and approaches based on Frequency Division Multiplexing (FDM) as in WiFi or coax are not suitable because only a relatively small band is available for PLC. It is therefore necessary to devise mechanisms to limit the harmful interference caused by non-inter-operable neighboring devices. Note that similar considerations can be made about the interference limited nature of many wireless networks, for example, WiFi, WiMAX, Zigbee, Bluetooth, and so on.

It is also important to ensure coexistence between Smart Grid and in-home PLC technologies, since the former have traditionally a much longer obsolescence horizon than the latter. It is likely that the number of homes fitted with energy metering and control devices that utilize Smart Grid technology will dramatically increase in the near future. On the other hand, in-home PLC technology continuously evolves, improving the transmission rate. The adoption of a coexistence mechanism will enable continued and efficient operation of Smart Grid devices in the presence of newly-deployed in-home PLC devices.

The issue of PLC coexistence was first raised two decades ago in CENELEC. Since CENELEC did not mandate a specific PHY/MAC technology, it was necessary to provide a fair channel access mechanism that avoided channel capture and collisions when non-interoperable devices operate on the same wires. In fact, if non-interoperable devices access the medium, then native CSMA and virtual carrier sensing do not work and a common medium access mechanism must be defined. CENELEC mandated a CSMA/CA mechanism only for the C-band, where a single frequency (132.5 kHz) is used to inform that the channel is in use.

Another approach to coexistence was introduced by the HomePlug Powerline Alliance to solve the issue of non-interoperability between HomePlug 1.0 and HomePlug AV devices. The HomePlug *hybrid delimiter* approach allows HomePlug AV/IEEE 1901 FFT-OFDM PHY stations to coexist with Home-Plug 1.0 (TIA-1113) stations by pre-pending to their native frame the HomePlug 1.0/TIA-1113 delimiter. This allows stations to correctly implement CSMA/CA and virtual carrier sensing.

The hybrid delimiter approach is a CSMA-based coexistence mechanism and, thus, does not eliminate interference caused by non-interoperable stations and cannot guarantee QoS when the traffic of at least one of the coexisting technologies grows. The use of hybrid delimiters is a somewhat inefficient approach if multiple technologies are to coexist as it would be necessary to prepend multiple delimiters (one for every non-interoperable technology) with increasing loss in efficiency.

The HomePlug hybrid delimiter method also exhibits security weaknesses as it is not a mechanism based on *fair-sharing*. In fact, HomePlug AV/IEEE 1901 FFT-OFDM PHY can defer indefinitely HomePlug 1.0 (TIA-1113) stations from accessing the medium so that, while HomePlug 1.0 (TIA-1113) stations cease all transmissions, HomePlug AV/IEEE 1901 FFT-OFDM PHY stations remain the only active ones on the medium. This capability may raise security concerns since HomePlug AV/IEEE 1901 FFT-OFDM PHY stations (either legitimate or rogue) can prevent Smart Grid devices based on HomePlug 1.0 (TIA-1113) from working.

Except for the CSMA mechanisms described above, the issue of coexistence between BB-PLC devices has been rarely addressed in the technical literature and the first published paper dates back only few years. The coexistence specification by OPERA was meant to ensure compatibility between access and in-home BB-PLC deployments. The BB-PLC coexistence scheme developed in by the Consumer Electronics Powerline Alliance (CEPCA) and UPA was to ensure coexistence between non-interoperable BB-PLC devices. In fact, since the lack of BB-PLC standards was causing a proliferation of proprietary solutions and since the industry did not seem to align behind any specific technology, coexistence seemed a necessary "evil" to ensure some etiquette on the shared medium and prevent interference. The CEPCA/UPA coexistence protocol is now included as an option in the IEEE 1901 standard.

For the specific case of coexistence between the two IEEE 1901 PHYs, Panasonic proposed to the IEEE 1901 WG a novel coexistence mechanism called the Inter-PHY Protocol (IPP). The IPP was designed initially to ensure compatibility with the CEPCA/UPA coexistence protocol but it was simpler, it allowed some distributed features, and it also allowed devices to perform Time Slot Reuse.[5]

Although the IPP was originally designed to enable efficient resource sharing between devices equipped with either the IEEE 1901 Wavelet-OFDM or the IEEE 1901 FFT-OFDM PHYs, it was soon recognized that the IPP could have been also an excellent tool for regulating simultaneous access to the channel of both IEEE 1901 and non-IEEE 1901 devices (e.g., the ones based on the ITU-T G.hn standard). Panasonic modified the IPP originally conceived to extend coexistence to G.hn devices and proposed this enhanced mechanism called Inter-System protocol (ISP) to both ITU and IEEE. The ISP is now a mandatory part of the IEEE 1901 standard (see [7], Chapter 16) and is also specified in ITU-T Recommendation G.9972, which was ratified by the ITU-T in June 2010. The approach followed in the design of the IPP/ISP is a radical conceptual departure from previous designs in CENELEC and in HomePlug which are both based on CSMA. Thus, none of the drawbacks mentioned above are present in the ISP.

As a result of the efforts of PAP 15, IEEE 1901 compliant devices implementing either one of the two IEEE 1901 PHY/MACs can coexist with each other. Likewise, ITU-T G.9960/9961 (G.hn) devices that implement ITU-T G.9972 can coexist with IEEE 1901 compliant devices implementing either one of the two IEEE 1901 PHY/MACs, and vice versa. A recent set of PAP 15 recommendations to the Smart Grid Interoperability Panel (SGIP) requires to "*Mandate that all BB-PLC technologies operating over power lines include in their implementation*" coexistence. In more detail, in order to be compliant with this recommendation, IEEE 1901 compliant devices must implement and activate ISP (i.e., be always on), ITU-T G.9960/G.9961 compliant devices must be compliant with and activate ITU-T G.9972 (i.e., be always on), and any other BB-PLC technology must be compliant with and activate coexistence (i.e., be always on) as specified in ITU-T G.9972 or as in the ISP of IEEE 1901.

As there are already PLC technologies deployed in the field that do not implement ISP, it is important to understand to what extent the existing installed base of legacy technologies can create interference issues when ISP-enabled devices are deployed. The first consideration to make is that this is a minor issue, as the installed base of BB-PLC devices is still very small when compared to other LAN technologies such as WiFi. Secondly, a lesser-known, but important, benefit of ISP is its capability of eliminating in many cases of practical interest the interference created by an installed base of devices that does not use ISP but can be still controlled in an ISP-compliant manner under some mild assumptions. For more information on the IPP/ISP, see [25].

13.5   Power Grid Topologies*

13.5.1   Outdoor Topologies: HV, MV, and LV

High voltage (HV) lines bear voltages in the 110–380 kV range and span very large geographical distances. These lines have been used as a communications medium for voice since the 1920s, via double and single-sideband amplitude modulation (power carrier systems) [2]. Nowadays, power line communications over HV lines comprises both analog systems (tele-protection) and digital systems (voice and data transmission).

Recent results on the topological characteristics of the transmission side (HV) of the power grid were reported in [9]. These results show that the transmission power grid topology has sparse connectivity, well emulated by a collection of sub-graphs connected in a ring, each of which closely matches the characteristics of a small-world network [10]. A key observation that follows is that the path distances separating nodes are relatively small when compared to the size of the network, which clearly has beneficial implications on the communication delay if the topology of the power delivery network

---

[5] Time Slot Reuse is the capability of nodes to detect when it is possible to transmit simultaneously to other nodes in neighboring systems, without causing harmful interference. Time Slot Reuse gains can be achieved also when the multiple stations sharing the medium are interoperable.

*Source: [Wiley 2010]. Reproduced with permission of Wiley.

matches the communication one – as is the case when PLC is used. Peculiarities in this section of the grid also include the exponential tail of the nodal degree distribution and the heavy tail distribution of the line impedances.

Typically, Medium Voltage (6–30 kV) and Low Voltage (100–400 V) power distribution lines are used for high-speed PLC communication [11]. Medium-voltage power systems are typically deployed in start, ring and mesh topologies. Sometimes they can also be found deployed as open-loop systems and tree systems, with radial arranged lines. European MV levels are in the 10–30 kV range, depending on distance, which typically ranges between a few kilometers (urban) and a few tens of kilometers (rural).

US and Asian distribution include an additional MV level around 6 kV before the LV section of the grid. A high number of small transformers is used on this MV level to supply a limited number of households – typically six household per LV transformer. Additionally, distribution lines consist of either underground or overhead cables where the former are used in urban environment and the latter in suburban ones. For the MV distribution power grid there are basically three topologies: star, ring and mesh.

In a single-phase configuration, a hot and a return (neutral) wire are fed to the premises main panel. Sometimes, a separate ground (earth) wire is also added. This configuration is typical of small residential buildings. Generally, the power company distributes three phases, and only one of these is fed to a house, whereas a neighbor may be served off another phase. In the US, voltage ratings are 60 Hz 120 V,[6] allowing a range of 114–126 V (ANSI C84.1). The new harmonized nominal voltage in Europe is 230 V[7] (range: 207–253 V) 50 Hz (formerly, 240 V in the UK, 220 V in the rest of Europe).

The two-phase configuration is not common in Europe, but is typical in the US in the split-phase configuration. In a typical home in the US three cables come into the premises panel from the service. Basically, there is a center-tapped step-down transformer on the electrical line pole, with the tap grounded and each socket connected across one side of the transformer. Larger devices (electric stoves, central air conditioning units, electric dryers, etc.) are wired across the entire transformer, receiving 240 V. In the US, sometimes apartment complexes are even fed with a 120/208 V "y" configuration.

Three-phase (three hot wires plus one return) configurations are common in Europe, but not in the US. In this case voltage ranges between 230 V and 400 V. The three-wire system that the user sees is typically derived from three-phase distribution, which uses a four-wire or five-wire system. In the five-wire system, there are three hot wires, one neutral wire, and one grounding wire. The common three-wire receptacle uses only one of the three hot wires. A few hundred households are typically connected to a single transformer.

13.5.2 Indoor Topologies

Residential power cables are comprised of two or three conductors in addition to the ubiquitous earth ground [12]. These include "hot" (black), "return" (white), safety ground and "runner" (red) wires, all confined by an outer jacket that maintains close conductor spacing. Most common cables are unshielded cables (NEMA type NM). Cross-sectional views of the two most common unshielded residential power line cable types, 2/ground and 3/ground, are shown in Figure 13.1, together with some important cable parameters directly measured using Time Domain Reflectometry (TDR) techniques. Note that power line cables differ from traditional twisted-pair cables in that the mutual capacitance of the black and white signal conductors $C_{BW} \approx 22$ pF/m is significantly smaller than the mutual capacitance between the signal wires and the central ground conductor C_{BG} or $C_{WG} \approx 70$–120 pF/m.

---

[6] The US has 110 V (60 Hz) because the original light bulb invented by Thomas Edison ran on 110 volts DC, and that voltage was kept even after converting to AC, so that it was not necessary to buy new light bulbs. The most prevalent frequency used was 60 Hz, supplied by Westinghouse-designed central stations for incandescent lamps. The development of a synchronous converter which operated best at 60 cycles encouraged convergence toward that standard.

[7] Europe's mains voltage is 230V (50Hz), because at the beginning of 1900 the German company AEG had a virtual monopoly on electrical power systems, and AEG decided to use 50 Hz.

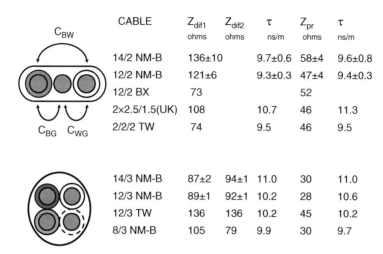

	CABLE	Z_{dif1} ohms	Z_{dif2} ohms	τ ns/m	Z_{pr} ohms	τ ns/m
	14/2 NM-B	136±10		9.7±0.6	58±4	9.6±0.8
	12/2 NM-B	121±6		9.3±0.3	47±4	9.4±0.3
	12/2 BX	73			52	
	2×2.5/1.5(UK)	108		10.7	46	11.3
	2/2/2 TW	74		9.5	46	9.5
	14/3 NM-B	87±2	94±1	11.0	30	11.0
	12/3 NM-B	89±1	92±1	10.2	28	10.6
	12/3 TW	136	136	10.2	45	10.2
	8/3 NM-B	105	79	9.9	30	9.7

Figure 13.1 Cross-section of typical three- and four-conductor cables found in residential power line networks. The table lists typical parameters for the differential and pair-modes that arise in multi-conductor transmission line modeling. Source: [12]. Reproduced with permission of IEEE.

Tree or star configurations (i.e., a single cable feeds all the wall outlets in a room only) are almost universally used. In Europe, both two-wire (ungrounded) and three-wire (grounded) outlets can be found. Interestingly, if a three-phase supply is used, separate rooms in the same apartment may be on different phases. The UK has its exceptions and may use special ring configurations: a single cable runs all the way around part of a house interconnecting all of the wall outlets, and a typical house will have three or four such rings. There are also some cases, especially in old buildings, where only two wires run around the house (neutral and ground share the same wire).

While the "white" return wires and safety grounds are isolated throughout all distal network branches, many national and international regulatory bodies today mandate that the "white" return and ground cables be connected together or "bonded" via a low resistance shunt R_{SB} (e.g., short) at the service panel (see Figure 13.2). There is substantial mode coupling created by the electrical path through R_{SB}, which has been largely ignored in the modeling of indoor power line links [12].

As described above, wiring and grounding practices come in many flavors, and this makes modem design much more challenging. However, international harmonization has been underway for the past 20–30 years, and many regulatory bodies, such as the US National Electric Code (NEC), have revised and mandated a harmonized set of practices. In particular, the following practices are now mandatory in most part of the world:

- Typical outlets have three wires: hot, neutral and ground.
- Classes of appliances (light, heavy duty appliances, outlets, etc.) must be fed by separate circuits.
- Neutral and ground are separate wires within the home, except for the main panel where they are bonded.

Residential and commercial premises power line networks now usually comprise a service panel feeding multiple branching paths that include receptacle or outlet circuits, fixed or embedded appliances and lighting circuits. Figure 13.3 illustrates some important differences between these three types of circuits. Receptacle circuits are typically dedicated 15 or 20 amp branching circuits that exhibit symmetry in the connections of black and white wires, which preserves differential balance with respect to the safety

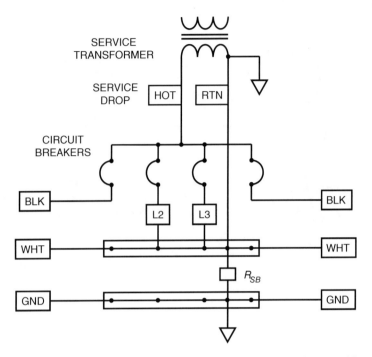

Figure 13.2 Diagram of a typical service panel with four circuit breakers. Two branch cables are explicitly shown, along with two additional loads. Source: [13]. Reproduced with permission of IEEE.

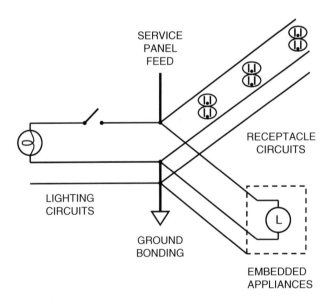

Figure 13.3 Layout of a typical residential or commercial premises power line network. A service panel feeds multiple branching paths that include receptacle or outlet circuits, fixed or embedded appliances and lighting circuits. Ground bonding at the service panel is also indicated. Source: [13]. Reproduced with permission of IEEE.

ground. Embedded appliances are typically dedicated non-branching receptacle circuits with a 20–50 amp rating. In contrast to receptacle circuits, lighting circuits have significant asymmetry between hot and return wires created by the insertion of dedicated switches in the "hot" side that produces considerable differential and pair-mode coupling.

Although complex network topologies can indeed exist, the abovementioned regulations greatly simplify the analysis of signal transmission over receptacle circuits.

13.6 Outdoor and In-Home Channel Characterization

The PLC channel is a very harsh and noisy transmission medium that is difficult to model: it is frequency-selective, time-varying, and is impaired by colored background noise and impulsive noise (for a comprehensive review of PLC channel characteristics, see [11]). Additionally, the structure of the grid differs from country to country, and also within a country, and the same applies for indoor wiring practices.

Every section of the grid has its own channel characteristics from a communications point of view. On the transmission side (HV), attenuation and dispersion are very small and can be well coped with. However, as we move towards the distribution side and towards the home, attenuation and dispersion grow considerably especially at higher frequencies. Furthermore, the PLC channel is characterized by a high noise level at all voltage levels. There are various kinds of noises, which are often time-, frequency- and weather-dependent.

As is well known today, signal propagation along power lines follows a multipath law. This is mostly due to mismatched impedances and bridged taps which, together with grounding, account for most multipath generation. A bridged tap with a length equal to a quarter of the wavelength λ causes a π-shifted reflection to add coherently with the main signal thus producing a notch at the frequency corresponding to wavelength λ. If the bridged tap length is a large multiple of the quarter wavelength, then the π-shifted reflection will be highly attenuated with respect to the main signal and this will cause just a dip in the frequency transfer function.

The PLC channel also shows significant differences in terms of path-loss in different sections of the grid. Though it is hard to give universal values for path loss, since many factors influence it (overhead or underground cables, type of cables, loading, weather, etc.), typical values of path loss for the PLC channel in dB/km are given in Table 13.1 [4].

Additional data collected in the USA suggest that the average in-home channel attenuation encountered by BB-PLC transceivers ranges between 40–50 dB for urban and sub-urban homes, respectively. Statistical values of measured channel attenuation and root-mean-square delay spread are given in Table 13.2.

In the next sub-sections, we will review the salient characteristics of the PLC channel with particular attention to the MV and LV sections of the grid where BB-PLC are typically used for internet access and LAN applications.

Table 13.1 Typical path loss values for PLC in dB/km. Values may vary depending on cable type, loading conditions, weather, and so on

Path loss (dB/km)	$f = 100\,kHz$	$f = 10\,MHz$
Low voltage (LV)	1.5–3	160–200
Medium voltage (MV) – Overhead	0.5–1	30–50
Medium voltage (MV) – underground	1–2	50–80
High voltage (HV)	0.01–0.1	2–4

Source: [4]. Reproduced with permission of IEEE.

Table 13.2 Statistical values in dB and in μs of measured channel attenuation and RMS-DS of US indoor links

	USA suburban		USA urban	
	$\overline{A}_{dB}$	$\sigma_\tau(\mu S)$	$\overline{A}_{dB}$	$\sigma_\tau(\mu S)$
Min	19.7	0.1	14.5	0.11
Max	68.1	1.73	65.1	0.47
Mean (μ)	48.9	0.52	41.5	0.23
Standard deviation (σ)	9.8	0.28	13.4	0.09
Kurtosis	3.9	7.6	2.3	3.82
Skewness	−0.6	1.7	−0.4	0.87
50%-percentile	49.4	0.46	44.3	0.23
90%-percentile	60.4	0.94	58.1	0.34

Source: [22]. Reproduced with permission of IEEE.

13.6.1   Characteristics of the HV Power Line Channel

HV lines are good waveguides as channel attenuation characteristics show a benign pass-band and time-invariant behavior. Noise is mainly caused by corona effect and other leakage or discharge events, and corona noise power fluctuations of some tens of dB can be observed due to change in weather conditions. Background noise is time variant and colored. Low-power impulse events periodical and synchronous with the mains are also present. These kinds of impulses are caused by discharges on insulators and other electrical substation devices. Compared to LV/MV, HV lines are a much better communications medium, characterized by very low attenuation (see Table 13.1).

Today, state-of-the-art HV digital modems support data rates of 320 kbit/s in a 32 kHz band and a reach of 100 km. Note that this is a very high spectral efficiency (10 bits/s/Hz), which is 50% higher than BB-PLC can achieve (<7 bits/s/Hz), or nearly an order of magnitude more than NB-PLC is capable of (1 bit/s/Hz). Today, the use of PLC over HV lines is well established, and thousands of links have been installed in more than 120 countries for a total length of some millions of kilometers. Digital PLC over HV lines has not yet been standardized but, a few years ago, the IEC TC57/WG20 started to work on updating the obsolete analog PLC standard IEC 60495 to include digital PLC for HV.

13.6.2   Characteristics of MV Power Line Channel

In the CENELEC band, the highest frequency is 150 kHz, so wavelengths are 1 km long or more. It is very rare to have a π-shifted reflection, and somewhat soft notches are appreciable because attenuation in this band is also low. Strong notches are often observable around 50–60 kHz, but these are not due to multipath but, rather, to resonance effects caused by lumped components at devices and appliances connected to the power line. In the FCC/ARIB bands that extend up to 500 kHz, wavelengths are of the order of 300 m and branches of some 100 m in length can therefore create strong frequency-selective fading with sharp notches. Channel modeling efforts have traditionally addressed the CENELEC band, as the vast majority of devices were operating in that band, and results are today well established [11].

There is still a basic disagreement about how to model MV lines at HF, and this disagreement is based on the underlining model of dissipative TLs above lossy ground. Of interest to this topic are recent results on the modeling of dissipative TLs above lossy ground reporting that previous classical models were not accurate at high frequencies, since they did not incorporate ground admittance [4]. More experimental results are needed to confirm the appropriate model for MV power line signal propagation.

The MV overhead and underground lines have in common the lack of bridged taps, but have very different attenuation characteristics. Overhead MV links offer moderate attenuation and mild ISI, exhibiting around 2–5 MHz, a large dip in the frequency response when a transformer is found along the line – as is usually the case for US and Asian topologies. Underground MV links offer substantial attenuation and moderate ISI, exhibiting often a pronounced low pass behavior. Examples for these two kinds of MV links can be seen in Figures 13.4 and 13.5.

13.6.3   Characteristics of LV Power Line Channel

A lot of work has been done for the modeling of the low voltage side of an outdoor power line network. Most approaches use a multipath or two-conductor TL formalism although, more recently, multi-conductor approaches have been suggested to account for ground.

Bridged taps with lengths equal to a small multiple of the quarter wavelength of the frequencies in the HF band ($10\,\text{m} < \lambda < 150\,\text{m}$) are common for indoor topologies, so that the indoor PLC channel in the HF region is characterized by many frequency notches and dips. Specifically, the PLC channel introduces both amplitude and delay distortion, and it is characterized by large excursions of the group delay in correspondence to notches – the deeper the notch, the larger the excursion.

Typical indoor transfer functions and corresponding group delay variations are shown in Figures 13.6 and 13.7.

The characteristic of the PLC channel in the VHF band (30–300 MHz) is very different from the one in the HF region, as indoor bridged taps usually have lengths that are much longer that a quarter of the wavelengths of VHF frequencies ($1\,\text{m} < \lambda < 10\,\text{m}$). As a consequence, bridged tap lengths are typically a large multiple of the quarter wavelength so that the π-shifted reflection will be heavily attenuated with respect to the main signal and this will cause just a small dip (not a notch) in the frequency transfer function. Also, as a consequence, also delay distortion will be small, because fewer and less pronounced group delay peaks would be present.

On the other hand, attenuation also increases, but not as much as one would imagine. For example, a quantitative analysis of channel characteristics between 30 MHz and 100 MHz reported that the median channel attenuation in the 30–100 MHz range was only few dB higher than in the HF band. Another characteristic of the VHF region is that, while channel attenuation increases with frequency, noise decreases following a Lorentzian $1/f$-shaped law. Thus, at higher frequencies, lower noise levels are encountered.

A peculiar aspect of the PL channel that differentiates it from other transmission line-based channels is that the PL channel is a Linear and Periodically Time Varying (LPTV) channel, as demonstrated by Canete *et al.* in [14]. The LPTV behavior is due to the fact that the electrical devices plugged in outlets (loads) contain non-linear elements such as diodes and transistors that, relative to the small and rapidly changing communication signals, appear as a resistance biased by the AC mains voltage. The periodically changing AC signal swings the devices over different regions of their non-linear I/V curve, and this induces a periodically time-varying change in their resistance. The overall impedance appears as a shunt across the hot and return wires and, since its time variability is due to the periodic AC mains waveform, it is naturally periodic.

An example of this behavior is seen in Figure 13.8, where the measured time variation of an IH PL channel transfer function is shown.

13.6.4   Power Line Noise Characteristics

There are various kinds of noises, which are often time-, frequency- and weather-dependent. In HV/MV networks, background noise is mainly caused by leakage or discharge events, power converters, transformer, etc. There are also impulsive events due to switching transients, lightning and other

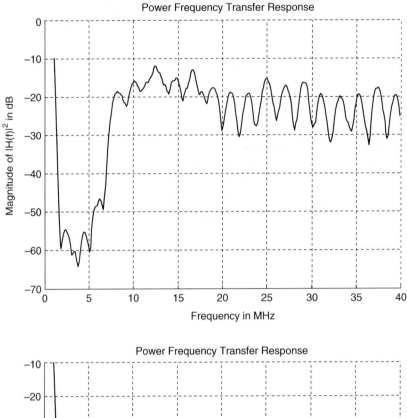

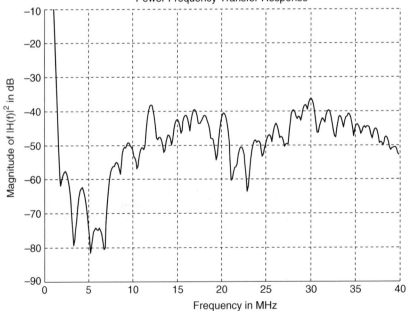

Figure 13.4 Measured frequency transfer functions of US MV overhead links

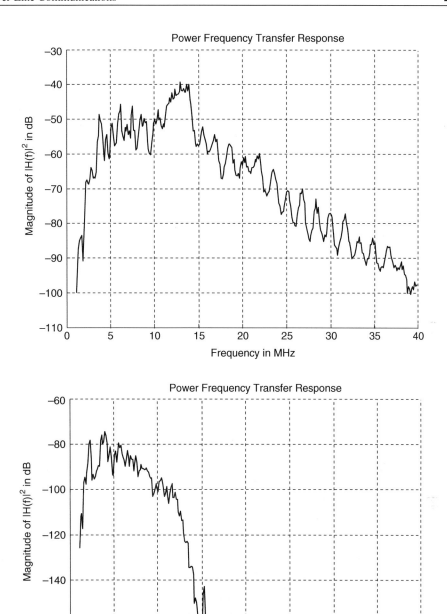

Figure 13.5 Measured frequency transfer functions of US MV underground links

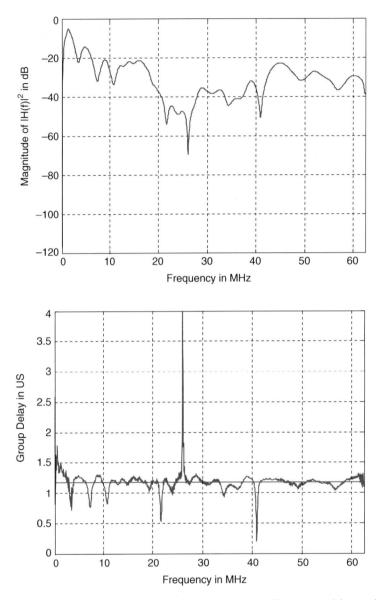

Figure 13.6 Measured frequency transfer function and corresponding group delay variation for a typical US indoor link with modest attenuation. Source: [3]. Reproduced with permission of John Wiley & Sons, Ltd.)

discharging events. In the LV and indoor environment, appliances become the cause of linear and periodically time-varying behavior in the channel impulse response, as well as sources of cyclostationary noise. Furthermore, electrical devices are noise generators and, in view of the Nyquist theorem, noise also appears to be cyclostationary. An example of a noise waveform generated by a halogen light is shown in Figure 13.9 (only half AC cycle is shown).

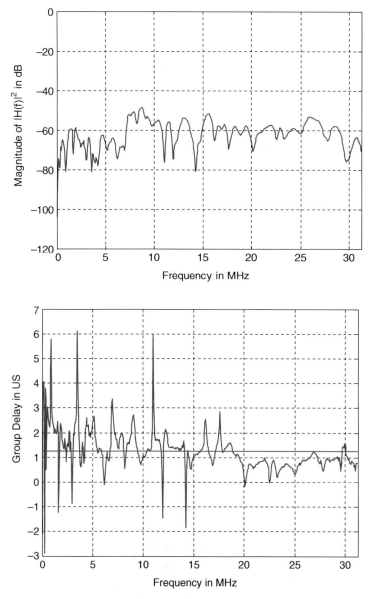

Figure 13.7 Measured frequency transfer function and corresponding group delay variation for a typical US indoor link with heavy attenuation

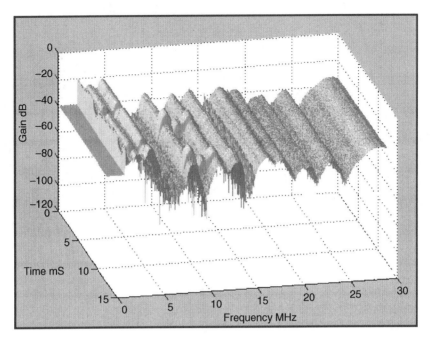

Figure 13.8 Measured time variation of indoor power line channel. Source: [5]. Reproduced with permission of IEEE.

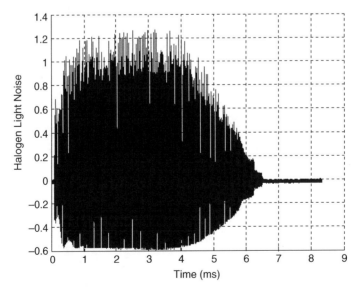

Figure 13.9 Measured noise waveform generated by a halogen light

As PLs are also both source and victims of electromagnetic interference, narrowband noise is also often present. Modem design is thus challenging, especially in dealing with the various sources of noise, which probably represents the most challenging problem in PLC.

13.7 Power Line Channel Modeling*

In the last decade, a lot of progress has been made in achieving a better understanding of signal propagation along power lines. Among the advances reported in the last decade, we point out the most prominent ones [4]:

- The multipath law [15].
- The classification of the several types of noise and their modeling [16,17].
- The quasi-isotropy of the PLC channel [18].
- The linear and periodically time-varying nature of the PLC channel [14].
- The relationship between grounded and ungrounded links, which now can be analyzed under the same formalism [12].
- The log-normal distribution of channel attenuation and RMS delay spread of the channel [19,22].
- The recent proof that block models similar to those used in wireless and wireline DSL channels can be used in the PLC context as well – an important result since key advances in wireless and DSL technologies were fostered by utilizing block transmission models and precoding strategies [20].

At first, PLC channel modeling attempts were mostly empirical and not necessarily tied to PLs *per se*. The first popular model that attempted to give a phenomenological description of the physics behind signal propagation over PLs is the multipath-model introduced in [15]. According to this model, signal propagation along PL cables is predominantly affected by multipath effects arising from the presence of several branches and impedance mismatches that cause multiple reflections. In this approach, the model parameters (delay, attenuation, number of paths, etc.) are fitted via measurements. The disadvantage of this approach is that it is not tied to the physical parameters of the channel. Furthermore, this approach is not even tied to the PLC channel *per se*, as it describes generic signal propagation along any TL-based channel (e.g., see [21] for the case of twisted pairs in DSL).

To overcome this drawback, classical two-conductor TL-theory can be used to derive analytically the multipath model parameters under the assumption that the link topology is known *a priori*. Unfortunately, the computational complexity of this method grows with the number of discontinuities and may become very high for the in-home case (see, for example, Section III.A in [12]).

TL-based channel models have today reached a good degree of sophistication, as they have been extended to include the multi-conductor TL (MTL) case. Pioneering work on the application of MTL theory to power distribution networks was made by Wedepohl in 1963, and tools on mode decoupling were successively introduced by Paul. Building on these results, a model for including grounding in LV indoor models was recently developed [12,13,18]. The MTL approach is a natural extension of the two-conductor modeling to include the presence of additional wires, such as the ground wire, and allows computation of the transfer function of both grounded and ungrounded PL links by using transmission matrices only. These results allow us to treat both grounded and ungrounded indoor PLC channels with the same formalism.

The transfer function of a TL-based channels can be calculated deterministically once the link topology is known. However, the variability of link topologies and wiring practices gives rise to a stochastic aspect of TL-based channels that has been only recently addressed in the literature. To encompass several potential scenarios and study the coverage and expected transmission rates of PLC networks, one needs to combine these MTL-based deterministic models with a set of topologies that are representative of the majority of cases found in the field. This approach is reminiscent of what has been done in xDSL context with the definition of the ANSI and CSA loops. Although this approach may be suitable for the outdoor

---

*Sources: [4] and [22]. Reproduced with permission of IEEE.

MV/LV cases, its applicability to the in-home case may be questionable, due to the wide variability of wiring and grounding practices.

An excellent approach is to generate random in-home topologies, where the appropriate national electric codes are used to set constraints on the topologies in terms of the number of outlets per branch, wire gauges, inter-outlet spacing, and so on. This is probably the most realistic and accurate way of generating randomly channel realizations, although a generalization of this approach requires the knowledge of the electric codes of every country.

A useful result for the modeling of the PLC channel and the calculation of its achievable throughput was the discovery that attenuation in LV/MV PLC channels is log-normally distributed [19,22]. Considering signal propagation along TLs as multipath-based, channel distortion is present at the receiver due not only to the low pass behavior of the cable, but also to the arrival of multiple echoes caused by successive reflections of the propagating signal generated by mismatched terminations and impedance discontinuities along the line. This is a general behavior and is independent of the link topology or, in the case of PLs, of the presence of grounding. According to this model, the transfer function is [15]:

$$H(f) = \sum_{i=0}^{N_{paths}-1} g_i(f) e^{-\alpha(f)v_p \vartheta_i} e^{-i2\pi f \vartheta_i}$$

where:

$g_i(f)$ is a complex number, generally frequency-dependent, that depends on the topology of the link
$\alpha(f)$ is the attenuation coefficient which takes into account both skin effect and dielectric loss
ϑ_i is the delay associated with the i th path
v_p is velocity of propagation along the power line cable
N_{paths} is the number of non-negligible paths.

Similarly, we can write in the time domain:

$$h(f) = \sum_{i=0}^{N_{paths}-1} e_{ep}^{(i)}(t - \vartheta_i)$$

where $e_{ep}^{(i)}(t) = \mathrm{FT}^{-1}[g_i(f)e^{-\alpha(f)v_p \vartheta_i}]$ is the signal propagating along the i-th path and its amplitude and shape are a function of the reflection coefficients $\rho^{(i)}$ and the transmission coefficients $\xi^{(i)} = (1+\rho^{(i)})$ associated to all the impedance discontinuities encountered along the i-th path, and of the low-pass behavior of the channel in the absence of multipath (for analytical expressions of $\rho^{(i)}$ and $\xi^{(i)}$, see [12] for the case of forward traveling signal paths and [21] for the case of backward traveling echo paths).

Thus, the path amplitudes are a function of a cascade (product) of several random propagation effects, and this is a condition that leads to log-normality in the central limit since the logarithm of a product of random terms becomes the summation of many random terms. Since log-normality is preserved under power, path gains are log-normally distributed as well. As the sum of independent or correlated log-normal random variables is well approximated by another log-normal distribution, we can finally state that also the PLC channel average gain (or attenuation) is log-normally distributed.

Empirical confirmation of this property of the PLC channel has been reported for indoor US sub-urban homes [19], indoor US urban multiple dwelling units [22], and for US outdoor MV underground PLs [22]. Furthermore, these PLC channel characteristics have also been observed in other wireline channels, such as coax and phone lines, so that a new generalized wireline statistical channel model has been recently formulated in [22]. The availability of these results greatly facilitates the study of coverage which is necessary for proper planning and deployment. Details on this new modeling approach is given in the next section.

13.7.1   Recent Results on the Modeling of Wireline Channels: Towards a Unified Framework

A recently published paper has confirmed that channel power gain and RMS-DS of not only LV/MV PLC channels but of also other wireline channels are correlated lognormal random variables [22]. An important consequence of this property is that channels characterized by large RMS-DS (severe ISI) are also characterized by small channel gains (large attenuation and, thus, also low SNR for fixed transmit power), and vice versa. Similar considerations may also be made in the wireless context, as several researchers have reported empirical results confirming that the correlation between channel gain and RMS-DS is sometimes observable in radio channels. On the basis of these results, it is possible to formulate a new approach to wireline statistical channel modeling, where the correlation between gains and RMS-DS is imposed by design.

A scatter plot of measured PLC channels is shown in Figure 13.10, together with a trend line calculated using a robust iteratively reweighted least squares algorithms with a bi-square weighting function that assigns a smaller weight to data points further from model predictions. The robust regression line exhibits a negative slope clearly indicating that channel power gain and RMS-DS are negatively correlated.

The property that channel gains and RMS-DS of PLC channels are correlated seems to be a natural characteristic of other physical communications channels. It is particularly interesting to verify that this

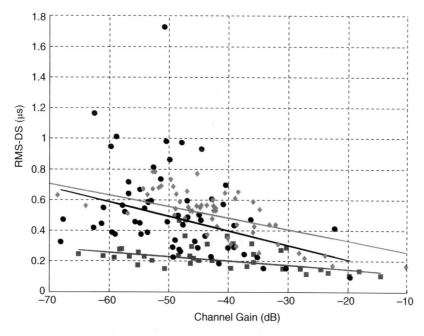

Figure 13.10 Scatter plot (channel RMS-DS versus channel power gain) of measured LV/MV PLC channels with least squares trend lines: measured underground MV PLC links (diamonds), indoor LV sub-urban (circles), and indoor LV urban (squares)

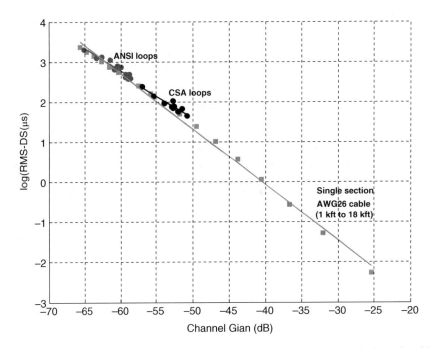

Figure 13.11 Scatter plot of log(RMS-DS) vs channel power gains – simulated channels with least squares trend lines: (circles) ANSI DSL loops; (circles) CSA DSL loops; (squares) single section AWG26 cables, from 1 kft to 18kft with 1 kft increments. Source: [22]. Reproduced with permission of IEEE.

holds true also for typical DSL channels where the functional relationship between channel power gain and the logarithm of RMS-DS becomes perfectly linear.

Figure 13.11 shows a scatter plot of power channel gain versus the logarithm of the RMS-DS for typical DSL loops (CSA and ANSI) and for a straight AWG26 cable with no bridged taps and length ranging between 1 kft and 18 kft, step 1 kft. The regression line shows a perfectly linear correlation with correlation coefficient equal to −0.99.

The scatter plots are shown together with a trend line given by the equation below:

$$\text{RMS-DS}^{(\mu S)} = \alpha G^{(dB)} + \beta$$

where RMS-DS$^{(\mu s)}$ is the RMS-DS expressed in μs and $G^{(dB)}$ is the channel power gain in dB. When a higher correlation is found between channel power gains and the logarithm of the RMS-DS, the trend line equation is:

$$\text{RMS-DS}^{(\mu S)} = \zeta G^{(dB)} + \xi$$

For the case of PLC channel, the robust equation parameters are:

$$\alpha = -0.094\mu s/dB; \quad \beta = 0.02\mu s \quad \text{(LV sub-urban)}$$
$$\alpha = -0.0028\mu s/dB; \quad \beta = 0.089\mu s \quad \text{(LV urban)}$$
$$\alpha = -0.0075\mu s/dB; \quad \beta = 0.183\mu s \quad \text{(MV underground)}$$

For the DSL case, the robust equation parameters are:

$$\zeta = -5.6\mu s/dB; \quad \xi = -0.139\mu s \quad (DSL\text{-}AWG26)$$
$$\zeta = -3.85\mu s/dB; \quad \xi = -0.109\mu s \quad (DSL\text{-}CSA)$$
$$\zeta = -3.81\mu s/dB; \quad \xi = -0.11\mu s \quad (DSL\text{-}ANSI)$$

13.8   The IEEE 1901 Broadband over Power Line Standard

The IEEE 1901 standard provides PLC specifications for both access and in-home related applications [5,6]. As opposed to in-home systems, access networks usually cover large areas, consisting of hundreds or thousands of nodes that are generally centrally controlled, and the networks have to be flexible enough to be deployed in diverse topologies such as tree, meshed, ring, or a combination of these. Access networks are also more dynamic than in-home ones, due to impedance changes, topology changes (electrical switches that open or close), and external interference that vary with the time of day.

13.8.1   Overview of Technical Features*

The standard offers a solution with a common MAC and the capability of supporting two PHYs, one based on Wavelet-OFDM, and one on windowed FFT-based OFDM.

A conceptual overview of the standard is shown in Figure 13.12. The common MAC handles the two different PHYs via an intermediate layer called the Physical Layer Convergence Protocol (PLCP). There are two PLCPs: the O-PLCP that handles the interaction between the common MAC and the windowed OFDM PHY; and the W-PLCP that handles the interaction between the common MAC and the Wavelet-OFDM PHY. Another key component of the standard is the presence of a mandatory Inter-System Protocol (ISP) that allows PLC devices based on the IEEE 1901 standards to share the medium efficiently and fairly regardless of the PHY differences. Furthermore, the ISP also allows IEEE 1901 devices to coexist with devices based on the ITU-T G.9960 (G.hn) standard (see Section 13.4.1.4 and [4,25] for more details on coexistence).

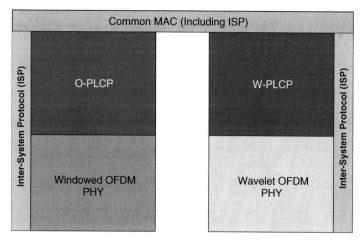

Figure 13.12   Architecture of the IEEE 1901 Standard. Example of functionalities present in each layer – Common MAC: frame formats, Addressing, SAP, SAR, Security, ISP; W-PLCP and O-PLCP: channel adaptation, PPDU format, FEC, etc.; PHY: Wavelet-OFDM PHY, Windowed FFT-OFDM PHY. Source: [5]. Reproduced with permission of IEEE.

---

*Source: [5]. Reproduced with permission of IEEE.

The decision to have multiple-PHY solution in IEEE 1901 is not the consequence of a technical necessity. There are certainly some advantages in using either the Wavelet- or the FFT-based OFDM PHYs, but those advantages alone do not really warrant the necessity of including both PHYs in the standard. Therefore, the multiple-PHY nature of the IEEE 1901 standard derives more from a political necessity than a technical one. On the other hand, this decision will also enable continuity and smooth migration from the currently deployed devices based on the HomePlug and the Panasonic technologies to the standardized IEEE 1901 devices. The ISP will also facilitate coexistence of IEEE 1901 and ITU-T G.9960 (G.hn) devices, avoiding any performance degradation due to the interference that devices based on these two non-interoperable standards may create.

13.8.2   The MAC and the Two PLCPs

The fundamental architecture used to coordinate the IEEE P1901 network is Master/Slave. The Master (QoS Controller) authorizes and authenticates the slave stations in the network, and may assign time slots for transmissions using either CSMA-based or TDM-based access. Network stations can communicate directly with each other (as opposed to a wireless access point that retransmits all traffic). This increases the efficiency of the network and reduces the load on the Master.

The MAC layer employs a hybrid access control based on TDMA and CSMA/CA by defining a Contention Free Period (CFP) and a Contention Period (CP) to accommodate data with different transmission requirements. The CFP is a portion of the total transmission cycle, during which stations that have low-delay/low-jitter requirements are allowed exclusive use of the medium. All streams requiring transmission in the CFP are managed by a QoS Controller. The CFP starts with a beacon, which is periodically sent by the QoS Controller, and ends when all reserved streams are transported. The rest of the Beacon Cycle is used for the CP. In the CFP, data streams that have a time allocated to them through a bandwidth reservation procedure managed by the QoS Controller are transported. Frequency Division Multiplexing (FDM) can also be supported to allow for co-existence between in-home and access networks. Fragmentation support, data bursting, group-ACK, and selective Repeat ARQ are also important features of the standard.

Intelligent TDMA is also defined in IEEE 1901. This is a dynamic bandwidth allocation mechanism that exploits information about the amount of traffic queued in each transmission station. This mechanism realizes stable transmission, which can cope with errors and IP/VBR traffic. In each transmitted data packet, each station inserts the number of frames pending to be transmitted. Since traffic information is directly obtained from data packets, the QoS Controller can perform accurate real-time operation. An option for line cycle synchronization is also present for coping with the periodically time-varying channel and cyclostationary noise.

13.8.2.1   The FFT OFDM based PHY*

FFT-based windowed OFDM is one of the two multichannel transmission techniques defined in the IEEE 1901 standard. Through the use of time-domain pulse shaping of the OFDM symbols, deep frequency notches can be achieved without the additional requirement of transmit notch filters. The OFDM PHY uses a maximum of 1893 carriers in the 1.8–48 MHz band for maximum data rates up to 400 Mbit/s. Frequencies above 30 MHz are optional and support up to 48 MHz may be included. Flexible spectral notching can support regional and application requirements. In addition, each OFDM tone can be loaded with 1, 2, 3, 4, 6, 8, 10 or 12 bits, using QAM on the basis of each carrier's SNR. This PHY uses Turbo Convolutional Coding for Forward Error Correction (FEC). Channel adaptation mechanisms, based on detecting zero crossings and understanding where noise is most likely to occur, have also been defined, as they significantly improve system performance in the presence of periodically time-varying noise.

The basic parameters of the FFT-OFDM PHY are given in Table 13.3.

---

*Source: [5]. Reproduced with permission of IEEE.

Table 13.3 Basic IEEE 1901 PHY parameters

	IEEE 1901 – FFT	IEEE 1901 – Wavelet
ISI Mitigation Technique	FFT-OFDM	Wavelet-OFDM
Discrete Transform Points	4096	512 (Mandatory)
		1024, 2048 (Optional)
Carrier Spacing [kHz]	24.414	61.035
Guard Interval [samples]	556, 756, 4712	None
Subcarrier Modulation	BPSK, QPSK	(2, 4, 8, 16, 32)-PAM
	(8, 16, 64, 256, 1024)-QAM	
Frequency Band [MHz]	1.8–30 (Mandatory)	1.8–28 (Mandatory)
	1.8–50 (Optional)	1.8–50 (Optional)
Forward Error Correction	Turbo Convolutional Coding	RS-CC (Mandatory)
		Convolutional LDPC (optional)
Maximum Data Rate [Mbit/s]	200	200

13.8.2.2 The Wavelet OFDM based PHY

Wavelet-OFDM [23] is the second multichannel transmission technique defined in the IEEE 1901 standard. The fundamental characteristic of Wavelet-OFDM is that the usual FFT-based transform and the rectangular/raised-cosine windowing used in conventional OFDM is replaced with critically decimated Perfect Reconstruction Cosine Modulated Filter Banks, which exhibit several desirable properties such as very low spectral leakage. One of the most interesting aspects of Wavelet-OFDM is that it is not necessary to introduce a guard interval between consecutive symbols.

The Wavelet-OFDM system specified here places 512 evenly spaced carriers into the frequency band from DC to around 30 MHz. Of these 512 carriers, 338 (approximately 2 MHz to 28 MHz) are used to carry information. With the use of an optional band up to 50 MHz, maximum PHY rates on the order of half a Gbit/s can be achieved. Every carrier is loaded with real constellations such as M-PAM (M = 2, 4, 8, 16, 32).

It is important to point out that the fact that Wavelet-OFDM employs real constellations does not mean that it has lower spectral efficiency than conventional OFDM employing 2D constellations, such as QAM. In fact, the frequency resolution of Wavelet-OFDM is twice that of windowed OFDM, because the use of non-rectangular windowing allows for a higher degree of spectral overlap. As a consequence, for the same total bandwidth and the same number of transform points K, Wavelet-OFDM uses K real carriers that employ PAM, whereas OFDM uses $K/2$ complex carriers that employ QAM. Thus, OFDM and Wavelet-OFDM have the same spectral efficiency. Specified FECs include a mandatory concatenated Reed-Solomon/convolutional code scheme and an optional convolutional LDPC code that allows easy scalability to high data rates at reasonable complexity. The basic parameters of the Wavelet-OFDM PHY are given in Table 13.3.

13.8.3 Access-Specific Features

IEEE P1901 subdivides the access network into the so called "access cells" [7]. An access cell consists of several components (an example of the BB-PLC access cell topology is shown in Figure 2 of [6]):

- a controlling station – the head end (HE);
- one or more repeaters (RPs);
- network termination units (NTUs);
- customer premises equipment (CPEs) stations.

Each access cell contains a group of access nodes administrated by a head end (HE), and there can be one or more access cells in an access network. The HE can also be seen as a gateway, as it establishes connections to the WAN and communicates to other HEs. All access nodes in the cell synchronize their local clocks with the clock of the HE.

End-to-end communication between the HE and a station in the access cell can be a point-to-point link, if the two nodes are in range, or the link can be established over multiple hops using a repeater (RP). A station decides which neighboring station to utilize as a RP (next hop) for communications with the HE based on link quality information. As the PLC environment is time-varying, topology information can change over time and must, therefore, be updated.

NTUs are RPs that can also bridge the access network with the in-home network or other non-PLC networks such as Wi-Fi and Ethernet. The functionality of the NTU is application-specific where the NTU may act as a proxy or a repeater.

Many IEEE 1901 features are common to both access and in-home, but the access part has some specific features. Specifically, we have:

- *Common features:*
  - The physical layer.
  - Channel adaptation.
  - MAC data plane.
  - Encryption.
- *Main Differences:*
  - MAC, specifically the support of distributed TDMA, point-to-multipoint and repeating functions.
  - Routing.
  - Key distribution.
  - Device authentication.

IEEE 1901 access uses a beacon-based periodic channel access mechanism, and the duration of each beacon period is twice that of the AC line cycle period (either 33.33 ms or 40 ms, depending on whether the mains frequency is 50 or 60 Hz). A beacon is transmitted every 1–2 seconds, and this is used for various purposes: selection of the best station to track; clock synchronization; TDMA period synchronization; and TDMA scheduling.

The access network synchronization is based on a centralized clock, similarly to the in-home case, but the procedure is more complex because of longer distances, and it uses a multihop mechanism.

All access cell stations synchronize on the clock of the HE. Every station that hears the HE beacon message synchronizes its clock to the HE. The multihop mechanism requires that the HE periodically transmits a beacon message with the time when the beacon was sent and the beacon level. The beacon level is always zero for the HE. When the beacon is re-transmitted by a station, it includes the station's transmitting time, based on the station's clock and the station's beacon level, which is 1 in this case. Only stations that are not synchronized to the HE synchronize to these beacons. This process continues for every beacon level until all the stations in the network have shared and synchronized the same clock value between all stations in the access cell according to their beacon level.

IEEE 1901 access nodes (or stations, STA) operate by default in CSMA/CA mode. This uses a modified binary exponential back-off, with four priority classes. The HE determines the contention window. CSMA grouping is also specified to make CSMA more efficient (e.g., making sure that there are no hidden stations in a group). Furthermore, CSMA groups are scheduled to avoid conflicts with other CSMA groups, and this is accomplished by scheduling different groups in different intervals for CSMA contention.

When a more deterministic channel access capability is needed, TDMA can also be used. Time resources are divided in time regions, and each region defines the interval for CSMA, for guaranteed TDMA allocations, and for stay-out regions (regions where a node cannot transmit because the TDMA allocations have been granted to neighboring access nodes). Two types of TDMA are specified [6]: centralized, when the HE allocates every region; and distributed, when each node defines its own allocations subject to authorization from the HE. Typically, a HE manages TDMA schedules for the first few hops thus enabling the HE to optimize the allocations in its vicinity, which is the most congested part of the access network. Stations further from the HE manage TDMA schedules in a distributed manner.

IEEE 1901 also specifies an FDM mode of operation. The access network is divided into clusters, and stations in a cluster operate in a particular frequency band. Each cluster operates as an independent access cell, and the HE of each cluster bridges traffic either to the backbone network or to another cluster. Network synchronization, channel access security, etc. are common to both TDM and FDM modes.

13.9   PLC and the Smart Grid*

It is broadly believed that the growth of energy demand has outpaced the rate at which energy generation can grow by traditional means [4]. Additionally, many governments agree that greenhouse gas emissions need to be contained to control or prevent climate change. The necessity of modernizing the electric grid infrastructure around the world is the consequence both of the limited investments made in it in the last decades, and of the result of new requirements that emerge in the safe integration of:

- utility scale Renewable Energy Sources feeding into the transmission system;
- Distributed Energy Resources feeding into the distribution system or the home;
- decentralized storage to compensate for the time varying nature of wind and photovoltaic sources;
- plug-in (hybrid) electric vehicles (PHEV) that may cause large load increases on sections of the grid;
- microgrids;
- allowing active participation of consumers via demand side management (DSM) and demand response (DR) programs.

All of the above are advocated as sustainable solutions to our energy crisis.

The power grid is a commodity delivery system where the commodity (electric power) has a production-to-consumption cycle time of zero: generation, delivery and consumption happen all at the same time! This creates unique challenges in sensing, communications, and control because electrical power moves just as fast as communication signals do. Balancing generation and demand of this "perfect just-in-time" system will, then, require the integration of additional protection and control technologies that ensure grid stability – not a trivial patch to the current power grid control network (Supervisory Control and Data Acquisition, or SCADA).

Furthermore, this problem also poses a fundamental design challenge because it is a combined problem of communications, sensing and control, and decoupling communications from control and management is problematic because the separation of time scales between control signals and controlled commodity is impossible. Hence, the concept of *Smart Grid* has emerged, encompassing the cyber-physical infrastructure including wide-area monitoring, two-way communications and enhanced control functionalities, that will bridge the present technological inadequacies of the SCADA system.

Since communications is such a fundamental element of the Smart Grid, the appropriate design for physical, data and network communications layers are today a topic of intense debate. The debate on what is the actual role of PLC in the Smart Grid is also still open and ongoing. While some advocate that PLC is a very good candidate for many applications, others express concerns and look at wireless as a more established alternative. There is no doubt that the Smart Grid will exploit multiple types of communications technologies, ranging from fiber optics to wireless and to wireline. Skeptics, however, contend that PLC has an unclear standardization status and offers data rates that are too small. Others also contend that PLC modems are still too expensive, and that they present electromagnetic compatibility (EMC) issues. Recent advances in PLC clear many of these concerns.

---

*Source: [4]. Reproduced with permission of IEEE.

Among the wireline alternatives, PLC is the only technology which has deployment cost that can be considered comparable to wireless, since the lines are already there. A promising sign, confirming that PLC has already exited the experimental phase and is a technology mature for deployment, is the extensive use of PLC over both the transmission and the distribution parts of the grid, including the wide penetration that PLC has gained for supporting Automatic Meter Reading (AMR) and Advanced Metering Infrastructure (AMI) applications.

There are many applications scenarios in the Smart Grid that require a diversity of communications technologies. Although it is expected that the Smart Grid will be supported by a heterogeneous set of networking technologies, PLC is an excellent and mature technology that can support a wide variety of applications from the transmission side to the distribution side and also to and within the home. There are many PLC technologies, either already available or currently under standardization, and it is very important to refrain from advocating a single PLC technology, rather than exercising a judicious choice in the selection of the right PLC technology for the right set of applications. There are many possible choices at the disposal of communications and utility engineers, and the vast majority of these technologies can find a suitable application within the Smart Grid.

Many of the available PLC technologies are well separated in frequency from each other, so that a good design of the analog front end would eliminate interference between non-interoperable technologies. However, there are also multiple non-interoperable technologies that operate in the same band so that coexistence mechanisms are required to alleviate the performance degradation due to mutual interference. Although it may be true, as some believe, that coexistence stands in the way of interoperability and may delay industry alignment behind a single standard, it is important to understand that usage of the PLC spectrum is not regulated, so that any PLC technology can use channel resources without having any legal obligation to protect other PLC technology from interference. Thus, any deployed PLC technology is a source/victim of interference to/from the installed base of PLC devices if a common coexistence mechanism is not supported, since there is not enough bandwidth to implement FDM efficiently, as in the WiFi and coax cases. Furthermore, the implementation of coexistence in PLC transceivers also allows that diversification of deployment that is today a necessary ingredient for achieving a better understanding of how to build the Smart Grid without having to pay the penalty of interference, performance degradation, and service disruption.

On the basis of the above considerations, coexistence can be seen as a transitory and necessary "evil" that will allow the industry to align behind the right PLC technology for the right application on the basis of field deployment data, and not on the basis of a pre-selection strategy. Furthermore, coexistence will also ensure that the operation of Smart Grid and home networking devices can be decoupled and allowed to mature at their traditional obsolescence rate, even if operating in the same band. Last but not least, coexistence will also allow utilities and other service providers to avoid having to resolve "service" issues caused by the interference between non-interoperable PLC devices supporting different applications.

Below, we summarize the fundamental benefits offered by PLC when it is employed for the Smart Grid and, more generally, for utility applications [4]:

- Utility applications almost always require redundancy in protection and control, and the need for redundancy must include the availability of redundant communications channels. PLC allows exploitation of the existing wired infrastructure, thus greatly reducing the cost of deploying a redundant communications channel.
- The use of PLC allows blurring together the traditionally separated functions of *sensing* and *communicating*, as a PLC transceiver can be designed to switch between functioning as a "sensor" and as a "modem".
- PLs often represent the most direct route between controllers and IEDs when compared to packet switched public networks, so that PLC offers substantial advantages when dealing with applications such as tele-protection, where ensuring a low and bounded latency is crucial.

- Power lines provide a communication path that is under the *direct and complete* control of the utility, which is a fundamental benefit when operating in countries where telecom markets are de-regulated.
- There is a wide variety of PLC technologies that can find a role in most Smart Grid applications, so that PLC can, indeed, provide a wide class of technologies that can be employed as a communications solution from the transmission side of the grid down to the HAN.

There are many examples of applications where PLC can be used for utility applications and these will be reviewed in the next sub-sections.

13.9.1   PLC for MV

An important requirement for future Smart Grids is the capability of transferring data concerning the status of the MV grid, where information about state of equipment and power flow conditions must be transferred between substations within the grid. Traditionally, substations at the MV level are not equipped with communications capabilities so the use of the existing PL infrastructure represents an appealing alternative to the installation of new communication links. Some substation automation functions need substation IEDs to communicate with one or more external IEDs. In the case of fault location, fault isolation and service restoration, substation IEDs must communicate with external IEDs such as switches, reclosers, or sectionalizers. In another example, implementation of voltage dispatch on the distribution system requires communication between substation IEDs and distribution feeder IEDs served by the substation. All of these communications require low-speed connectivity that is well within PLC capabilities.

A large portion of current MV equipment throughout the world was installed more than 40 years ago. Consequently, fault detection, as well as monitoring for ensuring longer lifespan to critical cable connections, is becoming a true operational, safety and economical necessity. Most techniques used today include on-site expensive truck rolls. For example, available power cable diagnostics are based today on partial discharge measurements (typically based on time domain reflectometry) on temporally disconnected connections which are externally energized. From an operational point of view, online diagnostic tools are preferable and will soon become the main trend. The coupling of PLC signals up to 95 kHz (European CENELEC A-band) for online diagnostic data transfer over MV cables has been studied, and it was reported that there is an advantage in integrating diagnostics tools that serve the dual purpose of sensing and communication devices [4].

DG systems can supply unintentional system islands, isolated from the remainder of the network. It is important to detect these events quickly, but passive protections based on traditional measures may fail in island detection under particular system operating conditions. The use of LDR NB-PLC (CENELEC A-band) for injecting a signal in the MV system has been analyzed and tested, and this appears to be less expensive, compared to other methods based on telephone cable signals. A similar approach has been investigated for the prevention of islanding in grid-connected photovoltaic systems, and it was found that PLC-based islanding prevention *"offers superior islanding prevention over any other existing method"* [4]. Other applications of PLC within the area of DG can also be found.

In addition to remote control for the prevention of the islanding phenomenon, other applications related to monitoring on the MV side (e.g., temperature measurement of oil transformers, voltage measurement on the secondary winding of HV/MV transformers, fault surveys, power quality measurement) have also been discussed and analyzed by various researchers.

13.9.2   PLC for LV

Most PLC Smart Grid applications on the LV side are in the area of AMR/AMI, vehicle-to-grid communications, DSM, and in-home energy management [4]. These applications will be addressed in the next sub-sections.

13.9.2.1 Automatic Meter Reading and Advanced Metering Infrastructure

In addition to basic one-way meter reading (AMR), AMI systems provide two-way communications that can be used to exchange information with customer devices and systems. Furthermore, AMI enables utilities to interact with meters and enables customer awareness of electricity pricing on a real-time basis. Although, today, smart meter deployment is getting a lot of attention worldwide, a smart meter is not really a necessary part of the Smart Grid, as there are several alternative ways to implement Smart Grid applications without smart meters. On the other hand, smart meters are important tools for utilities to reduce their operational costs and losses because they provide capabilities that go beyond simple AMR, such as remote connect/disconnect and reduction of so-called non-technical losses (e.g., losses due to energy theft).

PLC technology is certainly well suited for AMR/AMI. There is a vast amount of field data about the performance of PLC-based smart meters, as a few hundred million UNB/NB-PLC devices have been deployed around the world.

UNB-PLC devices were the first ones to be used for AMR/AMI. Although UNB-PLC systems are characterized by very low data rates, UNB-PLC signals propagate easily through several MV and LV transformers. Furthermore, UNB-PLC does not require any kind of PL conditioning, as other PLC technologies operating at higher frequency would often require due to the low pass effect of shunt power factor correction capacitors and series impedances of distribution transformers. As a consequence, these systems are able to cover very large distances (150 km or more).

In the last couple of decades, UNB-PLC systems have experienced good success in the market. The Turtle System has found good applicability in those areas served by US rural cooperatives, which are characterized by low population density and wide geographical spread. Several million TWACS-based end-points have been deployed in rural as well as in urban and sub-urban areas located in the US and in Latin America, and provide meter reading at 15-minute intervals.

Also, NB-PLC technologies are gaining interest for AMI applications, an interest exemplified by the recent standardization of both G3-PLC and PRIME as ITU-T Recommendations and the development of the IEEE 1901.2 standard. The capability of HDR NB-PLC of delivering substantially higher data rates with respect to UNB-PLC comes at the price of reduced range and, sometimes, transformer conditioning. Not all PLC technologies offer the same reliability in passing the distribution transformer and, often, this capability strongly depends on the transformer itself.

The equivalent circuit of a transformer contains both capacitors and inductances, where the capacitors appear as shunts and this produces the well-known low-pass behavior. However, the combination of various capacitors and inductances should give rise to a resonant behavior at various frequencies thus adding frequency selectivity on top of the low-pass trend. This should be a general behavior at all frequencies although, at high frequencies, other effects such as RF coupling may also appear. This behavior was recently confirmed in the literature, where it was reported that transformers offer several narrowband windows of low attenuation from the low frequency up to the very high frequency regions. Thus, even though BB-PLC signals do not pass through (or around) the distribution transformer, and broadband connectivity between MV and LV necessarily requires the installation of coupling units to by-pass the transformer, low data rate communication between two BB-PLC nodes located on the two sides of a distribution transformer may sometimes be possible without by-pass couplers.

These windows of low attenuation are present also at lower frequencies thus allowing NB-PLC technologies to pass the transformer in some cases – most likely when there is high frequency diversity, such as in HDR NB-PLC so that multiple windows of low attenuation fall in the communication. These characteristics call for sub-banding techniques and frequency agility capabilities in PLC transceivers. Although these results are encouraging, it is difficult to draw at this time general conclusions on this matter, since there is no statistical model for transformers that allows a more quantitative assessment of the capability of PLC signals to pass through (or around) the distribution transformer.

The architectural consequence of MV/LV connectivity is that many more meters could be handled by a single concentrator located on the MV side. This concentrator node would then send the aggregated data from many meters back to the utility, using either PLC or any other networking technology available *in situ*. This capability also heavily impacts the business case when there is a very different number of customers per MV/LV transformer. In North America, the majority of transformers serves between five and ten customers; in Europe, the majority of transformers serves 200 customers or more. Thus, especially in the US, it is economically advantageous to avoid coupler installation and resort to technologies that allow connectivity between the MV and LV sides – and possibly also between meters served by different distribution transformers (LV-to-MV-to-LV links).

When there are very few end-points (meters) per distribution transformer, as in the US, it is convenient to push the concentrator up along the MV side (and even up to the substation) and handle multiple LV sections, so that more end-points can be handled per concentrator. The low number of end-points per transformer in the US makes UNB-PLC solutions like TWACS attractive, as the concentrator is located in the substation and can handle a large number of meters with no additional communication infrastructure (e.g., repeaters or couplers) between substation and meters. On the other hand, the large number of end-points per transformers in Europe does not really require location of the concentrator up in the substation or on the MV side, as it can be conveniently located on the LV section of the grid. Thus, the capability of a PLC technology to pass through distribution transformers may be more appealing in the USA rather than in Europe.

In emergency situations, it is often the case that conventional networking technologies encounter congestion due to a spike in the collision rate, i.e., when all meters tend to access the channel at the same time (blackout, restoration, etc.), or when multiple DR signals requiring immediate action are sent to households. In these challenging scenarios, traditional networking approaches, including wireless sensor networks fail due to the network congestion and competitive channel access mechanism.

Unlike wireless solutions based on ZigBee or WiFi, PLC-based AMI have a proven track record of being able to avoid network congestion when cooperative schemes are employed (see the REMPLI project) [4].

13.9.2.2 Vehicle-to-Grid Communications

A PHEV charges its battery when connected to electric vehicle supply equipment (EVSE) which, in turn, is connected to premises wiring or to distribution cables (airport, parking lots, etc.). A variety of applications scenarios can be envisioned in enabling a communication link between the PHEV and the utility (e.g., for the control of the localized peak load that the increasing penetration of PHEVs would inevitably create). The availability of a communication link between the car and the EVSE (and even beyond the EVSE to the meter, the internet, the HAN, the appliances, the utility, etc.) will be the key enabler for these applications.

The first distinctive advantage of PLC for vehicle-to-grid communications is the fact that an unambiguous physical association between the vehicle and a specific EVSE can be established, and this is something that is not possible to accomplish with a wireless solution, even if short range. A physical association has advantages, especially in terms of security and authentication. Although the PLC channel in this scenario is impaired by several harmonics present due to the inverter, there are, today, several ongoing tests on both BB-PLC and NB-PLC solutions within the "PLC Competition" being conducted by the Society of Automotive Engineers (SAE). In terms of cost, worldwide regulations, and ease of upgrade, NB-PLC solutions offer some advantages with respect to BB-PLC. Since NB-PLC is also an excellent choices for meters and appliances, the availability of a single class of PLC technologies for the inter-networking of different actors is, of course, tempting.

13.9.2.3 Demand Side Management (DSM)

One of the primary DSM applications on the LV side is DR, which has been receiving growing interest, especially in the US. DR refers to the ability to make demand able to respond to the varying supply of generation that cannot be scheduled deterministically, for example, solar and wind. Thus, DR is a means to alleviate peak demand and to bring more awareness on energy usage to the consumer. It is believed that DR will allow a better control of peak power conditions, maximize the use of available power, increase power system efficiency through dynamic pricing models, and allow customers to participate more actively to energy efficiency. Implementation of DR requires establishing a link (either direct or indirect, e.g., via a gateway) between the utility and household appliances.

The largest direct load control system in the world has been operating in Florida for over 20 years, using a UNB-PLC technology (TWACS). Florida Power and Light manages via TWACS over 800 000 Load Control Transponders installed at the premises of over 700 000 customers, and it can shed up to 2 GW of load in a matter of a few minutes. Florida Power and Light has also deployed 1.4 million TWACS-enabled end-points for AMI. It is interesting to verify that such large scale DR/AMI systems can operate successfully using a communications system characterized by very low data rate.

Due to the higher attenuation that PLC signals experience over the LV side, BB-PLC solutions may not always be ideal for DR applications when direct load control is implemented, since the distance between appliances and the utility signal injection point (the smart meter, the MV/LV transformer) may be, in some cases, too large. On the other hand, when DR is implemented with indirect control via a gateway (e.g., a home/building energy management system (HEMS/BEMS)), then BB-PLC solutions are technically adequate and would provide the added benefit of being able to transfer securely data from Smart Grid applications to the HAN, and vice versa. Although technically adequate, other considerations related to cost may arise as BB-PLC technologies may be overly-dimensioned for carrying out DR. Due to the much lower path loss at lower frequencies, NB-PLC solutions are also good candidates for DR applications for both direct and indirect load control.

13.9.2.4 In-Home Environment

There are intriguing possibilities of tying Smart Grid applications with HEMS, and there is a strong belief that these applications will help to foster a behavioral change in how consumers address energy consumption. The home is a natural multi-protocol and multi-vendor environment, and it is unrealistic to suppose that this will change any time soon, even though there is a lot of pressure by some industry sectors to reduce the number of allowed networking choices. A variety of BB-PLC solutions will continue to be installed by consumers, regardless of any convergence in the networking choices for the Smart Grid.

From this point of view, segregating Smart Grid applications in one band (CENELEC/FCC/ARIB), and separating them from traditional entertainment and internet access ones running on BB-PLC (but also with the capability of linking these applications securely via the HEMS), seems a good engineering solution that balances efficiently the various requirements of these very different applications. However, the use of NB-PLC in the in-home environment may require special attention to cope with reduced cross-phase connectivity, since the capacitive nature of cross-phase coupling yields higher attenuation at lower frequencies than at the higher ones used in BB-PLC.

Although there is evidence that a HEMS does not provide compelling financial benefits to residential customers, there is substantial evidence that a HEMS can yield substantial benefits to utilities in terms of improving grid reliability, as well as reducing peak demand. In fact, a HEMS can serve the function of "sensor" in a much more complete and effective way than what a smart meter would be capable of doing. While the smart meter is a low-cost sensor and can only report instantaneous demand, a HEMS could actually report to the utility (or third party energy service provider) the forecasted demand of energy, and provide more complex sensing functions. For example, the forecasting capability of a HEMS could be very accurate, as it would be based on the "state" of the home and on the behavioral model built on consumer activity.

The state of the home tracked by a HEMS could include: the present and predicted energy demand of an appliance as it goes through its service cycle; storage levels of batteries; amount of consumer shifted demand (service queue); and so on. If a utility had at its disposal the knowledge of the state of every home (or of a set of homes or microgrids via aggregators), forecasting and scheduling of generation and DSM would be possible with more relaxed communications requirements. Furthermore, storage levels and queued demand could also become part of pricing models.

We also point out that there is, today, a growing interest in hybrid AC/DC wiring infrastructure. Within the home, the development of a DC infrastructure yields great benefits to energy generation (photovoltaic, fuel cell) and storage (rechargeable battery). Both NB and BB-PLC greatly benefit from operating over DC lines, as the channel is time-invariant and appliance cyclostationary noise disappears – with the exception of impulsive noise caused by AC/DC inverters.

13.10   Conclusions

The success of PLC for in-home and in-vehicle applications is growing and very promising, and its use for Smart Grid applications has been successful for years and will certainly continue to be so for more years to come. There are also several strong, and sometimes unique, motivations to consider PLC as an important enabler of telecommunications services in the area of broadband access:

- The power grid is ubiquitous, constituting an already existing high-penetration telecom infrastructure that is much more extensive and pervasive than any other wired/wireless alternative. Being already in place, it is potentially cheaper than other forms of local telecommunications access that are comparable in scale.
- The fine-grained size of the power grid allows PLC to be an excellent candidate for last-mile connectivity.
- As virtually every line-powered device can become the target of value-added services, PLC may be considered as the technological enabler of a multitude of future applications that would otherwise probably not be available.

On the other hand, there has also been a lot of skepticism about the technology and its commercial viability – a skepticism amplified by the slow pace at which standardization activities in IEEE, ETSI, and ITU-T have been progressing. Today there is only one standard for PLC as an access technology (IEEE 1901), but this has seldom been used in the field. There are, today, very few PLC deployments in the world for broadband access, and its use in industrialized countries, where the availability of other broadband access technology is abundant and cheaper, has made PLC a marginal technology for access applications.

The main reasons for skepticism can be summarized as follows:

- The PL channel is a very harsh and noisy transmission medium and is difficult to model, so it poses unique challenges to the modem designer.
- The multitude of wiring and grounding practices, as well as topologies, pose unusual robustness requirements.
- Regulatory issues naturally arise due to the unshielded nature of power line cables, which are both the source and the target of electromagnetic interference.
- Failure to deliver the desired bandwidth at the right cost has compromised the commercial viability for broadband access applications, especially in those areas where there is already a lot of broadband penetration.

Perhaps the area where broadband access via PLC may still have some possibility of success is in third world countries, where access to the internet is essential to economic growth but there is no or very little telecom infrastructure. Similarly, rural areas in industrialized countries, where it is very uneconomical to provide broadband services at competitive prices, could also benefit of from the deployment of PLC, as most of these areas lack traditional telecom infrastructure but nevertheless have access to power.

References

1. Schwartz M. History of communications – carrier-wave telephony over power lines: Early history. *IEEE Communications Magazine*. 2009; **47**(1): 14–18.
2. Dostert K. *Power Line Communications*, Prentice Hall PTR; 2001.
3. Ferreira H, Lampe L, Newbury J, Swart T. (eds) *Power Line Communications*, 1st edn. New York, NY: John Wiley & Sons; 2010.
4. Galli S, Scaglione A, Wang Z. For the grid and through the grid: the role of power line communications in the smart grid. *Proceedings of the IEEE*. 2011; **99**(6): 998–1027.
5. Galli S, Logvinov O. Recent developments in the standardization of power line communications within the IEEE. *IEEE Communications Magazine*. 2008; **46**(7): 64–71.
6. Goldfisher S, Tanabe S. IEEE 1901 access system: An overview of its uniqueness and motivation. *IEEE Communications Magazine*. 2010; **48**(10): 150–157.
7. IEEE Standard for Broadband Over Power Line Networks: Medium Access Control and Physical Layer Specifications, IEEE Std. 1901–2010; Sep. 2010.
8. Galli S. On the fair comparison of FEC schemes. IEEE Int. Conf. on Commun. (ICC), Cape Town, South Africa, May 23–27, 2010.
9. Wang Z, Scaglione A, Thomas R. Generating statistically correct random topologies for testing smart grid communication and control networks. *IEEE Transactions on Smart Grid*, 2010; **1**(1): 28–39.
10. Watts D, Strogatz S. Collective dynamics of small-world networks. *Nature*. 1998; **393**: 393–440.
11. Amirshahi P, Cañete F, Dostert K. *et al.* Chap. 2: channel characterization, in *Power Line Communications*, 1st edn (eds H. Ferreira, L. Lampe, J. Newbury, and T. Swart). John Wiley & Sons; 2010.
12. Galli S, Banwell T. A deterministic frequency-domain model for the indoor power line transfer function. *IEEE Journal on Selected Areas in Communications*. 2006; **24**(7): 82–83.
13. Banwell T., Galli S. A novel approach to the modeling of the indoor power line channel – Part I: Circuit analysis and companion model. *IEEE Transactions on Power Delivery*. 2005; **20**(2): 655–663.
14. Cañete F, Cortés J, Díez L, Entrambasaguas J. Analysis of the cyclic short-term variation of indoor power line channels. *IEEE Journal on Selected Areas in Communications*. 2006; **24**(7): 1327–1338.
15. Zimmermann M, Dostert K. A multipath model for the powerline channel. *IEEE Transactions on Communications*. 2002; **50**(4): 553–559.
16. Katayama M, Yamazato T, Okada H. A mathematical model of noise in narrowband power line communication systems. *IEEE Journal on Selected Areas in Communications*. 2006; **23**(7): 1267–1276.
17. Zimmermann M, Dostert K. Analysis and modeling of impulsive noise in broad-band powerline communications. *IEEE Transactions on Electromagnetic Compatibility*. 2002; **44**(1): 249–258.
18. Galli S, Banwell T. A novel approach to the modeling of the indoor power line channel – Part II: Transfer function and its properties. *IEEE Transactions on Power Delivery*. 2005; **20**(3): 1869–1878.
19. Galli S. A simplified model for the indoor power line channel. IEEE International Symposium on Power Line Communications and Its Applications (ISPLC), Dresden, Germany, Mar. 29–Apr. 1, 2009.
20. Galli S, Scaglione A. "Discrete-Time Block Models for Transmission Line Channels: Static and Doubly Selective Cases," Cornell University Library. Available online: http://arxiv.org/abs/1109.5382; 2011.
21. Galli S, Waring D. Loop make-up identification via single ended testing: beyond mere loop qualification. *IEEE Journal on Selected Areas in Communications*. 2002; **20**(5): 923–935.
22. Galli S. A novel approach to the statistical modeling of wireline channels. *IEEE Transactions on Communications*. 2011; **59** (5): 1332–1345.
23. Galli S, Koga H, Kodama N. Advanced Signal Processing for PLCs: Wavelet-OFDM. IEEE International Symposium on Power Line Communications and Its Applictions (ISPLC), Jeju Island, Korea, April 2–4, 2008.
24. Biglieri E, Lee YW, Poor HV, Vinck H. Power Line Communications. *IEEE Journal on Selected Areas in Communications (JSAC)*, vol. 24, no. 7, pp. 1261–1266, July 2006.
25. Galli S, Kurobe A, Ohura M. The Inter-PHY Protocol (IPP): a simple co-existence protocol for shared media. IEEE International Symposium on Power Line Communications and Its Applications (ISPLC), Dresden, Germany, Mar. 30–Apr. 1, 2009.

Further Reading

Most of the contributions on the topic of PLC can be found in the IEEE Transactions on Consumer Electronics, Power Delivery, Industry Applications and Industrial Electronics, whereas fewer papers on PLCs have appeared in publications traditionally dealing with communications problems.

A collection of "Best Readings" has been recently put together by the IEEE Communications Society Technical Committee on Power Line Communications (TC-PLC). Furthermore, the IEEE TC-PLC has also created a search tool for PLC papers that includes the Proceedings of the International Symposium on Power Line Communications and Its Applications for the years 1997 to 2004 which are not on the IEEE Xplore database.

For more information, see "Best Readings" and "PLC DocSearch" on the IEEE TC-PLC website http:// committees.comsoc.org/plc/.

14

Wireless Broadband Access: Air Interface Fundamentals

14.1 Introduction

In this chapter, we lay down the groundwork needed to understand the technologies that have evolved for wireless broadband access, by discussing various concepts that form the basis of modern wireless communication systems. By nature, the wireless medium is not a closed medium, and it is therefore susceptible to interference and may need to support shared access. Furthermore, due to the propagation characteristics of electromagnetic waves, wireless links can be very unreliable unless several steps are taken to mitigate this shortcoming. In the following sections, we first discuss duplexing techniques used to support two-way communications. We next explore the basic building blocks that comprise the physical layer of a robust wireless communication system, and concepts pertaining to shared access of the medium. Finally, we discuss algorithms such as link adaptation and scheduling, which control the operation of the physical layer and are vital to the performance of the system. As an example application of these concepts, we describe a satellite-based broadband access system.

14.2 Duplexing Techniques

Duplex communications allow the two ends of a communication link to transmit and receive simultaneously. This mode of communications is also referred to as full-duplex, and is in contrast to simplex communications, which is only unidirectional (e.g., broadcast TV). Most broadband access systems tend to use a central node that terminates the link for several access nodes that receive the broadband service. Communications from this central node (e.g., a base station or access point) towards the access node, such as a user device, is usually called the downlink or forward link. In the opposite direction, the link from the user to the base station is called the uplink or reverse link.

In order to establish simultaneous communications on both links in such a way that the transmissions in either direction do not interfere with each other, the downlink can either be separated from the uplink in frequency or in time. This gives rise to the two most widespread techniques for duplexing – frequency-division duplex (FDD), and time-division duplex (TDD). Recently, there has been promising research in the area of full-duplex communications, where both links use the same time and frequency resources. In such systems, advanced interference cancellation is employed in both the analog and digital domains, and the downlink is separated from the uplink by using two different antenna systems that are carefully designed. However, such systems are not yet ready for commercial deployment.

Broadband Access: Wireline and Wireless – Alternatives for Internet Services, First Edition.
Steven Gorshe, Arvind Raghavan, Thomas Starr and Stefano Galli.
© 2014 John Wiley & Sons, Ltd. Published 2014 by John Wiley & Sons, Ltd.

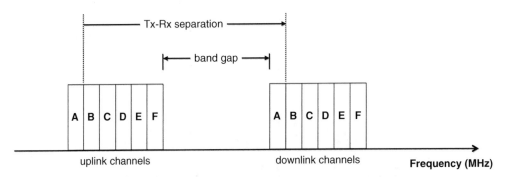

Figure 14.1 Frequency division duplexing

14.2.1 Frequency-Division Duplex

In FDD systems, two disjoint portions of the spectrum are allocated for the uplink and downlink, as shown in Figure 14.1.

Such a spectrum allocation is usually referred to as "paired" spectrum, where the two disjoint sub-bands of the spectrum have the same bandwidth and are assigned a common band number. The width of the band between one uplink channel and its corresponding downlink channel is sometimes called the Tx-Rx frequency separation – from the perspective of the user device – as it determines the RF front-end filter requirements. The space between the aggregate uplink and downlink portions of the band is called the band gap. For example, band II, commonly known as the Personal Communication Services (PCS) band, uses 1850–1910 MHz for the uplink and 1930–1990 MHz for the downlink. In this example, the band gap between the uplink and the downlink is 20 MHz, and the Tx-Rx separation is 80 MHz.

With FDD, the distribution of capacity between the uplink and downlink is static, because the pairs of sub-bands have the same bandwidth. As a result, the ratio of the uplink capacity to downlink capacity depends on the respective spectral efficiencies. Typically, the uplink requires transmissions from a battery-power constrained user device, so therefore the uplink is usually less spectrally efficient than the downlink. Taking this into account, the uplink sub-bands are usually allocated below the downlink sub-band, because of better propagation characteristics at lower frequencies.

FDD deployments form the vast majority of deployments around the world, with GSM being the most widely used FDD technology. As a result, technologies that followed as an evolution of GSM also tended to be FDD, and FDD systems still vastly outnumber TDD systems.

14.2.2 Time-Division Duplex

In contrast to FDD spectrum, TDD spectrum is usually called "unpaired" spectrum, as the uplink and downlink share the same contiguous spectrum allocation. Simultaneous operation in both links is achieved by switching the active link between the uplink and downlink at a very fast rate, as shown in Figure 14.2. In order to allow each end of the link to switch between transmitting and receiving, a short time gap is provided between each switch, analogous to the duplex gap in FDD systems.

The fraction of time that each link is active can be controlled, allowing the capacity of each link to be more closely matched to the expected traffic volume in each direction. However, in a network with several high-power base stations and user devices, such as a cellular network, this fraction must be the same across the entire network. If the uplink/downlink ratio is not the same, some high-power base stations can be transmitting when other base stations are trying to receive low-power transmissions from user devices, causing unacceptably high interference.

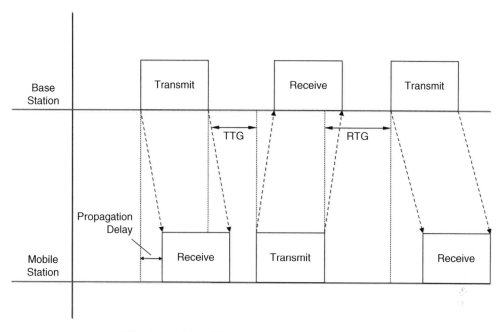

TTG – transmit transition gap   RTG – receive transition gap

Figure 14.2 Time division duplexing

14.3 Physical Layer Concepts

The propagation of signals over a wireless link can be unreliable, because of the time-varying quality of
the link. In the sub-sections that follow, we describe the characteristics of a wireless channel and several
techniques such as diversity transmissions, channel coding, and interleaving, that are applied to improve
reliability of the link. We also introduce multi-antenna systems and discuss how such systems provide a
way to extract maximum efficiency out of limited spectrum resources.

14.3.1 The Wireless Channel

In tethered communications, the propagation medium – copper or optical fiber – is structured, and the
signal loss can be estimated or predicted with relatively high accuracy. On the other hand, in a wireless
channel, the medium of propagation is highly unstructured, and signal loss is much more difficult to model
and predict. The attenuation of the signal, usually referred to as *path loss*, is usually much greater than that
for a wired medium for a given distance. In addition to the distance, the path loss is influenced by two key
factors: the frequency of the signal and the type of environment. Free space is usually the most benign
environment for radio propagation and, as the "clutter" increases, as in dense urban environments, the path
loss worsens.

The path loss also generally increases with frequency. This makes the lower frequency bands, in the
few hundreds of megahertz (MHz), very desirable because of their improved range and wall
penetration characteristics for indoor coverage. Lower frequencies, however, require larger antennas
for efficient transmission and reception. For example, a two-antenna system designed for operation at
700 MHz can require almost twice as much antenna surface area as a four-antenna system designed for
operation at 2100 MHz. Antenna towers are expensive to install and maintain, and the "real estate"

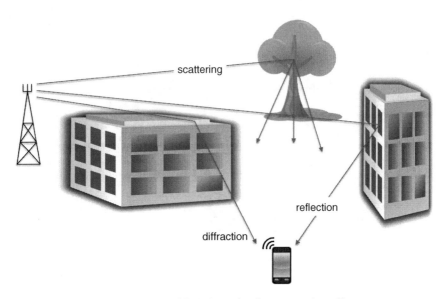

Figure 14.3 Simplified view of radio propagation effects

available at the top of the tower is very valuable. As we shall see later, increasing the number of antennas can provide important system performance benefits. Therefore, there exists a trade-off between increased path loss and lowered antenna sizes, and this must be taken into account for wireless system design.

In addition to path loss, the wireless signal strength can vary with time and frequency. Due to the presence of large obstacles between the transmitter and receiver (e.g., hills, buildings, and trees), the signal undergoes large-scale variations in strength, referred to as *shadowing* or *shadow-fading*. Wireless signals propagate from the transmitter to the receiver via various paths due to reflection, scattering, and diffraction caused by various objects in the environment, as shown in Figure 14.3. This characteristic of radio propagation is referred to as multi-path propagation. As a result of these multiple paths, the sinusoidal signals arrive at the receiving antenna with different phase-shifts, and these can combine constructively or destructively, resulting in the phase and amplitude of the signal varying in an unpredictable manner. This phenomenon called *fast fading*. These are small-scale variations and can be both time-varying and frequency-dependent. Note that fading can occur even if the mobile is stationary, due to changes in the environment from moving obstructions such as vehicles and trees.

In addition to impairments such as path loss and fading that are introduced by the medium, the wireless channel also suffers signal corruption due to the presence of interference and noise. Interference is usually due to signals from similar systems operating in the vicinity. Noise is typically thermal noise induced by the electronics at the receiver.

14.3.2 Diversity

On one hand, path loss and fading result in greatly attenuated and unpredictable signals that make communications difficult. On the other hand, the dynamic multi-path nature of the propagation presents an opportunity to improve the quality of the link. In order to exploit channel variations to improve the overall link, the system can be designed so that the channel is used (or "sampled") multiple times across a diverse set of channel conditions. The resulting link has better performance, because such a system increases the probability that the channel will be sampled at least once when the channel quality is good.

Time diversity is a technique in which the transmission is repeated at multiple instances separated in time, so that the channel quality between transmission instances is sufficiently uncorrelated to make the probability of success independent from one instance to the next. Similarly, if a wide bandwidth is used, the properties of the environment may make one part of the band fade independently from another, allowing the use of *frequency diversity*.

If multiple antennas are used at either end of the link, and if there is sufficient separation between the antennas at the same end of the link, then the signal may take independent paths across the link from different antennas, due to the multi-path nature of radio propagation. Therefore, channels with low correlation may also be exploited using the *spatial diversity* provided by multiple transmit or receive antennas. We will discuss spatial diversity in more detail in a subsequent section.

14.3.3   Channel Coding

The presence of severe impairments can result in communication errors, unless steps are taken to mitigate the effects of the wireless channel. Channel coding is an indispensable technique employed to improve the reliability of digital communications.

The easiest way to explain how channel codes work is by using the example of a simple coding scheme called the repetition code. This code is constructed by simply repeating the bit that needs to be sent. A repetition-3 code would therefore send 111 to represent 1 and 000 to represent 0. At the receiver, any sequence of bits other than 111 or 000 would therefore indicate that an error has occurred. In addition to detecting errors, if it is known that the channel is most likely to introduce only one bit-error, then it is intuitively clear that if a 110, 101, or 011 is received, 111 was most likely the transmitted code, and one bit-error can be corrected. Therefore, by adding redundant bits to the transmission, channel codes provide the capability to detect and correct errors. These redundant bits are sometimes referred to as parity bits. Because channel codes can correct errors without the data being retransmitted, they are called Forward Error Correction (FEC) codes.

In practice, there are several more sophisticated coding schemes that are available for both detection and correction of codes. Cyclic-redundancy check (CRC) codes are typically used for error detection. Turbo codes and low-density parity check (LDPC) codes are considered to be the most powerful channel codes in use today, although convolutional codes and Reed-Solomon codes are also used in several systems.

14.3.4   Interleaving

As explained earlier, fading can cause rapid variations in the signal, both in time and frequency. If a deep fade occurs, either in time or in a portion of the frequency band, then all the bits transmitted in that portion of the band can get corrupted, resulting in a high probability of error for a sequence of consecutive bits. If all of the corrupted bits come from part of the same FEC codeword then, with no reliable bits in the code, the FEC is rendered ineffective. In order to avoid this kind of decoding failure, the bits from a several codewords can be *interleaved*, so that consecutive bits in time or frequency belong to different codewords. If a sequence of bits are corrupted in transmission, then the process of de-interleaving at the receiver will ensure that these corrupted bits get distributed among various codewords, with the FEC able to effectively correct a few bits in error.

14.3.5   Multi-Antenna Techniques and Multiple-Input Multiple-Output (MIMO)

In a previous subsection, methods to harness diversity were shown essentially to involve distributing access along a physical dimension, such as time or frequency. In this subsection, we discuss how wireless systems exploit yet another dimension – space – to achieve more efficient resource utilization. Multi-antenna techniques allow us to explore this "final frontier" for improving peak speeds and capacity of wireless systems.

Increasing the number of antennas can be useful at both the transmitter and receiver. There are three key benefits of multiple antennas:

- Spatial diversity.
- Coherent combining or power gain.
- Spatial multiplexing.

The basic idea is that different antennas at the same end of the link can sample the spatial variations in the channel and provide diversity. In addition, these antennas can also increase the strength of the composite signal by combining the samples smartly. Finally, if the transmit antennas are such that the channel from each antenna is not correlated with the channels from other antennas, then a different stream of data can be sent from each antenna and multiplexed "in space" to increase the data rate of the link.

The term Multiple-Input Multiple-Output (MIMO) is usually used to denote spatial multiplexing. When multiple antennas are used at the transmitter to send multiple streams in the manner described, then multiple antennas are required at the receiver to separate the streams. If N streams are spatially multiplexed using N transmitting antennas, then at least N receiving antennas are required to effectively separate these streams.

To see this, it is best to think of each receive antenna forming a composite received signal that can be expressed as a linear equation, with the N signals being the unknowns, and the channel from each transmit antenna to a given receive antenna forming the coefficients. In order to solve for the N unknown signals, we need N equations, which can be obtained by using N receive antennas. See Figure 14.4 and the Equation 14.1 for a more precise mathematical representation of a MIMO system. The transmitted symbols are x_1 and x_2, and the received symbols are r_1 and r_2. Given that the channel is known (usually by sending known transmitted symbols called pilots or reference symbols), the h and r values are known constants, and one has to only solve for x.

$$r_1 = h_{11}x_1 + h_{21}x_2$$
$$r_2 = h_{12}x_1 + h_{22}x_2$$

(14.1)

The gains one can extract from the spatial dimension depend on the properties of the channel and the type of antenna configuration employed. If the antenna spacing is large relative to the surrounding multi-path propagation environment, then the channels from each antenna are likely to be uncorrelated. It also helps if the environment is conducive to scattering. Such an environment is likely to exist in dense urban areas with a lot of clutter in between the transmitter and the receiver. In an outdoor environment without much clutter – say, in rural areas – it is possible that the multi-path propagation has a strong line-of-sight (LoS) component. In this situation, the scattering components may be weak compared to the strong LoS path, and the channels at widely spaced antennas will still be quite correlated due to the presence of this strong LoS path. A strong LoS path can usually deliver a high SNR, but it does not provide spatial

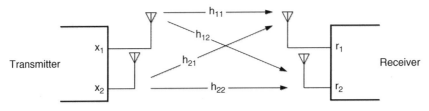

Figure 14.4 A simple MIMO system

diversity. In order to still exploit diversity, cross-polarized antennas must be used, with each antenna transmitting on a different polarization, resulting in channels that have low correlation.

We will first look the types of techniques used, and the benefits of adding antennas separately at each end of the link, and that will be followed by a discussion on the techniques used in a true MIMO system with multiple antennas at both ends of the link.

14.3.5.1 Multiple Antennas at the Receiver

When multiple antennas are used at the receiver, with only a single antenna at the transmitter, multiple copies of the transmitted signal are received, with each copy distorted differently based on the channel to the different receive antennas. Given that the channel to each receive antenna is known, this presents an over-specified system of equations, so different techniques can be used to estimate the transmitted signal, depending on the amount of information available. One popular technique is called Maximal Ratio Combining (MRC), and this simply tries to maximize the received signal energy in the presence of thermal noise by adding up the received samples coherently, weighted by the signal strength of the individual paths to each receive antenna. Using MRC maximizes the SNR of the received signal.

Another technique relies on the use of information about the interference, and tries to maximize signal-to-interference-and-noise ratio or SINR. This technique is called either Minimum Mean Squared Error (MMSE) or Interference Rejection Combining (IRC), depending on the implementation and the level of interference knowledge that is incorporated into the algorithm. IRC usually performs best in the presence of a few strong interfering users and, if designed well, it will not result in any degradation in performance when interference is absent. However, IRC is usually more complex to implement than MMSE or MRC.

14.3.5.2 Multiple Antennas at the Transmitter

When multiple antennas are used at the transmitter, with only a single antenna at the receiver, then the composite signal received should be "predistorted" at the transmitter, so that the receiver perceives a better signal than it would have if only one transmitting antenna were used. In other words, it is important to use a technique whereby the multiple copies from the transmitter do not cause interference at the receiver, which is clearly an incorrect way to use multiple transmit antennas. Here we discuss three different techniques that are available at the transmitter end to improve the quality of the link through the use of multiple transmit antennas.

One straightforward scheme is basically a reciprocal of the MRC and MMSE/IRC. This is called *transmit beamforming*, where the transmit signals are predistorted so that they combine coherently at the receiver of interest and, optionally, combine destructively at other receivers that may otherwise perceive interference from this link. The latter optional technique is sometimes called *nulling*. The predistortion is accomplished by rotating and scaling the signal with a different phase and amplitude for each transmit antenna. These amplitudes and phases, applied to the antenna array, are often called *beamforming weights*.

Note that for transmit beamforming and nulling, the transmitter must have up-to-date knowledge of the channels from multiple antennas to multiple receivers. In general, this knowledge is hard to collect without significant feedback overhead in FDD systems. In TDD systems, channel reciprocity (i.e., the fact that the channel is essentially the same in both directions) can be exploited to obtain channel state information. However, given that the path through the RF electronics is not the same in both directions, the antennas have to be calibrated to use channel reciprocity. This can make the system complex and expensive.

If channel information is not available at the transmitter, then Space-Time Block Coding (STBC) or Space-Frequency Block Coding (SFBC) can be used, which provides diversity benefits at the receiver. These transmit diversity schemes send multiple copies of the same signal, either at successive time instances (STBC) or on adjacent frequencies (SFBC), under the assumption that the channel does not change significantly across the time or frequency dimension. The transmission of signals across the space and time/frequency dimension is done in such a way that the copies from different antennas can be separated and combined in an optimal way through simple matrix manipulation. STBC and SFBC allow the user to harness spatial diversity even if the receiver has only one antenna. However, note that this increased spatial diversity does not deliver the type of increase in rate or capacity that comes with the use of spatial multiplexing.

One other technique that is available for use in Orthogonal Frequency Division Multiplexing (OFDM) systems is cyclic-delay diversity (CDD). We will look at OFDM in more detail in the next section, but it is sufficient for now to know that, in OFDM, the signal consists of a few hundred frequency "subcarriers" sent together on the same time symbol. In CDD, mulitple antennas transmit cyclically shifted versions of the OFDM signal. In the frequency domain, this is equivalent to a different phase shift being applied to each subcarrier and, therefore, it is as if each subcarrier is using its own "beam". This manipulation of the signal results in the channel appearing to be more frequency-selective at the receiver, and it therefore allows the system to harness more frequency diversity. In addition, CDD is also useful when multiple antennas need to be virtualized to look like a single antenna. For example, when a four-antenna system needs to look like a single-antenna system for legacy mobile devices that are not equipped to handle a four-transmit antenna system, the same signal can be sent out of all four antennas, with a progressively increasing cyclic delay applied to the signal out of each antenna. The composite signal appears to be arriving from a single virtual antenna after passing through a channel with increased frequency selectivity.

14.3.5.3 MIMO and Spatial Multiplexing

In a system where multiple antennas are used at both ends of the link, it is possible to send muliple streams of information, as explained earlier. For spatial multiplexing, the number of transmit antennas and receive antennas should be greater than one, and the maximum number of streams that can be spatially multiplexed is equal to the minimum number of antennas present at either end of the link. A spatially multiplexed stream is often called a spatial *layer*, and the number of multiplexed streams is called the *rank* of the transmission. For example, a transmission with two spatial layers is called a rank-2 transmission.

There are generally three types of MIMO techniques that one can configure. The first is single-user MIMO or SU-MIMO, where multiple streams are sent to the same user. Another technique is multi-user MIMO or MU-MIMO, where the spatially multiplexed streams are intended for multiple users. For example, four transmit antennas can be used to send a total of four layers, as long as there are a total of four receive antennas. However, if appropriate beamforming is used, the four layers can be sent as two rank-2 transmissions to two seperate receivers, each with two receive antennas. It is important to realize that all transmit antennas can be (and usually are) involved in the transmission of each layer, with beamforming weights being used across all antennas to separate the layers. Finally, the reciprocal of this system, where multiple users transmit to a single receiver (usually in the uplink) is called Virtual MIMO or Collaborative Spatial Multiplexing (CSM). For example, two users with a single transmit antenna each can transmit collaboratively to a receiver with two receive antennas.

Clearly, in all cases, the coexistence of multiple streams in the spatial dimension requires either the channels originating at different transmit antennas to be uncorrelated, or beamforming and nulling to be used to reduce the interference between the streams. This process of attempting to separate various spatially multiplexed streams for improved performance via beamforming is also called precoding, and this is typically performed using a *codebook* of *precoder matrices*. The precoder matrix is nothing but a set of fixed beamforming weights. Note that the weights form a single dimensional vector when a single spatial layer (or rank-1 beamforming) is used (one weight for each antenna), and form a matrix with one vector for each spatial layer when spatial multiplexing or transmit diversity is used.

The receiver selects the precoding matrix from the codebook that best matches the channel conditions that it estimates to exist between the transmitter and itself. It then sends only the index of this particular matrix within the codebook of matrices, along with a recommendation for the rank of the transmission, back to the transmitter. This index is usually called the precoding matrix index (PMI), and the rank is called the rank indicator (RI). The tradeoff in system design is to design a codebook that is large enough to have sufficient entries to match the channels that may exist over the link, yet small enough so that the bandwidth used for feedback does not result in unnecessary overhead.

14.4   Access Technology Concepts

In this section, we discuss several general concepts that apply to wireless access. Wireless communications use a shared medium and, as such, mechanisms are required for coordinated use of the spectrum. This coordination includes *multiple access* techniques that permit multiple communication links to coexist without interference. In addition, *medium access control* (MAC) protocols are also required to specify how users can access the resources required for the link to operate.

14.4.1   Frequency Division Multiple Access (FDMA)

In this multiple access technique, each link is assigned its own frequency channel, so therefore the available spectrum is divided into several channels for multiple users to simultaneously access the network. This is one of the most basic methods but, as wireless systems have evolved over the years, more complex mechanisms have been built on top of this fundamental division of resources. In the early days of cellular networks, the Advanced Mobile Phone System (AMPS), a second generation cellular standard, relied solely on FDMA. Some frequency channels were set aside for access control, but most of the spectrum was divided into single-user channels. As new systems were developed, multiple users were assigned to the same channel, thereby increasing the capacity of the underlying FDMA system.

14.4.2   Time Division Multiple Access (TDMA)

Another technique to create multiple links is to take a fixed period of time, usually referred to as a *frame*, and divide it into several time-slots or *subframes*, giving each user a different slot or subframe. In its most basic implementation, the time-slot assigned to a user repeats in a periodic fashion, frame after frame, until the user no longer needs the link. When the AMPS system was extended to increase capacity, each user channel was divided into three time-slots, and transmissions were digitized and compressed, thereby allowing three users to use the channel instead of one. This new system was commonly referred to as TDMA, although the formal standard that specified this system is called IS-54, which was later enhanced in IS-136. In addition, most standards that came later, including GSM, HSPA, and even LTE, have TDMA as a basis, in addition to more complex technologies, as we will see in subsequent sections.

14.4.3   Code Division Multiple Access (CDMA)

In the most general sense, CDMA refers to the ability of separating users based on orthogonal or near-orthogonal codes that modulate the signals from or to different sources. CDMA systems are also sometimes called spread-spectrum systems, as the code usually takes a narrowband signal and spreads it to occupy a much wider band, in which several other narrowband signals have been spread in such a way that they coexist with little or no interference. One method to accomplish this is frequency-hopping (FH), where the location of the narrowband signal is changed according to a predefined code that defines a hopping pattern as shown in Figure 14.5. If the hopping codes are selected so that different users rarely, or never, hop to the location at the same time, then multiple access is achieved. The original PHY specification of 802.11 includes an FH option that is almost obsolete.

The more common method used for CDMA is Direct Sequence (DS), wherein a narrowband signal is multiplied with a higher-frequency spreading sequence or code. If the spreading codes are orthogonal, then different users can coexist at the same time in the same frequency channel. In order to recover a

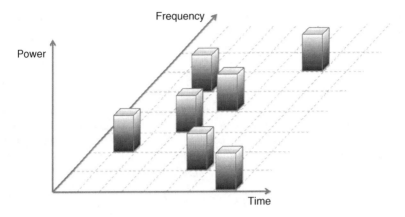

Figure 14.5 Frequency hopping code division multiple access (FH-CDMA)

specific user's signal, one has to simply multiply the user's received signal with the code used to spread it, as shown in Figure 14.6. Many systems, including the 802.11b PHY and cellular systems like cdma2000 and WCDMA, use DS-CDMA for multiple access. DS-CDMA is so much more widely used than FH-CDMA that DS-CDMA is simply called *CDMA*.

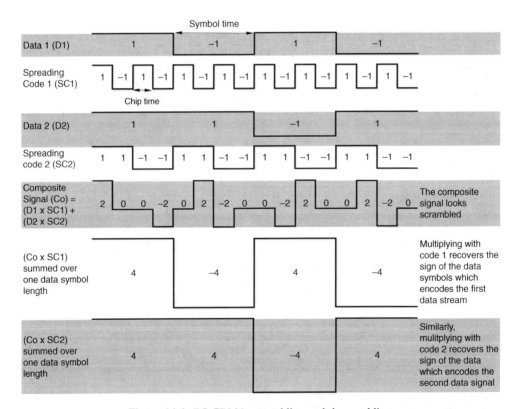

Figure 14.6 DS-CDMA: scrambling and descrambling

14.4.4   Orthogonal Frequency Division Multiplexing (OFDM)

Although FDMA allows multiple frequency carriers to be used within a given spectrum band, it is inefficient, because each frequency is given its own guard band, and filters at the front end of the receiver are tuned to separate each carrier. OFDM provides a much more efficient alternative to pack frequency carriers into a given spectrum band. It achieves this efficient packing by means of the mutually orthogonal property of the spectral signature of each frequency component of the Discrete-time Fourier Transform (DFT). This orthogonal property of the spectrum is illustrated in Figure 14.7. As shown in the figure, the samples of interest are evenly spaced every 1/T Hz, where T is the duration of the composite OFDM symbol. At these sample points, there is only one carrier with a signal peak, and no contribution from other carriers.

In practice, the DFT is implemented by means of the Fast Fourier Transform (FFT) algorithm. The actual OFDM signal is obtained by the inverse FFT operation, which produces samples of a time domain signal. Several low-rate bit streams are transmitted in parallel – one per frequency carrier – by transforming them to a single time-domain symbol. Each low-rate stream is a sequence of complex numbers that usually represent points of a quadrature amplitude modulation (QAM) constellation, which maps the input bit streams to QAM symbols. To distinguish between the composite signal carrier, the individual carriers that carry these low-rate bit streams are usually referred to as subcarriers or tones.

Figure 14.8 shows the block diagram of a typical OFDM system. At the receiver, the information carried by each subcarrier (the complex numbers originally transmitted) is simply retrieved by means of an FFT operation, and the steps taken at the transmitter are reversed to retrieve the original bit stream. Due to the dispersive properties of the channel, inter-symbol interference (ISI) is possible, where different propagation paths can cause copies of the same signal to interfere mutually. An OFDM system has a very elegant way to combat ISI, which is one reason why this is such an effective and widely deployed technology.

The periodic nature of the FFT operation can be exploited by copying a portion of the time domain signal and inserting it at the beginning of the symbol. This process is called cyclic-prefix (CP) insertion. At the receiver, samples corresponding to this CP are removed, leaving only an ISI-free block of samples for processing. Moreover, removal of ISI also greatly simplifies the equalization process to account for channel distortions. A greatly simplified equalizer is a key distinction when comparing it to the technologies that preceded it, like CDMA.

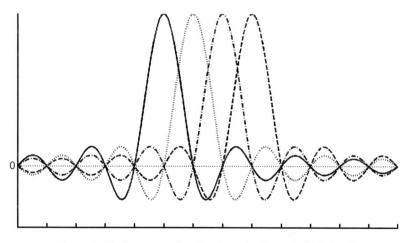

Figure 14.7   Spectrum of each sub-carrier in an OFDM signal

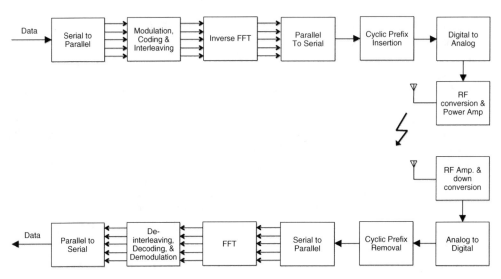

Figure 14.8 Block diagram of a simple OFDM system

In order to protect a particular set of subcarriers from being corrupted by a deep fade in a frequency selective channel, Forward Error Correction (FEC) could be applied, followed by an interleaver to spread the bits over a large number of subcarriers separated in frequency. By doing so, a set of corrupted subcarriers will result in bit errors that are not consecutive at the input to the FEC decoder, making it easier for the decoder to correct the errors. Another technique that may be used to exploit frequency-selective fading is frequency-selective scheduling, where a subset of active users is selected, such that each user is allocated resources in a region of the band where the user's signal is the strongest. We will discuss channel-dependent scheduling in more detail in a subsequent section.

14.4.4.1 OFDMA

The orthogonal-subcarrier property is exploited in a straightforward manner in order to enable multiple access in a multi-user system. The available subcarriers can be partitioned into subsets of subcarriers based on different criteria, and thus assigned to different users. The criteria may include the presence or absence of interference in different portions of the spectrum for specific users, or the frequency-selective nature of a specific user's fading channel. In the downlink, the base station can take data from several users, split it into parallel data streams modulated on different subcarriers in the OFDM symbol, and signal the allocation by means of a control channel. At the receiver, the user only demodulates the subcarriers of interest to receive the data intended for it. In the uplink, different users can transmit on only the portion of the subcarriers allocated to them in a mutually exclusive fashion, and the uplink receiver at the base station follows the same allocation partition to collate data from different users.

The actual allocation of subcarriers can either be contiguous in frequency, or distributed over the entire band. Distributed allocations work better for cases where only the average channel conditions over the entire band are known, so the user cannot be effectively mapped to the portion of the spectrum that is most suitable. In these situations, distributed allocations are still able to provide beneficial frequency diversity.

14.4.4.2 SC-FDMA

The OFDMA signal from multiple users, as described in the previous section, suffers from one practical problem. When subcarriers from a single user are spread over the entire bandwidth in an unstructured

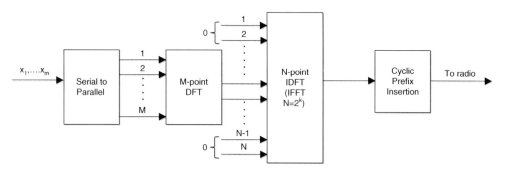

Figure 14.9 SC-FDMA transmitter

manner, large variations are introduced in the instantaneous transmit power. As a result, the efficiency of the power amplifier at the front end of the transmitter decreases, as it is not always used at the optimal operating point. This problem is exacerbated in mobile devices which are power-limited. As a result, a modified version of the basic OFDM system, called single-carrier FDMA (SC-FDMA), is more suitable for the uplink.

A simplified block diagram of the SC-FDMA transmitter is shown in Figure 14.9. In the first step, a sequence of M complex modulation symbols (QPSK or 16-QAM, for example) is transformed by means of a size-M DFT operation. In the next step, an inverse DFT operation is performed over N samples, where N > M, with zeros inserted to make up the required samples. The actual location of the M samples among the zero padding determines the frequency shift of the final signal. Therefore, for multiple access, different users have their signals shifted in an non-overlapping fashion.

Note that for efficient implementation, the inverse DFT operation, in practice, actually uses an IFFT operation over a set of points that is an exponent of 2, similar to a OFDMA system. Also note that when M = N, the DFT and its inverse cancel each other out, generating a wide bandwidth single-carrier signal. It is as if the serial stream of data were directly modulated onto a single frequency. Even when M < N, the combination of the DFT and IDFT operation in tandem produces a signal that is single carrier in nature, and therefore SC-FDMA has more desirable instantaneous-power properties than OFDM, with similar spectral efficiency.

14.4.5 MAC Protocols

The multiple-access techniques described above assume that there is a separate entity that controls and assigns the resources (time slots in TDMA, frequency channels in FDMA, etc.) to various users with traffic demands. In other words, the multiple access may be centrally coordinated. However, in some systems, such a central entity may not be present to coordinate access and, in such systems, a mechanism is required to coordinate access to the channel in a distributed manner.

One simple example of such a channel access mechanism is the Aloha protocol, in which a user simply transmits on the channel when the need arises, assuming that the channel is available. If an acknowledgment is not received from the other end of the link, then a collision is assumed to have occurred and a retransmission is initiated. More sophisticated protocols are obviously possible when it is not simply assumed that the channel is free when the need to transmit arises. WiFi systems use a protocol that belongs to this class of channel access techniques, as we shall see in a later chapter.

On the other hand, even when a central entity such as a base station may be present, in order for users to dynamically enter the system, a mechanism must be provided for initial access. After the initial access, a low-bandwidth control channel can be used to assign resources for user traffic, leaving most of the bandwidth for user traffic. For initial access, certain system resources are usually set aside for random access by users that wish to join the system for the first time. Again, this random access channel can follow an Aloha-style protocol or something more sophisticated, in order to improve the probability of successful access. Cellular systems typically use random access channels in conjunction with control channels. We shall explore the operation of these channels in more detail in subsequent chapters that describe cellular systems.

14.5 Cross-Layer Algorithms

14.5.1 Link Adaptation

Due to the time-varying nature of the channel, most modern wireless systems use knowledge of the current state of the channel to adapt the parameters of the transmission in order to maximize the rate of information transfer, with a constraint on the maximum permissible error rate. The typical parameters that the system uses to adapt the information rate are the modulation scheme and the code rate of the FEC, jointly referred to as the modulation and coding scheme (MCS). For example, for users located at the edge of cellular coverage, a robust modulation scheme is used, along with a low rate code; while, for users close to the cell site, a higher-order modulation scheme (carrying more bits per transmitted symbol) can be used, with very little FEC protection. In addition to the MCS, the link adaptation algorithm can also be used to optimize MIMO transmissions. The choice of whether spatial multiplexing or spatial diversity should be used, and the number of spatial layers that should be used depends on the channel conditions. We will see how link adaptation works for spatial multiplexing in more detail when we discuss LTE systems.

The knowledge of the channel state needed to adapt the transmission can be obtained from explicit link information feedback on the reverse link. This feedback consumes valuable bandwidth that otherwise can be used for data traffic. Design of the feedback scheme so that the maximum benefits of link adaptation can be achieved with minimal overhead cost is an active area for research and innovation. Note that, in the case of TDD systems, as the wireless channel characteristics are reciprocal, the channel state can be obtained by directly observing the signals directly on the reverse link. However, as the signal uses different paths to propagate through receive and transmit hardware at the front end, calibration may be required before the channel measured in one direction can be used in the other direction. In cellular systems, the task of link adaptation usually lies with the base station, as it usually has access to a wider set of channel state metrics and can make decisions in a centralized fashion to optimize the performance of the system as a whole.

14.5.2 Channel-Dependent Scheduling

Given that the knowledge of the channel state is usually available at the transmitter for link adaptation, this information can be further exploited to provide channel-dependent scheduling gains. There are at least two ways of achieving these gains. One method is to schedule users at time instances when their channel conditions are most favorable. The other method is to schedule users in that portion of the band where their channel conditions are most favorable. Both methods work best when there are a large number of users in the system.

The first method is sometimes referred to as *multi-user diversity* transmission, and the corresponding gains are *multi-user scheduling* gains. To understand this method, let us assume for a moment that only one user can be scheduled in a given time slot. The basic idea then is to select, from a pool of users, the user with the best *scheduling metric* in that time slot. The scheduling metric can be composed of different criteria for user selection. In *maximum-rate* scheduler, the user whose data rate is the highest in the pool of users is selected for scheduling in a given time-slot. This algorithm ensures that the system is always transmitting at the maximum rate it can achieve and, therefore, it maximizes the capacity of the system. This idea is pictorially depicted in Figure 14.10.

However, in a system where users are distributed at various distances from the transmitter, this policy can be unfair to users that are far from the transmitter. If the users close to the transmitter always have data

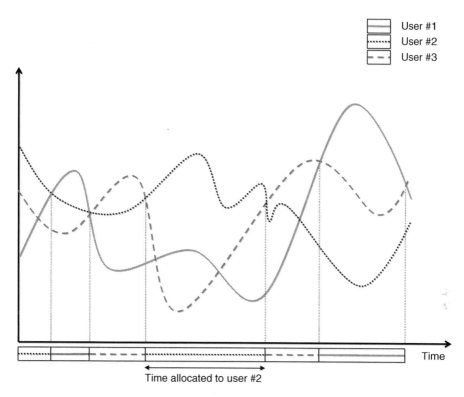

Figure 14.10 Channel-dependent multi-user scheduling

to receive, they will always be selected over users that are further away, thereby starving the under-scheduled users.

At the other end of the fairness scale is the *equal-rate* scheduler, where users are selected on the basis of the history of data rates achieved based in previous scheduling instances. Users who have not received sufficient allocations are thus prioritized over users who have been allocated resources recently. The system behaves in a fair manner to all users, but the capacity of the system is greatly reduced, because the scheduling is not optimized to use the channel conditions effectively to provide an overall improvement to system performance.

A popular technique that tries to balance per-user fairness with the need to maximize system capacity by serving users with good channel conditions is the *proportionally-fair* (PF) scheduler. The scheduling metric used to select users in this scheduler has one component that reflects the instantaneous data rate that the user can achieve on the next time slot, based on the channel conditions of a given user, and another component that is a reflection of the average data rate that the user has been experiencing. Equation 14.2 below shows one typical formulation of this metric, where α and β are weights that can be tuned to achieve the desired performance:

$$Scheduling\ Metric = \frac{[Expected\ user\ rate\ in\ next\ timeslot]^{\alpha}}{[Average\ rate\ received\ by\ user]^{\beta}} \qquad (14.2)$$

The idea is to prioritize a user whose current conditions are much better than the average conditions that the user normally experiences. As a result, in a system with a large number of users, the scheduler picks users who are in relatively good channel conditions, thereby improving the capacity of the system. Users closer to the transmitter will experience a data rate that is higher than for users further away, but no user starvation will occur, as the metric will pick users that are further away as long as their current channel conditions are much better than their average channel conditions. Note that such a scheduler provides capacity gains mainly when the there is fast-fading, otherwise the instantaneous channel conditions will not differ much from the average channel conditions. Also note that one can increase or reduce the amount of fairness (or conversely, system capacity gains) by increasing or decreasing the weight given to the component in the scheduling metric that reflects the average data rate (or conversely, instantaneous data rate).

In systems where resources are available in a two-dimensional time-frequency grid, an additional step can be added to the scheduling process in which the frequency characteristics are exploited in the same manner as the time dimension. When multiple users are scheduled in a given time-slot, the frequency allocations given to each user can be determined on the basis of the frequency-specific channel conditions of the subset of users selected for scheduling in that slot. This method is called *frequency-selective scheduling*, and it can provide significant gains in OFDMA systems. Note that for such systems (e.g., wide-bandwidth OFDM systems), the channel bandwidth is usually wide enough for frequency-selective fading to occur. Frequency-selective fading takes advantage of the fact that the deep fades and peaks in frequency do not occur at the same portion of the band for all users.

14.5.3 Automatic Repeat Request (ARQ) and Hybrid ARQ (HARQ)

It is important to present an error-free channel to higher layer protocols such as TCP, because such protocols were originally designed to run over highly reliable wired links. As such, TCP has an error recovery mechanism which detects errors and requests a retransmission very slowly, because it expects packets to be dropped only occasionally due to congestion. Furthermore, in response to the congestion, the rate of transmission is reduced by TCP whenever a packet drop is detected.

Wireless systems therefore tend to build in their own error recovery mechanisms. The first level of error recovery is called ARQ. This uses a CRC to detect errors and request a retransmission from the other end of the link, and the erroneous packet is dropped. Unlike TCP, which relies on missing packets in a sequence to detect an error, the error detection in ARQ is immediate and the recovery is much faster. ARQ usually operates in the MAC or Radio Link Control (RLC) sub-layer above the physical layer.

HARQ operates at both the physical layer and the MAC layer and assists the link adaptation algorithm. Like ARQ, it uses a CRC to detect errors but, unlike ARQ, it does not discard the erroneous packet. HARQ combines the erroneous packet with retransmitted copies of the same packet to improve the probability for decoding the packet correctly. The retransmitted packet may contain the same FEC as the original packet, in which case the process is called *Chase combining*. In this method, the additional packets help by providing two benefits:

1. Additional signal energy, due to multiple received copies of the packet.
2. Time diversity, because the copies pass through the channel at different times.

An alternate HARQ method, called *incremental redundancy (IR)*, adds new parity bits to a retransmitted packet instead of sending an exact copy, with the result that the combined packets successively contain more and more parity bits when compared to the original. This technique enjoys the additional benefit of coding gain, due to the combined packet utilizing a lower-rate FEC that is more robust than the FEC in the original packet.

The ability of HARQ to reuse older copies of the packet results in a greatly increased probability of decoding with each successive try, because of the benefits indicated above. This feature of HARQ is exploited by the link adaptation algorithm. The link adaptation algorithms target a fairly high packet error rate (usually 10%) by recommending the use of a less robust MCS that contains more data. This results in a higher throughput over the link and, for the packets that are in error, the subsequent retransmissions are much more likely to be successful when combined, due to HARQ.

14.6   Example Application: Satellite Broadband Access

In this section, we provide a brief overview of the technology used for satellite-based internet services, in order to illustrate how the concepts discussed in this chapter come together in the design of a complete system. In remote rural areas, the cost of providing internet services either via terrestrial wireless systems or via wired technologies is cost-prohibitive, because of the extremely low density of users served, compared to the magnitude of the infrastructure that must be put in place to serve them. Geostationary satellites can use *spot beams* to provide a strong, focused signals over an extremely wide geographic area, and they are therefore ideally suited for covering remote areas. Satellite-based broadband access systems employ a star topology that consists of a Very Small Aperture Terminal (VSAT) at the user's location that communicates with the satellite, plus a gateway or earth station that communicates with the satellite to play the role of the hub for all the data being transferred to and from the users. The gateway provides the connection to the internet, while the satellite serves as a bent-pipe transponder, because it merely serves as a relay in the conduit that connects the user's VSAT to the gateway.

The duplexing for satellite-based broadband services is based on FDD. For example, services operating in the Ku-band tend to use 14 GHz for the uplink and 10.9 GHz - 12.75 GHz for the downlink. The multiple access is a combination of TDMA and FDMA that is sometimes called MF-TDMA (for multi-frequency TDMA). The band in use for the service is divided up into multiple frequency carriers or channels, and TDMA is used within each channel.

The most advanced of these commercial systems which are being deployed today is based on the Digital Video Broadcasting – Satellite – Second generation (DVB-S2) standard, which supports downlink access, and the DVB Return Channel via Satellite (DVB-RCS), which supports uplink access. In addition to a sophisticated TDMA system, the DVB-S2 technology has support for a dynamic demand-based channel allocation mechanism called Demand Assigned Multiple Access (DAMA), link adaptation via a feature called Adaptive Coding and Modulation (ACM), and a forward error correction coding scheme based on LDPC.

Since 2011, a new generation of satellites have been launched, based on the DVB-S2/DVB-RCS technologies. One such system, called Excede®, utilizes the ViaSat-1 satellite in the Ka-band and provides a total capacity of 134 Gbps, with user download rates of 12–15 Mbit/s. It should be noted that the Achilles heel of satellite-based broadband services using geostationary satellites has always been the high latency associated with the propagation of signals to and from the satellite. These services usually have a latency of around 500 ms, which is not sufficiently small for fast-response applications such as gaming. Even services like VoIP cannot support high quality voice that usually requires ear-to-mouth latency of 250 ms or less, but the latency is not high enough to render VoIP services unusable. However with support for high data rates of over 10 Mbit/s, most other data-intensive applications that are relatively insensitive to latency should perform comparably to terrestrial systems.

14.7   Summary

In this chapter, we have covered the concepts essential for understanding wireless broadband access technologies. In order to enable bidirectional communications on a wireless link, duplexing mechanisms are used, separating the directions of each link in frequency or time, in the case of frequency division duplexing and time division duplexing, respectively. The phyiscal layer of a wireless link has special challenges compared to wired links, due to higher signal loss, and because of the potential for interference from other sources. Furthermore, propagation of signals over multiple time-varying paths generates fading, which causes the received signal to vary rapidly over time and frequency. In order to mitigate (and even benefit from) the effects of these impairments, diversity, interleaving, channel coding, and multi-antenna techniques must be used.

Diversity is a technique whereby the signal is sent wireless link in such a way that different channel conditions are sampled. The expectation is that the portion of the signal that experiences good conditions will help to improve the overall quality of the signal. Channel coding, or forward error correction (FEC) coding, has the ability to correct errors that may occur in a sequence or block of bits sent over a wireless link. Interleaving ensures that the errors on a link are not all clumped together, making the channel coding more effective at correcting the errors that occur. Multi-antenna techniques allow the system to harness or deliver more useful signal energy by using multiple antennas at one or both ends of the link. At the receiver, multiple antennas help to improve the quality of the received signal and, potentially, suppress interfering signals. At the transmitter, multiple antennas help by providing spatial diversity and, potentially, capacity gains, through the use of spatial multiplexing. Multiple-antenna systems, especially those that employ spatial multiplexing, are usually called multiple-input multiple-output (MIMO) systems.

In order for multiple users to transfer data over the link concurrently, the physical layer must support mechanisms that provide multiple access. Time-division multiple access is the method that is most widely used, where time is divided into slots or subframes, and these are assigned to different users. Frequency-division multiple access uses the same concept, with frequency being divided into different channels to support multiple users. In code division multiple access, a special code is used to spread each users signal in such a way that they do not interfere much, and the receiver can separate different users based on the codes applied at the transmitter.

Finally, the most current systems use orthogonal frequency division multiplexing (OFDM), which relies on the spectral properties of the Discrete Fourier Transform (DFT) to pack frequency subcarriers efficiently. OFDM has several beneficial properties in addition to the efficient packing, and these include a simpler receiver structure, efficient implementation using Fast Fourier Transforms (FFTs), the ability to harness frequency diversity, and frequency-selective scheduling. In addition to physical layer support for multiple-access, there is a need for an entity that coordinates or allocates the use of the physical layer between different users. This entity is the medium-access control (MAC) layer. It provides the structure that is needed for controlling access to the channel, either in a distributed manner (e.g., WiFi) or in a centralized manner (e.g., cellular systems).

Finally, a key component of all wireless systems today is a cross-layer function, where the MAC layer uses information about the state of the physical links in the network to optimize system efficiency. The three key inter-related and inter-dependent cross-layer sub-functions are link adaptation, scheduling, and HARQ. The link adaptation function is responsible for choosing the appropriate modulation and coding scheme for transmission. In addition, for systems that support MIMO, link adaptation includes selection of the number of spatially multiplexed layers and the beamforming configuration. The scheduling function takes into account the quality-of-service requirements of different users and services, and thereby allocates resources in a fair and efficient manner, utilizing knowledge about the links to various users to select those users who are in the best situation to be served at a given time. Lastly, Hybrid Automatic Repeat reQuest (HARQ) is used to enable fast retransmission of packets in error, and it also provides a mechanism to improve the probability of retransmission success by combining all received copies of a packet.

These concepts apply to all the broadband access technologies discussed in this section of the book. As an example, the architecture of a satellite-based broadband system was outlined in the last section. In the subsequent chapters, we will see how all of these techniques have been combined in different ways to develop elegantly designed systems for high-speed internet access.

Further Reading

1. Tse D, Vishwanath P. *Fundamentals of Wireless Communication*, Cambridge University Press; 2005.

15

WiFi: IEEE 802.11 Wireless LAN

15.1 Introduction

We begin our discussion of broadband wireless access technologies with WiFi. WiFi has been the most widely adopted wireless access in the world, both in terms of devices and infrastructure. WiFi chipsets have been part of the standard network interfaces in laptops and smartphones for many years now. Even before all laptops and smartphones were WiFi enabled by default, there was a plethora of WiFi cards that were common in the marketplace, and WiFi networks had begun to become widespread both in offices and homes.

WiFi is the commonly used term for technology that is standardized by the Institute of Electrical and Electronics Engineers (IEEE) under their 802 umbrella of standards for Local Area Networks (LANs). Strictly speaking, "WiFi" or "WiFi certified" is the trademark name given to products that are certified to be interoperable by the Wi-Fi Alliance, which is a trade association that promotes WiFi and performs the certifications. The IEEE 802 standards development organization, which focuses on local and metropolitan area networks, has several working groups under it to develop standards for specific access environments. The 802.11 WG was charged with the responsibility to create a standard for wireless LAN (WLAN) in 1992, and the first 802.11 standard was adopted in 1997.

In 1985, the Federal Communications Commission (FCC) deregulated the spectrum band from 2.4–2.5 GHz and made it available for use by the industrial, scientific, and medical (ISM) communities without any licence requirements. However, the FCC did impose some rules, which mainly restricted the maximum transmit power of radiators in this spectrum band to ensure that the band did not become unusable because of unrestrained interference. Another requirement was that devices should use "good engineering design" to ensure that they did not cause unnecessary and harmful interference, and that a given system could co-exist and operate in the presence of other interfering systems or sources. The first stipulation has resulted in the proliferation of several low-range wireless systems, of which WiFi is the prime example. The second requirement has influenced the design of the WiFi protocols to be interference-friendly, as we shall see in subsequent sections. At any rate, the deregulation of the ISM band spectrum has spurred decades of innovation, and has been the key enabler of ubiquitous wireless access today.

Another reason for the proliferation of WiFi deployments is the ease of configuration, and compatibility with the wired Ethernet networks which preceded WiFi. In the home environment, self-install of WiFi networks has been relatively hassle-free; even for those who are not technically savvy, a professional installation is not very costly, and usually can be provided by the internet service provider. Furthermore, these home networks require almost no maintenance. In enterprise and public environments, as we shall

Broadband Access: Wireline and Wireless – Alternatives for Internet Services, First Edition.
Steven Gorshe, Arvind Raghavan, Thomas Starr and Stefano Galli.
© 2014 John Wiley & Sons, Ltd. Published 2014 by John Wiley & Sons, Ltd.

see, there are more architectural elements and management requirements, which result in a bigger design and operational burden, but this is still a far cry from the installation and operational complexity of cellular networks, making WiFi the obvious choice for broadband access in areas with dense user populations and low mobility.

15.2 Technology Basics

15.2.1 System Overview

A wireless LAN system generally consists of two types of nodes: clients and access points (APs). In the infrastructure-based topology, several clients associate directly with an access point, and they use the AP to connect to the internet. This topology is the most commonly used one, and it is not unlike the topology of a cellular network. Unlike cellular standards, the 802.11 standard does allow other topologies. For example, it is possible for clients to establish peer-to-peer connections directly with other clients that are in range in an *ad hoc* topology. Furthermore, the standard also specifies a mesh topology, in which clients can use multiple peer-to-peer connections in tandem to reach an AP that may be out of range for a distant client. These topologies are depicted in Figure 15.1.

The access protocol for wireless LANs consists of specifications at the two lowest layers of the protocol stack, with the interface to the higher layers (IP and above) being identical to that of an Ethernet-based wired LAN. In the following subsections, we describe the MAC layer and the physical layer, which have some unique features when compared to cellular systems. Unlike cellular systems, where the uplink and downlink access are very different, WiFi networks have links that are much more symmetric. This fundamental symmetry comes from the requirement to support both *ad hoc* (or mesh) and infrastructure

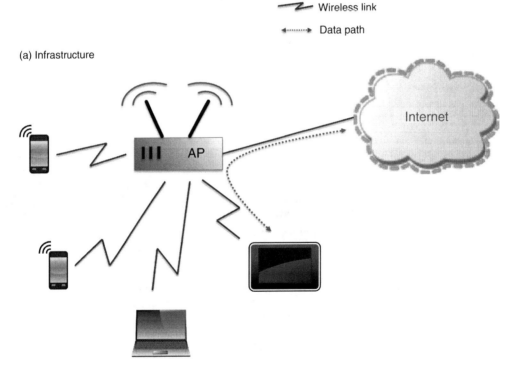

Figure 15.1 Wireless LAN topologies

(b) Ad-hoc

(c) Mesh

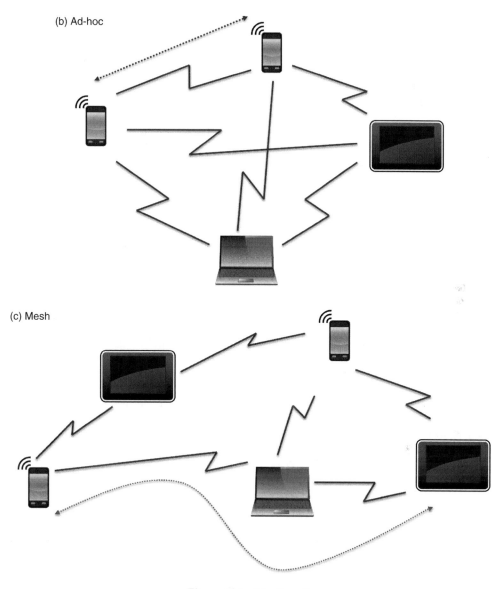

Figure 15.1 *(Continued)*

modes of operation. As a result of this requirement, the basic MAC protocol uses a carrier-sensing mechanism that is essentially the same at both the AP and the client, resulting in no significant difference between the air-interface protocols in the upstream and the downstream.

Furthermore, the transmit power difference between a client and an AP in WiFi systems is not as great as that between mobile stations and the base station in cellular systems, allowing the WiFi physical layer to be very similar in both directions. The symmetry of the upstream and downstream links has resulted in a system which is less complex and allows more flexibility in topology. The lower complexity and associated low cost has been a key factor in the widespread proliferation of WiFi networks.

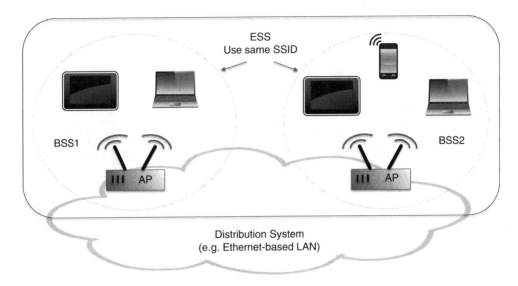

Figure 15.2 Extended service set

Note that in terms of capabilities, the AP and clients may still be quite different. For example, as we shall see in a subsequent section, the MIMO capabilities of the AP are usually much better than those of a client, in terms of the number of antennas for transmit and receive and the MIMO modes supported. However, the standard does not explicitly preclude any of the typical AP capabilities from also being enabled at a client.

A set of nodes consisting of an AP and clients that are controlled by a common coordination function is called a basic service set (BSS). We shall discuss coordination functions in more detail in a subsequent section. Each BSS is identified by a service set identifier (SSID). Multiple BSSs, even in close proximity, can use the same SSID. A set of APs that advertise the same SSID form an extended service set (ESS). To clients, the entire ESS appears to be a single logical network, as depicted in Figure 15.2.

15.2.2 MAC Layer

The MAC layer consists of two major functional entities:

- The MAC sublayer.
- The MAC layer management entity (MLME).

The core functions of channel access, retransmissions, packet fragmentation and encryption are handled by the MAC sublayer. The MLME handles higher MAC functions such as synchronization, power management, and connection management, which include association and authentication. In the following subsections we will take a closer look at some of these functions.

15.2.2.1 Channel Access

As mentioned earlier, a central controller is not mandated for a WiFi system, therefore access to the channel must be controlled in a distributed manner. The core access mechanism is referred to as the Distributed Coordination Function (DCF), and this forms the basis for other access mechanisms that may require a central coordination point (usually the AP), as shown in Figure 15.3. Access functions that use DCF as its basis are extended to support additional features such as QoS or mesh access.

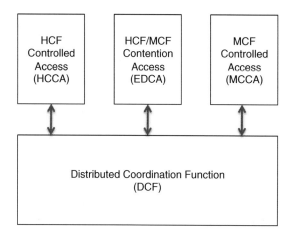

HCF: Hybrid Coordination Function
MCF: Mesh Coordination Function

Figure 15.3 WLAN MAC protocol architecture

DCF uses a listen-before-transmit protocol called Carrier Sense Multiple Access with Collision Avoidance (CSMA/CA), as shown in Figure 15.4 and explained below. Carrier sensing is performed to determine whether the channel is free or busy before initiating any transmission. The behavior of the node is different in the two cases. In the first case, where the channel is determined to be free, a node (client *or* AP) waits for a fixed amount of time and then transmits only if the channel is still free. This fixed amount of time is called the inter-frame space (IFS). By specifying IFSs of different lengths, different types of transmissions are prioritized. For example, for a node that is transmitting an ACK immediately following the reception of a packet, the short IFS (SIFS) is used. The SIFS is not long enough to sense the channel, and it is used to switch the front end from receive to transmit mode. The SIFS is used any time the node needs to respond to a received transmission and, therefore, it is the shortest IFS and does not require channel sensing. A node that needs to initiate a new transmission waits longer, for a span called the distributed IFS (DIFS), before it transmits.

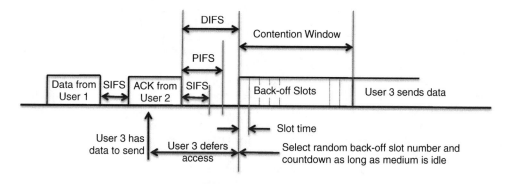

Figure 15.4 Basic CSMA/CA protocol

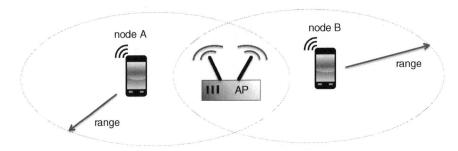

Node A and B are within range of the AP, but not within range of each other.

Figure 15.5 Hidden node problem

On the other hand, in the second case, where the channel is determined to be busy, it is assumed that there may be other nodes that may also want to transmit. In this case, transmitting directly after the DIFS can result in a collision. To avoid these sorts of collisions, a backoff mechanism is used. Each node selects a random backoff interval, computed as a random number of backoff slots of fixed duration, during which it continues to sense the channel. The duration for which a node must sense the channel before it can transmit is called its *contention window*, and this is initially set to the selected random backoff interval. If the node senses that the medium has stayed idle for the duration of its contention window, the node can initiate a new transmission.

If the node detects another transmission during its contention window (most likely due to another node that picked a shorter contention window), it stops the countdown of its backoff slots. Once the medium is free again, it sets its contention window to equal the number of slots remaining after the countdown. In this manner, nodes that pick a large number of random slots initially keep increasing their probability of getting to transmit next, as nodes with shorter backoffs finish their transmissions. The backoff mechanism constitutes the "collision avoidance" part of the CSMA/CA protocol.

There are two methods used for carrier sensing: physical and virtual. Physical sensing is performed by the RF front end of the modem, and it relies on signal power thresholds to determine whether the medium is busy. The 802.11 frames contain a preamble which aids physical sensing by permitting a device to distinguish between a valid signal and noise. In addition, frames contain a duration field which aids in virtual sensing. Using the duration field, a node is able to construct a network allocation vector (NAV) that allows it to defer transmissions to a future time without physically sensing the channel.

Channel sensing can be unreliable in a network where the clients are spread over a large area and, although every client may be within range of the AP, two clients may not be within range of each other. In this situation, an ongoing transmission from one client may not be sensed by the other client, resulting in a collision of frames at the AP receiver. This problem is sometimes called the *hidden node* problem, as shown in Figure 15.5. The same situation can also arise when two APs are in range of a client but not within range of each other, causing a collision at the client.

Note that a transmission is considered unsuccessful if either the data frame or the ACK is lost. To alleviate this problem, which can occur with high frequency in a congested network, two short frames, request-to-send (RTS) and clear-to-send (CTS), sent by the transmitter and receiver respectively, are used to notify all nodes of an upcoming data frame or ACK. The RTS is received by all nodes within range of the transmitter, and the CTS is received by all nodes within range of the receiver. Both messages contain the duration of the upcoming exchange and permit the nodes to update their NAV. Therefore, the RTS protects the transmitter from an ACK collision, and the CTS protects the receiver from data frame collisions. The RTS/CTS exchange adds overhead to the protocol, which can reduce the throughput in a lightly congested, or closely clustered, network where all nodes are in range of each other.

DCF, as described in the preceding paragraphs is the primary mode of channel access in WiFi networks. However, as should be evident, there is no inherent guarantee on latency or quality-of-service differentiation that is provided by this mechanism. In order to address these issues, the network can be operated with a contention-free period, where the allocations are controlled by the AP. The contention-free period alternates with a contention period, and this access mechanism is called the Hybrid Coordination Function (HCF). During the contention-free period, the node waits for an IFS called the point coordination function IFS (PIFS) before initiating a transmission. The PIFS is smaller than the DIFS, thereby implicitly giving priority to transmissions that are coordinated by the AP.

Quality-of-service differentiation is possible even within the contention period, by exploiting the inherent priority of access that is dictated by IFS of different lengths. An improved version of DCF called enhanced distributed channel access (EDCA) uses an additional set of IFSs that are associated with different *access categories* (ACs), to provide quality-of-service differentiation between the ACs. The HCF or EDCA will not be discussed here in greater detail as they have not been widely deployed. Similarly, in the mesh mode, the coordination function is called the mesh coordination function (MCF), and an optional access method has been specified that permits nodes to access the medium with lower contention than would be possible with EDCA. This mechanism, called the MCF controlled channel access (MCCA), is based on one node becoming the *owner* of this function, and the neighboring nodes becoming the *responders*. This will not be discussed here further, due to lack of commercial interest in this mode.

15.2.2.2   Connection Management

The three main steps that the client must follow in order to establish a connection with an AP are as follows:

- Scan for an AP.
- Authenticate with the AP.
- Associate with the AP.

In order for a client to join a BSS, it must first identify the BSSs that are within range, and this is accomplished by scanning. Scanning can be passive, in which case the client node simply listens for beacon frames; each BSS advertises its SSID in a beacon frame that is sent periodically. Alternatively, a client can actively query an AP for BSS information by sending a probe frame, to which an AP must respond with information similar to that contained in a beacon. The actual algorithm used by a client to select a BSS is left to the implementation, but the scanning process allows a client to create a list of potential BSSs.

Once a BSS is selected, the client follows a security protocol to authenticate with the AP. After authentication, the client associates with the AP by exchanging capability information that includes the bandwidths and data rates supported. The AP also conveys additional configuration parameters, such as beacon interval, in its response. Reassociation is a process in which a client switches its connection from one AP to another (one BSS to another) in the same ESS. This process consists of dissociating with the old AP and associating with the new AP. This process permits WiFi clients to roam seamlessly among APs in an ESS.

15.2.3   Physical Layer

Although the MAC sublayer has seen several major enhancements since the original specification was published in 1997, the basic DCF, with a few minor enhancements, is still the only form of the MAC that continues to be widely deployed. On the other hand, the physical layer, which provides all the raw throughput enhancements, has always been on the forefront of products roadmaps and deployments.

In contrast to the MAC enhancements, physical layer enhancements have found their way into products even before the standard has been fully ratified!

The physical layer consists of four major functions:

- Physical-layer dependent framing.
- Channel sensing (CS) or Clear Channel Assessment (CCA).
- Transmission of frames.
- Reception of frames.

The first function is part of the Physical Layer Convergence Procedure (PLCP) sublayer, and the last three functions are part of the Physical Medium Dependent (PMD) sublayer of the physical layer. The PLCP ensures that the correct frame format is used for any data that must be transmitted by the PMD. For example, in the presence of legacy devices, a legacy preamble is prepended to the frame, so that legacy devices can decode the frame duration information that is required for correctly updating the NAV.

As indicated earlier, most of the major throughput improvements have been brought about through modifications in the physical layer. As such, the physical layer has gone through radical changes, in the form of new transmission technologies, from one product cycle to the next. We shall take a detailed look at these physical-layer enhancements, as well as MAC enhancements, as we discuss the technology evolution in the next section.

15.3   Technology Evolution

The original specification of the standard had three options for the physical layer: infrared (IR), frequency hopping (FH), and direct-sequence (DS). The regulatory authorities either specified strict spectrum mask requirements on transmit power or actually mandated the use of spread spectrum technologies for use in unlicensed spectrum, so that no one system generated too much interference for other systems nearby. This is clearly reflected in the choice of FH and DS spread spectrum waveforms of the original standard. IR systems were never produced, because their poor propagation characteristics made them significantly less usable than the RF-based options like FH and DS. These initial FH and DS systems supported a peak throughput of 1 and 2 Mbit/s and operated in the ISM band at 2.4 GHz. The FH-based systems did not see widespread deployment, but the extension to the DS-based system which specified higher speeds was the first WiFi system that saw widespread deployment, as we shall see in the next section.

15.3.1   802.11 b

The first set of enhancements to the 802.11 standard – the 802.11a and 802.11b amendments – were ratified in September, 1999. The 802.11b provided higher speeds for operation in the 2.4 GHz band by adding two higher rates – 5.5 Mbit/s and 11 Mbit/s – to the already existing rates of 1 Mbit/s and 2 Mbit/s. The spreading code used in the DS specification is the 11-bit Barker code, transmitted at 11 million chips/second. The 1 Mbit/s signal uses DPSK modulation at 11 chips/bit, and the 2 Mbit/s signal uses DQPSK modulation to double the number of bits per Barker code.

The high rate improvements were made by adding a new modulation, Complementary Code Keying (CCK), to support 5.5 and 11 Mbit/s. CCK uses a spreading code length 8 that is used in conjunction with the Barker code to maintain backward compatibility. The 2.4 GHz unlicensed spectrum is divided into 14 channels that have center frequencies which are 5 MHz apart, with the first operating channel specified to have a center frequency of 2.412 GHz (see Table 15.1). The null-to-null bandwidth for the spreading waveform at 11 million chips/second is 22 MHz. Therefore, in order for the channels to be non-overlapping and not to interfere with each other, they should be at least five channels apart, as shown in Figure 15.6, making channels 1, 6, and 11 popular choices for deployment. Products based on the 802.11b standard were the first to be certified by the Wi-Fi Alliance, and this resulted in the establishment of a robust eco-system for WLAN based products and services.

Table 15.1 802.11b channel plan

Channel ID	Frequency (MHz)	Availability
1	2412	Global
2	2417	Global
3	2422	Global
4	2427	Global
5	2432	Global
6	2437	Global
7	2442	Global
8	2447	Global
9	2452	Global
10	2457	Global
11	2462	Global
12	2467	Except N. America
13	2472	Except N. America
14	2484	Japan Only

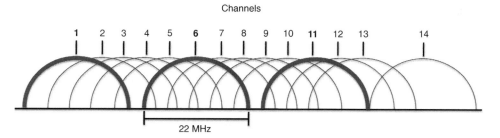

Figure 15.6 Channel overlap and planning for 802.11b

15.3.2   802.11 a/g

The next speed boost for WiFi systems was provided by the introduction of OFDM to the physical layer. The 802.11a standard supports rates ranging from 6 Mbit/s to 54 Mbit/s at a bandwidth of 20 MHz. Irrespective of the signal bandwidth, the actual OFDM signal consists of 52 subcarriers, of which four are pilot tones for channel estimation. The higher bandwidth signals simply send these OFDM symbols at a faster rate by shrinking the symbol time. For example, the 20 MHz signal has a 4 µs symbol time, which is a quarter of the 16 µs symbol time used by a 5 MHz signal. It is noteworthy that other OFDM systems, such as WiMAX and LTE use the same symbol time irrespective of the bandwidth but squeeze in additional subcarriers in proportion to the bandwidth, as we shall see in a subsequent chapter. The FEC used by 802.11a systems is a convolutional code, with constraint length 7, and coding rates of 1/2, 2/3, and 3/4 are supported. The modulation used is QAM, with 64-QAM being the highest order modulation supported. The parameters of the OFDM PHY are shown in Table 15.2.

Although the 802.11a standard was ratified in 1999, along with the 802.11b standard, availability of products lagged behind that of 802.11b, mainly due to the higher complexity and cost of the early implementations of OFDM chipsets. In addition, the 802.11a standard was specified for operation only in the 5 GHz U-NII band for the United States, with consecutive channels being 5 MHz apart, starting at 5.18 GHz. Support for European and Japanese frequencies in the 5 GHz bands were ratified as part of 802.11 h and 802.11j in 2003 and 2004 respectively.

Table 15.2 OFDM PHY parameters for 802.11 a/g

Parameter	Value (20 MHz channel spacing)	Value (10 MHz channel spacing)	Value (5 MHz channel spacing)
Number of data subcarriers	48	48	48
Number of pilot subcarriers	4	4	4
Number of subcarriers total	52	52	52
Subcarrier frequency spacing	0.3125 MHz	0.15625 MHz	0.078125 MHz
Inverse Fast Fourier Transform (IFFT)/Fast Fourier Transform (FFT) period	3.2 µs	6.4 µs	12.8 µs

With the proliferation of 802.11b devices that enjoyed a higher range in the 2.4 GHz ISM band, it was felt that there was also a need to provide the higher rates specified in 802.11a in this lower band. The result was the 802.11 g standard, which was ratified in 2003. This version of the standard takes the OFDM PHY as specified in 802.11a and adds support for backward compatibility with legacy 802.11b systems already operating in the 2.4 GHz band. Recall that the "legacy" 802.11b physical layer use DSSS so, therefore, steps have to be taken to ensure that the OFDM physical layer used by 802.11 g devices does not disrupt the operation of the legacy devices.

One mechanism used to allow the OFDM transmissions to occur safely in the presence of DSSS devices is the "CTS-to-self" mechanism. As the name indicates, a node autonomously sends a CTS, not with the intention of responding to an RTS, but with the intention of ensuring that no other node transmits in the area surrounding it. Legacy devices can decode this CTS, as it is sent using the DSSS interface, and it prevents them from using the channel for the duration specified in the CTS. The 802.11 g devices are then free to use OFDM transmissions on the channel without any disruption to legacy devices.

15.3.3 802.11 n

The next major step taken to improve access speed for 802.11 was spearheaded by the task group "n" (TGn), which worked on improving the OFDM PHY developed by task group "a" for the 802.11a standard. The major enhancement introduced by 802.11n was the inclusion of MIMO with support for up to four spatially multiplexed streams. In addition, the bandwidth supported was doubled to 40 MHz. Over this bandwidth, with four spatial streams, 802.11n boasts a top speed of 600 Mbit/s. The highest order QAM modulation stays the same (64-QAM) as 802.11a. However, TGn added support for one higher FEC rate: 5/6.

An optional FEC scheme, Low Density Parity Check (LDPC) codes, was specified to provide vastly improved error correction capabilities over convolutional codes, thereby improving the performance in poor signal-to-noise (SNR) ratio conditions, and improving range. In addition, changes were made to the MAC to increase efficiency by specifying support for frame aggregation. 802.11n has been specified for operation in both the 2.4 GHz ISM band, and the 5 GHz U-NII band. A list of mandatory and optional features are shown in Table 15.3, and are explained further in the following subsections.

15.3.3.1 MIMO Enhancements

Although the standard specifies operation for up to four spatially multiplexed streams, supporting four streams would require four transmit antennas at the source, and also four receive antennas at the destination. For client nodes that are typically power-limited, driving four transmit or receive chains may be difficult to support while still ensuring adequate battery life for other functions. Moreover, in phones and USB dongles, space is also at a premium, making multiple antennas difficult to accommodate. As such, the standard specifies that only a single spatial stream is mandatory for a clients to transmit whereas, for an AP, support for transmitting two spatial streams is mandatory.

Table 15.3 Mandatory and optional features in 802.11n

Feature	Mandatory	Optional
Bandwidth	20 MHz	40 MHz
FEC	Convolutional code	Low density parity check codes
MIMO	Spatial multiplexing with 1 or 2 streams	Spatial multiplexing with 3 or 4 streams
		Transmit beamforming
		Space-time block codes

Although it has been a few years since 802.11n chipsets hit the market, smartphones with 802.11n capability today still use only one receive antenna. For example, Apple's iPhone 5 is capable of receiving only a single spatial stream over its WiFi interface, but supports 40 MHz operation and can use both the 2.4 GHz and 5 GHz bands. It is worth noting here that iPhone 5, and other smartphones that support LTE, do support two spatial streams over their cellular interface. As we shall see later, currently deployed LTE systems predominantly use 10 MHz bandwidths in a given direction. Therefore, the additional complexity and battery drain can be justified for greater throughputs for LTE whereas, for WiFi, the use of larger bandwidths can deliver speeds similar to LTE without the need for high-order MIMO.

Although spatial multiplexing requires a minimum 2×2 (2 transmit, 2 receive) configuration of antennas at the two ends of the link, space-time block codes (STBC) can be supported from an AP which has two transmit antennas to a client which has one receive antenna. STBC, as noted earlier, is able to harness spatial diversity without compromising on the rate of transmission. Additionally, the presence of two antennas at the AP can be used to improve the quality of signal reception through the use of receive diversity. Note that receive diversity can be implemented even for reception of signals from legacy clients using 802.11 b/a/g, so multi-antenna APs provide a boost to all "uplink" transmissions.

The 802.11n standard also specifies support for transmit beamforming, but this is not mandatory. The key to effective beamforming is obtaining good channel estimates. As the WiFi system is a sort of TDD system, the channel reciprocity can be exploited to obtain channel estimates by utilizing signals received from a node towards which beamforming needs to be applied. However, the transmit and receive circuitry have hardware paths that are different, so a correction needs to be applied to make the transmit path correspond to the channel observed on the receive path. This correction is referred to as "calibration".

The standard provides support for this type of implicit beamforming by specifying signaling for exchange of calibration information. In addition, the standard also specifies support for explicit beamforming, where the receiver measures the channel on the transmit path directly and feeds it back to the transmitter, obviating the need for the transmitter to calibrate or use channel reciprocity. In practice, calibration is possible using proprietary techniques, without the help of any signaling from the receiver, and therefore vendors can implement transmit beamforming without any support from the receiver. This permits AP vendors to differentiate their products, and additionally to provide better performance for all clients, without expecting implementation of this optional feature at the client side.

15.3.3.2 Frame Aggregation

Each time the 802.11 MAC sublayer receives a packet (MSDU) for transmission, it adds a MAC header and a PHY header to the MSDU. In addition, it has to sense the channels, defer for the appropriate IFS, and backoff if necessary, before actually transmitting the packet. Therefore, there can be a significant amount of dead time and overhead bits associated with the transmission of a single packet of useful MSDU data, as shown in Figure 15.7. In order to make the MAC more efficient, and thereby increase the throughput, two types of packet aggregation have been defined.

Aggregate MSDU (A-MSDU) permits a node to concatenate several MSDUs together in one single MPDU, containing only one PHY header and one MAC header. In order to define the boundaries of each

Figure 15.7 MSDU transmission without aggregation

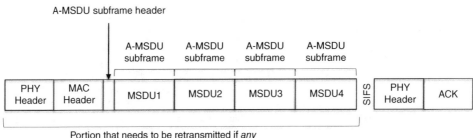

Figure 15.8 MSDU aggregation

MSDU in the concatenation, a MPDU subframe header is added (see Figure 15.8). A-MSDU is clearly very efficient in terms of overhead reduction but, if the link quality is poor and there are errors in the transmission, then the entire A-MSDU will need to be retransmitted, resulting in wasted bandwidth and reduced throughput.

Another aggregation technique specified in 802.11n is the Aggregate MPDU (A-MPDU). An A-MPDU contains a concatenation of MPDUs, with each MPDU subframe containing a MPDU delimiter, a MAC header, and an MSDU, as shown in Figure 15.9. Notice that the MAC header is repeated multiple times, whereas there is only one PHY header. The A-MPDU presents a tradeoff between overhead and reduced throughput due to retransmissions because, if there are errors in the aggregated frame, those MSDUs that have errors can be detected by using the CRC check in the MAC header and, therefore, only those MSDUs need to be retransmitted.

15.3.4   802.11 ac

The latest and greatest version of the WLAN standard, 802.11ac, was finalized in task group ac (TGac), and ratified in December 2013. TGac has taken the 802.11n standard and made three key enhancements to increase the raw throughput that it provides:

- The channel bandwidths supported have been increased from 40 MHz to include both 80 MHz and 160 MHz. The maximum speed being directly proportional to the bandwidth given the same OFDM signal structure, 160 MHz bandwidth results in a 4× boost to 802.11n speeds.

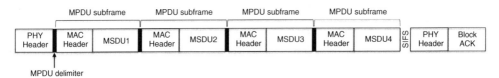

Figure 15.9 MPDU aggregation

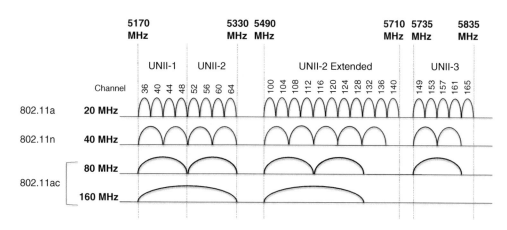

Figure 15.10   5 GHz channel overview

- The highest order modulation has been increased from 64-QAM to 256-QAM. Each QAM symbol can now carry 8 bits of data, 2 more bits than 802.11n supports with 64-QAM. This represents a 33% increase in rate.
- The maximum number of spatially multiplexed streams has been increased from 4 to 8. Nominally, this also represents a proportional increase in the top speed, like the expansion in bandwidth. However, in practice, both channel conditions and hardware limitations may make this increase hard to achieve.

Unlike 802.11n, the only band specified for use with 802.11ac is the 5 GHz band. This avoids all the interference and coexistence issues that arise with the use of the 2.4 GHz band. For example, coexistence with Bluetooth, and older versions like 802.11b and 802.11 g, are no longer a consideration. This permits the design to be cleaner. Moreover, there is insufficient spectrum to support 160 MHz in the 2.4 GHz band, and not many options to support 80 GHz, either. An overview of the channel configurations possible in the 5 GHz band is shown in Figure 15.10.

In applications where multiple APs may need to operate in close proximity, such as enterprise deployments, the 5 GHz band provides more options for non-overlapping channel plans. The initial products in the market are likely to support only 80 MHz, with chips supporting 160 MHz expected to come later. The peak speed per spatial stream is 433 Mbit/s for 80 MHz, and 867 Mbit/s for 160 MHz. This gives a theoretical maximum of just under 7 Gbps if all eight spatial streams are used with 160 MHz bandwidth.

At the physical layer, multiplexing of eight spatial streams will require both ends of the link to use eight antennas in an 8×8 configuration. For small form factor devices, such as smartphones, the use of eight antennas, the associated front end processing and power consumption will make this configuration unlikely. Indeed, as we saw earlier, smartphones today still use only one receive antenna, and only high end chipsets support even 3×3 MIMO, with 4×4 MIMO being very rare.

On the other hand, TGac has taken steps to make other multi-antenna techniques, such as transmit beamforming, a lot easier to deploy. From the plethora of beamforming options specified by TGn, none of which saw any significant productization, TGac has picked one option that they have refined for use in the new standard. Transmit beamforming in 802.11ac uses the explicit compressed feedback mode. Being the only supported option, interoperability testing should be easier. Similarly, the options for STBC have also been restricted to two (2×1 and 4×2) of the four permitted in 802.11n. Two additional STBC modes are available for systems with greater than four antennas, but beamforming is likely to be the mode of choice over STBC for all but the 2×1 configuration. Finally, multi-user MIMO (MU-MIMO), yet another technique that is being standardized by TGac, can be used to improve the efficiency of transfers from the AP to clients that have fewer antennas than the AP.

At the MAC sublayer, steps have been taken to improve the CCA mechanism via an enhanced RTS-CTS exchange. Note that several legacy nodes using 20 MHz bandwidth may be present in the vicinity of an 802.11ac system that intends to use 80 MHz of bandwidth. These legacy nodes can be full systems, with APs and clients on a separate BSS, or simply 802.11a or 802.11n clients that wish to associate with the 802.11ac AP. The enhanced RTS-CTS mechanism permits the nodes at both ends of the link to determine when and how much of the bandwidth is free, thereby allowing it to restrict the transmission to a bandwidth that does not cause interference to legacy systems. In addition, a primary channel that is 20 MHz wide is designated for use by legacy clients associated with an 802.11ac AP, so that they can receive beacons and perform virtual carrier sensing for all transmissions that overlap this primary channel.

15.4   WLAN Network Architecture

In the previous sections, we discussed the basic WLAN topology options, the MAC and PHY layers and their evolution. In a small office or home environment, with just one or two APs deployed, configuration and use of the APs is fairly straightforward. In this case, the architecture consists simply of an AP or two connected to the internet via a switch, and no additional layers or nodes are needed for management of the network. The APs usually host a web server for basic configuration. However, in bigger deployments, where tens or hundreds of APs are deployed across multiple floors, buildings, campuses, and even cities, additional software or hardware is needed to manage and operate the network efficiently.

The typical solution introduces an additional node called the WLAN controller, which is dedicated hardware that can be used to offload a lot of the processing needed for user data, control, and management of the network, thereby making the APs "lightweight". Such a WLAN controller-based architecture, along with the paths taken for data and control, are shown in Figure 15.11. The APs mainly play the role of the RF interface to the network. User data, control signaling, and management

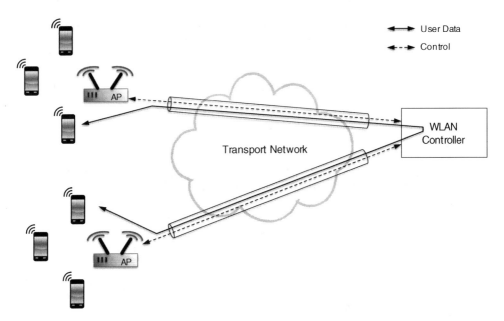

Figure 15.11   WLAN controller based architecture

directives are sent between the controller and APs via a tunnel that can traverse other networks transparently to get to the controller.

WLAN controllers make it possible to scale up a WLAN deployment with centralized control, management, and configuration. WLAN controllers usually provide the following features:

- *Radio resource management*: as seen in previous sections, the physical layer can use several different spectrum bands and, within a band, there may be multiple choices of channels. In order to minimize interference between APs and optimize performance, frequency planning is needed, and this is one of the key functions that WLAN controllers provide. Additionally, in dense deployments, controllers may adjust the transmit power and antenna beam-shapes of nearby APs in order to reduce interference and maximize coverage.
- *Faster roaming*: by caching security keying material, the WLAN controller is able to permit faster roaming between APs connected to the controller, and even between controllers, via layer-3 tunneling between controllers.
- *Security and authentication*: WLAN controllers can enforce centrally defined firewall policies for security. In addition, they can implement VLANs and active directory-based authentication, so that "guest" access is separated from employee access at enterprises. They can also be used to implement "captive portal" based authentication, which is frequently seen at hotels and airports, where a user wishing to access the network is redirected to a page via a web browser where authorization may be performed.
- *AP discovery and provisioning*: the IETF standard Control and Provisioning of Wireless Access Points protocol (CAPWAP) is used to communicate between the controller and other lightweight access points on the network. The WLAN controller uses CAPWAP for discovery and provisioning of the APs in the deployment.
- *Quality-of-Service, traffic monitoring, filtering, and shaping*: traffic shaping and filtering can be used to implement bandwidth control and user-access control (e.g., to ensure that no single user dominates the bandwidth usage on the network). In addition, voice services can be prioritized by enabling QoS policies. Other options for QoS and traffic grooming include optimizations for delivery of video services such as streaming, progressive downloads (e.g., YouTube), and video calling.
- *Load balancing and mesh networking*: in order to manage load and connectivity better, controllers can manage the load on various APs by reassigning users in overlap areas to APs with lower load. Furthermore, in case there is loss of access, the APs can form a mesh network to communicate back to an AP with a reliable backhaul connection to the internet.
- *Various management functions*: these include centralized scheduling of firmware updates, dashboard-based troubleshooting using various performance statistics, and status checks. Security monitoring is also possible from a central console via a WLAN controller.

More recently, the increasing deployment of cloud-based services have given rise to controller-less architectures, where more capable hardware at the APs can be used to reduce the need for a hardware-based controller. By shifting some of the processing back to the AP, one can design a purely software-based "controller in the cloud". This can still handle all the management functions and, in addition, can still handle several control functions. There are several benefits of cloud-based architecture with less-lightweight APs:

- Rapid deployment with self-provisioning and self-optimizing hardware.
- Future-proof and always up to date, with seamless over-the-web firmware updates.
- Arbitrary aggregation and grouping of APs for configuration management.
- Autonomous APs can still be deployed without any cloud support.

15.5   TV White Space and 802.11 af

To meet the growing need for spectrum to satisfy the increasing demand for broadband access, unused analog-TV spectrum is being made available in the United States and around the world. This spectrum is called TV *white space*, as it refers to the frequencies that available between TV bands that are still in use by broadcasters. These white space frequencies lie in the VHF and UHF bands, for example, between 54–698 MHz in the US and 470–790 MHz in Europe. In these bands, the radio signals experience a much lower path loss than they do in the 2.4 GHz band. In addition, the regulation surrounding this spectrum, at least in the US, allows for higher transmit powers. As a result, WiFi can be used in TV white space to deliver broadband access over a much wider area. This is especially attractive in rural areas, where installing wired access may be expensive, but availability of white space spectrum may be high because only a few TV channels may be in use in these regions.

The actual frequencies available for use in any local region depends on the broadcasters who own and operate TV channels in the region. In order to use the white space frequencies effectively, geolocation databases are being created that list the available frequencies for a given location, and the duration for which these frequencies are available. White space devices (WSDs) are required to consult the database when using these frequencies.

The availability of white space channels can vary with time and, therefore, a set of mechanisms needs to be specified for WSDs to satisfy the regulatory requirements associated with the use of these frequencies. The IEEE 802.11af standard currently under development aims to provide the protocols required to enable TV white space access on a global basis, and this is expected to be completed in the first half of 2014.

15.6   Summary

In this chapter, we discuss the technology evolution of IEEE 802.11, more popularly known as WiFi. IEEE 802.11 develops the wireless LAN protocols standards for the physical layer and MAC layer, and the WiFi Alliance certifies interoperability between devices manufactured by different vendors. The main reasons for the rapid proliferation and ubiquity of WiFi are its use of unlicensed spectrum bands, its ease of configuration and deployment, and smooth integration with Ethernet-based wired networks.

The most commonly used wireless LAN topology consists of two types of nodes: clients and access points (APs) that are used in the infrastructure mode, with all direct communications occurring only between an AP and a client. The AP is usually provides connectivity to the rest of the network, usually including the internet. However, the protocol used between the AP and the client is symmetric, unlike in a cellular network, where the uplink and downlink are distinct in terms of the physical layer and MAC protocols.

The fundamental access mechanism in wireless LANs is based on Carrier Sense Multiple Access with Collision Avoidance (CSMA/CA). This access mechanism operates in a distributed manner, without the need for a central coordinating entity. The basic idea of CSMA/CA is to listen to the medium before transmitting and, if the medium is free, then the node defers transmission for a short gap (called the inter-frame spacing, IFS), in order to allow higher priority transmissions to take precedence. The node transmits if the medium is still free after this gap. On the other hand, if the medium is busy, then the node initiates a random backoff procedure to ensure that its transmission does not collide with those from other nodes also waiting to for the medium to become free. Therefore, protocol enables collision avoidance via a random backoff mechanism. The MAC protocol supports Quality-of-Service (QoS) levels by the use of different IFS values for different QoS levels.

The first widely deployed version of the standard was 802.11b, which supported data rates up to 11 Mbit/s, and used roughly 22 MHz of bandwidth in the 2.4 GHz ISM band. This technology was soon supplanted by 802.11 g, which used OFDM in the physical layer, and provided data rates up to 54 Mbit/s over 20 MHz bandwidth. Another OFDM-based standard, 802.11a, which was developed for deployment in the 5 GHz U-NII band, did not get much traction in the market. The next step in the evolution of the physical layer to provide higher data rates occurred with the ratification of the "11n" standard. The major enhancements provided by 11n include support for MIMO, 40 MHz bandwidth, and MAC efficiency features. With 802.11n, a peak data rate of 600 Mbit/s is possible over a 40 MHz band. With the wider bandwidth, operation has been specified for both the 2.4 GHz ISM band, and the 5 GHz U-NII band, where more spectrum is available for multiple non-overlapping 40 MHz bands.

The most recent version of the physical layer is based on 802.11ac, which was ratified in December 2013. 802.11ac is specified for operation only in the 5 GHz U-NII band, with up to 160 MHz wide channels. The highest order modulation possible is increased from 64-QAM to 256-QAM, and the number of spatial streams is increased from four to eight. Although the peak aggregate capacity of an 160 MHz MU-MIMO system can be very close to 7 Gbps, that would require eight spatial streams. For the initial systems which are likely to be 80 MHz, the peak speed per spatial stream is 433 Mbit/s.

The widespread proliferation of WiFi networks, especially in large enterprises, and as major commercial deployments for public use, has given rise to an entire industry catering to the design, control, operation, and management of networks with tens to hundreds of APs. In these networks, additional hardware and software may be needed to configure, optimize, and manage these APs effectively. Dedicated hardware, in the form of WLAN controllers, may be used, and the APs may themselves be made "lightweight" in terms of functionality, with the majority of the work offloaded to these controllers. On the other hand, the APs may still retain a significant portion of the functionality, with a cloud-based management service used to control and optimize the APs. At any rate, for any deployment with more than a few APs, some form of centralized control and management is needed to simplify the operation of the network.

To meet the growing need for spectrum, the 802.11 af working group is working on specifying mechanisms to use unused TV spectrum, called TV white space. White space devices will be required to consult a geolocation database when operating in these portions of the spectrum. The database contains information about where and when such white space frequencies are available for unlicensed use.

Further Readings

1. IEEE802.11-2012. Part 11: Wireless LAN Medium Access Control (MAC) and Physical Layer (PHY) Specifications, 29 March 2012.
2. 802.11ac-2013 - IEEE Standard for Information technology–Telecommunications and information exchange between systems—Local and metropolitan area networks–Specific requirements–Part 11: Wireless LAN Medium Access Control (MAC) and Physical Layer (PHY) Specifications–Amendment 4: Enhancements for Very High Throughput for Operation in Bands below 6 GHz, Dec. 18 2013.

16

UMTS: W-CDMA and HSPA

16.1 Introduction

With the rapid expansion of the internet in the mid-1990s, applications and services that used packet data saw explosive growth. There was widespread proliferation of mobile phones around the same time, mainly for voice services, and the need was felt to enable data services on mobile phones. There were two paths available to provide data access on mobile devices: enhance the existing second generation (2G) systems to support higher data rates, or develop a new system designed with support for both data and voice services in mind from the outset.

The 2G systems, which are still in wide use today, are based on a technology called GSM (Global System for Mobile) which was developed only for voice services. It was enhanced to support data services, first by the introduction of the GSM Packet Radio System (GPRS), and next by the addition of Enhanced Data-rates for GSM Evolution (EDGE). However, these systems were designed to depend fundamentally on the GSM air interface structure and they were, therefore, limited in their ability to deliver true broadband speeds. The second alternative – development of a new system designed to support both data and voice from the outset – resulted in the third generation of cellular systems, which based on an entirely different air interface technology.

In this chapter, we explore third generation (3G) systems which represent the first cellular technology that could truly provide broadband speeds of several megabits-per-second (Mbit/s) on a consistent basis. As an introduction to 3G systems, we first take a brief look at the various organizations responsible for developing requirements for standardization, the standards themselves, and harmonization of spectrum for the technology.

The International Telecommunication Union (ITU) is the United Nations specialized agency for information and communication technologies. In the mid-1980s, the concept for International Mobile Telecommunications (IMT-2000) was conceived at the ITU. The framework for the development of 3G systems was established in 1992 at ITU's World Administrative Radio Conference (WARC-92) where, among other regulatory provisions, the radio-frequency spectrum bands were identified on a global basis for use by countries when deploying IMT-2000 systems. This spectrum, known as IMT-2000 bands, includes allocations for terrestrial and satellite access. The spectrum for terrestrial access using FDD has the uplink band from 1920–1980 MHz and the downlink band from 2110–2170 MHz.

The Universal Mobile Telecommunication System (UMTS) was developed as the 3G cellular standard that meets the IMT-2000 requirements. UMTS is the umbrella term used for 3G systems that are based on the wideband-CDMA (WCDMA) air interface and a core network that has evolved from the GSM-based

Broadband Access: Wireline and Wireless – Alternatives for Internet Services, First Edition.
Steven Gorshe, Arvind Raghavan, Thomas Starr and Stefano Galli.
© 2014 John Wiley & Sons, Ltd. Published 2014 by John Wiley & Sons, Ltd.

core network for circuit-switched and packet-switched services. The standardization of 3G UMTS systems is carried out under the auspices of the Third Generation Partnership Project (3GPP). This is a standards development organization (SDO) in which the partners and participation are drawn from regional standards organizations including ATIS (North America), ETSI (Europe), ARIB (Japan), TTA (Korea), TTC (Japan), and CCSA (China). Individual member companies are represented through these regional organizations.

Another system based on relatively narrowband CDMA air interface, called CDMA2000, represents a parallel technology evolution of the IS-95 standard. It was developed by a sister organization of the 3GPP called 3GPP2. Note that both 3GPP and 3GPP2 developed standards for IMT-2000 based on the recommendations from the ITU, and the standards that were developed were submitted to ITU and ratified as IMT-2000 standards.

Typical speeds for 3G systems (as measured by speed tests) tend to be in 2–4 Mbit/s range for the downlink, and 0.75–1.5 Mbit/s in the uplink. The term "3G", or third generation systems, originally applied to all CDMA-based systems that replaced GSM-based systems, with the term "4G", or fourth generation systems, reserved for the next major update to the air interface that introduced OFDM. However, operators in the USA that deploy W-CDMA-based systems supporting peak rates of 42 Mbit/s, including AT&T and T-mobile USA, started marketing their service as 4G, thereby blurring the distinction between the CDMA and OFDM systems as far as the generation of the technology is concerned.

The rest of the chapter is organized as follows. We first explore the overall network and protocol architecture for UMTS systems. Next, we look at the inner working of the various layers that comprise the basic WCDMA-based access technology which is the foundation for future enhancements to the system. The following section then covers the step-by-step evolution of the technology to provide true high-speed broadband access by describing the extensions that were introduced to the original system. Note that the focus of the chapter is UMTS because, not only is it the more widely deployed system, it is also conceptually very similar to CDMA2000 systems. In the last section of this chapter, we briefly discuss CDMA2000 and how it differs from UMTS.

16.2 Technology Basics

16.2.1 Network Architecture

Compared to WiFi systems, 3G systems are more complex, as they are designed to support high-grade voice services, with seamless mobility, over a much greater area. Moreover, when these systems were designed, the expectation was that, at least initially, the predominant source of traffic would be voice. Therefore, the architecture contains two distinct parts to support the two modes of operation: circuit-switched mode for voice, and packet-switched mode for data (the eventual reality was that, in markets with high penetration of smartphones, data usage overwhelmed voice usage in terms of volume).

An overview of the network architecture of UMTS-based 3G systems is shown in Figure 16.1. The system may be divided into three sub-systems:

- The mobile devices or user equipment (UE).
- The fixed infrastructure portion that handles the access over the air, formally called the UMTS Terrestrial Radio Access Network (UTRAN).
- The back-end network, called the core network (CN), that supports the access network consisting of the UE and UTRAN.

Radio Access Network (RAN) is the commonly used contraction for UTRAN. The various nodes and interfaces that make up the architecture are briefly described in the following paragraphs.

The UE contains the subscriber identity module (SIM) that is used for authorization and authentication, in addition to the hardware and software that implements all the end-to-end communication protocols. There are two nodes that comprise the UTRAN: the Node-B and the radio network controller (RNC). The

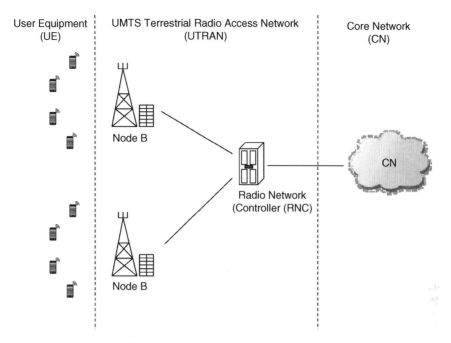

Figure 16.1 UMTS network architecture

Node-B hardware consists of the antennas, the radio unit, and the baseband unit. The baseband unit generates signals that are up-converted to radio frequency and amplified to high power levels before transmission over the antenna. A Node-B may support multiple sectors, using directional antennas to cover the area surrounding the antenna tower. Typically, three sectors are used, with three antenna systems, each covering a 120° sector, with some overlap. The RNC controls several Node-Bs (typically hundreds of them), and is responsible for implementing the mobility management and admission control functions. As the technology evolved, some of the functions implemented at the RNC were moved to the Node-B, as we shall see in a subsequent section.

The CN contains all the switching and routing functions required for global network connectivity as shown in Figure 16.2. It consists of two nodes that support the circuit-switched (CS) mode of operation, the mobile switching center (MSC) and the gateway MSC (GMSC). The two nodes that support the packet-switched (PS) mode of operation, the serving GPRS support node (SGSN) and the gateway GSN (GGSN), were nodes originally developed for GPRS, and subsequently enhanced with support for the added 3G functionality. In addition, it contains the Home Location Register (HLR), which supports both modes of operation. The MSC plays the role of the telephone switch in the mobile network by switching calls in the CS mode. The GMSC serves as a gateway node to route calls to the external telephone network. The SGSN provides IP connectivity to the mobile for packet-switched services by routing packets to and from the mobile. The GGSN serves as the gateway router to the internet for a given service provider's 3G network.

16.2.2   Protocol Architecture

In this section, we take a closer look at the protocol architecture of the RAN. The different protocol layers and the channels that form the interface between these layers are shown in Figure 16.3. The protocols terminating at the RAN are divided into three layers, numbered starting from the lowest layer, the physical

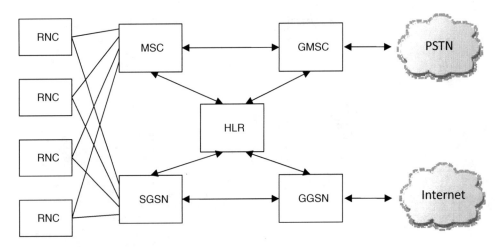

GGSN – Gateway GPRS Support Node
GMSC – Gateway Mobile Switching Center
HLR – Home Location Register
MSC – Mobile Switching Center
PSTN – Public Switched Telephone Network
RNC – Radio Network Controller
SGSN – Serving GPRS Support Node

Figure 16.2   Core network architecture

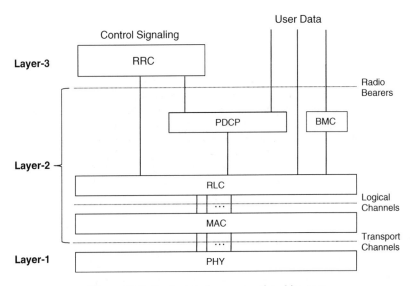

Figure 16.3   Radio interface protocol architecture

layer, that is responsible for the actual transmission of signals over the air. Layer-1 (L1), or the physical layer, provides *transport channels* for use by layer-2 (L2). It is the structure of these transport channels and how they are mapped to *physical channels* in the physical layer that define the air interface technology, and we will look more closely at these channels and their evolution in subsequent sections. Transport channels are concerned with the timing and format of the data transfer, and therefore they define the characteristics of data transfer, including latency and data rate. On the other hand, the physical channels actually specify the structure of the signals transmitted over the air interface, and they deal with interleaving, multiplexing, FEC, modulation and spreading of the data.

One of the sub-layers in L2 is the MAC sub-layer. This provides *logical channels* that interface to the other sub-layers in L2 and Layer-3 (L3) above it for transfer of both control and user data. Logical channels only specify what type of data is transmitted (e.g., broadcast, control, user traffic, etc.). In the RAN, L3 contains only the radio resource control (RRC) function. The service provided by L2 in the RAN to the core network is called a *radio bearer*, and each radio bearer can be used to transport a different user service. For example, there can be separate radio bearers for voice and data services. Radio bearers that are used by the RRC do not carry user data, and are called *signaling radio bearers*.

The protocol layers that are present in the UTRAN are part of what is called the *access-stratum (AS)*. The UE can also have peer protocol entities that do not terminate in the UTRAN, and these entities are part of what is called the *non-access stratum (NAS)*. The protocols that transport user data are called *user-plane* protocols, whereas those that transfer control signaling are called *control-plane* protocols. To make these terms clear by means of examples, the TCP/IP suite, and all applications or mobile apps, are part of the NAS protocols in the user plane. The RRC in the L3 sends several messages to configure and tear down connections between the RAN and the mobile, and these messages use AS protocols in the control-plane. We shall see the functions and operation of these three AS layers, L1, L2, and L3, in more detail in the following sections.

16.2.3   Physical Layer (L1)

As indicated earlier, the physical layer of UMTS systems is based on WCDMA. In this section, we discuss various aspects of this air interface, starting with a description of how CDMA is applied to the downlink and uplink, the modulation and coding used, and the configuration of various transport channels (for an introduction to CDMA concepts, see Chapter 14). Physical layer procedures and techniques such as power control and handover are described to complete the overview of the physical layer.

16.2.3.1   Channelization, Scrambling, and Modulation

Conceptually, CDMA allows multiple users to utilize the same frequency band at the same time by using different spreading codes, as explained in Chapter 14. In order to build an entire cellular communication system using this concept, the system design should specify how these codes are allocated to separate uplink and downlink users within a cell, and across different cells. Furthermore, the design should permit a user to transmit or receive multiple streams in parallel. To accomplish this, two types of codes are used in tandem: a channelization code, and a scrambling code.

- Channelization codes are based on Orthogonal Variable Spreading Factor (OVSF) codes. These codes, as the name suggests, are orthogonal, and have the interesting property that they can be picked in such a way that they continue to be orthogonal even when the codes are of different lengths. The spreading factor of the code is directly proportional to its length, and the codes can be organized in a tree-like structure as shown in Figure 16.4. Two codes of different lengths are orthogonal as long as the longer code is not a direct descendent of the shorter code on the code tree. The channelization code is transmitted at the chip rate, irrespective of its length, so the shorter codes carry the modulated symbols at a higher rate. Therefore, higher rate transmissions have a smaller spreading factor and lower processing gain. The data stream is first spread using the channelization code to obtain a stream at the

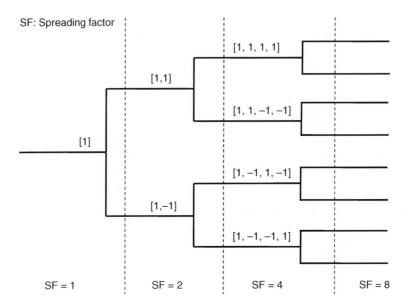

Figure 16.4 Structure of Orthogonal Variable Spreading Factor (OVSF) codes

chip rate, and the scrambling code is then applied to it at the same rate. As a result, the scrambling code does not spread the signal any further.

- Scrambling codes are usually much longer in length and are based on Gold codes that are known to have good cross-correlation properties, although they are not orthogonal. The chip rate of WCDMA systems is 3.84 MChips/sec, with each WCDMA carrier occupying around 5 MHz of bandwidth, including the guard bands on either side of the signal.

In the downlink, channelization codes are used to separate different users belonging to the same cell. They can also be used for multi-code transmissions to the same user. Being orthogonal, these different transmissions do not interfere with each other. However, it should be noted that, due to multipath propagation, the codes are no longer perfectly orthogonal when they are received at the UE. To overcome this degradation, equalization must be used at the UE receiver. The same scrambling code is used for all users in a given cell, but different scrambling codes are used by different Node-Bs to separate users belonging to one cell from another.

In the uplink, channelization codes are used to separate different channels from a UE to the Node-B. For example the control channel and data channel from a given UE use different channelization codes. Furthermore, the scrambling code is used to separate different users being received simultaneously by the Node-B. Note that, unlike the downlink, the uplink is not orthogonal across users and, therefore, intra-cell interference can be a major problem unless fast power control is used to ensure that all users are received at their target SINR.

In addition to the spreading operation, the physical layer is also responsible for modulation and coding. The FEC used is usually a rate 1/3 Turbo code, but convolutional coding is also supported. Rate-matching, interleaving and multiplexing of different channels is also handled by L1. The early releases of the specification only supported QPSK modulation in the downlink and BPSK modulation in the uplink. However, higher order modulations were introduced as the standard evolved, as we shall see in a subsequent section.

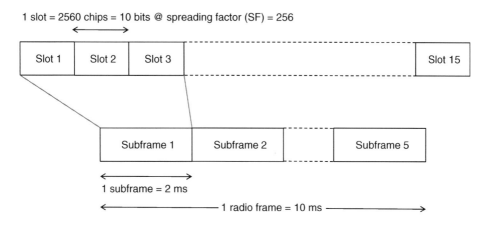

1 slot = 2560 chips = 10 bits @ spreading factor (SF) = 256

Figure 16.5 Basic slot structure for WDCDMA

16.2.3.2 Transport Channels and Physical Channels

The physical layer serves L2 by mapping transport channels to physical channels that actually carry the data over the air. In addition, there are several physical channels that do not directly serve L2 but are, rather, used by L1 itself to aid synchronization, acquisition, demodulation, and so on. Transport channels are either dedicated to a single user or shared between users. There is only one dedicated transport channel, simply called Dedicated CHannel (DCH) in each direction. The DCH is used to carry user data for either circuit-switched (voice) services or packet-switched services.

In addition to the DCH, there are several shared channels in the uplink and downlink. In the downlink, the Broadcast Channel (BCH) is used to convey important system configuration information to the UEs in a cell. The Forward Access Channel (FACH) is used to transfer small amounts of user data in the downlink. The Paging Channel (PCH) is used in the downlink to notify the UE of an incoming voice call, SMS, or data session. In the uplink, the random access channel (RACH) is used to permit new UEs to initiate a connection to the network.

These transport channels are mapped onto one or more physical channels, which have a specific frame structure that is based on slots, and sub-frames that are 2 ms long, as shown in Figure 16.5. Each slot carries 2560 chips, and there are three slots in a sub-frame. The DCH is mapped to two physical channels: the dedicated physical data channel (DPDCH) and the dedicated physical control channel (DPCCH). The DPDCH carries the data payload and the DPCCH carries the signaling information associated with the payload, which includes the transport format, power control, and pilot for channel estimation.

The structure of the uplink DPDCH and DPCCH are shown in Figure 16.6. The DPCCH carries the transmit power control (TPC) bits, feedback information (FBI), pilot for coherent demodulation, and the transport-format combination indicator (TFCI). In the uplink, these two channels are transmitted together by a process called I-Q code multiplexing. In I-Q code multiplexing, the two channels are separated using different channelization codes, and are modulated on to the in-phase and quadrature portions of a QPSK constellation with different gains, as shown in Figure 16.7. The spreading factor for the DPCCH is fixed at 256, allowing 10 bits to be transmitted in each slot, while the spreading factor of the DPDCH is variable between 4 and 256, allowing for different rates.

In the downlink, the DPDCH and DPCCH are time-multiplexed, as shown in Figure 16.8. The spreading factor may be varied from 512 down to 4. The actual number of data bits carried depends on the spreading factor. The DPCCH carries a known pilot, TPC bits, and an optional TFCI. In addition to

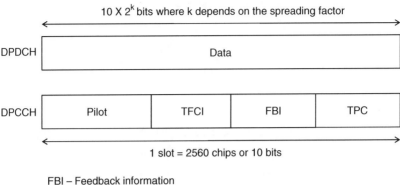

FBI – Feedback information
TFCI – Transport-format combination indicator
TPC – Transmit power control

Figure 16.6 Structure of the uplink DPDCH and DPCCH

the DPDCH and DPCCH, several other physical channels are specified. Some of these are L1-specific signaling channels that assist in the operation of the physical layer, while others carry shared transport channels.

Two important signaling channels in the downlink are the common pilot channel (CPICH) and synchronization channel (SCH). The CPICH carries no data modulation, is scrambled using a different code than the one used to multiplex all other downlink channels, and is used to assist with channel estimation at the UE receiver. The SCH is used for the cell search process by providing signals that assist with timing acquisition and the identification of cell-specific scrambling codes. In addition, the downlink also transmits the primary and secondary common control physical channels (PCCPCH and SCCPCH). The PCCPCH carries the BCH, and the SCCPCH carries the FACH. In the uplink, the RACH is carried on the physical RACH (PRACH) channel, on which the UE can transmit a preamble followed by a data payload. Random access is a two-step process, where the UE first sends a preamble and then listens to a

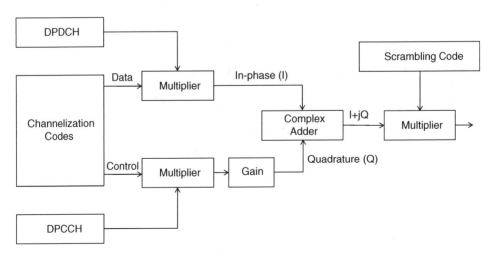

Figure 16.7 I-Q code multiplexing

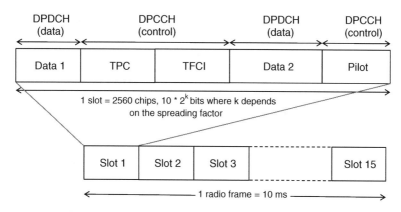

TPC – Transmit power control
TFCI – Transport-format combination indicator

Figure 16.8 Structure of the downlink DPCH

downlink channel called the acquisition indicator channel (AICH), which informs the UE whether its random access attempt was successful before it can transmit the payload.

16.2.3.3 Power Control

One key element of the WCDMA air interface is power control. Although power control implementation is not restricted to the Node-B where the physical layer implementation resides, it is discussed in this section because it is mainly a physical layer procedure. Power control comes in different flavors, and applies both to the uplink and downlink.

Let us first discuss why power control is needed. In the uplink, as the scrambling codes from different UEs are not orthogonal, significant interference can be caused at the Node-B. This is the classic "near-far" problem, where a nearby UE may be received at a much higher power than a UE far away, making it difficult to detect the far UE. The solution to this problem is to control the transmit power of all UEs in the cell, so that the near UEs transmit at lower power than the far UEs, thus resulting in all UEs being received at similar power-levels. In the downlink, the strongest signal is usually from the serving Node-B, so the near-far problem does not apply as much. However, the signal from the Node-B can vary quite a bit due to fast fading. To ensure adequate link performance, the transmit power can be increased to combat fading, but this can cause unnecessary interference to other cells. Instead, power control may be used to compensate for fading, and this will result in a lower power increase on average, lowering the inter-cell interference and improving capacity.

The flavor of power control used for the uplink and downlink DCH is fast, closed-loop power control. Power control commands are sent once per slot, resulting in 15 commands every 10 ms, for a power control frequency of 1500 Hz. The power control commands ask the transmitter (UE or Node B) to increase or lower the power by a known quantum. Another flavor of power control, open-loop power control, is used in the uplink by the RACH. In open-loop power control, no explicit commands are sent from the Node-B. Instead, the UE estimates the path-loss based on the downlink received power, and uses this estimate to set its initial power. If the transmission is not successful, it increases its power in steps until it is heard by the Node-B.

Finally, there is a slow, closed-loop power control mechanism that is referred to as "outer-loop power control". The purpose of this mechanism is to provide a target SINR for the fast power control. The target

SINR is usually based on the target BLock Error Rate (BLER) of decoding a transport block. For example, for a voice service, the target block error rate may be set to 1%, and the corresponding target SINR may be set to 6 dB. However, the actual channel conditions may be such that an SINR of 7 dB is required for ensuring a BLER of 1%. In this situation, when the outer loop tracks the BLER and it realizes that the BLER is larger than what is acceptable, it increases the SINR target for the fast power control algorithm.

The outer-loop algorithm varies the target SINR at a rate that is much slower than the operation of the fast power control loop, because it takes many cycles of the fast loop for the outer loop to form a good estimate of the BLER needed to adjust the target SINR. Outer-loop power control is needed both in the uplink and downlink, as the DCH uses fast power control in both directions. A general overview of the closed loop algorithms is shown in Figure 16.9.

16.2.3.4 Handover

Handover is usually understood to be the process by which a mobile UE can stay connected to the network without perceptible interruption when it moves across coverage areas of multiple cells, with some Node-Bs going out of range and other Node-Bs coming into range. However, more generally speaking, handovers can be performed to move the UE from a given carrier signal another carrier signal. The other carrier can be at a different frequency but may still belong to the same Node-B, in which case it is an *inter-frequency* handover. The other carrier can also be using a different Radio-Access-Technology (RAT), in which case it is called an *inter-RAT* handover. Most commonly, however, handovers tend to be *intra-frequency*, which are either performed at the same carrier frequency between different Node-Bs or different sectors of the same Node-B.

Loss of coverage due to mobility is clearly the primary reason to enable handovers, but the process can also be motivated by other considerations, such as service availability, load balancing, and interference management. For example, certain services may be available only on a specific RAT (e.g., data services are not available on 2G). It is also possible that the coverage footprint of a given frequency band may not be as large as that of a different (lower) frequency band, causing an inter-frequency handover with the same sector of the connected Node-B. Furthermore, the load across different carriers on a Node-B may be different, and handovers may have to be performed to balance the load across different frequencies. Another situation in which a handover may need to occur is in the presence of interference from a femtocell, which is a small-form-factor Node-B installed in a residence. When a femtocell is serving only a closed subscriber group (CSG) – typically tenants of the residence – then a UE that does not belong to the CSG may have to be handed off to a different frequency to avoid interference from the femtocell.

A unique handover feature of 3G technologies, which sets it apart from both previous technologies and more recent technologies, is soft handover – the situation in which the UE is connected to multiple Node-Bs simultaneously. Each UE maintains an *active set* that lists all the Node-Bs with strong, yet similar, signal strengths. Soft handover occurs at the border of coverage or cell-edge. In the downlink, multiple Node-Bs transmit the same data to a UE, and the UE combines multiple signals to improve the quality of the link. This type of diversity combining is called *micro-diversity* combining, as the signals are combined at the physical layer to improve the SINR before decoding the FEC.

The diversity provides protection against fast fading, as the probability of at least one of the links having a strong signal improves as the number of links in the active set increases. In the uplink, if the signals are received by different Node-Bs, then the physical layer processing is completed at the Node-B first and, if decoding is successful, the packet is forwarded to the RNC. The RNC discards any duplicates if multiple Node-Bs send decoded packets, or requests a retransmission if none of the Node-Bs is successful in decoding the packet. This process is called *macro-diversity* combining. If two different sectors of a Node-B are in the active set of a UE, then these sectors can perform micro-diversity combining before forwarding the data to the RNC. This situation is called a "softer handover". Micro-diversity combining fares better than macro-diversity combining, as the signals are directly combined to improve the SINR, thereby improving the probability of successful decoding.

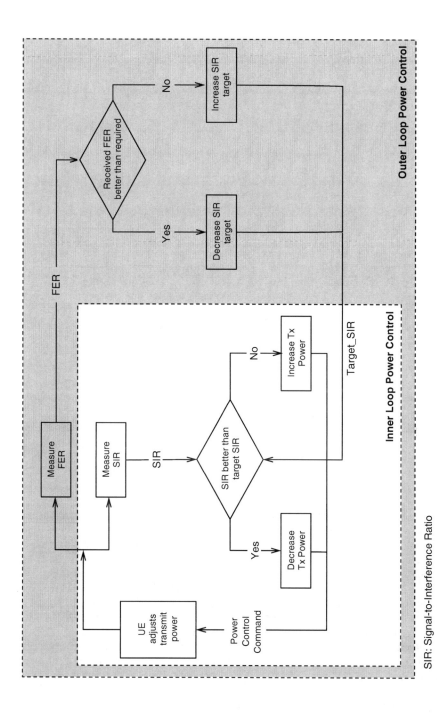

SIR: Signal-to-Interference Ratio

FER: Frame Error Rate

Figure 16.9 Overview of closed loop power control algorithms

16.2.4 Layer-2

Layer-2 (L2) in the protocol architecture shown in Figure 16.3 consists of several sub-layers:

- Medium Access Control (MAC).
- Radio Link Control (RLC).
- Packet Data Convergence Protocol (PDCP).
- Broadcast/Multicast Control (BMC).

We shall examine the services and functions of these sub-layers, and their interactions.

16.2.4.1 Medium Access Control (MAC)

The MAC sub-layer presents several logical channels for use by the sub-layers above it (RLC, PDCP, and BMC), and RRC (L3). These logical channels can be divided on the basis of the type of data they carry (control signaling or user traffic) and whether they are common (shared) or dedicated. As such, the following logical control channels are available:

- *Broadcast Control Channel (BCCH)*. This channel is used for broadcasting system control information.
- *Paging Control Channel (PCCH)*. This channel is used to notify a UE about an incoming call or data session when the UE cell location is not known, or when the UE is utilizing sleep mode.
- *Dedicated Control Channel (DCCH)*. This is a point-to-point bidirectional channel for transfer of control information.
- *Common Control Channel (CCCH)*. This channel is also bidirectional, and it is used before a DCCH is available, typically when a UE has just connected to the cell after cell-reselection.

There are two logical traffic channels:

- *Dedicated Traffic Channel (DTCH)*. This channel is a point-to-point channel dedicated to user data transfer for one UE. This channel can exist both in the downlink and uplink.
- *Common Traffic Channel (CTCH)*. This channel is a point-to-multipoint, downlink-only channel, for transfer of user data to all or a group of UEs.

The main functions of the MAC sub-layer are summarized below:

- **Logical-to-transport channel mapping**. There is some flexibility in mapping logical channels to transport channels. For example, the BCCH can be mapped to the BCH or the FACH, depending on the type of broadcast data being sent.
- **Transport format selection**. The transport format defines the instantaneous rate used over the radio link, and it depends on the source rate for efficient use of the transport channel.
- **Priority handling and scheduling of flows**. This function is implemented both at the UE and the UTRAN. At the UE, multiple radio bearers may be active, and the UE MAC uses the priority attributes of the radio bearer to map the data flows on to transport channels. At the UTRAN, dynamic scheduling of resources on the downlink is used for bursty data transfers. We will take a more detailed look at scheduling as we look at the evolution of UMTS to support high-speed data.
- **Multiplexing/demultiplexing of PDUs to/from transport blocks**. Higher layer PDUs are delivered to the MAC and these are formatted (e.g., CRC and MAC header are added) into a transport block for transfer on a transport channel over the physical layer. In the reverse direction, transport blocks are de-multiplexed into higher-layer PDU.
- **HARQ functionality**. To support high-speed data, HARQ functionality is required, and this is implemented in the MAC sub-layer at both the UE and UTRAN.

- **Segmentation/reassembly**. Depending on the dynamic scheduling of resources, the higher layer SDUs may not fit into the available radio resources, necessitating segmentation and reassembly at the MAC sub-layer.
- **Ciphering**. Ciphering secures the data from eavesdropping, and is performed by the MAC for transfers from the RLC where the RLC is a pass-through (i.e., transparent mode, as we shall see in the next section).

16.2.4.2   Radio Link Control (RLC)

The main function of the RLC protocol is ARQ, to mitigate the errors that may occur on the radio link. However, this function may not be needed for all services. As such, the RLC can be operated in three modes:

- *Acknowledged Mode (AM)*. This mode is the "full-featured" operation of the RLC, and it supports retransmissions for packets in error, flow control, in-sequence delivery, and segmentation/reassembly. In this mode, the reliability of the link is greatest, but the latency may increase in order to ensure reliability. As such, it is typically used for TCP-based transfers like web browsing and e-mail, which are not as delay-sensitive but are sensitive to errors. Acknowledgments required for the ARQ operation require a reverse channel and, therefore, like TCP, AM-RLC is bidirectional. Recall that, in macro-diversity combining with soft handovers, packets from multiple Node-Bs need to be received at a central location before an acknowledgment can be issued. This central node is the RNC, and it acts at the termination point for the AM-RLC connection with a UE. Moreover, it is possible that the limited bandwidth link between a RNC and Node-B may get congested, so flow control is also part of the AM-RLC protocol. Functionally, the AM-RLC is similar to TCP, but it can react to errors at a much smaller timescale than TCP. Whereas TCP has congestion control as its primary objective, AM-RLC has error recovery as its primary objective. As a result, they work well in tandem. In AM, the data is encrypted to prevent unauthorized access.
- *Unacknowledged Mode (UM)*. In this mode, the ARQ operation of the RLC is not possible. However, segmentation and reassembly are supported, and so is in-sequence delivery. As such, the RLC can operate with lower latencies, but it cannot mitigate packet errors. This mode is well suited to latency-sensitive, but somewhat error-tolerant, operations such as Voice-over-IP (VoIP). The RLC is responsible for encrypting the data in this mode, too.
- *Transparent Mode*. In this mode, the RLC can be configured to be a virtual pass-through, with no protocol overhead added. However, if needed, a limited form of segmentation and reassembly functionality can be enabled at radio-bearer setup. It can be used for broadcast messages such as system information or paging.

16.2.4.3   Packet Data Convergence Protocol (PDCP)

The main function of the PDCP sub-layer is data compression. IP packets carry a 20-byte header and, when carrying a voice payload, the RTP/UDP header is another 20 bytes. For a given connection, these headers can be compressed significantly to reduce the capacity used on the air interface.

PDCP is present only in the user-plane, and only for packet-switched services. It is also responsible for user data transfer, by forwarding packets between the connected RLC entity (UM or AM) and the non-access stratum.

16.2.4.4   Broadcast/Multicast Control (BMC)

This protocol also exists only in the user-plane. Its main function is the storage, scheduling, and transmission of cell-broadcast messages. Emergency services, such as the Commercial Mobile Alert System (CMAS), can be carried using this protocol. In this service, the US Federal government can send a broadcast text message to all subscribers via mobile operators who enable this service in their networks.

16.2.5   Radio Resource Control (RRC)

Layer-3 consists of the RRC protocol which operates solely in the control-plane. Radio resource management algorithms such as admission control, handovers, outer-loop power control, etc. rely on the signaling enabled by the RRC protocol for their operation. We shall first take a brief look at the key functions enabled by RRC:

- **Broadcast of system information**. System information from both the access and non-access stratum (core network) is sent periodically over the air interface to allow UEs entering the network to understand the configuration of the network. The RRC is responsible for scheduling, segmentation and periodic repetition of this information.
- **RRC connection management**. Before a UE can start communicating with the network, an RRC connection has to be established with the network. The establishment of this connection includes an optional cell-reselection, admission control, and L2 signaling link setup. The RRC connection can exist in one of several RRC states, and we shall take a closer look at these states later in this section.
- **Radio bearer management**. Upon request from the non-access stratum, the RRC can setup and teardown radio bearers. Radio bearers can also be reconfigured. Radio bearers contain QoS attributes that define the type of service carried by the radio bearer, and we shall take a closer look at these QoS classes later in this section. The RRC layer is responsible for control of QoS requested by radio bearers, and for allocation of resources required to ensure that the QoS requirements are met.
- **Radio resource allocations**. The assignment, reconfiguration, and release of channelization codes to RRC connections is one of the key functions of the RRC layer. It handles the allocation of resources to the uplink and downlink, and between multiple radio bearers assigned to an RRC connection.
- **Mobility functions**. The RRC layer is responsible for evaluation of handover options, making handover decision, and executing the signaling required to complete a handover. It also handles cell selection and reselection in idle mode (one of the RRC states we shall discuss shortly). The measurements required from the UE to enable these decisions are also orchestrated by the RRC layer.

16.2.5.1   Quality of Service (QoS)

There are four classes of QoS defined in UMTS:

- Conversational class.
- Streaming class.
- Interactive class.
- Background class.

The main difference between these classes is the latency requirement. The traffic carried by the conversational class is typically voice and video telephony, and this is the most delay-sensitive. Streaming class is used for applications such as real-time video streaming, which can use some buffering and are, therefore, not as delay-sensitive as telephony applications, but still require tight bounds on the delay. The interactive class includes web browsing, database/server access, etc., which require some guarantee of response time, but which also require low error rates. Finally, the background class is suitable for applications such as SMS, file download, email, etc., that do not have strict delay requirements but need error-free payload delivery.

The RRC layer is responsible for the admission control functionality that is required to manage QoS in the network. This function must maintain information about the level of resource utilization in the network, and must also be able to estimate the resource requirements of radio bearers with different QoS requirements. It should be able to reserve resources, configure and modify radio bearers as needed. Resource management is a key function and includes scheduling, bandwidth management and power control.

16.2.5.2 UE Modes and RRC States

As indicated earlier, the UE can be in one of two modes: *idle mode* or *connected mode*. When the UE first connects to the network, it comes up in idle mode, where it selects a cell and "camps" on it. In this situation, the UTRAN does not maintain any state information about the UE and does not know the exact cell where the UE is located. The UE only registers its UTRAN Registration Area (URA) with the NAS. This allows the UE to be paged when it is in idle mode, because all Node-Bs in the registration area will carry a page for the UE known to be in that registration area. In the idle mode, the UE power consumption is minimal, as it responsible to execute very few functions. It uses a discontinuous reception (DRX) cycle that allows it to wake up occasionally to monitor the paging channel, measure the strength of the cell it is camped on and reselect if necessary.

In the connected mode, data may be transferred between the UE and the UTRAN. The connected mode is divided into four RRC states: Cell_DCH, Cell_FACH, Cell_PCH, and URA_PCH. These states, and the transition between the states and modes, are shown in Figure 16.10.

- In the Cell_DCH state, the UE maintains a dedicated physical channel to the UTRAN. It is power-controlled, and can be in soft handover. The UE power consumption in this state is the highest of any RRC connected state, and the amount of network resources consumed is also the greatest.
- In the Cell_FACH state, the UE has no dedicated physical channel, and is not power-controlled. It can use the FACH and RACH to exchange small amounts of data with the network. The power consumption in this state is the second highest of all RRC states.
- In Cell_PCH, the UE cell location is known at the serving RNC. Its power consumption is less than that of the Cell_FACH state as it can use DRX.
- Finally, in the URA_PCH state, the UE cell location is not known, but only the URA is known at the serving RNC.

The Node-B uses timers to decide which RRC connected state to keep the UE. When an active data transfer is in progress, the UE is kept in the Cell_DCH state. If the data transfer ends, it is transferred to the Cell_FACH state after a certain amount of time. If there is no activity in the Cell_FACH state, it may be

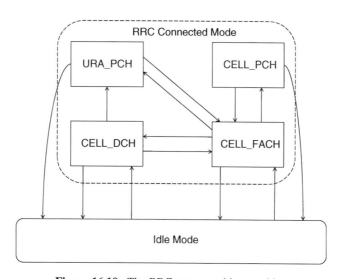

Figure 16.10 The RRC state transition machine

transferred to one of the PCH states, or to the idle state, depending on the implementation. There are RRC signaling messages exchanged with each state transfer. For efficient operation of the network, it is important to manage the UE in such a way that the UE does not stay in the Cell_DCH state for longer than necessary (to minimize battery and resource consumption), yet the RRC signaling load should be kept within check.

16.3 UMTS Technology Evolution

The initial specification of the UMTS WCDMA air interface and core network elements was completed in early 2000, with the publication of Release 99. All the basic technology elements discussed in the previous section were put in place as part of this first release. The next release was called Release 4, and subsequent releases were numbered sequentially as Release 5, 6, 7, and so on. Release 11 is currently the latest release and was completed in the second half of 2012. Release 12 is slated for release in late 2014. In this section, we will follow the evolution of UMTS by looking at the features that were included in various releases over time. The key releases were Release 99 as already indicated, followed by Release 5, which introduced high-speed downlink packet access (HSDPA), followed by Release 6, which included high-speed uplink packet access (HSUPA).

16.3.1 Release 99

The peak data rate supported by the first release in the downlink was actually in the excess of 2 Mbit/s which, in theory, can be achieved by using three channelization codes, each with spreading factor 4, for a per-code rate of 768 kbit/s. In the uplink, a peak rate of 384 kbit/s could be achieved by using a single code with spreading factor 8. However, in practice, a single code with spreading factor 8 was configured in the downlink for a maximum speed of 384 kbit/s, and initial uplink speeds were limited to 128 kbit/s. Although there were two shared transport channels defined – DSCH in the downlink and CPCH in the uplink – only dedicated transport channels (DCH in both directions) were used in practice. The shared channels were never implemented, and were superseded by high-speed versions in subsequent releases. As a result, they have been removed from the specifications.

The architecture for Release 99, showing the key transport channels, along with the functional breakdown between the RAN elements is illustrated in Figure 16.11. As shown in the figure, the Node-B is mainly a modem (modulator/demodulator), responsible for physical layer processing, including fast power control, error correction coding/decoding, and spreading/despreading. Functions like scheduling, admission control, outer-loop power control, handover management, etc., are handled by the RNC. Note

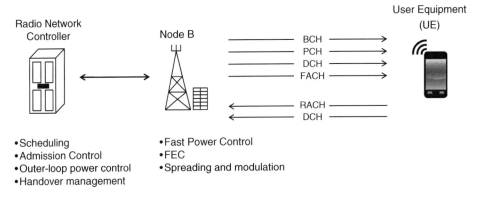

Figure 16.11 Release 99 transport channels and UTRAN functional architecture

that the spreading factor and codes are assigned by the RNC via higher layer signaling in a semi-static manner, and the smallest transmission time interval (TTI) over which a transport block is sent is 10 ms, with support for TTIs that are 20, 40, and 80 ms long.

16.3.2   Release 5: High-Speed Downlink Packet Access (HSDPA)

The release immediately following Release 99, called Release 4, was published in 2001, and did not contain any major enhancements to the air interface. The main changes in Release 4 were to the core network, so Release 4 will not be discussed further here. Release 5 was the next major release, published in early 2002. It introduced HSDPA, ushering in the era of fast wireless broadband access in a truly mobile environment supporting high vehicle speeds. The peak speed claim of Release 5 HSDPA is around 14 Mbit/s, with typical speeds of 2–3 Mbit/s, an order of magnitude higher than was possible before on UMTS. Most of the modifications needed to enable HSDPA were restricted to the air interface and, as such, the major upgrades were applied to the Node-B and RNC. The key features of HSDPA that enabled significantly higher speeds over the air interface are:

- HARQ.
- Link adaptation.
- Dynamic scheduling over short TTI (2 ms).
- Enhanced multi-code operation.
- Support of 16-QAM.

These enhancements were incorporated into the existing WCDMA architecture by defining new transport and physical channels, adding more functionality to the Node-Bs and moving some of the control from the RNC over to the Node-B. An overview of the functional breakdown between the Node-B and the RNC are shown in Figure 16.12, which is in contrast to Figure 16.11. The Node-B in the enhanced architecture is responsible for dynamic scheduling, link adaptation, and HARQ operation.

The transport channel added to support high-speed data is called the High-Speed Downlink Shared Channel (HS-DSCH). To support the operation of the HS-DSCH, three new physical channels were added: the High-Speed Shared Control Channel (HS-SCCH) and the High-Speed Physical Downlink Shared Channel (HS-PDSCH) were added in the downlink, and the High-Speed Dedicated Physical Control Channel (HS-DPCCH) was added in the uplink. The functional mapping of these channels is shown in Figure 16.13.

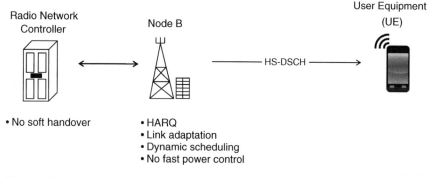

Figure 16.12   Release 5: transport channel inclusion and functional changes to UTRAN

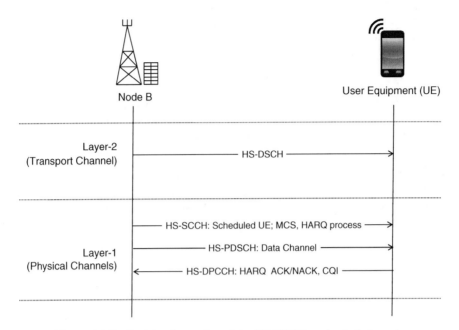

Figure 16.13 Functional mapping of the HS-DSCH to physical channels

The HS-PDSCH is the physical channel that actually carries the user data traffic using 2 ms TTI. To support the operation of the HS-PDSCH, the HS-SCCH carries control information such as the scheduled UE, the MCS, and the HARQ process information. The HS-DPCCH carries the HARQ ACK/NACK and Channel Quality Information (CQI). Note that in addition to these new channels, the DCH needs to continue to operate in both directions. In the next few subsections, we shall delve further into how the features listed above were designed by examining these new channels that were added as part of Release 5 in more detail.

16.3.2.1 HS-DSCH/HS-PDSCH

As indicated earlier, the transport channel that carries high-speed data is the HS-DSCH, and it is mapped to the physical channel HS-PDSCH. The HS-PDSCH has the following characteristics (see Figure 16.14):

- Short TTI of 2 ms duration.
- Fixed spreading factor (SF) of 16.
- Multi-code transmissions.
- Variable modulation and coding scheme (MCS) via link adaptation.
- QPSK or 16-QAM.
- Turbo-coding for FEC.
- HARQ for fast error recovery.
- No soft handover.
- No fast power control.
- Dynamic scheduling over the TTIs (time) and codes.
- Carries only user data – no control information.

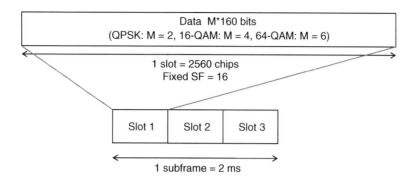

Figure 16.14 Subframe structure for the HS-PDSCH

The Node-B plays a key role in the operation of the HS-PDSCH by implementing HARQ, link adaptation, and scheduling. The DCH channels still support soft handover, and therefore need to be controlled by the RNC. In order for the Node-B to have complete control over the HS-PDSCH to implement these features, soft handover is not supported on the HS-DSCH. Furthermore, as the uplink DCH may still use soft handover, its power control algorithm may potentially take into account the signals from multiple Node-Bs in its active set, so the HS-PDSCH cannot depend on the power control commands received in the uplink for its single Node-B operation. On the other hand, the lack of power control is more than compensated by the introduction of dynamic scheduling and link adaptation with HARQ.

The scheduling algorithm at the Node-B operates over two dimensions – time and code – as shown in Figure 16.15. Every TTI, the scheduler can allocate code resources to one or more users, depending on the availability of data in the buffers of active RRC-connected users. The code resources are all equal in terms of their suitability to users (i.e., no codes can carry more data than other codes). Therefore, if one user has

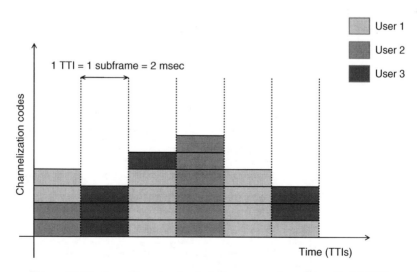

Figure 16.15 Two-dimensional scheduling over code and time for HSDPA

better instantaneous channel conditions than others, relative to its own average channel conditions, it might make sense to give all the code resources to that user opportunistically, so that it can take advantage of a higher rate. On the other hand, if a user only has a small amount of data in its buffers, multiple users can be scheduled on the same TTI in order to ensure that the available capacity is not wasted. The scheduler takes into account the CQI feedback from the UE to decide how much data can be scheduled in each TTI, by using a link adaptation algorithm.

Note that scheduling algorithms usually give preference to HARQ retransmissions of transport blocks for which NACKs were received over new user data (please refer back to Chapter 14 for a discussion on HARQ, link adaptation and scheduling algorithms). The actual algorithms used for scheduling can make a significant difference in the quality of service and user experience. Therefore, these algorithms are not specified in the standard, but rather left to the equipment manufacturers to bring in their own innovations.

16.3.2.2 HS-SCCH

The HS-SCCH is the downlink control channel at the physical layer that accompanies the HS-PDSCH. Like the HS-PDSCH, it is a shared channel that is monitored by all capable UEs in the RRC-connected state. It carries the following information about the transport blocks sent on the HS-PDSCH:

- The UE ID to which the information is directed.
- The number of codes allocated to the UE.
- The modulation and coding format (transport block size).
- HARQ process information.
- New data/retransmission indicator flag.

As the HS-PDSCH may send transport blocks to multiple UEs via code-multiplexing, multiple HS-SCCHs are needed, and these can be configured by the Node-B if required. The HS-SCCH uses a fixed spreading factor of 128, and carries 40-bits for information transmitted over three slots (2 ms) (see Figure 16.16). The HS-SCCH is sent two slots ahead of the user data on the HS-PDSCH, to allow the UE that is going to receive the data to have prior knowledge of the channelization codes and modulation format in order to reduce processing time and implementation complexity. Accordingly, the HS-SCCH data is sent in two parts. The first part occupies one slot and contains the channelization codes and modulation format. The second part contains HARQ-related information. The UE ID is implicitly included in both parts by using UE-specific scrambling for part-1 and a UE-specific CRC in part-2.

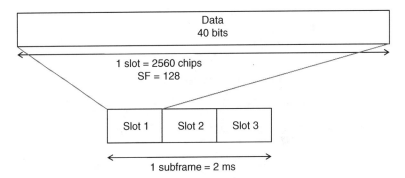

Figure 16.16 Subframe structure for the HS-SCCH

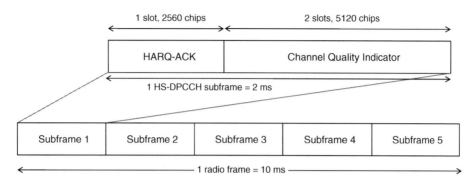

Figure 16.17 Subframe structure for uplink HS-DPCCH

16.3.2.3 HS-PDCCH

In the uplink, a channel is needed to carry feedback to the Node-B for operation of downlink HARQ and link adaptation. This uplink control channel, the HS-PDCCH, needs to be separate from the uplink DPCCH, because the DPCCH needs to continue to work with Node-Bs that may not support HSDPA operation. More importantly, unlike the HS-PDCCH, which is only needed at the Node-B originating the HS-DSCH, the DPCCH may operate in soft handover and be received by multiple Node-Bs in the active set. An example where this may occur is when circuit-switched voice calls may be simultaneously carried over the DCH along with packet data on the HS-DSCH. (Recall that the DCH is mapped to the DPDCH/DPCCH).

The HS-PDSCH carries the following feedback information:

- ACK/NACK status of the downlink HARQ transmission,
- Instantaneous channel quality information (CQI).

Much like the HS-SCCH in the downlink, the HS-DPCCH is carried over three slots that are subdivided into two parts, as shown in Figure 16.17. The first part carries the HARQ ACK/NACK information over a single slot, and the second part carries the CQI information over the remaining two slots. A spreading factor of 256 is used for the HS-PDCCH. The CQI information does not carry an explicit value of SINR that indicates channel quality but, rather, it encodes one of 30 fixed MCS levels that can be used in the downlink for a pre-specified block error rate of 10%. The MCS, in effect, corresponds to the data rate that can be supported by the UE. Therefore, this type of feedback incorporates the UE receiver capability, along with an indication of the actual channel conditions. More capable receivers can report feedback higher data rates than less capable receivers for the same channel conditions, thereby allowing the Node-B to maximize the capacity of the HS-DSCH channel.

Clearly, each HS-PDCCH transmission is tied to a corresponding downlink HS-PDSCH transmission, and therefore there needs to be a timing alignment between the two transmissions. This is achieved by ensuring that the HS-PDCCH is symbol-aligned and sent approximately after 7.5 slots of the HS-PDSCH transport block. The HS-PDCCH need not be time-aligned to the other uplink DCH channels (DPDCH/DPCCH), but it does need to be chip-aligned with all uplink channels to keep all uplink channelization codes orthogonal to each other.

16.3.3 Release 6: Enhanced Uplink

As one might naturally expect, a high-speed uplink packet access (HSUPA) feature was developed to follow the development and deployment of HSDPA. Both of these features are frequently referred together as High Speed Packet Access (HSPA), and they form the complete 3G mobile broadband access technology part of UMTS. Although the enhanced uplink has some features in common with HSDPA, the uplink access is fundamentally different from the downlink access, resulting in some key differences in the design of HSUPA. In this section, we shall first examine the peculiarities of uplink access to motivate some of the design choices made for HSUPA, and then describe the new channels introduced to support enhanced uplink access.

Uplink access differs from downlink access in two fundamental respects. In the downlink, the UE is usually served by the strongest Node-B, and is not expected to demodulate signals from interfering Node-Bs that are typically weaker than the serving Node-B. However, in the uplink, the Node-B receives non-orthogonal transmissions from different UEs within the cell that are distributed at different distances and, therefore, it may have to contend with near UEs drowning out the far UEs. This *near-far* problem can be particularly debilitating in the uplink and, therefore, fast power control is essential to keep the intra-cell interference under control. In fact, the intra-cell interference level is the key parameter that must be controlled by sharing the power allocated to various UE by employing a scheduling algorithm. Furthermore, fast power control must operate at the serving Node-B in conjunction with other surrounding Node-Bs to manage the inter-cell interference, which motivates the need for soft handover.

The other key difference between the downlink and uplink is in the ability to coordinate transmission/reception from multiple Node-Bs. In the downlink, the use of soft handover implies that transmissions must be coordinated across different Node-Bs, which becomes a difficult proposition, given fast dynamic scheduling over 2 ms TTI. However, on the uplink, each UE can be received by multiple Node-Bs, and the packets can be combined at the RNC without the same tight coordination constraints, because the scheduling is mainly controlled by the serving Node-B and the UE. Therefore, soft handover is enabled in the uplink.

Techniques such as link adaptation, fast scheduling and HARQ have been utilized in the uplink, for the same reasons that they are used in the downlink. However, the key difference in the uplink is that these techniques must now be capable of operating over simultaneous links to multiple Node-Bs, due to the possibility of soft handover. The other difference in the uplink arises from the fact that the resource allocation occurs at the Node-B, and it does not have access to the exact current status of the uplink buffers at the UE. As a result, mechanisms are needed to transfer scheduling information to the UE and buffer information from the UE back to the Node-B. We shall delve deeper into how the design accounts for these differences as we discuss the channels that have been introduced to support the enhanced uplink.

HSUPA introduces a new transport channel called the E-DCH to support high-speed access. Unlike the HS-DSCH, which is a shared channel, E-DCH is a dedicated channel like its lower-speed precursor, the DCH. In fact, as the name suggests, it shares several features with the DCH, and is better understood to be a faster version of the uplink DCH with HARQ, link adaptation, and dynamic scheduling, rather than the uplink version of HS-DSCH. The transport channel E-DCH is mapped to physical channels E-DPDCH and E-DPCCH. The user data is sent over one or more E-DPDCHs and the data rate is a function of the number of E-DPDCHs and their spreading factors. The E-DPCCH is the control channel that accompanies the E-DPDCH, and it carries the transport format, buffer and data rate modification indication, and new-data/retransmission information.

To support dynamic scheduling and HARQ, three new downlink physical channels are required to be part of the enhanced uplink. The E-DCH HARQ Indicator Channel (E-HICH) carries the HARQ ACK/NACK information in the downlink. The E-DCH Absolute Grant Channel (E-AGCH) and E-DCH Relative Grant Channel (E-RGCH) carry scheduling grants. Finally, a more efficient version of the DPCH, called the Fractional DPCH (F-DPCH) has also been specified to carry fast power control information only to replace the downlink DCH.

16.3.3.1   E-DCH/E-DPDCH

The E-DCH is an advanced version of the DCH transport channel, because it supports the HARQ, link adaptation, and fast scheduling. The user data on the E-DCH is transported over the E-DPDCH on the physical layer. The E-DPDCH is an enhanced version of the DPDCH that uses shorter TTIs, and supports a lower spreading factor. The TTI is typically 2 ms long, although the specification also supports 10 ms TTIs for use in large cells, where the uplink is power-limited and can support relatively lower data rates. The longer TTI is beneficial as it uses less overhead. The E-DPDCH is also permitted to use a spreading factor of 2, which in effect doubles the allowable channel data rate. Moreover, the multiple E-DPDCHs can be used at the same time via multi-code transmission.

The E-DCH supports fast power control and soft handover. For fast power control signaling, the downlink control channel used can be either the DPCH, based on Release 99, or the new F-DPCH channel that is described later. Scheduling on the E-DCH is mainly controlled by the serving Node-B. The serving Node-B controls the uplink interference by allocating different power levels to UEs, based on their channel conditions and buffer status. The power levels are specified as a ratio of the E-DPDCH power to the DPDCH power. The UE scheduler must allocate power to the DCH first, before allocating the remaining power to the E-DPDCH to ensure that power allocated for any delay-sensitive circuit-switched services is not affected by the presence of data bursts.

16.3.3.2 E-DPCCH

The E-DPCCH carries three pieces of information in support of the E-DPDCH:

- Transport block configuration.
- Retransmission sequence number.
- The "happy" bit.

The transport block format is referred to as the Enhanced Transport Format Combination Indicator (E-TFCI), and is encoded using seven bits. There is only one E-DPCCH for potentially multiple E-DPDCHs, so the E-TFCI encodes the transport block size, which can be used to infer the number of E-DPDCHs and their spreading factors. The E-DPCCH itself uses a spreading factor of 256 and carries a total of ten information bits sent over three slots (2 ms).

The retransmission sequence number is two bits, supporting four different values. This field is similar to the new-data flag used in the downlink. However, as the HARQ operation in the uplink can occur during soft handover, error scenarios are more likely, because all Node-Bs in soft handover may not have good channels to the UE. A single bit used in toggle mode can more easily lead to corruption of data in the HARQ buffer, because of incorrect interpretation of the new-data flag by any one of the Node-Bs in the active set that goes through a sequence of errors due to a weak link. To provide additional protection against this eventuality, the retransmissions are protected by a sequence number which has four values, and takes longer to "wrap around".

The "happy" bit is a simple toggle indicator sent by the UE to the Node-B to indicate whether it is happy with the data rate that it is receiving, based on its channel conditions and data in the buffer. If this bit is set, then the Node-B will try to increase the power ratio allocated to the UE if the scheduling constraints permit it to do so.

16.3.3.3 E-AGCH

The E-DCH absolute grant channel is a shared channel that carries the following information:

- The actual value of the E-DPDCH/DPDCH power ratio.
- An activation flag that applies to an individual HARQ process.
- The UE identity.

Each UE-specific message carries a total of six bits of information, of which five bits are used to encode the granted power ratio and one bit is used to enable or disable HARQ processes. The UE ID is encoded as part of the CRC calculation, so that a UE can determine whether the message is directed towards it. The E-AGCH is transmitted over three slots (2 ms), with a spreading factor of 256. Each UE monitors only one E-AGCH – the one sent by its serving cell.

16.3.3.4   E-RGCH

In contrast to the E-AGCH, the EDCH Relative Grant CHannel (E-RGCH) is a dedicated channel from the serving Node-B. It carries one of three messages: UP, HOLD, or DOWN, which indicates to the UE whether the E-DPDCH/DPDCH power ratio must be increased, kept the same, or reduced by a preset amount that is signaled separately. Also, in contrast to the E-AGCH, UE must monitor the E-RGCH from all the Node-Bs in its active set, but the non-serving cells transmit a common physical channel that carries an "overload" indicator. The overload indicator can only take the values DTX or DOWN. The DTX command indicates that there is no overload, and the UE is free to follow the grant from the serving cell. The DOWN command indicates that there is an overload and the UE should not increase its power.

A situation may arise in which the serving cell may allow the UE to increase its power by sending an UP command in response to the lowered interference that may result when a UE responds to a DOWN command from a non-served cell. In this situation, the UE will not let itself increase its power for one HARQ round-trip cycle, to ensure that the conflicting commands from the cells do not end up in a repetitive cycle.

16.3.3.5   E-HICH

The E-DCH HARQ indicator channel is used to send HARQ ACK/NACKs in the downlink for uplink HARQ transmissions. This channel is a dedicated physical channel that carries an ACK, a NACK, or a DTX. The DTX corresponds to a situation where the Node-B did not receive any HARQ transmission in the expected TTI on the uplink. Non-serving cells in the active set also transmit this channel as a dedicated channel, although they only send an ACK or DTX. The idea is that the all non-serving Node-Bs can save downlink power because, if the serving cell sends the NACK, it is very likely that they will also send a NACK. This unnecessary repetition of NACKs can be avoided by stipulating that non-serving cells only send ACKs. Following this protocol, the UE will keep retransmitting the data as long as it does not receive an ACK from any Node-B in its active set. Only the serving Node-B sends an explicit NACK to the UE.

The E-HICH channel uses 40-bit orthogonal sequences that permit 40 different single-bit messages to be transmitted over a single channelization code. The E-RGCH shares this same structure with the E-HICH in the serving cell. Usually, the E-RGCH and E-HICH share 20 sequences each, so that a UE can monitor a single channelization code to receive both the E-HICH and E-RGCH. The E-HICH and E-RGCH are sent using a spreading factor of 128 and can occupy three slots (2 ms) or 12 slots (8 ms).

16.3.3.6   F-DPCH

When the downlink employs an HS-DSCH, a downlink DCH may still be enabled to carry circuit-switched data and RRC signaling messages. In this case, a DPCCH is required to carry dedicated pilots and TFCI for the associated DPDCH. However, if all the DPDCH traffic is carried over the HS-DSCH, then a full blown DPCCH is not needed, as the only control information that still needs to be carried is the fast power control signaling for the uplink. Keeping this in mind, a new channel, called the Fractional Dedicated Physical Channel (F-DPCH), was designed. Instead of allocating a unique channelization code for each DPCH, the F-DPCH shares the same channelization code among ten UEs. Each UE receives two TPC bits that are time-multiplexed with other UEs on the same channelization code in a single slot. The spreading factor used is 256. Note that the F-DPCH cannot be used when a Release 99 DCH is required for circuit-switched services that cannot be mapped to the HS-DSCH.

16.3.4   Release 7

From release 7 onwards, the HSPA channels put in place in the previous releases have been progressively improved, with the addition of several more advanced features. There are three main directions that wireless standards tend to extend into for increased peak rates and capacity:

- Increasing the spatial dimension (increasing the number of transmit or receive antennas).
- Higher order modulation.
- Increasing bandwidth.

The key features added in release 7 are support for spatial multiplexing (MIMO), continuous packet connectivity, enhanced FACH, and higher-order modulation. These exploit the first two directions mentioned above. In Release 8, we shall see how the third dimension is brought into play. While most of these features have pertained to the air interface, some features enhance the higher layers, too. For example, the RLC operation has been streamlined with the use of larger PDU sizes, which can become segmented by the MAC, based on instantaneous channel conditions. An enhanced FACH operation has been defined to permit the use of HSDPA in cell_FACH state. As more features are developed in the standard, a note about *UE categories* is required here.

The concept of UE categories is used to classify UEs, based on their capabilities to support various subsets of optional features in the standard. Among other parameters, the UE categories specify the number of simultaneous codes that can be sent or received, the length of the transmission (in terms of number of consecutive TTIs or the length of a TTI), the maximum transport block size, and so on. The category of the UE usually determines the maximum throughput it can support (based on the TTI and the maximum transport block size). Therefore, it is important to keep in mind that, although a specific release may have support for a certain maximum throughput, in practice, one would need vendors to produce devices of the UE category that supports the maximum throughput. Release 7 is sometimes also called *HSPA+*, and it increases the maximum supported throughput in the downlink from around 14 Mbit/s to over 42 Mbit/s.

16.3.4.1   MIMO

Use of two transmit antennas to achieve transmit diversity in the downlink has been supported from Release 99. Employing two receive antennas at the UE for receive diversity combining does not require standards support so, in theory, it is a feature that UEs could have exploited from the early days of UMTS. In practice, UEs with receive diversity did not appear in the market until recently.

Note that using two receive-chains at the UE does increase its battery drain, so it is always a consideration in small-form factor devices such as smartphones. However, neither transmit or receive diversity increase the peak rate achievable by the link, as they still permit only a single stream to be used per TTI (i.e., only one transport block can be sent per TTI).

In order to increase the peak rate, spatial multiplexing is required to permit two streams of data to be transmitted at the same time. This requires a 2×2 MIMO system of two transmit antennas at the Node-B and two receive antennas at the UE. Furthermore, in order for the link to effectively support spatial multiplexing, high signal-to-interference and noise ratio (SINR) is required, and the wireless channel must have sufficient scattering or polarization diversity to support two spatial channels.

The flavor of MIMO defined in Release 7 is dual-stream, multiple codeword, pre-coded spatial multiplexing. The use of two codewords in a dual stream transmission implies that each spatial stream can use its own MCS based on the strengths of the individual spatial streams. Moreover precoding is used to form two orthogonal beams to reduce the self-interference between the streams at the receiver. The selection of pre-coding weights is assisted by feedback from the UE. Note that, although enabled, each transmission does not need to be a spatially multiplexed transmission. The system allows for the use of a single stream or two streams, dynamically based on the channel conditions.

In order to support MIMO, the HS-SCCH has been expanded to carry MIMO configuration information to the UE. In the first part (first slot), the number of streams and the pre-coding matrix are sent; in the second part (last two slots), the transport block size and HARQ information for the second stream is added if the HS-DSCH is spatially multiplexing two streams. In addition, the HS-DPCCH is also enhanced to carry pre-coder indicator (PCI) and CQI information, along with transmission rank (number of spatial streams) for up to two streams.

Note that the Node-B can decide to use a single stream even if the UE indicates that two streams can be supported. This may be due either to power constraints or to lack of data in the buffer for the UE. In this situation, it is not possible to infer the PCI/CQI accurately for a single stream, given two-stream feedback. To assist the Node-B in this situation, two types of reports are sent – one that reports PCI/CQI for two streams if the channel conditions are right, and a second report that contains the PCI/CQI for a single stream in case the Node-B should choose to use only one stream. The Node-B can configure how many of each report type the UE should send.

16.3.4.2 Continuous Packet Connectivity (CPC)

As discussed in the section on RRC, there is a trade-off between latency and resource consumption in the various RRC states. In the Cell_DCH state, the UE link is very responsive to traffic demands and can service a request for bandwidth with very low latency. However, this responsiveness comes at the price of increased overhead, interference to the network from the control channels, and greater battery drain. Ideally, one would like to be able to marry the fast access time of the HS-DSCH with the low overhead and power consumption of the Cell_PCH or Cell_FACH state.

Continuous Packet Connectivity attempts to accomplish this ideal through the use of three features:

- Discontinuous Transmission (DTX). The UE does not transmit the DPCCH continuously when the E-DCH is enabled; rather, the DPCCH is turned on two slots prior to the E-DPDCH/E-DPCCH activity and turned off one slot after the activity stops.
- Discontinuous Reception (DRX). The UE does not monitor the HS-SCCH, E-AGCH, and E-RGCH continuously, but moves into a cycle of reception of these channels, followed by a sleep period where it turns its receiver off. The cycle is initiated when there is no activity on the HS-DSCH.
- HS-SCCH-less HS-DSCH transmission. This mode of operation is used to transmit small transport blocks to the UE for services such as VoIP, without the high overhead associated with frequent signaling of small packets. The modulation used is restricted to QPSK, and one of four pre-defined transport block formats are allowed. This permits the UE to perform blind decoding to detect a transmission for itself. The CRC used is UE-specific, allowing the UE to decide which packets belong to it.

It is interesting to note that in order to achieve an "always-on" experience on HSPA, with low resource cost, one has to resort to discontinuous channels via a feature called continuous packet connectivity!

16.3.4.3 Higher-Order Modulation

The size of the signal constellation has been extended in both the uplink and the downlink to enable higher rates when the channel conditions are good (i.e., under high SINR conditions). In the downlink, 64-QAM is supported to permit six bits of information to be transmitted per symbol. In the uplink, 16-QAM transmissions carrying four bits of information per symbol is permitted.

16.3.5 Release 8 and Beyond

Release 8 of UMTS/HSPA also coincides with the first release of LTE, the next generation standard that is beginning to see rapid deployment in some parts of the world. Although HSPA continues to evolve, it is very attractive for operators to switch to deploying LTE, because of the greater spectral efficiency provided by LTE and the rapid proliferation of smartphones that have LTE chipsets. As such, there are not many deployments of HSPA that extensively use features developed beyond Release 8. At any rate, it is still early days, in terms of device availability, for the latest HSPA and LTE releases, so one cannot predict with certainty to what extent the newer releases of HSPA will be implemented in devices and deployed by operators. The main thrust for capacity improvement in HSPA, starting with Release 8 and beyond, has been the use of multiple carriers. The use of 64-QAM was permitted with MIMO in Release 8, and the enhanced FACH operation was further improved to support HSUPA in cell_FACH.

16.3.5.1 Fast Dormancy

In the quest to optimize responsiveness and battery life, a feature called fast dormancy was fine-tuned in Release 8. Fast dormancy allows the UE to switch quickly to idle mode when it detects inactivity on the PS connection. However, in releases before Release 8, the actual algorithm used by the UE was left to implementation, and poor implementations could result in excessive signaling over the network and poor response times. In Release 8, the control of the fast dormancy was given to the Node-B through the use of a timer specified by the Node-B. In addition, the transition from cell_DCH was made to URA_PCH instead of the Idle mode, so that the UE was still RRC connected and the HSPA connection could be revived faster.

16.3.5.2 Advanced Receivers

One option to improve the performance of a wireless link is to employ better receiver algorithms. Several basic approaches, and combinations thereof, are available for the reception of WCDMA signals. The basic approach used in CDMA receivers is the use of a "rake"-like structure, with different fingers or taps of the rake synchronized with different multipath signals. By synchronizing the taps to different delays of the desired signal and combining the different copies in a coherent way, the signal-to-noise ratio (SNR) can be increased.

A more complex approach involves using some fingers in the rake to collect information about the interfering signals (which could include self-interference from other symbols or interference from other users) that arrive at different delays. The SINR of the received signal can be maximized by appropriate selection of combining weights, such that these interfering signals are canceled and delayed copies of the desired signal are combined coherently. As the different path delays are resolved at the chip level, these receivers are also called chip-level equalizers. Another approach is to use multiple receive antennas. The use of multiple antennas at the receiver provides some diversity (depending on the correlation between the antennas) and coherent combining gain. As such, four receiver types have been specified:

- Type 1: Receive diversity.
- Type 2: Chip-level equalizers.
- Type 3: Receive diversity with chip-level equalization.
- Type 3i: Receive diversity with interference cancellation.

The 3GPP specification does not mandate any particular receiver structure; rather, it specifies receiver requirements that are derived using a baseline receiver structure. Vendors are free to choose any implementation that allows them to meet the performance specified. It is also worth noting that it is optional for a UE to meet the performance specified for a given type of receiver, as the advanced receiver performance is not necessary for correct operation of the network from a protocol and signaling perspective. However, competitive pressures in the market usually push vendors to implement receivers that meet these requirements anyway. Alternatively, operators can specify mandatory requirements for devices that will be certified for use in their networks, as a way to ensure a certain level of performance in their network.

16.3.5.3   Multi-Carrier and Multi-Flow

A multi-carrier system works by coordinating the scheduling of packets over multiple carriers. As the system has more carriers that can be used simultaneously, the peak rate that is supported is increased in proportion to the number of carriers used. Without multi-carrier operation, UEs are assigned to a specific carrier and, depending on the traffic distribution, all carriers may not be equally utilized. Load balancing across carriers requires inter-frequency handovers, which is signaling-intensive and occurs at much larger timescales than a TTI. On the other hand, if all carriers are available to the pool of all UEs, the resources are more efficiently utilized by allocating resources dynamically to UEs every TTI. This increase in efficiency is called "trunking gain." Finally, upgrading a system to support multi-carrier transmissions can be less expensive than MIMO, as the towers do not need to be updated with additional hardware (antennas, radios, cables) and provide benefit over a wider range of SINR conditions than MIMO. Of course, multi-carrier assumes the availability of spectrum to deploy additional carriers, but MIMO extracts additional efficiency with the same amount of spectrum.

In Release 8, dual-carrier support was specified for HSDPA, with both carriers residing in the same frequency band. Gradually, over various subsequent releases, support has been added for more carriers, with more features and less restrictions. In Release 9, support was added for MIMO and 64-QAM on dual-carrier HSDPA. In Release 10, multi-carrier was further extended to allow two carriers on two different bands to be aggregated, thereby supporting four carriers in total. Most recently, in Release 11, 8-carrier HSDPA was standardized, with four carriers each on two different bands.

Multi-flow is conceptually different from multi-carrier, because it allows the same carrier transmitted from different cells carrying different information streams. In a sense, it is a combination of multi-carrier and soft handover that operates like a spatially separated MIMO system. By receiving two different streams on the same carrier, the system exploits spatial multiplexing gains, which improves the peak rate. Furthermore, at the cell-edge, other-cell interference in multi-flow can actually be a useful stream towards the UE, which it extracts using advanced interference cancellation receivers that are able to suppress self-interference. Thus, multi-flow also improves the cell-edge rate. Multi-flow was standardized in Release 11, and it supports transmission of two carriers each from two different sectors with 2×2 MIMO.

16.4   CDMA2000

In parallel with the development of UMTS in 3GPP, a sister organization – 3GPP2 – developed another access technology, CDMA2000, which is conceptually very similar to WCDMA. In terms of worldwide subscribers, in early 2012, the number of CDMA2000-based 3G users represented about a fifth of the all 3G users, who were predominantly users of UMTS/HSPA. The subscribers were mainly in Asia/Pacific (Japan and Korea) and North America (subscribers of Verizon Wireless and Sprint in the United States).

One of the fundamental differences between CDMA2000 and UMTS is in how it evolved to support high-speed data. The original CDMA2000 specification, called 1×RTT for 1 carrier Radio Transmission Technology, was published in 1999. A new uplink and downlink structure was developed to support high-speed data, but was designed only to carry packet-switched services. This evolution was originally called 1×EV-DO for Data Only. Later, EV-DO was modified to represent Data Optimized. However, the key difference between UMTS and CDMA2000 evolutions was that separate carriers were needed for voice and data services in CDMA2000, whereas the same carrier could support simultaneous voice and data on UMTS.

The first release of 1×EV-DO, Rev 0, and its subsequent releases Rev A and Rev B, are all based on fast scheduling, link adaptation, short TTI, Hybrid ARQ, and higher-order modulation. Rev A focused on uplink enhancements. Rev B added support for multiple carriers. Note that CDMA2000 was defined to operate on a carrier bandwidth of 1.25 MHz. Although the spectral efficiency of the two technologies 1×EV-DO and HSPA are similar, the peak rates are much lower for 1×EV-DO, because of the lower bandwidth. As a result, the multi-carrier support added in Rev B was important to boost peak rates.

16.5   Summary

This chapter discusses the third generation of cellular systems, popularly known as 3G, the first cellular technology that can be considered to truly provide broadband access, with typical speeds of around 2–4 Mbit/s in the downlink. Two 3G systems were developed in parallel: UMTS and CDMA2000. With around 80% of the deployments being based on UMTS today, the focus of this chapter is UMTS. CDMA2000 systems are very similar to UMTS systems in the fundamental access technology used, the main differences being the bandwidth per carrier, which is smaller for CDMA2000, and the need for separate carriers in CDMA2000 to support data and voice services. UMTS was designed from the start to consist of two separate modes of access: circuit-switched for voice services, and packet-switched for all other data services. The specifications for UMTS was, and continues to be, developed by the 3GPP, a consortium of standards development organizations (SDOs) from around the world. Individual companies participate through their regional SDOs.

The basic UMTS network architecture consists of three nodes in the access network: the user equipment (UE), the base station with the RF front end called the Node-B, and the radio network controller (RNC) that manages several Node-Bs. The protocol architecture consists of three layers: the physical layer (layer-1); layer-2, comprising the medium-access control (MAC), radio link control (RLC), and the packet data convergence protocol (PDCP); and radio resource control at layer-3.

The physical layer uses a CDMA-based design called Wideband CDMA (WCDMA) over a bandwidth of around 5 MHz, with a chip rate of 3.84 MChips/sec. Channelization codes are sent at the chip rate, and are used to separate users in the downlink and different channels from the same user in the uplink. Scrambling codes are used to separate users in the uplink and different sectors and Node-Bs in the downlink. The physical layer implements physical channels which carry transport blocks from layer-2 transport channels. The main transport channel that carries traffic in the initial release of WCDMA is called the dedicated channel or DCH, and it is a bidirectional channel, mapped to a control channel and a data channel in the physical layer. Power control is needed for these physical channels to ensure that the received signal power and the interference power are balanced correctly for a good quality link. Power control is also used to combat fast fading. At layer-2, the MAC layer is mainly responsible for scheduling of flows, ink adaptation, and HARQ. The RLC implements ARQ and flow control. The PDCP performs header compression of RTP/UDP/IP headers. Radio resource control operates only in the control plane, handling admission control, handovers, power control, and connection management.

The first release of UMTS was in late 1999, and was called Release 99. This contained the specifications for the first WCDMA-based air interface, but did not really support high speed broadband access. However, it formed the baseline for subsequent releases, which were sequentially numbered releases, starting with Release 4. High speed downlink packet access (HSDPA) was introduced as part of Release 5. The key features of HSDPA include HARQ, link adaptation, dynamic scheduling over 2 ms TTI, and support for 16-QAM. A new transport channel, called High Speed Downlink Shared Channel (HS-DSCH) was added as part of this release, and this channel implements HSDPA. HSDPA was designed as an extension to the existing WCDMA systems by mapping the HS-DSCH on to new physical channels that handle both user data and control signaling. The HS-DSCH is different from the DCH because it does not support soft handovers or power control, and is dynamically scheduled. The initial release of HS-DSCH supports a peak speed of 14.4 Mbit/s in the downlink.

In the next release of UMTS, Release 6, High Speed Uplink Packet Access (HSUPA) was introduced to go along with HSDPA. Together, these are called High Speed Packet Access (HSPA), and they form a complete high-speed broadband access system for mobile users. The HSUPA system is designed differently from the HSDPA, because of the fundamental asymmetry between a cellular uplink and downlink. In HSUPA, a new transport channel called the Enhanced-DCH (E-DCH) is introduced to provide higher speeds than are possible with DCH. Due to the greater significance of the near-far problem in the uplink, power control is still part of E-DCH, unlike the HS-DSCH in the downlink. Moreover, it is actually possible to maintain UEs in soft handover in the uplink because, for uplink reception, different

Node-Bs can be coordinated at the RNC without the tight delay requirements that would be needed in the downlink case. Like the HS-DSCH, the E-DCH also uses link adaptation, HARQ, and dynamic scheduling. New uplink physical channels have been added to support the operation of the E-DCH and, again, these are extensions of the basic WCDMA-based uplink.

Starting with Release 7, HSPA has been enhanced to support several new features that increase the capacity of the network and provide higher peak speeds. These enhancements include the introduction of MIMO, higher-order modulation, support for advanced receivers, and optimizations to the RRC state machine operation for increased battery life and reduced latency. In addition, one key feature first introduced in Release 8 was the simultaneous use of multiple carriers. This allows the effective bandwidth of a connection to be increased greatly, with a proportional increase in the peak throughput. The number of simultaneous carriers was progressively increased in subsequent releases, with Release 11 permitting the use of four parallel carriers, each in two disjoint frequency bands. Another related enhancement is multi-flow, in which the same carrier is transmitted from different cells, carrying different data. This can be seen as a type of MIMO system that exploits spatial multiplexing gains and improves peak rate without the need for additional spectrum.

The WCDMA-based UMTS system continues to evolve, with new releases from 3GPP. However, with LTE systems becoming mature and providing higher capacity, the enhanced HSPA features may not see wide deployment as operators start investing in LTE systems, which are more efficient, to maximize their returns.

Further Readings

1. 3GPPTS 25.201. Physical layer – general description.
2. 3GPPTS 25.211. Physical channels and mapping of transport channels onto physical channels (FDD).
3. 3GPPTS 25.213. Spreading and modulation (FDD).
4. 3GPPTS 25.214. Physical layer procedures (FDD).
5. 3GPPTS 25.301. Radio Interface Protocol Architecture.
6. 3GPPTS 25.401. UTRAN Overall Description.
7. 3GPPTS 25.321. MAC Protocol Specification.
8. 3GPPTS 25.322. RLC Protocol Specification.
9. 3GPPTS 25.331. RRC Protocol Specification.
10. 3GPPTS 25.308. High Speed Downlink Packet Access (HSDPA) Overall Description.
11. 3GPPTS 25.319. Enhanced Uplink Overall Description.

17

Fourth Generation Systems: LTE and LTE-Advanced

17.1 Introduction

Following the natural progression that was seen in wired broadband access, where voice-centric connections were succeeded by wide-bandwidth connections with voice being just another service, LTE was developed as a true mobile broadband access technology to support generic always-on broadband access. As such, unlike the UMTS architecture, where voice is carried over an explicitly designed circuit-switched portion of the network, the LTE architecture is designed only for packet-switched services, with voice carried as Voice over Internet Protocol (VoIP). Furthermore, all LTE networks to date have been deployed as an overlay to the existing UMTS/HSPA and CDMA2000 networks, providing only generic data services, with mobile voice calling still relying on the underlying UMTS CS technology. This is a clear testament to the explosion in demand for data and the shrinking bandwidth consumed by voice as a proportion of the data consumption. Major operator-managed voice-over-LTE (VoLTE) roll-outs are only expected in 2014, more than three years after LTE data services were turned on.

17.1.1 LTE Standardization

Following the successful development of global standards for GSM and UMTS under the auspices of the 3GPP, the standardization process for LTE was kicked off at the RAN Future Evolution Workshop in Toronto in November 2004. Several operators and equipment vendors from around the world presented their views on the long-term requirements for technology. In addition, these companies also discussed their proposals for the technology selection in the uplink and downlink. The formal standardization process was started in 2005, and the first version of the specification, Release 8, was frozen in December 2008.

The standardization work in 3GPP is handled by four technical specification groups (TSGs), as shown in Figure 17.1:

- GSM Edge Radio Access Network (TSG-GERAN).
- Radio Access Network (TSG-RAN).
- Services and System Aspects (TSG-SA).
- Core Network and Terminals (TSG-CT).

Broadband Access: Wireline and Wireless – Alternatives for Internet Services, First Edition.
Steven Gorshe, Arvind Raghavan, Thomas Starr, and Stefano Galli.
© 2014 John Wiley & Sons, Ltd. Published 2014 by John Wiley & Sons, Ltd.

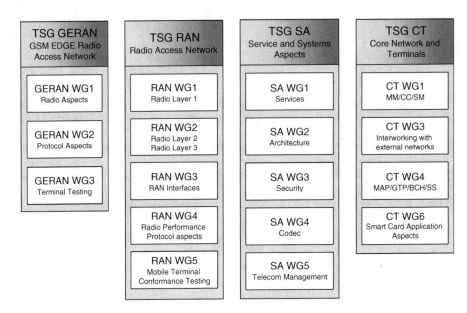

Figure 17.1 TSG structure

The air interface aspects of the system are standardized under the heading "Long Term Evolution" and the non-radio aspects of the system are part of the "System Architecture Evolution" (SAE). Together they form the "Evolved Packet System" (EPS). LTE standardization is discussed in TSG-RAN, which is divided into five working groups:

- RAN-WG1: physical-layer specifications.
- RAN-WG2: radio layer-2 and layer-3 specifications.
- RAN-WG3: RAN interfaces between nodes within the RAN and to the core network.
- RAN-WG4: radio performance specifications.
- RAN-WG5: mobile terminal conformance testing.

17.1.2 LTE Requirements

For the first LTE version, Release 8, the following key requirements were put forth to ensure that the new standard would be sufficiently forward-looking and could remain competitive in the long term:

- Downlink peak rate of 100 Mbit/s in a bandwidth of 20 MHz. This translates to 5 bits/sec/Hz in spectral efficiency.
- Uplink peak rate of 50 Mbit/s in a bandwidth of 20 MHz. This translates to 2.5 bits/sec/Hz in spectral efficiency.
- Downlink cell-edge user throughput (fifth-percentile) of 2–3 times that of the Release 6 HSDPA system, and average user throughput 3–4 times that of the same baseline.
- Uplink cell-edge user throughput (fifth-percentile) of 2–3 times that of the Release 6 Enhanced Uplink system, and average user throughput 2–3 times that of the same baseline.
- Latency: in the control plane, the latency of transitioning from the idle state to the connected state should be less than 100 ms. In the user plane, the latency of transferring an IP packet across the air interface in either direction shall be less than 5 ms in low load conditions.

Some of the key requirements for the architecture are as follows [REF]:

- The architecture shall be packet based, although provision should be made to support real-time and conversational traffic.
- The architecture shall minimize the presence of "single points of failure" where possible without additional cost for backhaul.
- The architecture shall simplify and minimize the introduced number of interfaces where possible.
- The architecture shall support end-to-end QoS. The core network shall provide the appropriate QoS requested by the RAN.

In the next section, we will take a closer at the overall architecture of the evolved UTRAN (E-UTRAN), and see how the requirements specified above are incorporated into the design. The resulting system design supports a peak rate of 73 Mbit/s in the downlink, assuming a 2×2 MIMO configuration with 10 MHz of bandwidth. Typical downlink speeds for bursty downloads are around 5–12 Mbit/s. Latencies are low enough to support smooth streaming of audio and video at speeds of around 400 kbit/s (sufficient for a smartphone screen form factor) even at the cell edge.

The rest of this chapter is organized as follows. We devote the next section to understanding the basics of LTE, including the overall network and protocol architecture, QoS support, the details of the OFDM-based physical layer, and higher layer functions in the RAN. The technology features covered in this section form the first release of E-UTRAN, Release 8. We then look at the main features in Release 9, which include enhanced multimedia broadcast and multicast service (eMBMS), and self-optimizing network (SON) enhancements. We next discuss Release 10, in which a collection of features including carrier aggregation, relays, enhancements to MIMO, and support for heterogeneous networks (HetNets), jointly referred to as *LTE-Advanced*, were standardized. Finally, we conclude the chapter with a discussion of Release 11 and beyond including a discussion of cooperative multi-point (CoMP).

17.2   Release 8: The Basics of LTE

17.2.1   Network Architecture

The architecture of the LTE network differs from that of the UMTS network in two important respects:

- It is an all-IP architecture.
- It is a flat architecture (no RNC to impose hierarchy).

The logical nodes that comprise the network are shown in Figure 17.2. The radio access portion of the network is called E-UTRAN for evolved-UTRAN, and the non-radio core network is called the Evolved Packet Core (EPC). As shown in the figure, the only infrastructure node in the RAN is the eNodeB (for evolved-Node-B), and the core network contains a new node called the mobility management entity (MME), along with the serving gateway (S-GW) and packet data network (PDN) gateway, or P-GW for short. In addition, there is a home subscriber server (HSS) and policy and charging rules function (PCRF) server. The functions of these nodes in an LTE network are described below:

- eNodeB: the eNodeB is the only node in the RAN portion of the LTE network and, as such, it is responsible for all the radio-access related functions. These include:
  - implementation of physical layer protocols over the air interface to communicate with the UE, including modulation, coding, link adaptation, HARQ, power control, and channel state feedback mechanisms;
  - radio resource management, including resource allocation via dynamic scheduling, mobility control for active UEs, admission control, and interference management;

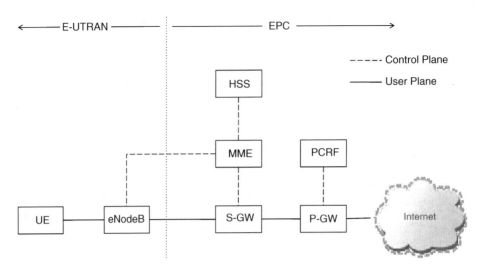

Figure 17.2 LTE logical nodes

- EPC connectivity by forwarding and receiving user IP packets to and from the S-GW and control messages to and from the MME for UE location management and security;
- encryption and header compression of user data to and from the UE.
- S-GW: the UE has only one S-GW, and this node serves as the router to forward IP packets to and from potentially different PDNs via different P-GWs. It acts as the mobility anchor for handovers between eNodeBs, and buffers packets for a UE in idle mode.
- P-GW: the P-GW connects the UE to the internet and acts as its default router. It is also responsible for IP address allocation for the UE and default EPS bearer setup (bearers are discussed in detail in a subsequent section). It handles the Differentiated Services Code Point (DSCP) marking for QoS differentiation in the core network and filters IP packets to map different IP flows to QoS bearers.
- MME: the MME is the node that handles most of the signaling, and is the key node in the control plane of the LTE network. As such, its key functions include:
- Non-access stratum (NAS) signaling, security, and authentication.
- Mobility management, including idle mode reachability, tracking area management, and roaming.
- Bearer management functions.
- Selection of the S-GW, P-GW, and MME selection during handover.
- HSS: the HSS is the master database that stores information about the user including subscription, identity, and service profile. It is involved in the authentication and authorization of the user. It generates security information for authentication, integrity checks, and ciphering, and it can also provide information about the user's location.
- PCRF: the PCRF delivers the QoS policy required for each user session to the MME. In addition, it also stores and serves the rules required to ensure that the user account is charged correctly.

In order to meet the requirement of minimizing single points of failures, and for load balancing possibilities, each eNodeB can be connected to one or more MMEs and S-GWs, as shown in Figure 17.3. The concept of an MME/S-GW *pool* has been introduced to facilitate the use of multiple MME/S-GWs as potential connection points for a given eNodeB. The MME/S-GW pool area covers multiple *tracking areas*, where a tracking area defines a set of eNodeB coverage areas, within which a mobile is paged by the

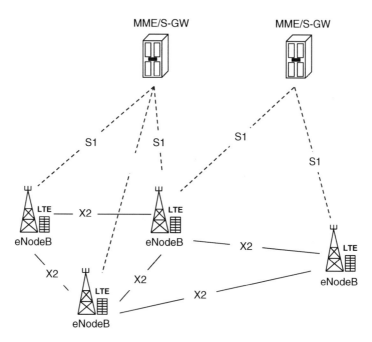

Figure 17.3 MME/S-GW pool

network. The network only retains knowledge of a UE's tracking area when it is in idle mode, so that the UE in idle mode does not have to send frequent updates to the network as it moves between the coverage area of different eNodeBs. As such, the UE does not need to change its association with an MME or an S-GW unless it moves to a pool area that does not cover the tracking area to which the UE belongs. The S1 interface connects the eNodeB to the core network and supports a many-to-many relationship to promote fault tolerance via redundancy. The MME, S-GW, P-GW, and so on are logical nodes and can be implemented on the same hardware in one physical node.

Note that the eNodeB itself is also a logical node, and it can support many instances of the air interface. The most common configuration of an eNodeB is one that supports three sectors. Other configurations include remote radio-heads (RRHs) where the antenna and the RF hardware (including filters and power amplifiers) are installed on the top of a tower (or at a good vantage point for coverage), with the baseband eNodeB hardware located at a central location supporting multiple RRHs. A distributed antenna system (DAS) is another variation on this theme, and is typically used in large buildings or venues to serve a dense user population restricted to a small area. In a DAS, the RF hardware could be centrally located with the eNodeB, with RF cables carrying the signal to a widely spread installation of antennas (e.g., in a stadium).

Finally, in Figure 17.4, the interfaces between these various nodes are depicted. The LTE-Uu interface is the air interface that we discuss in detail in this chapter. It carries both user-plane and control-plane data between the UE and the eNodeB. The S1-u interface carries user-plane data between the eNodeB to the S-GW. The S1-MME interface carries the control plane signaling between the eNodeB and the MME, and includes the bearer set-up messages and idle mode update messages. The X2 interface between eNodeBs is used for handover signaling and for exchange of load and interference information.

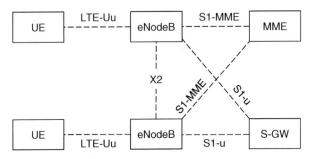

Figure 17.4 E-UTRAN interfaces

17.2.2 PDN Connectivity, Bearers, and QoS Architecture

The EPS provides connectivity between an external PDN like the internet and the UE. It achieves this connectivity via the use of EPS *bearers*. An EPS bearer uniquely identifies the traffic flows that receive a specific QoS treatment between the UE and the P-GW. Therefore, separate EPS bearers are needed for a UE connecting to different P-GWs, or if the UE supports multiple QoS levels. For example, a UE may have one EPS bearer for regular internet service, and another for VoIP service. The first EPS bearer that is established when a PDN connection is initiated is called the *default* EPS bearer, and this provides "always on" IP connectivity to that PDN. Any additional EPS bearer that is established to the PDN is called a *dedicated* bearer.

In order to differentiate traffic flows based on QoS or different PDN considerations and map them to different EPS bearers, traffic flow templates (TFTs) are used. TFTs are packet filters used at the UE and at the PGW, based on IP source or destination addresses, port, protocol, etc. to sort incoming IP packets (both UL and DL) into different EPS bearers, as shown in Figure 17.5. Typically, each TFT identifies a Service Data Flow (SDF), and a set of SDFs representing an SDF aggregate is mapped on to an EPS bearer. As shown in Figure 17.5, an EPS bearer is actually three low-level bearers in tandem: a radio bearer over the air interface, an S1 bearer between the S-GW and eNodeB and, finally, an S5/S8 bearer between the S-GW and P-GW. There are one-to-one mappings at the eNodeB and the S-GW to ensure that the incoming bearer over an external interface is mapped correctly to the outgoing bearer over another external interface.

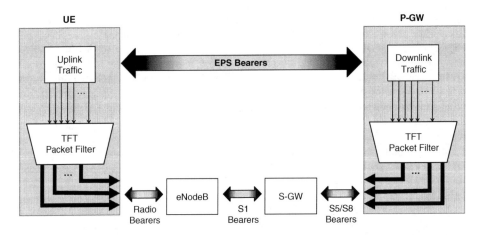

Figure 17.5 EPC bearer architecture and QoS mapping

Each EPS bearer is associated with two QoS parameters:

- QoS Class Identifier (QCI); and
- Allocation and Retention Priority (ARP).

QCI is the key value that controls the forwarding treatment received by the packets in the traffic flow, which includes the eNodeB scheduling treatment, queue management thresholds, DSCP mapping, etc. During EPS bearer setup (both for dedicated and default bearers), the eNodeB and UE are informed of the QCI associated with the bearer.

The QCI mapping defines the following parameters:

- Resource type: guaranteed bit rate (GBR) or non-GBR. In case of GBR, as the name suggests, the flow will enjoy, or at least perceive, guaranteed availability of resources.
- Priority.
- Packet delay budget: this parameter specifies the delay measured from the UE to the P-GW.
- Packet error loss rate.

It should be noted that the implementation of scheduling and queue management to ensure that the packet forwarding treatment meets the requirements specified by the parameters above is left to the vendor's proprietary algorithms. In fact, vendors try to differentiate themselves by implementing algorithms that implement QoS treatment so that, in addition to meeting requirements for a given flow, other considerations such as battery usage, the bandwidth left over for best-effort flows, and overall resource utilization are also optimized. For example, although GBR flow nominally expects resources to be reserved, in practice the eNodeB scheduler will not let resources get wasted if the flow does not actually need these resources. Furthermore, as explained earlier in the discussion on channel-dependent scheduling, even the guaranteed resources will actually be assigned as far as possible (taking delay constraints into account) only when the channel conditions are favorable to maximize the spectral efficiency.

The QCI is an 8-bit field permitting 255 values, of which nine standardized QCI mappings are defined (for values 1–9). The standard mappings, shown in Table 17.1, include QCIs that can be used for

Table 17.1 Standardized QCI mappings

QCI	Priority	Packet delay budget	Packet error loss rate	Example services
GBR QCIs				
1	2	100 ms	10^{-2}	Conversational voice
2	4	150 ms	10^{-3}	Conversational voice (live streaming)
3	3	50 ms	10^{-3}	Real-time gaming
4	5	300 ms	10^{-6}	Non-conversational video (buffered streaming)
Non-GBR QCIs				
5	1	100 ms	10^{-6}	IMS signaling
6	6	300 ms	10^{-6}	Video (buffered streaming) TCP-based (e.g., www, e-mail, chat, FTP, p2p file sharing, progressive video, etc.)
7	7	100 ms	10^{-3}	Voice, video (live streaming), interactive gaming
8	8	300 ms	10^{-6}	Video (buffered streaming), TCP based (e.g., www, e-mail, chat, FTP,
9	9			p2p file sharing, progressive video, and so on

well-known services such as voice, video conferencing, and so on. By defining such standard mappings, it is possible for the service to receive the same minimum level of QoS in multi-vendor deployments and in case of roaming.

The ARP affects admission control during bearer establishment, and also the preemption of bearers in congested situations. Note that once a bearer is established, ARP has no impact on the packet forwarding; the QCI is used for that purpose. For example, ARP can be used to treat the video and the voice portion of a video call differently in a case of congestion. By mapping the video portion of the call to an EPS bearer with a different ARP value, the bandwidth-heavy video streams can be discontinued to permit the voice calls to possibly face less congestion. Also, in disaster situations, capacity can be freed up for emergency responders and other similar high priority users, by having other user sessions preempted via use of ARP.

In addition to QCI and ARP, GBR bearers are associated with two parameters:

- Guaranteed Bit Rate (GBR);
- Maximum Bit Rate (MBR).

The GBR value denotes the rate expected to be provided to the bearer, and it usually matches the data rate of the service it carries. For example, voice may use a GBR of 12.65 kbit/s when it uses the AMR wideband codec to provide high quality voice, while video for a smartphone form factor may use a GBR of around 256 kbit/s. The MBR is used to limit the maximum rate that a GBR bearer may use by shaping any offered traffic that falls above that threshold. For example, for video which uses a variable bit rate based on the content, the MBR can be set to be equal to the maximum codec rate.

Finally, two additional parameters are used to limit the data rate offered to non-GBR bearers: per Access Point Name (APN) Aggregate Maximum Bit Rate (AMBR), or APN-AMBR, and UE-AMBR. The APN represents a PDN, and the aggregate bandwidth used by all connections to a PDN can be limited by the APN-AMBR.

One example where this may be used is a PDN that provides access to a popular video streaming site. In this case, the APN-AMBR can be used to ensure that all of the traffic in the network is not overwhelmed by one specific type. UE-AMBR represents the maximum data rate that a UE can achieve in either the UL or DL across all of its non-GBR bearers. This parameter can be used, for example, to throttle "abusive" users, so that the rest of the user population can get their fair share of the bandwidth.

17.2.3 Protocol Architecture

The protocol architecture, in terms of layers and peer protocol entities, is very similar to that seen in the UMTS system. The major difference is in the location of the layers, because LTE has a much flatter architecture. As such, the E-UTRAN has only the eNodeB on the infrastructure side of the link, and all the radio interface protocols in the user plane are implemented there. Similarly, the core network support for the control plane has also evolved to encapsulate all the functionality in the MME. The explicit protocol stack for the user plane is shown in Figure 17.6.

The user plane and the control plane have several access stratum protocols that are shared. The PDCP layer is responsible mainly for header compression, handling of in-sequence delivery and retransmissions during handover, and ciphering. The main function of the RLC layer is mitigating packet errors by means of an ARQ algorithm. It also handles the important function of segmentation and concatenation of higher layer packets for use by the MAC layer below. Also included in this protocol are functions for in-sequence delivery, discarding duplicates, and timer-based discard of SDUs in error. The MAC layer handles multiplexing of different RLC connections, dynamic scheduling between UEs, HARQ processing, and link adaptation by means of transport format selection.

Between the layers, as shown in the figure, the RLC provides radio-bearers to the PDCP layer for data transfer. Between the RLC peers, the MAC layer provides logical channels, and the MAC transfers data to

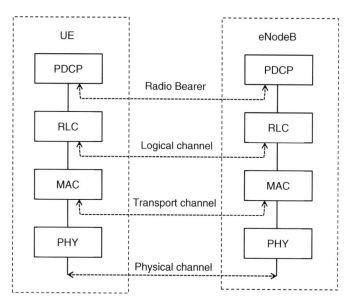

Figure 17.6 User plane protocol stack

its peer MAC layer by means of transport channels instantiated by the physical layer (layer-1 or L1). The physical layer itself uses physical channels to transfer data over the air interface.

The RRC layer is responsible for generating all the control signaling messages required for correct operation of the radio link. These include system broadcast information, UE-specific signaling for handover and measurement reporting, radio bearer establishment and tear-down, and RRC connection management. In the control plane, the PDCP, RLC, and MAC perform the same functions as they do for the user plane, with the main difference being that the PDCP for the control plane does not use header compression. There exists one PDCP and RLC entity for each radio bearer, and these entities can be thought of as being independent from each other. However, the MAC layer is common across all radio bearers, because different bearers are multiplexed and scheduled by the MAC layer. In the next several sections, we will delve much deeper into the structure and operation of these layers and their sub-layers.

17.2.4 Layer-1: The Physical Layer

In this subsection, we provide a detailed description of the physical layer by first looking at the time-frequency structure of the OFDM signal, then looking at how different uplink and downlink channels are mapped onto this structure. Next, we look at some physical layer procedures like power control, scheduling, link adaptation, and handover. Finally, we look at the MIMO transmission modes supported in Release 8 to get a sense for the baseline technology in initial deployments that forms the basis for enhancements in future releases of LTE.

17.2.4.1 Physical Layer Resource Structure

The basic multiple access system for the LTE OFDM physical layer is a combination of TDMA and OFDMA in the downlink and TDMA and SC-FDMA in the uplink. In the time dimension, the resources are divided as shown in Figure 17.7. The resources are made up of slots, with two equal-sized slots making up a subframe of duration 1 ms. A radio frame consists of ten subframes and forms the basic unit of an LTE signal which is repeated every 10 ms.

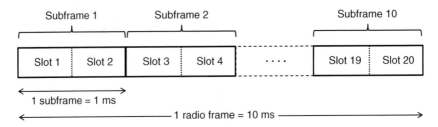

Figure 17.7 Basic LTE frame structure

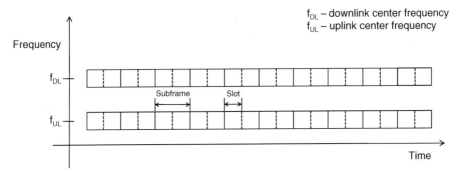

Figure 17.8 FDD frame structure

FDD

In an FDD system, the uplink and downlink frames are time aligned and the full set of ten subframes is available for use in both directions, as shown in Figure 17.8. Typically, both the mobile device and the eNodeB transmit simultaneously on the uplink and downlink in a *full-duplex* fashion. However, in special cases, it may be possible to configure operation in the *half-duplex* mode, where the uplink transmissions do not overlap with the downlink transmissions. Clearly, given that there are two separate bands for the uplink and downlink, this latter mode is not spectrally efficient. However, device complexity considerations may dictate this mode of operation is some cases.

TDD

In a TDD system, the radio frame is broken down into three types of subframes – uplink, downlink, and special subframes – as shown in Figure 17.9. The special subframes are designed to permit the mobile and eNodeB to transition from the uplink to the downlink or vice versa. For example, in an uplink-to-downlink transition, the receiver is first turned off and then the transmitter is powered up. In addition, due to

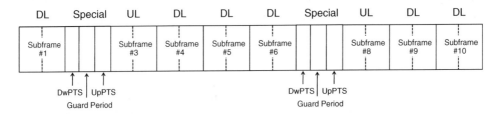

Figure 17.9 TDD frame structure

Table 17.2 TDD subframe configurations

Configuration	Periodicity	Subframe number									
		1	2	3	4	5	6	7	8	9	10
0	5 ms	D	S	U	U	U	D	S	U	U	U
1	5 ms	D	S	U	U	D	D	S	U	U	D
2	5 ms	D	S	U	D	D	D	S	U	D	D
3	10 ms	D	S	U	U	U	D	D	D	D	D
4	10 ms	D	S	U	U	D	D	D	D	D	D
5	10 ms	D	S	U	D	D	D	D	D	D	D
6	5 ms	D	S	U	U	U	D	S	U	U	D

D = Downlink subframe; U = Uplink subframe; S = Special subframe.

propagation delays, additional time gaps must be allowed for the signal from the furthest transmitter to be received properly. The special subframe consists of three parts: the downlink part, called the "DwPTS," the guard period (GP), and the uplink part, called the "UpPTS." The DwPTS and UpPTS terms are the legacy of the low chip-rate UTRA TDD system, also known as Time Division Synchronous CDMA (TD-SCDMA), which was deployed in China.

The first and sixth subframes are constrained to be downlink subframes, and the third subframe is constrained to be an uplink subframe. The standard provides for seven configurations of uplink and downlink subframes, supporting both 5 ms and 10 ms periodicities, as shown in Table 17.2. The special subframe is typically located in the second subframe and, for a 5 ms period, the next special subframe occurs in the seventh subframe. The arrangement of subframes in configuration #2 can be made to align with the transition points in the uplink and downlink subframes of other technologies, like WiMAX and TD-SCDMA, thereby permitting coexistence and compatibility with these technologies.

Resource Blocks: The Basic OFDM Resource Unit

We now discuss the actual structure of the OFDM signal within a subframe. Within each subframe, the OFDM subcarriers are divided into resource elements and resource blocks in a time-frequency grid, as shown in Figure 17.10. Recall that there are two slots in each subframe. For the most commonly used OFDM parameters, with 15 kHz frequency spacing between subcarriers, and the normal cyclic prefix, each slot is divided into seven OFDM symbols in the time dimension. In the frequency dimension, each time slot is further divided into frequency blocks of 12 subcarriers each. The 12×7 two-dimensional structure, consisting of 12 subcarriers across seven OFDM symbols, is called a *resource block* (RB). A resource block contains $12 \times 7 = 84$ subcarriers, each of which is also called a *resource element* (RE). The layer-2 scheduling is based on a unit containing two consecutive RBs in a subframe, which are usually allocated across the same set of subcarriers in frequency, as shown in Figure 17.10. In common parlance, this unit of allocation containing a pair of physical RBs (PRBs) is sometimes simply (and loosely) referred to as a "PRB".

Note that this resource structure based on resource elements, and unit of allocation based on PRBs, is common to both the uplink and downlink, although the actual processes of OFDM signal generation (OFDMA for the DL and SC-FDMA for the UL) are different. Also note that when spatial multiplexing is used, this resource structure applies to the allocation of each spatial stream. In other words, multiple PRBs are sent in parallel, one for each spatial stream.

17.2.4.2 Physical Layer Reference Signals

The structure of the OFDM frame, as described in the previous section, is essentially a set of REs or subcarriers that form a pair of RBs or PRBs. This set of REs is further partitioned into subsets, and each subset of REs has a special significance. One set of REs is set aside as reference signals, to aid the receiver in channel estimation for data reception and channel state feedback.

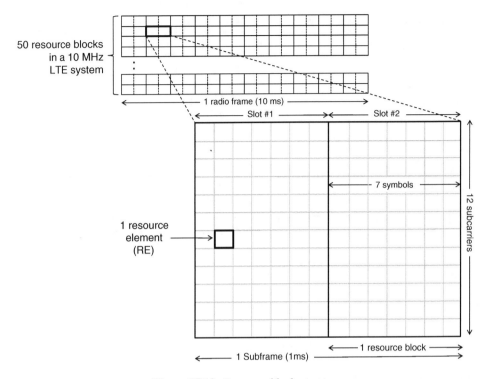

Figure 17.10 Resource block structure

Shown in Figure 17.11 are the REs that are used as common or cell-specific reference signals (CRS) in the downlink for a system with two antenna ports. These CRS are present in every PRB. In addition to CRS, UE-specific reference symbols are also defined in Release 8, as shown in Figure 17.11. These reference symbols are pre-coded using the same weights that are applied to the data-carrying symbols and, therefore, can be used with proprietary beamforming schemes.

In addition to reference symbols, as shown in Figure 17.12, some REs are set aside in specific PRBs within the frequency band and are used as synchronization signals. There are two types of synchronization signals: the primary synchronization signal (PSS) and the secondary synchronization signal (SSS). These two signals aid the UE in the cell-search and acquisition procedure, providing cell identity and information about whether the cell is in TDD or FDD mode, in addition to time and frequency synchronization information. The PSS and SSS are sent in the six central PRBs and occupy one symbol each at the end of the first slot of the first and sixth subframes. By locating the PSS and SSS at the center of the frequency band, the UE can detect these signals without knowledge of the actual system bandwidth. Note that the six REs at the top and bottom of the PSS and SSS allocation are actually not used, but form a small guard band to the adjacent PRBs.

In the uplink, there are two types of reference signals, as shown in Figure 17.13:

1. the Demodulation Reference Signals (DM-RS); and
2. the Sounding Reference Signals (SRS).

The DM-RS are primarily for channel estimation to aid coherent detection of the uplink transmission from a UE, and they are sent by the UE if and only if it is transmitting data in the PRB. The DM-RS are

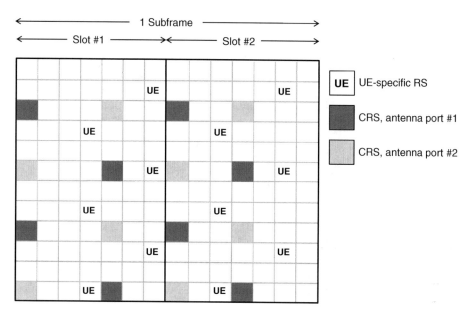

Figure 17.11 Downlink reference symbols

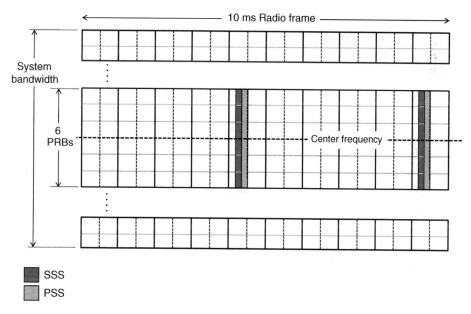

Figure 17.12 Downlink synchronization signals

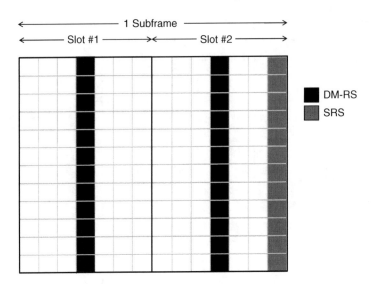

Figure 17.13 Uplink reference symbols

transmitted on the fourth symbol of each slot in a PRB for a system using the normal cyclic prefix, and occupy all the tones in that symbol.

In addition, the eNodeB needs to track the UE's channel for link adaptation and for channel-dependent scheduling, even when the UE may not have transmitted data recently. The SRS is configured by the eNodeB to be periodic, with a period that is a tradeoff between the overhead, the number of UEs that need to be sounded, and the speed of the UE. The faster the UE, the more frequent the SRS transmission needs to be for it to track the channel effectively. In Release 10, aperiodic or "one-shot" SRS transmissions are also supported. The SRS is located in the last symbol of the second slot of a PRB, as shown in Figure 17.13, and may occupy all the tones in that symbol, or every alternate tone to allow multiplexing of two UEs in the same PRB. In addition, up to eight cyclic shifts of the transmitted sequence are allowed to permit eight UEs to transmit on the same set of tones for further packing of UEs on the SRS symbols to minimize overhead.

17.2.4.3 Physical Layer Channels

In addition to subsets of REs set aside as reference and synchronization signals, there are other RE locations in the PRBs, which are assigned to various channels that are used to transport both user data and control information. In the downlink, three channels are available for transporting data (unicast, multicast, and broadcast):

- *Physical Downlink Shared CHannel (PDSCH).* This channel carries all the user data, and it is dynamically allocated by the scheduler and shared between all UEs with data to send. It makes up the majority of the REs in a PRB, as would be expected. After all, this is the channel that generates revenue for the service provider, and therefore a good design would minimize all the other overhead channels to maximize the resources available for the traffic channel. As shown in Figure 17.14, it occupies at least 11 of the 14 symbols in a PRB pair. We will discuss the different multi-antenna schemes used for data transmission in this channel in a separate section on transmission modes.

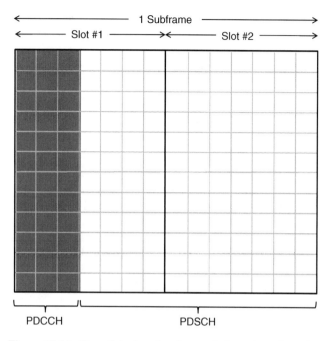

Figure 17.14 Downlink shared and control channel configuration

- *Physical Broadcast CHannel (PBCH)*. This channel carries essential system information that a UE needs to decode first before it can use the system, including the system bandwidth and frame number. As the system bandwidth is unknown before the PBCH is decoded, the PBCH is located in the central six PRBs of the band, as shown in Figure 17.15, in the four symbols that occur right after the PSS and

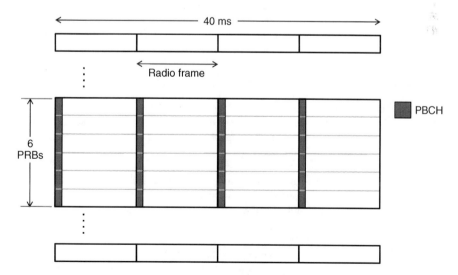

Figure 17.15 Structure of the physical broadcast channel

SSS in the first subframe. In addition, as no UE can connect to the system without decoding the PBCH, it uses a very robust repetition code and is transmitted over four radio frames spanning 40 ms, thereby providing time diversity against deep fades.

- *Physical Multicast CHannel (PMCH)*. Although listed in this section for completeness, the PMCH, used for multicast data to users was introduced only in Release 9 of the specification. We will look at the details of the multicasting feature of the LTE system in a subsequent section that discusses the evolution of LTE in Release 9.

In addition to the data channels discussed above, there are three control channels used by the physical layer that carry vital information required for the operation of the data channels, especially the PDSCH. These are discussed below:

- *Physical Downlink Control CHannel (PDCCH)*. This channel carries the dynamic allocation information generated by the scheduler, so that each UE knows which PRBs to receive in the downlink, or which PRBs to transmit on the uplink. It is located in the first few symbols of the PRB pair, as shown in Figure 17.14. In addition, this channel also carries the transport format used by the eNodeB in the downlink, or to be used by the UE in the uplink, and power control information. Depending on the type of traffic, and the number of active users, the number of symbols needed for the PDCCH may be varied between one and three symbols. It is important for this channel to be decoded correctly by the UE, otherwise all the data carried in the PDSCH or in the uplink is lost. As such, one strategy is to encode the information for all UEs in the most robust manner permitted, which uses eight times the bandwidth of the least robust scheme. However, this would imply that the number of UEs that can receive allocation is constrained. PDCCH capacity may not be a problem for data-heavy applications, in which very few users occupy all the bandwidth in a given subframe. However, PDCCH capacity may become a concern for VoIP, where there may be a large number of users, each requesting only a small amount of resources on a given subframe. In this situation, link adaptation can be used on the PDCCH, where users with strong signals are allocated fewer PDCCH resources with less robust coding to pack more allocations in a subframe.
- *Physical Control Format Indicator CHannel (PCFICH)*. In order to inform the UE how many symbols are used for PDCCH, and by extension the remaining symbols that decide the size of the PDSCH, a few REs are set aside for carrying these two bits of information that are needed to convey whether one, two, or three symbols are used for PDCCH. However, as the decoding of the entire subframe depends on the correct decoding of this critical configuration data, a very robust rate 1/16 block code is used and 32 encoded bits are used to map this channel to 16 REs that are all distributed within the first OFDM symbol in a subframe.
- *Physical Hybrid-ARQ Indicator CHannel (PHICH)*. The PHICH is used for transmission of HARQ acknowledgments in the downlink in response to uplink HARQ transmissions. It is important to encode this channel robustly, so that an ACK is not confused for a NACK or vice versa. Clearly confusing a NACK for an ACK is more serious, as it results in data loss, and so the error rate requirements for a NACK transmission are typically lower. The single bit of information is mapped onto 12 REs that are located in the PDCCH region of the subframe.

In the uplink, given the infrastructure-based topology, broadcast and multicast transmissions do not apply, and therefore there is only one uplink traffic channel. In addition, there is one control channel and one random-access channel, as described below:

- *Physical Uplink Control CHannel (PUCCH)*. The PUCCH is the only channel available in the uplink that exclusively carries control information. The three key pieces of information required at the eNodeB, and sent by the UE, are: (i) HARQ ACK/NACK information for PDSCH transmissions, (ii) Channel state information, including Channel Quality Information (CQI) of the downlink channel

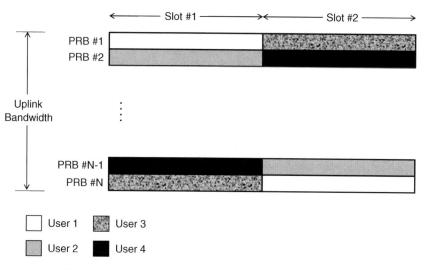

Figure 17.16 Structure of the physical uplink control channel

for downlink link adaptation, and (iii) scheduling requests when the UE has data to send. The PUCCH occupies the PRBs located at the edges of the configured uplink band, as shown in Figure 17.16, Within a subframe, a UE uses slot-hopping, and transmits one slot at either end of the band, as shown in Figure 17.16. One reason for this arrangement is to exploit frequency diversity and provide robustness to the control signals. Another important reason is to leave the band used for PUSCH unsegmented, so that SC-FDMA allocations which are required to be contiguous in frequency can utilize the entire remaining bandwidth for a single UE if needed.

- *Physical Uplink Shared CHannel (PUSCH)*. The PUSCH is the only traffic channel available in the uplink. In Release 8, only non-MIMO single antenna-based transmission is specified, so only a single spatial stream is allowed. As indicated earlier, this channel uses SC-FDMA transmissions to improve power efficiency at the UE. As shown in Figure 17.13, all the symbols and associated REs not used for the DM-RS and SRS are available for PUSCH transmission. Furthermore, the PRBs occupied by the PUCCH, as shown in Figure 17.16, are not available for PUSCH transmissions. In addition to user data traffic, the PUSCH is also used to carry control information in two different scenarios. The first situation occurs when a PUCCH transmission coincides with a PUSCH transmission from the same user. In such subframes, in order to maintain the SC-FDMA property of the uplink signal, the PUCCH data is multiplexed onto the PUSCH transmission. The other possibility is that the eNodeB can request aperiodic PUSCH-based CQI reports, which are allocated by the eNodeB using the PDCCH. Such reports can carry more detailed information regarding the channel state. This CQI reporting mode can be used when the number of active UEs is relatively small, thereby keeping the overhead of carrying control signaling on the PUSCH – used predominantly for data traffic – down to a reasonable level.
- *Physical Random Access CHannel (PRACH)*. When a UE does not have pre-configured periodic resources on the PUCCH, there is need for an alternate mechanism that permits the UE to communicate with the eNodeB. Such situations may occur either at the time of initial network entry, or during handover, or when the UE is not actively connected to the eNodeB (connected and idle states of a UE will be discussed in a subsequent section). These situations are common to all cellular systems and, therefore, a random access mechanism is a fundamental requirement in the uplink. In addition, random access is also useful to obtain timing synchronization with the eNodeB, so that the composite OFDM symbol formed by PRBs arriving from different UEs are all aligned to within the duration of the CP.

The PRACH uses six contiguous PRBs in one slot. There are two flavors of random access: contention-based and contention-free. The random access procedure consists of the UE sending a preamble, and the eNodeB responding to the preamble with timing corrections and allocation information if appropriate. If the preamble is selected at random, it can collide with that of another UE attempting random access, and contention resolution is required. On the other hand, in situations where the UE is connected to the network – for example, during handover – it can be asked to use a specific preamble, thereby avoiding contention.

17.2.4.4 Downlink Transmission Modes in Release 8

For PDSCH transmissions in Release 8, three antenna configurations are supported: one, two and four antenna transmissions. Although two-transmit antenna systems are the predominant configuration in most currently deployed LTE systems, use of four antennas will gain popularity as the need for greater capacity becomes apparent with the increase in demand. As such, in Release 8, several "transmission modes" have been defined to support different MIMO configurations. These are explained briefly below (please refer back to Chapter 14 for a general discussion of MIMO techniques and terms used below):

- **Transmission Mode 1**: single antenna transmission.
- **Transmission Mode 2**: transmit diversity for two and four antennas systems. In LTE, SFBC is used for two antenna systems. For four-antenna systems, a combination of SFBC and *Frequency-Switched Transmit Diversity (FSTD)* is used. A simple way to describe this scheme is that one pair of antennas uses SFBC for half the subcarriers, and the other pair of antennas uses SFBC for the other half of the subcarriers. The subcarriers are distributed between the two pairs of antennas, such that each group of four consecutive subcarriers is split into two separate groups and assigned to different antenna pairs.
- **Transmission Mode 3**: open-loop spatial multiplexing. For rank-1 transmissions, transmit diversity is used as described in mode-2. For rank-2 transmissions, spatial multiplexing is used with large-delay CDD.
- **Transmission Mode 4**: closed-loop spatial multiplexing. Codebooks have been defined in the standard for both two-antenna and four-antenna systems. The precoding is based on these codebooks for both rank-1 and rank-2 transmissions.
- **Transmission Mode 5**: multi-user MIMO (MU-MIMO). In release 8, there is support for a basic version of MU-MIMO that uses the same codebooks as in transmission mode-4. Each user in the pair of MU-MIMO users receives a rank-1 transmission. More sophisticated MU-MIMO techniques are introduced in subsequent releases.
- **Transmission Mode 6**: closed-loop spatial multiplexing restricted to rank-1 only. This mode is a subset of mode-4 with applicability to low-SINR UEs, where rank-2 transmissions are unlikely, and therefore feedback of the rank indicator from the UE can be avoided.
- **Transmission Mode 7**: UE-specific beamforming. In this mode, UE-specific RS are used for proprietary beamforming techniques. One possible use of this mode involves the use of a closely spaced linear array of antennas to form narrow beams to UEs, based upon angle-of-arrival estimation processing.

17.2.5 Layer-2 and Cross-Layer Algorithms

In this section, we will discuss in detail the function of the three sub-layers that comprise layer-2 of the LTE protocol stack:

- Medium Access Control (MAC).
- Radio Link Control (RLC).
- Packet Data Convergence Protocol (PDCP).

In addition, there are several algorithms and procedures that operate at the eNodeB and UE that can be loosely associated with the MAC, although these are better characterized as *cross-layer*, as they deal with the optimization of the physical link by using knowledge of the state of the physical layer in procedures implemented at the MAC layer. These algorithms are usually left to vendor implementation and differentiation, and they form the basis for the innovative research and development that has resulted in LTE being a major step forward over the previous technologies in terms of performance. We will discuss these algorithms that form the "brains" of the MAC layer after a brief description of the basic structure of the MAC layer and its functionality.

17.2.5.1 Medium Access Control (MAC)

The MAC sub-layer in layer-2 is asymmetric because of the centralized cellular architecture of the LTE system. The implementation of the MAC at the eNodeB is very complex, because all the algorithms and resource allocations functions are centrally located there. On the other hand, in order for the eNodeB algorithms to function reliably, there needs to be a clear specification of the operation of the MAC at the UE, on which the eNodeB can depend. As a result, the MAC layer at the UE is relatively simple, with precise definition of the procedures to be used. In this section, we will discuss the high-level functions of the MAC and the UE procedures, followed by the algorithms used at the eNodeB in the next section. The key functions of the MAC layer are as follows:

- Provide logical channels for use by the higher layer (RLC).
- Map the logical channels to transport channels provided by the physical layer.
- Multiplex SDUs from different logical channels to a transport block sent on a transport channel.
- Demultiplex SDUs from a transport block received to different logical channels.
- HARQ operation.
- Transport format selection (*eNodeB only*).
- Prioritizing between different UEs, and between flows of a given UE (*eNodeB only*).
- Logical channel prioritization based on received scheduling grant (*UE only*).
- Reporting of scheduling information (*UE only*).

We will examine some of these key functions in the following subsections.

Logical Channels and Transport Channels
Logical channels are provided by the MAC layer as a service to the layer above (RLC). They can be classified as control or traffic channels, as shown in Table 17.3.

The control channels are used by the control plane protocols implemented in the RRC (layer-3), and the traffic channel is used for user-plane data transfer. In order to provide these logical channel services and, in addition, for operation of the MAC itself, the MAC layer uses transport channels that are provided by the physical layer as shown in Table 17.4.

Note that these transport channels are unidirectional, as indicated in the table. The shared channels carry the bulk of the data, and they may be used for both control signaling and user traffic. Figures 17.17

Table 17.3 LTE logical channels

Logical channel	Uplink	Downlink
Broadcast control channel (BCCH)	No	Yes
Paging control channel (PCCH)	No	Yes
Common control channel (CCCH)	Yes	Yes
Dedicated control channel (DCCH)	Yes	Yes
Dedicated traffic channel (DTCH)	Yes	Yes

Table 17.4 LTE transport channels

Transport channel	Uplink	Downlink
Broadcast channel (BCH)		Yes
Paging channel (PCH)		Yes
Random access channel (RACH)	Yes	
Downlink shared channel (DL-SCH)		Yes
Uplink shared channel (UL-SCH)		Yes

and 17.18 summarize the mapping of logical channels to transport channels, and transport channels to physical channels for the downlink and the uplink, respectively. Again, it is noteworthy that the physical shared channels in the uplink and downlink are the most versatile and are used to carry different types of transport channels for the MAC layer.

HARQ Operation
As discussed in Chapter 14, HARQ distinguishes itself from ARQ by its ability to retain erroneous packets and combine them with retransmitted copies to greatly improve the probability of decoding the packet correctly. The actual protocol used to schedule transmissions, receive the ACK/NACK and schedule retransmissions follows a "stop-and-wait" protocol. This latter terminology is somewhat dated, and is more accurately described as a parallel-process stop-and-wait protocol. In a traditional stop-and-wait approach, the sender stops and waits for an ACK after sending a packet. The next packet (new or retransmission) is sent only after the ACK is received. This approach is inefficient, because the link is not utilized when the sender stops and waits.

In order to keep the link utilized, sufficient stop-and-wait processes are active in parallel so that either a new transmission or a retransmission can be scheduled on every subframe. The number of parallel processes depends on the ACK reception and processing delay which, in the case of LTE, is eight frames. In addition, for rank-2 transmissions, two spatial streams use a separate HARQ processes. Therefore, a total of 16 HARQ processes are required (and are permitted in the standard) for concurrent operation. The operation of parallel HARQ processes with the eight-frame ACK processing delay is depicted in Figure 17.19.

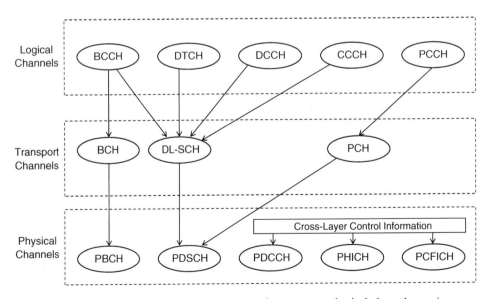

Figure 17.17 Downlink logical-to-transport and transport-to-physical channel mapping

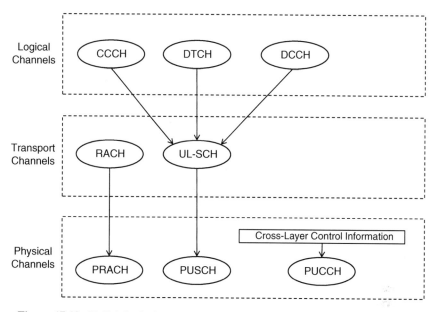

Figure 17.18 Uplink logical-to-transport and transport-to-physical channel mapping

Based on the timing of the retransmission and recombination strategy, HARQ in LTE is classified as follows:

- *Asynchronous* vs. *synchronous*. After an eight-frame delay, the ACK has been received and processed at the sender, and there is an option to then send a new transmission or retransmission. In the downlink, there may be special frames called MBSFN frames (which will be described later) that occur on the eighth frame and prevent the retransmission from being scheduled. As a result, the DL retransmissions have to be explicitly scheduled, since they may not follow an eight-frame cycle and are therefore deemed *asynchronous*. In the uplink, there are no special frames, so *synchronous* operation is possible. If a retransmission – signaled by the reception of a NACK – is required, it is implicitly scheduled and

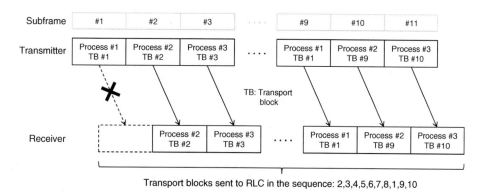

Figure 17.19 Parallel HARQ process operation

autonomously sent by the UE on the eighth frame following the original transmission. Synchronous operation saves the overhead required to signal the allocation for a retransmission.

- *Adaptive* vs. *non-adaptive*. In LTE, both modes of soft combining, chase combining and incremental redundancy (IR), are supported. Recall, as discussed in Chapter 14, that the same encoded packet is sent again when chase combining is used. In the non-adaptive mode for HARQ in LTE, in addition to the same encoded packet format as in chase combining, there is an additional constraint, in that the same frequency resources used in the original transmission must be allocated for the retransmission. In the adaptive mode, incremental redundancy can be used, and the actual location of the retransmission in terms of frequency resources can be modified. Both modes are supported in the uplink and the downlink, and the choice is left to the implementation. Adaptive HARQ can provide better performance at the cost of increased complexity and potentially greater signaling overhead.

If the link conditions are very poor, then decoding may still fail after repeated attempts. In this situation, the eNodeB can signal a discard of the old buffer, and a new PDU is attempted on the same HARQ process.

Scheduling Procedures

Another key function of the MAC sub-layer is to manage the scheduling procedures so that the scheduling information is exchanged between the eNodeB and the UE in an efficient manner. Note again that the actual algorithms used for scheduling are not specified, but these procedures convey information about DL allocations, UL allocations, how an UL allocation should be partitioned between multiple uplink flows, and how the UE should convey information about its bandwidth requirements to the eNodeB.

DL grants: the DL grants are signaled on the PDCCH as mentioned before, and the grant is valid for the same subframe that carries the PDCCH message. There are several formats used for PDCCH messages, depending on the amount of information to be conveyed. All formats carry, at a minimum, the following information:

- Allocation of PRBs: these can be either contiguous or noncontiguous, depending on whether the eNodeB uses frequency selective scheduling or not.
- HARQ process number and a "new data" indicator that is used to flush the old data from the HARQ buffer if needed.
- Modulation and coding scheme (MCS), which contains information about the modulation, the code rate, and the transport block size. As we shall see, the MCS is a direct output of the link adaptation algorithm.
- Multi-antenna transmit format: this includes the number of layers used and precoding information.

UL grants: the UL grants are also signaled on the PDCCH. However, they apply on the subframe that occurs four subframes after the one carrying the PDCCH message. For TDD, if the fourth subframe is not an uplink subframe, then the grant applies to the first uplink subframe following the fourth subframe after the grant is received. In addition, the grant in the uplink is per terminal, not per logical channel that has data in the uplink buffer. This type of signaling limits the overhead on the PDCCH. The actual grant information carried in the uplink PDCCH message is very similar to that in the downlink, with the exception that only contiguous allocations are allowed in the UL to maintain the SC-FDMA properties of the signal. Also, in Release 8, only single antenna transmission is supported in the uplink, so no multi-antenna information is needed.

Uplink logical channel prioritization: as the uplink grants are per terminal, in order to ensure uniform logical channel allocation behavior at the UEs, the standard specifies the algorithm to be used to partition the allocated bandwidth between the various uplink logical channels. Each logical channel is served according to its priority. However, an absolute priority is not used, because a high demand channel with high priority can starve other lower priority channels. To avoid starvation, an additional parameter called prioritized bit rate (PBR) is used in conjunction with the priority as follows. Each logical channel is served in order of priority, with the allocation size set based on the PBR. If all channels are served, and there is bandwidth left over, then the channels are served again, but this time it is based on absolute priority. In this manner, each logical channel is assured of getting bandwidth equal to its PBR at a minimum.

Scheduling requests and buffer status reports: in order for the eNodeB scheduler to assign resources to a UE, the UE must signal that data has arrived in its buffer. In order to do so, it uses a single bit flag that is assigned on the PUCCH on a dedicated basis to the UE. This flag is called a scheduling request (SR) and it is sent any time new higher-priority data arrives, or if no grant has been received since the last SR. Another option to request a grant from the eNodeB scheduler is to use the random access channel (RACH). However, UEs may face contention and backoff if many UEs that frequently have data to transmit resort to using the RACH. Therefore, allocating dedicated resources for the SR is advantageous in situations where the probability of SR collision is high when the RACH is used. On the other hand, if there is a very large number of UEs with only infrequent requests for bandwidth, dedicated PUCCH resources for SRs may not be available for all UEs, so they can fall back on the RACH mechanism for SRs. In addition to the SR, it is important for the eNodeB to be informed of the actual amount of data in the buffers of various logical channels. To get this information, the UE sends a buffer status report as part of the MAC PDU header periodically, or each time there is a need to add padding data on the PDU.

17.2.5.2   Cross-Layer Algorithms

In this subsection, we discuss cross-layer algorithms that operate between the MAC layer and the physical layer. These algorithms play a major role in optimizing the performance of the LTE system. In general, the details of these algorithms tend to be closely guarded secrets by infrastructure vendors, as their performance characteristics are used as a differentiating factor between their systems.

Power Control

Although power control may be considered a physical layer procedure, it is closely tied with the link adaptation and scheduling algorithms, and thus it merits discussion in this section. Unlike UMTS, which was designed initially for fixed-rate voice, LTE has been designed from the beginning to be a high-throughput system for general data traffic. In UMTS, power control was used to combat fast fading and to maintain an almost constant SINR at the receiver to support a fixed data rate. In addition, the CDMA air interface generates interference both from within and without the cell. LTE differs from UMTS in both these respects, and the emphasis is on maximizing data throughput and minimizing other-cell interference.

Downlink: in the downlink, the objective for the use of the PDSCH is to maximize data throughput and, as such, rate control by means of link adaptation is better suited for the job than power control. On the other hand, in the PDCCH, where important control messages are sent, the goal is maximizing coverage and minimizing error rates. As such, a combination of link adaptation and power control may be used for the PDCCH. The idea of using power control in the PDCCH as follows. Given a total transmit power budget, the power assigned to users with low path loss can be reduced, and this surplus power can be assigned to users facing high path loss conditions at the cell edge. The power control mechanism can be used in conjunction with suitable adjustments to the aggregation level (or effective code rate) to achieve the desired error rate on the PDCCH.

Uplink: in the uplink, the situation differs from the downlink in a couple of important respects:

1. the transmitters are power-limited UEs; and
2. the transmitters are distributed over the area of the cell, unlike the eNodeBs.

This results in the interference environment being more dynamic, and some measure of power control can help to control the interference. As such, the LTE standard specifies a power control procedure for the uplink that can be operated either as a purely open-loop system, or as a closed-loop system with an open-loop component. The basic idea is to set the UE transmit power such that the eNodeB receiver achieves a certain target SINR. The basic equation that is applied is:

$$\text{Target SINR(dB)} = \text{UE transmit power(dBm)} - \text{path loss(dB)} - \text{interference-and-noise power(dBm)}$$

In this equation, the target SINR is lumped together with the interference-and-noise power into one system parameter called P_0, which is broadcast by the eNodeB. In addition, two terms are added to adjust the target SINR for different MCS values and different allocation bandwidths. Note that higher target SINRs are needed for higher MCS values, and the transmit power is divided equally among the PRBs in the allocation. This gives us the equation:

$$\text{UE transmit power} = P_0 + \text{path loss} + \text{MCS factor} + 10 \log_{10}(\text{number of PRBs})$$

Furthermore, in order to reduce the interference from UEs at the cell-edge, it may be desirable to reduce the transmit power in proportion to the path loss; this is called *fractional power control*. To achieve this, the path loss (PL) is reduced by a factor α, where α takes on values between 0.7 and 1. Finally, in order to introduce closed-loop control, a UE-specific offset term is introduced so that it can be controlled by the eNodeB. The final equation is:

$$\text{UE transmit power} = P_0 + \alpha \, \text{PL} + \text{MCS factor} + 10 \log_{10}(\text{number of PRBs})$$
$$+ \text{accumulated closed-loop offsets}$$

This generic equation allows an eNodeB to exercise as much or as little control over the UE transmit power as it chooses. On one end, the eNodeB may set $\alpha = 1$ and not apply any closed-loop offsets for a non-fractional open loop power control system. At the other extreme, the eNodeB may use fractional power control with closed-loop adjustments. The adjustments from the eNodeB are sent on the PDCCH, and they can be based on uplink measurements on received subframes.

Note that the equation above is applicable to power control for both the PUCCH and PUSCH except that, for the PUCCH, the number of PRBs is always set to 1. In addition, the PUCCH algorithm also checks to make sure that the power allocated does not exceed the per carrier maximum power, P_{CMAX}. Although, in Release 8, no PUCCH transmission is allowed when PUSCH needs to be sent, simultaneous PUCCH and PUSCH transmissions are allowed in Release 10. In this case, the PUSCH algorithm checks to make sure it does not exceed P_{CMAX} minus the power assigned to the PUCCH, which ensures priority of control signaling on PUCCH over data transmissions on PUSCH.

Link Adaptation

Link adaptation is the process by which the eNodeB selects transmission parameters to match link conditions, and thereby extract the optimal performance from the link. In essence, the link adaptation algorithm selects the optimal transport block formats and the MIMO transmission parameters. The MIMO transmission parameters include the transmission modes discussed earlier and, if appropriate for the transmission mode, the number of spatial layers, or rank of the transmission, and the precoding weights. The information that the link adaptation algorithm needs to make these decisions is obtained using the channel state feedback. In an FDD system, the uplink and downlink are separated in frequency, so information about the downlink channel is only available to the UE, and the eNodeB must depend on measurements at the UE to get channel state feedback. In TDD, it is possible for the eNodeB to use the reciprocity of the channel to obtain some channel state information, although it can still rely on the UE for feedback.

There are three key pieces of information that the UE provides to the eNodeB by means of measurements:

- *Channel Quality Indicator (CQI)*. The channel quality indicator is not actually a measure of the signal strength or SINR, but is an indication of the highest MCS that the UE can decode successfully with a block error rate of 10% or less. The reason for this metric is that this information implicitly includes the UE receiver capabilities and is a true reflection of the rate that the UE can support, given the channel conditions. It is also noteworthy that a somewhat high block error rate has been chosen as the target for MCS selection. The reason for this error rate lies in the way HARQ operates. Recall that the probability of error drops significantly as the number of copies of retransmitted signals available to combine increases. As a result, keeping the error rate on the first transmission high (by picking a suitably high-rate MCS) allows the overall throughput to be high even after retransmissions have been taken into account. The CQI can be either *wideband*, with one value of CQI computed for the entire band, or *subband*, where the bandwidth is divided into subbands with a few PRBs per subband, and a separate CQI is reported for each subband. Subband CQI is necessary for frequency-selective scheduling.
- *Rank Indicator (RI)*. The rank indicator is used by the eNodeB to decide whether spatial multiplexing should be used or not. The UE uses the knowledge of its channel conditions to decide which mode is likely to support the highest overall throughput – a single-layer transmission that supports a relatively high rate, or a dual-layer transmission, where each layer is possibly at a lower rate than that of the single layer, but whose sum exceeds the single-layer rate. Note that if rank-2 transmission is supported, then it measures the CQI for each stream and reports two CQI values.
- *Precoding Matrix Index (PMI)*. The precoding matrix index reported by the UE corresponds to the rank indicator reported by the UE and, therefore, is suitable for both rank-1 and rank-2 precoding in the closed-loop MIMO mode (transmission mode 4). Just like the CQI reports, PMI reports can also be wideband or subband, depending on the MIMO algorithms implemented at the eNodeB.

It should be noted that it is not mandatory for the eNodeB to apply the channel state information from the UE "as-is" when choosing the transport block format and MIMO parameters. In fact, the CQI is usually "smoothed out" by an outer-loop link adaptation algorithm that tracks the block error rate to make adjustments to the instantaneous CQI reported by the UE. This is akin to the outer-loop power control algorithm discussed in the UMTS context. Furthermore, depending on the type of traffic and amount of data to be sent, the eNodeB may also overrule the rank recommendation from the UE (e.g., if small VoIP packets are being sent, rank-1 transmissions may be sufficient).

There are procedures defined for the actual feedback mechanism from the UE. As noted earlier when discussing uplink physical channels, there are two types of feedback: periodic feedback, that is usually sent on the PUCCH; and aperiodic feedback, that is sent on the PUSCH. The periodic feedback is sent on the bandwidth-limited PUCCH so, therefore, can carry less information than the PUSCH. One strategy is to augment the information from the periodic feedback with aperiodic feedback that can carry more information. Another strategy may be to use PUCCH for VoIP-like flows where the data throughput requirements are low but a large number of users need to be supported. Aperiodic PUSCH feedback can be limited to a few heavy data users where optimal link adaptation by getting granular information results in more efficient use of the link.

Scheduling

The scheduler is considered to be the "brains" of the eNodeB, because it is responsible for coordinating the operation of all the uplink and downlink channels. A typical (but not exhaustive) list of functions of a scheduler is:

- Decide the MCS and MIMO configuration based upon UE channel state feedback.
- Select the appropriate transmission mode.
- Apply the appropriate scheduling algorithm for different QoS requirements.
- Manage HARQ retransmissions.
- PDCCH power control and uplink power control.

Based on the traffic intensity, user population size, and mobility conditions in the cell, the eNodeB must decide how the periodic and aperiodic feedback mechanisms are to be configured for different sets of UEs in the cell. In a highly loaded cell with a small set of users and a large volume of traffic, it is possible to extract significant gains from frequency-selective scheduling. In LTE frequency-selective scheduling, users are assigned those PRBs in a subframe where their expected data rate is highest. This will require using a feedback mode where subband CQI is reported.

On the other hand, if there is a large number of users, each with moderate amounts of bursty data, it may be hard to justify the high overhead of subband-CQI reporting, and frequency-diverse scheduling may look more attractive. In frequency-diverse scheduling, wideband CQI reports are sufficient, and the PRBs assigned to the user can be distributed over the entire bandwidth in order to maximize frequency diversity gains.

If the cell is supporting a large number of VoIP users, then the capacity of the PDCCH channel may not be sufficient to support the large number of UL and DL grants required every subframe. LTE supports an alternative to dynamic scheduling called semi-persistent scheduling (SPS), where the user allocation can be set to repeat periodically and signaled only once. This type of scheduling is very suitable for use with VoIP frames, which are of known size and arrive with fixed periodicity. For the duration of a talk spurt, only one message needs to be sent on the PDCCH to start the periodic allocation pattern, instead of a PDCCH message every 20 ms which corresponds to the VoIP frame period. Note that asynchronous retransmissions in the downlink have to still be signaled dynamically, but SPS may be used for synchronous retransmissions in the uplink.

The scheduling strategy may not always hinge on maximizing the cell capacity. It is possible to design a scheduling strategy around providing an improved user data rate at the edge of the cell, where the SINR is poor. This would make sense if, for example, one wants to ensure a certain QoS for all users in the network. LTE is usually deployed as a *reuse-1* system, where the same frequency is used in all cells, because the system is robust enough to support cell edge users by adapting their link to match the prevailing conditions. However, if it is desirable that the link should support a certain minimum data rate, then the strategy should be to reduce interference at the cell edge. This can be done by coordinating transmission between different eNodeBs, such that portions of the bandwidth are reserved for cell edge users and these portions of the band do not overlap between neighboring cells, thereby reducing the interference for cell-edge users. This mechanism is called *Inter-Cell Interference Coordination (ICIC)*.

Messages have been defined in the standard that eNodeBs can exchange over the X2 interface to convey information about the PRBs on which they intend to transmit, and also regarding the interference they are perceiving on different PRBs. These messages can be used for coordination purposes, but the exact algorithm to be used is not specified and is left to the vendor's implementation.

In summary, the scheduler is a very complex entity in an eNodeB, and has the task of optimizing the system performance over various parameters. An overview of the types of options available to the scheduler are:

- Frequency-selective vs. frequency-diverse scheduling.
- Dynamic scheduling vs. SPS.
- Closed-loop power control vs. fractional open-loop power control.
- Open-loop MIMO vs. closed-loop MIMO (transmission modes 3 and 4).
- Wideband feedback vs. subband feedback.
- Periodic feedback vs. aperiodic feedback.
- ICIC or full reuse.

Note that even within a single cell, the scheduler may have to use different strategies, depending on the load, user mobility, and channel conditions. Furthermore, some of these strategies may have to adapt dynamically, making the scheduler a fertile area for innovation.

17.2.5.3   Radio Link Control (RLC)

Very much like the RLC sublayer in UMTS, the RLC, which is a sublayer of layer-2 in LTE, also operates in three modes: transparent (TM), unacknowledged (UM), and acknowledged (AM).

The main functions of the AM RLC are as follows:

- Segmentation and concatenation of higher layer SDUs into RLC PDUs.
- In-sequence delivery and duplicate detection.
- Retransmission of erroneous PDUs.
- Reassembly of RLC PDUs into SDUs for the higher layer.

Note that there is no retransmission of erroneous PDUs for UM-RLC. In transparent mode, the RLC is just a pass-through and does not introduce any overhead (or do anything useful to the flow, for that matter).

Unlike UMTS, where the RLC function and HARQ function are in separate nodes (RNC and nodeB), in an LTE system, the RLC and HARQ are implemented in the eNodeB, along with the scheduler. Therefore, the RLC PDU sizes are dynamic and are decided by the scheduler on the basis of the transport block sizes chosen by the link adaptation algorithm. As shown in Figure 17.20, an RLC PDU may be constructed by either segmentation of an SDU in multiple chunks, or by concatenation of several SDUs. In the figure, we see a general case, where both concatenation and segmentation occur in the same PDU.

Although, at first glance, it may seem that the RLC and HARQ mechanisms are both achieving the same goal, this is not so. HARQ is first responsible for maximizing the throughput of the link by starting out with a high error rate on the first transmission, then successively improving upon it by means of a fast feedback mechanism to request retransmissions. The residual error after retransmissions is low (typically 0.1–1%), but not low enough to allow TCP to operate smoothly. This is where the RLC protocol steps in to clean up the remaining errors, so that TCP perceives an almost error-free channel, with packet error rates well less than 0.01%. The RLC protocol feedback is much slower, but it needs only to correct errors remaining from the HARQ operation, so the feedback delay is only an occasional cause for increased latency in the actual end-to-end user experience. Note that AM-RLC is used with TCP-based flows (e.g., uploads and downloads) that are error-sensitive and latency-insensitive services. On the other hand, UM-RLC is usually used with UDP-based flows (like VoIP) that can tolerate a small amount of errors in exchange for lower latency.

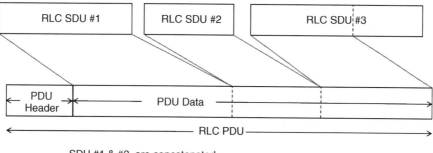

SDU #1 & #2 are concatenated
SDU #3 is segmented

Figure 17.20   RLC PDU construction via segmentation and concatenation

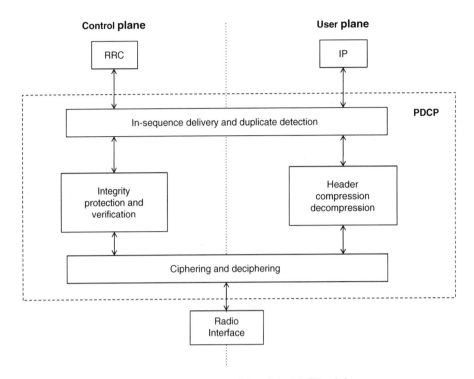

Figure 17.21 Functional view of the PDCP sub-layer

17.2.5.4 Packet Data Convergence Protocol (PDCP)

The PDCP layer acts as an interface between the user plane IP data and the control plane RRC signaling messages coming down to the access layer, as shown in Figure 17.21. As such, it performs the following functions at this interface:

- Applies the RObust Header Compression (ROHC) protocol to compress the higher layer headers and performs decompression at the receiving end. For example, in VoIP flows, several short packets are sent repeatedly. These packets have headers with IP source/destination information, UDP headers, etc. that do not change from packet to packet. Header compression greatly improves the efficiency of the air interface by eliminating transfer of repeated higher-layer headers in such flows.
- Maintenance of sequence numbers for in-sequence delivery of packets and elimination of duplicates. The PDCP plays a key role in ensuring lossless handover by reordering packets, and by retransmitting packets if necessary. It also discards packets that have timed out, preventing them from congesting the air interface when they are no longer needed.
- Ciphering, deciphering, integrity protection and verification of data.

17.2.6 Layer-3: Radio Resource Control (RRC)

As in the case of WCDMA-based systems, LTE systems also have a third layer above the physical medium that contains the Radio Resource Control (RRC) functions, solely dealing with control-plane protocols. Some of the key functions implemented here include connection management, mobility, and security. After we describe an overview of the functions in this layer, we will delve into two important topics – RRC states and DRX – which affect the battery life of devices and the mobility functions.

17.2.6.1   Overview

The RRC function is located at the eNodeB, and it is responsible for control signaling between the E-UTRAN and the UE. The following key procedures are handled by the RRC:

- Broadcast of system information required by the UE to communicate with the network. This information is needed not only by UEs entering the network, but also by UEs that are already connected to the network
- Connection management: as discussed earlier, signaling and data radio bearers are established before a UE can exchange information with the network. These bearers are set up, maintained, and torn down through signaling procedures managed by the RRC function.
- Mobility: RRC handles all mobility-related procedures, which includes cell selection and reselection when the UE is not active, and handovers when the UE is active. Mobility between LTE and other radio interface technologies are part of the RRC function. The measurement configuration and reporting that is required from the UE for mobility decisions are also managed by the RRC.

17.2.6.2   RRC States

Much like the connected mode and idle mode for UEs in a UMTS system, the RRC in LTE is in one of two states: RRC_CONNECTED or RRC_IDLE. In contrast, there are no sub-states based on the transport channel being used (DCH/FACH/PCH), as was the case in UMTS. However, these RRC states do play a very key role in determining the power consumption of the UE and, therefore, the battery life of the device. As such, we will take this opportunity to discuss the different discontinuous reception (DRX) cycles as they relate to these RRC states. See Figure 17.22 for an overview of the different DRX and RRC state transitions in the connected mode.

In the RRC_CONNECTED state, the UE has an RRC context established with the network, which implies that the parameters needed to establish a link are known to both sides. As a result, data transfer can occur immediately without any need for additional control-message exchanges. In this state, the UE can conserve power by judicious use of DRX cycles. There exists a trade-off between latency and power savings. When the UE is in active data transfer mode, it monitors the PDCCH on each subframe and, therefore, its radio is effectively turned on all the time. However, when there is a break in the data transfer, the UE can conserve power by turning off its radio, according to a DRX cycle that is known to both the eNodeB and UE. During the DRX cycle, the UE turns on its radio for a certain period of time when the eNodeB can signal additional data transfer, and it is asleep for a certain period of time when the eNodeB cannot immediately contact the UE.

Normally, for traffic like web browsing or file downloads, a longer sleep-cycle is not a problem from a latency perspective because, once a transfer is completed, there is usually a gap of several hundreds of milliseconds, or even several seconds before the next transfer begins. However, for low-rate, latency-sensitive flows like voice or video telephony, the link may need to carry short bursts of data repeatedly at periodic intervals. In this case, the sleep cycle needs to be in sync with the periodicity of the voice codec for an optimal trade-off between latency and power consumption. As such, there are two DRX cycles– the long DRX and short DRX cycles, as shown in Figure 17.22. After the active state, the UE goes into a short DRX cycle to see if there is any periodic transfer and, if there is none, it proceeds to the long DRX cycle and then eventually to the RRC_IDLE state if there still no activity. If any data arrives during one of the DRX cycles, the UE goes back to the active state.

Note that even in the RRC_IDLE state, there is some power consumption, as the UE needs to wake up periodically to listen for paging messages. Paging messages either indicate the presence of downlink data for the UE, or a change in the system information which the UE must receive to be in sync with the

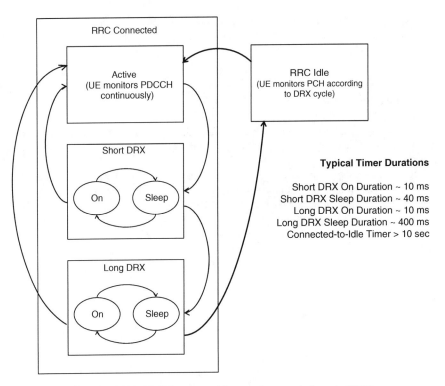

Figure 17.22 RRC State transitions and connected-mode DRX

eNodeB. However, this power consumption is much lower than when the UE is in the active state. Also note that moving the UE out of active state very quickly may reduce power consumption, but can have an adverse impact on both the latency and the number of control signaling messages that are exchanged to bring the UE back to the connected state. Therefore, it is important to tune the timers so that the correct balance is struck between latency, power consumption, and signaling overhead.

17.2.6.3 Mobility

Depending on the RRC state of the UE, the mobility is either controlled by the UE or the eNodeB. In the idle state, mobility is controlled by the UE by the process of cell selection and reselection. The UE has the task of selecting between different cells on the basis of frequency and signal strength. To select between frequencies, the UE must adhere to different network-wide priorities configured for different frequencies by the eNodeB via broadcast messages. If the priorities are equal, signal strength is used. In addition, the UE capabilities are also taken into account by the network to configure UE-specific priorities. Finally, the UE also selects between different RATs, based on priorities configured by the network.

In the connected state, the final decision on mobility rests with the eNodeB, but the process is assisted by measurements from the UE, as shown in Figure 17.23. In Release 8 of LTE, there are no soft handovers as in UMTS, so the UE is connected to only one eNodeB at any given time. The eNodeB decides, on the basis of measurements triggered from the UE, when the UE must to be handed over to a suitable "target eNodeB". At this time, the eNodeB, also called the "source eNodeB", initiates the process by transferring the RRC context of the UE.

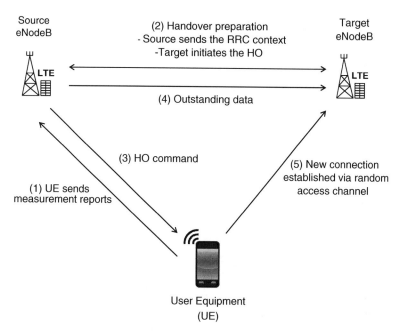

Figure 17.23 LTE hard handover procedure

When the target eNodeB is ready, it initiates the handover and the source eNodeB sends the handover command message to the UE, along with any system information to be used in the target cell. At this point, the source eNodeB also transfers any outstanding data in the buffers to the target eNodeB, and stops communicating with the UE. The UE then proceeds to connect to the target eNodeB by performing a random access procedure and establishing a new RRC connection. Communication between the source and target eNodeB can occur over the S1 interface or the X2 interface but, from the perspective of the UE, the process looks the same in either case.

17.3   Release 9: eMBMS and SON

After the completion of Release 8 in 2008, 3GPP RAN working groups continued to evolve the standard, and they finished the next major release, Release 9, in 2009. The major feature added in Release 9 was evolved Multimedia Broadcast Multicast Service (eMBMS), which allowed data transfer to move away from a strictly unicast format to a more efficient multicast or broadcast format for content that lent itself to this type of transfer. We will look at this feature in more detail in this section.

Another feature that is worth discussing in the context of Release 9 is *Self Optimizing Networks (SON)*. Although there were some SON-related features developed in Release 8, they pertained mainly to self-configuration of the eNodeB. Configuration of the eNodeB parameters, such as cell IDs, IP addresses, eNodeB neighbor lists, etc. could be automated as a result. The major emphasis on optimization in SON first occurred only as part of Release 9. We will provide an overview of the SON features later in this section.

In addition, support was added for an additional MIMO mode, transmission mode 8. Recall that support for a single-layer transmission, with non-codebook-based beamforming, is available via transmission mode 7. The UE-specific demodulation reference signals (DM-RS) specified for this mode were enhanced to support two mutually orthogonal codes to separate the DM-RS, thereby supporting the ability to transmit two beams to the UE. This mode is sometimes called dual-layer beamforming. Finally, location-based services were given a boost via added support for determining UE location. A typical implementation measures the time difference of arrival of signals from at least three different eNodeBs to determine the UE position via triangulation methods. In order to measure these time differences accurately, special reference signals were introduced in Release 9.

17.3.1   Evolved Multimedia Broadcast Multicast Service (eMBMS)

In situations where there is demand for the same information by a large UE population, using unicast links to transmit this information becomes inefficient, so methods to combine these links into bundles, or even to broadcast this information to all UEs, start to become very attractive from a system capacity standpoint. The demand for common information streams can arise in several situations:

- Live event broadcast: in venues for sporting events, there can be demand for live video that is presented from different camera angles, and also for time-sensitive but delayed content that may represent instant replays and promotions, or even for advertisements. In such densely populated venues, there is insufficient bandwidth to support on-demand, unicast delivery of such content, and eMBMS can be used very effectively. Note that, in general, not only sporting events but also live telecasts such as keynotes, public service announcements, etc. can benefit from eMBMS.
- Synchronized download of content: in several instances, content that has a large subscriber base is made available at a predetermined time, and this can result in a flurry of unicast download requests. Examples of such content include operating system updates, the availability of a much-awaited new version of a gaming app, publication of a new book by a popular author, new music releases, or even simple magazine subscriptions. Such content can be bundled into a synchronized "push" transfer to mobiles via eMBMS.
- Off-peak network offloads: customers can be incentivized to accept delayed download of content, such as TV shows or movies, and this content can be sent via eMBMS if a sufficient percentage of the user base all accept off-peak delivery of such content to their devices.

In the following, we describe how eMBMS is implemented in an LTE system, and how it provides system benefits beyond just reduction of unicast traffic. Recall that the cyclic-prefix in OFDM signals is used to combine multiple propagations paths from a single eNodeB with a simple equalizer. The idea used in eMBMS is to transmit the same content in a synchronized manner, over the same frequency carrier, from all eNodeBs in an area where such content is in demand. The multiple paths now arriving at a UE may actually be from different eNodeBs, but the same principle applies for equalization. The only difference in this situation is the potentially large difference in delay between the various paths, dictating that a larger CP be used. Such eMBMS transmissions can occur in six of the ten subframes in a frame and, in these subframes, the larger CP results in a decrease in the total number of non-PDCCH symbols from seven to six. This frame and subframe arrangement for FDD is shown in Figure 17.24.

The subframes on which eMBMS transmissions are permitted are called multicast broadcast single frequency network (MBSFN) subframes. MBSFN subframes can be configured to be carried on any or all of the six subframes shown in Figure 17.24. The other subframes carry special signaling channels and synchronization signals, so they are not available for use by eMBMS. Note that MBSFN subframes were defined in Release 8, and they represent special subframes that can be configured in a backward compatible manner for uses that future releases of the standard may specify. Therefore, a network can use eMBMS without any disruption to Release 8 UEs in the network, by scheduling such legacy UEs only in non-MBSFN subframes.

For the purposes of synchronized broadcast on a single carrier, the eNodeBs can be grouped into an *MBSFN area*, which can functionally represent the region in which it is desired to broadcast a certain type of content. Different MBSFN areas can overlap, and each eNodeB can belong to as many as eight areas. In

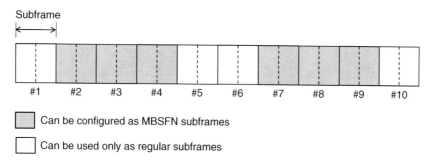

Figure 17.24 MBSFN subframe arrangement

terms of network architecture, three new *logical* nodes are introduced to support eMBMS, as shown in Figure 17.25:

- *Broadcast Multicast Service Center (BMSC)*. This node supports membership services (authentication, authorization, and charging), content synchronization, and configuration of data flow through the network.
- *MBMS Gateway (MBMS-GW)*. This node is responsible for configuring IP multicast to distribute the content efficiently to all the eNodeBs in the eMBMS region. In addition, it is also responsible for the control signaling required for a session set up via the MME.
- *Multi-cell/multicast Coordination Entity (MCE)*. This entity can be co-located at the eNodeB, and it is responsible for admission control and radio resource allocation. It controls the MBSFN frame configuration and transmission parameters such as MCS. One MCE can control multiple eNodeBs in the MBSFN area.

Note that the design of eMBMS, as discussed in the preceding paragraphs, provides several benefits:

- eMBMS makes very effective use of the OFDM properties of an LTE signal to improve greatly the coverage area of broadcasts by allowing a UE to combine signals from different eNodeBs that would otherwise cause interference for unicast transmissions.

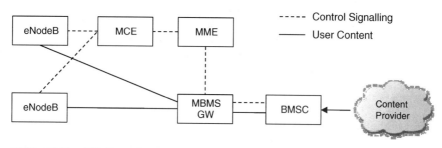

MCE – Multi-cell/Multicast Coordination Entity
MBMS-GW – MBMS Gateway
BMSC – Broadcast Multicast Service Center

Figure 17.25 eMBMS network architecture

- eMBMS leverages the TDMA structure of the OFDM frames to allow for a very flexible configuration of eMBMS, this and can be deployed in existing carriers, given that spectrum resources tend to be scarce. The service can be configured to carry 10–60% of the traffic as multicast.
- The design reuses LTE network elements and permits dynamic service deployment in only those regions of the network where there is demand or need for such services.

17.3.2 Self-Organizing Networks (SON)

As mentioned earlier, the focus of SON features defined in Release 9 is optimization, and this includes optimizing the capacity, coverage, interference, handovers, and random access procedure. The features can be divided into four broad categories as follows:

- *Mobility Load Balancing* (*MLB*). The main focus of this function is to optimize the capacity of the network by redistributing the UEs across different cells to rectify load imbalances between cells. In order to ascertain that load balancing is needed, the standard provides for several load measurements to occur. These include the radio resource load in the uplink and downlink, measured in terms of the allocation for GBR and non-GBR traffic and the PRB utilization. In addition, parameters such as the hardware load and transport network load can also be reported. The actual load balancing is accomplished by modifying the handover thresholds to change the area covered by different cells.
- *Mobility Robustness Optimization* (*MRO*). The other aspect of radio access operation that is addressed by SON is optimization of the actual process of handovers by introducing the automatic detection and correction of handover failures. Handover failures can occur in various situations, including:
  1. *late handover*, where the UE is not handed off soon enough for it to still have a usable signal from the source eNodeB to receive the handover command;
  2. *early handover*, where the UE is handed off to the target eNodeB before it has completely left the coverage area of the source eNodeB; and
  3. *handover to incorrect cell*, where there is a small overlap in coverage between the correct target eNodeB and another eNodeB that is unsuitable for handover. In this last case, handover to the unsuitable eNodeB will result in either a failure, or two successive handovers in a very short span. In all these cases, the standard defines signaling that can be used to report details of the erroneous conditions back to the eNodeB where corrections can be made to the handover parameters.
- *Coverage and Capacity Optimization* (*CCO*). In addition to load balancing, the coverage and capacity of a network of cells can be optimized by controlling the transmit power and antenna tilts of the eNodeBs. Modifying these parameters has the effect of changing the coverage boundaries of a cell and thereby closing a coverage "hole", or changing the capacity of the cell by improving the overall SINR distribution. Support for this feature was actually introduced in Release 10, but it may be possible for vendors to use proprietary algorithms to accomplish CCO without standards support.
- *RACH Optimization*. It is important for UEs to be able to perform network entry with minimum delay, by using as few attempts as possible. This also has the added benefit of reducing interference. The standard provides support for reporting of information related to RACH failures to the eNodeB, which can be used by the eNodeB to optimize the RACH parameters.

17.4 Release 10: LTE-Advanced

As discussed at the beginning of Chapter 16, the ITU developed specifications for IMT-2000, which formed the basis for the development of several 3G systems, the foremost being the WCDMA-based UMTS. With the wide deployment and maturity of 3G systems, and the explosive growth in user demand for data services, ITU developed specifications for the next generation of radio interface technologies (RITs), called IMT-Advanced. These requirements were set forth in a "circular letter" that spurred standard development organizations worldwide to respond with their proposals for IMT-Advanced.

Table 17.5 Peak requirements for LTE, LTE-Advanced and IMT-Advanced

		LTE (Release 8)	IMT-Advanced	LTE-Advanced (Release 10)
Peak data rate	DL	300 Mbit/s	1 Gbps (low mobility) 100 Mbit/s (high mobility)	1 Gbps
	UL	75 Mbit/s		500 Mbit/s
Peak spectrum efficiency (bps/Hz)	DL	15	15	30
	UL	3.75	6.75	15

Within 3GPP, these requirements were used to develop an evolution path for LTE (in keeping with its name). This evolution is called LTE-Advanced, and the requirements that 3GPP defined for LTE-Advanced were beyond those specified by IMT-Advanced. The development of the actual specifications for LTE-Advanced were rolled into the next release of the LTE specifications, Release 10. For comparison purposes, the peak throughput and spectral efficiency requirements for LTE, IMT-Advanced and LTE-Advanced are shown in Table 17.5.

In addition to these requirements, 3GPP also added backwards compatibility as a requirement for the development of LTE-Advanced. The implication of backward compatibility is that UEs based on Release 8 can continue to operate in a network based on Release 10. Similarly, UEs based on Release 10 can continue to operate in a network based on Release 8. We will see how and why this is possible as we explore the technologies that make up LTE-Advanced.

There were four key technologies that were developed in Release 10 to meet the requirements of LTE-Advanced:

- Carrier aggregation.
- Support for heterogeneous network (HetNet) deployments.
- Enhancements to uplink and downlink MIMO techniques.
- Relays.

Of these technologies, carrier aggregation and HetNets are starting to be deployed, and are the two highly touted technologies in LTE-Advanced that will drive large-scale capacity and coverage improvements in existing LTE systems. We will take a deeper look at these technologies in subsequent sections. In addition to carrier aggregation and HetNets, another key enhancement that was developed as part of Release 10 is the support for additional layers of spatial multiplexing, both in the uplink and the downlink, and the introduction of relays to extend coverage. Ultimately, LTE-Advanced cannot considered to be one major technological shift introduced into the existing LTE system but, rather, a series of enhancements that can be mixed and matched and introduced into an existing LTE network with differing timelines, based on a service provider's spectrum assets, tower assets, and service priorities.

UL MIMO enhancements: in previous releases, the uplink did not support spatial multiplexing. However, to meet the aggressive requirements for uplink in LTE-Advanced, it was deemed essential to develop UL MIMO capabilities as part of Release 10. Moreover, with the rise of social networking and camera-enabled smartphones, demand has exploded for uploads of photos and video, especially during events at popular venues. In order to meet this demand, one approach is to improve MIMO technologies at the eNodeB. These include improved interference cancellation algorithms and the use of additional receive antennas for the uplink only. Note that it is cheaper to deploy additional receive antennas, as they do not incur the considerable expense associated with the power amplifiers required for additional transmit paths in the downlink. The other alternative is to improve the uplink transmission itself to support a higher peak rate.

In Release 10, the PUSCH supports up to four layers of spatial multiplexing, with a maximum of two transport blocks. Only closed-loop precoding using codebooks is specified. For the PUCCH, however, there is an open-loop transmit diversity mode specified, as it is not practical to specify an auxiliary control channel required for signaling of dynamic closed-loop precoding the main control channel.

DL MIMO enhancements: there were two main improvements introduced to the downlink MIMO techniques:

1. support for up to eight layers of spatial multiplexing with a new dual-codebook structure; and
2. a new transmission mode (TM), TM-9, that uses a new set of reference signals, called Channel State Information Reference Signals (CSI-RS).

With support for eight spatial layers, it was possible for eNodeBs to use eight transmit antennas in the downlink. The codebook structure for closed-loop precoding contains two parts. The first codebook is considered the wide-band and long-term codebook, and it contains subsets of beams ("grid of beams") that are selected from a superset of all the beams allowed. The second codebook is narrow-band and short-term, and it permits selection of a beam from this subset in a frequency-selective manner.

In TM-9, the UE uses a combination of DM-RS and CSI-RS. The DM-RS allows the UE to perform the channel estimation needed for decoding, while the CSI-RS plays the role of the CRS and is used for channel-state feedback. Note that with additional antenna ports, use of CRS becomes prohibitively expensive in terms of overhead, necessitating the development of a low-overhead CSI-RS structure for channel state feedback.

Relays: one simplified way to understand a relay is to think of it simply as a signal repeater that amplifies a weak signal to extend the link. However, a brute-force signal repeater will amplify the interference along with the useful signal. A more elegant relay can be designed only to transmit the useful signal, and to prevent the interference from propagating. In practice, the LTE-Advanced relays are even more sophisticated. Two types of relays are possible:

- a *Type-1* relay, which appears to the UE to be a fully functional eNodeB with its own cell ID; and
- a *Type-2* relay that is transparent to the UE because it uses the same cell ID as the eNodeB to which it is connected (called the *donor* eNodeB).

In Release 10, only Type-1 relays are specified. In Type-1 relays, the relay node transmits its own control channels and traffic channels on the *access link*, allowing a UE to connect to it like it would connect to a regular eNodeB. However, the relay node does not have any backhaul transport mechanism, and it connects to the donor eNodeB via a *backhaul link*. A half-duplex relay cannot use the access link to the UE at the same time as it uses the backhaul link to the eNodeB. Only half-duplex relays are specified in Release 10, but a vendor who can provide sufficient isolation between the backhaul link and the access link can enable full-duplex operation.

As a final categorization, the backhaul link can either use the same frequency band as the access link, in which case the relay is called an *in-band* relay, otherwise it is an *out-band* relay. Because relays use precious LTE spectrum for the backhaul link, they are mainly seen as good options for coverage extension. For example, in rural areas, where the cost of running a fiber or DSL backhaul may not justify the revenue potential, coverage can be provided by means of in-band relays. Another potential use case is indoor coverage via a relay with an antenna on the outside of the building, but this approach can be expensive in terms of the installation cost to connect the external antenna with the indoor relay system.

17.4.1 Carrier Aggregation

17.4.1.1 Introduction

In order to support the LTE-Advanced requirement of a peak throughput of 1 Gbps, up to 100 MHz of bandwidth may be needed, and this amount of spectrum is usually very hard to obtain in a single contiguous allocation. Furthermore, most operators around the world have spectrum holdings that span several bands, with a varying amount of bandwidth per band.

Carrier aggregation is a feature that allows a network to be configured with a composite carrier that is stitched together using different *component carriers* (*CCs*). Currently, operators deploy multiple carriers, but they have to use inter-frequency handovers to balance the load between different carriers as users join and leave the network. Furthermore, a user can be scheduled on only a single carrier so, when other carriers are free, the bandwidth is not efficiently utilized. In addition to the gain in peak rate, trunking gains are also obtained, by aggregating multiple carriers to schedule traffic simultaneously for a larger pool of users. In the next two subsections, we shall see how CA can be configured, what restrictions are imposed by the standard, and how the protocol is designed to build upon Release 8 mechanisms and support CA for Release 10 UEs.

17.4.1.2   Configuration

As mentioned earlier, the spectrum holdings of an operator can be quite diverse in terms of how close or spaced-out the various frequency bands are in relation to one another. LTE permits carriers to be aggregated in all three possible situations:

- Contiguous intra-band.
- Non-contiguous intra-band.
- Inter-band.

These three configurations are shown in Figure 17.26. Nominally, the standard permits up to five CCs to be used in each direction. However, in practice, early deployments will use far fewer CCs and, in fact, the first few deployments are likely to use only two CCs in the downlink. Furthermore, in the uplink, each CC consumes a significant amount of power in an already power-limited device. Therefore, initial deployments are likely to enable CA only in the downlink, with a single carrier used in the uplink, in an asymmetric configuration. The standard also stipulates that the number of downlink CCs should be equal to or greater than the number of uplink CCs.

Another noteworthy aspect of carrier aggregation is that, although a network may have the capability to support a certain number of CCs in the downlink and the uplink, the actual configuration of CA is necessarily UE-specific. This is needed to support both legacy UEs that do not support any CA, and also UEs with the capability to support differing numbers of CCs. Furthermore, depending on the load and UE

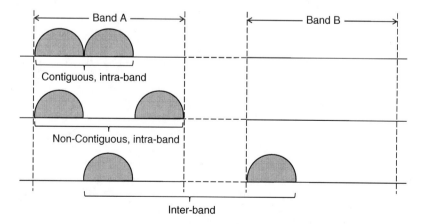

Figure 17.26   Carrier aggregation configurations

demand, the network may want to enable CCs selectively in different UEs. The final point to note about CA configuration is that, for TDD, all CCs must use the same UL/DL subframe split.

17.4.1.3   Protocol Design and Operation

A key design decision in the development of CA is the support of backward compatibility of carriers for legacy mobiles. Therefore, all CCs in a CA system are also Release 8 carriers that can support legacy UEs joining and leaving the system, and therefore can carry all the control channels and reference signals as described in a previous section.

The CA design comes into play only when a CA-capable UE joins the network. The initial CC to which the UE connects is called the Primary CC (PCC), and the corresponding cell with an uplink PCC is called the primary cell, or *PCell*. After the PCell RRC connection is established, a secondary cell, or *SCell*, is configured by means of further RRC control signaling. The SCell configuration may include only a downlink CC, or may contain both a downlink CC and an uplink CC. The UE obtains PCell information from the system information messages carried on the BCH. For SCells, however, in order to avoid the latency involved with decoding infrequently broadcast system information messages, the eNodeB unicasts the system information to the UE, along with the RRC signaling used to configure the SCell.

From the perspective of the RLC sub-layer, the operation of CA is completely transparent, in the sense that there need not be any modifications to the RLC to support CA. The MAC sub-layer, on the other hand, needs to be aware of multiple CCs in order to support transport block multiplexing on multiple HARQ processes. There is one HARQ process active per subframe for each spatial layer and for each CC.

Another key aspect of CA support is the ability to use a single PUCCH for CSI and HARQ ACK/NACK for multiple downlink CCs. For example, if there are five downlink CCs, then the capacity of the HARQ ACK/NACK channel on the PUCCH has to be increased fivefold if there is only one UL CC enabled. This is accomplished by defining a new PUCCH scheme to map the additional ACK/NACK bits. For CSI, when possible, the PUSCH is used to support the expanded payload, or the reporting intervals are controlled by the eNodeB so that there is no overlap of reports for different downlink CCs.

To convey scheduling grants on the downlink and uplink, the PDCCH can be used in the same manner as in Release 8, with the allocation information for a given PCell or SCell carried on the downlink CC belonging to that PCell or SCell. In addition, there is a more flexible technique called *cross-carrier scheduling*, which permits the scheduling grants carried on the PDCCH of one CC to apply to the PDSCH or PUSCH of a different CC. In order for the grant on one PDCCH to point to the PDSCH/PUSCH of a different carrier, the grant is augmented with an optional field called the *carrier indicator field (CIF)*, which can be configured to be part of the grant in a semi-static fashion via RRC signaling when needed.

There are some restrictions on how cross-carrier scheduling can be done. A given CC can be scheduled from the PDCCH of one, and only one, CC (either the same CC itself, or a different CC). In addition, the PCell PDSCH and PUSCH must be scheduled from the PDCCH of the PCell. Finally, the PDSCH and PUSCH must be scheduled from the same CC. Cross-carrier scheduling permits the PDCCH resources to be pooled on one CC, thereby allowing other CCs to have fewer PDCCH symbols. We will look at another benefit of cross-carrier scheduling in the next section on HetNets.

When CA is enabled in the uplink, the subcarriers transmitted from two different carriers are not contiguous so, therefore, the single carrier property of SC-FDMA is lost. Given that the UE implementations will now need to account for this increased inefficiency at the power amplifier, the specification relaxed the contiguous PRB allocation rule for UEs even within a single carrier. UEs that support CA-capability can be scheduled to transmit on PUCCH and PUSCH at the same time. In addition, the UL allocation can be in two non-contiguous clusters, in order to allow for increased scheduling flexibility.

17.4.2   Heterogeneous Networks with Small Cells

17.4.2.1   Introduction

The term "heterogeneous networks" (or HetNets) can, in the most generic sense, encompass a wide variety of networks under one umbrella. In this generic sense, it is sometimes used to refer to the fact that, in devices today, the user has several options for network access, with WiFi and cellular being the predominant ones. The management of access in such devices – either autonomously by the devices themselves, or in concert with the network infrastructure – can be considered to be part of the HetNet paradigm, and there are several interesting challenges to be solved, including the development of SON algorithms, such as mobility optimization and load balancing.

However, in this section, we will look at HetNets in a more narrow, but commonly used sense, wherein a network of low-power eNodeBs is deployed within the coverage area of the regular high-power or *macro*-cellular network. In contrast to the macro-cells, the low-power nodes are called *small cells*, because their coverage footprint is small compared to that of a macro-eNodeB. The term used for small cells in the LTE standard is *pico* cell. Figure 17.27 is a pictorial representation of a HetNet with two small cells in the footprint of a macro-cell.

In terms of frequency planning with respect to small cells, one can envisage several scenarios:

- A *co-channel* deployment with a single carrier, where both the macro-network and small cells share the same carrier.
- A non-co-channel deployment with multiple carriers, where the small cells and macro-network are on different carriers.
- A carrier-aggregation based scenario, where both the small cells and macro cells share two or more carriers.

A co-channel deployment has the advantage of reusing the spectrum efficiently. On the other hand, the use of different carriers between the small cell layer and the macro-cell layer makes for interference-free operation between the layers. As we shall see in a subsequent section, in a co-channel deployment, special techniques may be needed to manage the interference between layers. Finally, we will describe how the CA-based scenario can benefit from interference coordination techniques that are unique to it.

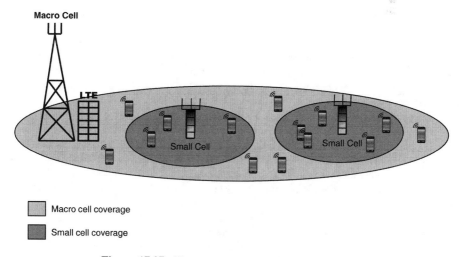

Figure 17.27   Heterogeneous network with small cells

Small cells are fairly similar to WiFi access points in terms of their physical size, range, and power consumption. However, they differ in a couple of important respects. First, they are deployed in licensed spectrum, and are operated by the service provider who owns the spectrum, unlike WiFi APs that can be privately owned and operated in unlicensed spectrum. Therefore, there can be better control over the interference environment in the case of small cells. Second, small cells use cellular air interface technologies in general, and LTE in our discussion here, so they are capable of providing high-quality voice services with seamless handovers between the macro-cells and small cells.

It should also be noted that, in our discussion, we will mainly focus on the open subscriber group configuration, where any user that is on the macro-network can associate with the small cell or pico cell. This is in contrast to the smaller "femtocell" or the home Node-B concept, where closed subscriber groups are typically used, and only a small set of subscribers – typically the subscriber who provides the backhaul for the femtocell and his trusted parties – can associate with the cell.

The main benefit of small cells is that they offer a cost-effective way to add capacity and coverage to an existing macro-cell network. They can be used to provide focused offload of traffic demand from the macro-network in dense user locations (or "hot-spots"), and their smaller size and power requirements typically make them easier to deploy, especially indoors. In this sense, they are a complementary solution to WiFi, and small cells may well support multiple radio access technologies, including LTE, WiFi, and UMTS. However, in outdoor settings, implementing backhaul links to a large number of small cells may prove quite challenging. In addition, outdoor deployments do not enjoy the natural protection that building walls can provide indoor small cells against strong macro-cell interference.

Overall, small cells are seen as an excellent option to augment capacity in a network. LTE is particularly suitable for small cells, because it was designed to operate well even in adverse interference environments, and small cells take advantage of this feature of the technology. In the next few subsections, we will try to shed some light on the use cases, performance characteristics of small cells, and interference mitigation techniques developed in Release 10 to make small cell deployment even more attractive.

17.4.2.2   Performance Considerations for Small Cell Deployment

In this section, we delve deeper into the different aspects of small cell deployment that can have a significant impact on their ability to deliver high throughput and capacity to an existing LTE macro-cellular deployment. Let us first consider the scenario in which the small cells and the macro-cells share the same carrier, without carrier aggregation. This is likely to be a very typical use case, because supporting multiple carriers in a low-cost, small-form-factor small cell may not be easy. In this scenario, the interference from the strongest macro-cell is one of the most important design considerations.

The key driving factor in small cell design is to provide a high SINR link to as many small cell users as possible, because this will enable the most offload of capacity from the macro-cell. Note that the capacity of a small cell is dependent on the average SINR that a user associated with the small cell enjoys. This average SINR is greatly dependent on the configuration of the small cell and, therefore, there can be a large variation between the capacity provided by different small cells.

One straightforward way to provide higher offload is simply to increase the power of the small cell. However, power increase can be limited by regulatory concerns for indoor deployments, as well as by the increased cost of power amplifiers and power delivery to the small cell. The choice of the small cell location with respect to the macro-cell and macro-cell size can also have a big impact on the performance of small cells. In a dense deployment of macro-cells, the coverage area of each macro-cell is not very large and the macro-cell power is likely to be strong in most parts of the network, making it difficult to create a large offload region by the use of small cells.

Similarly, the location of the small cell – whether it is placed close to the macro-cell or at the edge of macro-cell coverage – has a significant impact on its coverage. Small cells are ideally placed at the edge of coverage, where the interference environment is most benign and, therefore, allows small cells to provide higher capacity. However, the choice of small cell placement is usually driven by user demand, and therefore small cells are likely to be deployed in user hot-spots, which represent areas with high user density. These may not necessarily be conveniently located at the edge of macro-cell coverage. This discussion motivates the need for techniques to expand the coverage and capacity of the small cell, and we shall look at the approach taken in this direction by Release 10 features in the next section.

So far, the discussion of small cell performance has focused on the impact of high macro-cell power on low small cell power in the downlink. Although the downlink tends to be the more important link from a user-demand perspective, small cells can also provide benefits in the uplink. As noted earlier, the UEs are power-limited in the uplink due to battery considerations and, thus, it is of great interest to provide options that will permit a UE to transmit at lower power. Small cells fit the bill nicely. Being closer to the user, they result in lowered uplink power consumption and, therefore, they provide improved battery life. Furthermore, the lowered transmit power from users in small cells results in an improved uplink interference environment for the macro-cell. Conversely, the users associated with the macro-cell might still be running at high transmit power and, consequently, steps must be taken to tune the uplink power control algorithms to avoid excessive interference to small cell users from macro-cell users.

17.4.2.3 Enhanced ICIC for non-CA Deployments

As mentioned in the previous section, the effectiveness of small cells as means of capacity augmentation depends on the number of users that can associate with the small cell over the macro-cell, as well as the quality of the signal that can be delivered to these users. One method to associate a larger number of users with the small cell is by the use of a *cell-selection offset* (*CSO*), which represents a *bias* that the user will apply to the signal strength of a specific eNodeB before comparing it with the signal strengths from other eNodeBs for the purpose of determining the best server. For example, a CSO of 3 dB for a small cell will mean that the user will associate with the small cell, even in areas where the macro-cell is twice as strong as the small cell. Clearly, the use of CSO will enable greater offload of users from the macro-cell by creating a *cell-range expansion* (*CRE*) region for these users, as shown in Figure 17.28. However, the

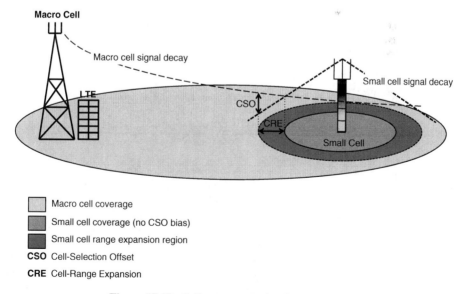

Figure 17.28 Cell-range expansion for small cells

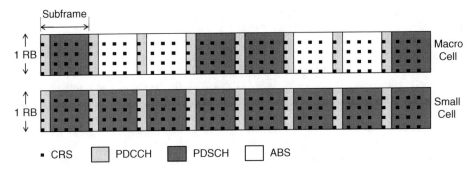

Figure 17.29 eICIC: Time-domain partitioning using almost blank subframes (ABS)

quality of the link for these users in the CRE region is poor because, by definition, they are associated with a cell that does not have the strongest signal.

In order to mitigate the high interference faced by UEs in the CRE region and provide a better quality link to the small cells, Release 10 enables an enhanced inter-cell interference coordination (eICIC) mechanism that is based on time-domain partitioning of resources between the macro-cells and the small cells. This partitioning is created by using *almost-blank subframes* (*ABS*), which are subframes transmitted by the macro-cell with no data. To be more specific, ABS contain only reference signals such as CRS, PSS, and SSS, to ensure that the macro-cell remains accessible to Release 8/9 UEs that depend on these signals for connecting to a cell and monitoring the link quality. Note that MBSFN can also be used for this purpose but, because they can be enabled only on specific subframes, it makes it very restrictive for use with synchronous HARQ, which operates on an eight-subframe periodicity.

As shown in Figure 17.29, by using ABS, the macro-cells create a set of subframes in time when the interference faced by the small cells is low, thereby greatly improving the average link SINR. The small cells can schedule users preferentially in the CRE region during ABS, so that they can realize higher throughputs compared to non-ABS operation. Note that the actual feature specified in the standards to enable ABS is actually the reporting mechanism at the UEs that takes into account the two levels of interference it will see, depending on the presence of ABS from the macro-cells.

Release 10 UEs are capable of restricting their measurements to different subsets of subframes for link adaptation purposes and radio-link monitoring. This feature allows UEs to continue to be connected with a small cell even with a large CSO bias, because they can be asked to monitor the link quality only on the ABS. Furthermore, the higher SINR enjoyed on the ABS can be used for link adaptation purposes by configuring the UEs to two sets of CQI – one for ABS and another for non-ABS – and adapting the link dynamically via the scheduler.

The use of CRE by means of CSO biases, and ABS, presents the system designer with some complex trade-offs and optimization opportunities when small cells are deployed. On one hand, off-loading users from a highly congested macro to several small cells with fewer users allows the spectrum to be reused multiple times, and this can produce attractive gains. On the other hand, in order for the small cell to capture sufficient users to justify the cost associated with installation and operation, CRE may have to be used. CRE reduces the quality of the link for users in the CRE region, so ABS may be required. However, the use of ABS reduces the resources available to the macro-cell, which may be highly congested. Therefore, the selection of CSO, configuration of ABS, the location of the small cells with respect to the user population, and macro-cell proximity, are all important design considerations that must be taken into account to optimize the performance of the network as a whole.

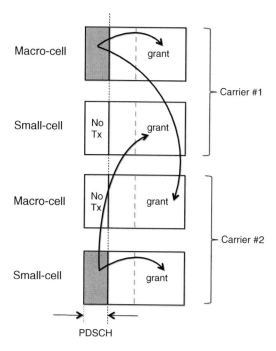

Figure 17.30 Cross-carrier scheduling for interference coordination in HetNets

17.4.2.4 Interference Coordination with CA

Small cells deployed with CA open up an option for interference coordination via the use of cross-carrier scheduling. Typically, the PDCCH requires additional protection, as errors in scheduling grants result in significant loss of throughput because large portions of the subframe, if not the entire subframe, may be rendered useless if the grant is not decoded correctly. When two carriers are aggregated in both the macro-cell and the small cell, then the PCell can be configured to be on different carriers in the small cells and the macro-cells. Using cross-carrier scheduling, the macro-cell and the small cell can enjoy interference-free PDCCH operation, as shown in Figure 17.30. The macro-cell uses its PCell for PDCCH transmission, and leaves the PDCCH symbols on its SCell interference free for the small cell to use as its PDCCH.

17.5 Future of LTE-Advanced: Release 11 and Beyond

As we discussed in the previous section, LTE-Advanced can actually be seen as a toolkit of features that service providers can use to enhance the performance of their networks. The evolution of LTE beyond Release 10 continues in the same manner, with more enhancements added to existing mechanisms, and the creation of new features by mixing related enhancements together in a specific system configuration. If one looks carefully at HetNets and the eICIC feature, for example, in terms of changes to the existing mechanisms in the standard, it is nothing but an enhancement to the UE measurement reporting procedures. However, when taken together with the concept of small cells, ABS, and cell range expansion (all of which are possible even in Release 8), it can be deployed in an implementation-specific manner to provide a robust interference coordination mechanism. In other words, with the completion of Release 10,

LTE can be seen as a platform for innovative performance enhancement techniques enabled by the rich set of features provided by the standard.

A word about terminology is appropriate here. LTE-Advanced is often abbreviated LTE-A, and the term LTE-A is now commonly used to describe Release 10 and Release 11. In order to be able to keep deriving new terms to describe the technology as it evolves, the term LTE-B is now being used to describe Release 12 and beyond. Of course, one can naturally expect that the community will move to LTE-C once LTE-B has been filled to the brim with features!

Release 11 follows in the footsteps of Release 10, and it provides enhancements to eICIC, control signaling, and measurement reporting of channel state information. The changes to the measurement reporting actually enable a feature called Coordinated Multi-Point, abbreviated CoMP, that we shall describe in detail in the following subsection. The enhancement to eICIC enables the CSO bias to be operated with values as high as 9 dB, thereby allowing for significant range expansion and the ability to offload more traffic from the macro-cells.

In order to operate in such severe interference conditions, the receivers implement interference cancellation algorithms that require knowledge of the reference signals used by the interfering eNodeBs, so the standard provides signaling support to convey this information to the UEs. Furthermore, at 9 dB bias, some broadcast channels become hard to decode correctly, so provisions have been made to permit the unicast transfer of some broadcast system information via RRC signaling. This package of enhancements is called further enhanced ICIC or FeICIC.

Another notable feature in Release 11 is the addition of an enhanced version of the PDCCH, called ePDCCH. This new channel is not restricted to being located in the first three symbols, like the PDCCH, but actually spans resource elements (REs) along the time axis at a given frequency subcarrier. The ePDCCH provides increased capacity to schedule more users in a subframe. In addition, it also permits beamformed transmissions of PDCCH messages for increased robustness, and it extends the benefits of frequency-domain ICIC to the PDCCH.

17.5.1 Cooperative Multi-Point (CoMP)

The major feature added to Release 11 is the ability of multiple eNodeBs to coordinate their transmission and reception to provide improved performance using a class of techniques called *Coordinated Multi-Point*, abbreviated as "CoMP." This uses signals from multiple sources in a coordinated fashion to improve the quality of the desired signal, which can be either through combining of useful signals in a constructive manner, or by the suppression of interfering signals in a synchronized manner. This processing usually results in the most improvement at the cell-edge where interference is of most concern, because it is in this region that coordination between different eNodeBs can provide the most benefit.

Like small cells, the benefit comes at the price of increased backhaul cost, because high-speed and low-latency backhaul links are needed to share the information needed for coordination. We will look mainly at downlink schemes in detail, where it will become clear that the support is needed from the standard. In the uplink, CoMP schemes can be implemented with little or no standards support.

The coordinating eNodeBs are usually called *transmission points* (TPs), or *reception points* (RPs), depending on whether the problem being studied is downlink or uplink respectively. The different configurations of CoMP can be studied by dividing them into four broad categories or scenarios, as shown in Figure 17.31:

A. Intra-eNodeB, or between the co-located sectors of an eNodeB.
B. Distributed eNodeB, or between high-power remote radio-heads (RRHs) or TPs connected to the same baseband processing system.
C. Heterogeneous network with an eNodeB connected to multiple low-power RRHs or TPs, all of which are different cells.
D. Heterogeneous network with an eNodeB connected to multiple low-power RRHs or TPs, all of which share the same cell ID.

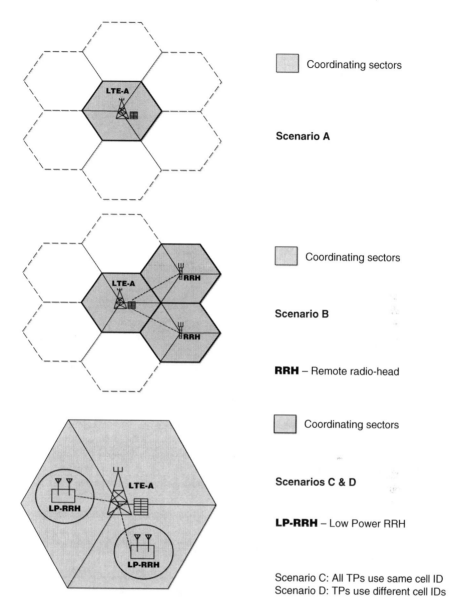

Figure 17.31 CoMP scenarios

Clearly, in scenario A, the TPs are co-located, and the backhaul is not of concern. In scenarios B, C, and D, the connection between the TPs and the baseband processing system must be low-latency and high-speed. Given that they actually connect the RF front-end back to the baseband, these connections are sometimes called "front-hauls". The actual mechanisms used for coordination in the downlink fall into two broad categories: joint processing (JP) and coordinated scheduling/coordinated beamforming (CS/CB). With CS/CB, the data for the UE is available for transmission from only one TP. The TPs schedule the use of beamforming and nulling in such a way that when one TP points a beam towards its intended receiver, the other TPs are pointing nulls at this receiver.

Under joint processing, which implies that multiple TPs coordinate their transmissions jointly with data for the UE available at multiple TPs, there are three different mechanisms discussed in the 3GPP study:

- The first of these is called Dynamic Point Selection (DPS) where, in any given subframe, a single TP dynamically is chosen from a set of TPs for the downlink transmission.
- The second mechanism is called Dynamic Point Blanking (DPB) where, in any given subframe, in addition to selecting a transmission point dynamically, the strongest interfering TP does not transmit, further improving the SINR of the link.
- Finally, in its most generic form, JP can be implemented via Joint Transmission (JT), where multiple TPs transmit the same information to a UE that can combine the signals constructively.

The development of the CoMP feature in Release 11 focused on enabling the UE reporting required for DPS and DBP, although it does not preclude a vendor's implementation from using other mechanisms. For the central coordinating entity in the RAN to select between TPs dynamically, it needs to be able to estimate the quality of the link under different hypotheses. For example, if there are two TPs, then the CQI for each TP is needed, assuming that the other TP is either transmitting or silent for a total of four hypotheses. As it is easy to see, the number of CQI measurements needed can quickly balloon as more TPs are considered for coordination.

In order for the UE to be able to estimate the CQI under different conditions, special measurement resources are configured that permit the UE to measure the interference and useful signal separately from each TP. These measurements can then be combined to calculate the CQI for different hypotheses. Note that in TDD systems, some of these measurements can be performed at the eNodeB directly, assuming channel reciprocity.

The implementation of CoMP in the uplink is much simpler, because the eNodeBs in the pool of coordinating eNodeBs do not have to depend on the UEs to make measurements corresponding to different hypothesis, but can do so directly by themselves. The two general mechanisms that can be employed in the uplink are joint reception and coordinated scheduling. In joint reception, the transport blocks from a UE are received by multiple reception points (RPs) and combined using interference cancellation algorithms to maximize the decoding probability. In coordinated scheduling, the central coordinating entity decides how to schedule uplink transmissions so that a UE's interference is minimized.

17.5.2   *Release 12 and the Future of LTE*

Work on the next release of LTE, Release 12, started in early 2013, and is slated for completion in mid-2014. Enhancements are planned for all major features developed in previous releases, including:

- DL-MIMO;
- HetNet mobility;
- carrier aggregation.

With regards to DL-MIMO, a new codebook has been developed for the 4-transmit antenna configuration. This codebook uses the dual-matrix structure used for 8-layer MIMO in Release 10, and provides sufficient gain over the Release 8 codebook to make deployment of 4-transmit systems attractive to service providers. In addition, 3GPP has a study under way to evaluate the benefits of "3D MIMO", which refers to the ability of the eNodeB to form beams that point not only in the azimuthal plane, but also in the elevation plane. Exploiting 3D-MIMO will eventually result in systems with antenna arrays having 16 or more elements, leading to systems sometimes called "massive MIMO". Other new areas being explored include machine-type communications for low cost devices, and also device-to-device communications for proximity services.

As small cells proliferate and become a dense capacity-and-coverage-augmentation layer of nodes underlying a macro network, the access provided by small cells will need to be optimized in several ways. The signaling for handovers will need to be reduced in order to prevent it from becoming a significant overhead on the operation of the network. As discussed before, the specifications provide a toolkit of tools that can be used in different combinations. This is particularly true when one considers the features provided by CoMP, CA, and densely deployed small cells together.

One generalization of these techniques is to have two carriers – one from the macro-cell and one from the small cell – which are used simultaneously by a UE. The macro-cell can provide an "anchor" carrier that is always available, and the small cell carrier can be used for short bursts of data based on demand. There are several ways that the carriers can be split. For example, a voice carrier that is sensitive to handover disruption can be anchored at the macro-cell, and web browsing or bulky download services can use the small cell. Another configuration may use the macro-cell for all low-bandwidth control-plane signaling, and the small cell to deliver high-bandwidth user-plane services. Even the uplink and downlink carriers can be terminated on different layers in order to manage interference and power consumption at the mobile.

Ultimately, the solid foundation provided by a high-performance, robust, and elegantly engineered LTE system can be evolved over many years to come and provide greater coverage and capacity at lower cost for a vastly improved user experience.

17.6   IEEE 802.16 and WiMAX Systems

About three years before the first LTE release in 2008, another OFDM-based system for mobile broadband access was standardized by the IEEE. Specifically, the 802.16e standard, which enhanced the 802.16 fixed wireless standard to support OFDMA and mobility, was ratified in late 2005. However, the 802.16e standard was not ready for implementation and deployment because of two shortcomings. The first of these had to do with the plethora of optional features in the standard, which made it hard to develop systems that could interoperate without additional definition of parameters and mandatory features. The second had to do with the fact that IEEE 802 standards only specify the MAC layer (layer-2) and below, usually assuming a flat IP-network architecture for the rest of the stack. However, for a mobile network, as we have seen, several other control-plane functions are needed for the system to operate. The WiMAX Forum, an industry-led consortium of companies interested in developing products and networks based on IEEE 802.16, stepped in to overcome both of these shortcomings.

The WiMAX forum developed a system profile by late 2006 that whittled down the smorgasbord of options in the original 802.16e specification, thereby making the standard suitable for implementation and interoperability testing. In parallel, working groups within the WiMAX forum began to develop the network architecture to support end-to-end data services over the WiMAX air interface and released the first version of the network architecture in late 2007.

WiMAX was seen as a competing technology to LTE by most industry experts, because both were designed to provide truly high-speed broadband access in a mobile setting, using a fourth-generation air interface based on OFDM. As such, WiMAX had a head start of at least a year before the Release 8 LTE specifications were ready. However, major industry players, such as Qualcomm and Ericsson, did not back the technology, and the leading operators in the USA, AT&T and Verizon, decided to deploy LTE. Only one US operator, Sprint, deployed WiMAX on a large scale, with a smattering of smaller deployments around the world.

At the time the first WiMAX deployments started, the smartphone industry was still in its infancy and, although a WiMAX-based smartphone was available before an LTE smartphone, smartphones based on HSPA were already in the marketplace. As such, the demand for WiMAX was never overwhelming, and a much larger eco-system started to flourish around LTE. Ultimately, the bigger WiMAX operators saw the writing on the wall and decided to switch to LTE in the long term, leaving WiMAX to enjoy only a small niche in the broadband access marketplace.

From an air interface technology standpoint, the original 802.16e standard is not as sophisticated as the LTE air interface. This is partly because its origins lie in a fixed wireless system which, in turn, was based on the cable-modem standard DOCSIS. As a result, the control signaling is not as efficient as that of the LTE system. Furthermore, the basic OFDM frame structure is based on resource blocks (called *subchannels*) that have subcarriers that are distributed over the entire spectrum. This made the first wave of deployed systems incapable of taking advantage of features such as frequency selective scheduling. Overall, LTE systems had a performance advantage over the first wave of WiMAX systems.

For submission as a candidate technology to IMT-Advanced, IEEE 802.16 developed a much better designed system that is popularly called "16m" after the working group "m" in which the standard was developed. However, with the original WiMAX system not seeing widespread adoption, it is not clear whether equipment vendors will develop chips and infrastructure for 16m-based systems. It seems more likely that the original WiMAX system will continue to see adoption in small-scale commercial and possibly military systems, without actually developing an eco-system with enough critical mass to justify the investment in the next generation of the technology. As such, we will not delve into the details of this technology any further.

17.7 Summary

This chapter discusses the fourth generation of cellular systems that have been developed with the goal of providing always-on access to broadband services. This represents a departure from all previous cellular technologies that were designed either primarily with voice services in mind, or with separate network and protocol architectures to serve voice and data. In these fourth generation systems, provisions are made to provide the appropriate quality-of-service to voice and other real-time services, but are otherwise treated in the same fashion as other generic broadband services.

There are two technologies based on OFDM that can be considered to be part of these fourth generation systems: LTE and WiMAX. Only LTE is dealt with in this chapter, because the WiMAX deployments started to decline even before the WiMAX ecosystem had a chance to mature. LTE on the other hand, has a very vibrant ecosystem, with all the major service providers worldwide either already deploying LTE systems or getting ready to do so in the near future. LTE standardization is carried out by 3GPP, the same body that developed the 3G standards. The culmination of these efforts has resulted in an LTE network that can support a peak speed of about 73 Mbit/s in 10 MHz of bandwidth with the baseline 2×2 downlink MIMO configuration, with typical download speeds ranging between 5 Mbit/s and 12 Mbit/s.

The radio access network architecture of an LTE system is a flat, all-IP architecture. It is flat in the sense that there is only a single type of node, the eNodeB, which forms the infrastructure of the network by connecting directly to the evolved packet core (EPC) network. In the EPC, the serving gateway (S-GW) and the packet gateway (P-GW) form the main user plane nodes to provide connectivity to the internet and other services, while the mobility management entity (MME) is the main hub in the control plane. Note that all these nodes are logical, in the sense that they can physically be co-located with other nodes or even implemented on shared hardware.

The other noteworthy aspect of this architecture is the concept of MME/S-GW pools, which allow an eNodeB to use an S-GW or MME from a pool of such nodes, with the benefit that single points of failure are avoided. As indicated earlier, service differentiation is provided in the architecture for real-time services by means of a powerful Quality-of-Service (QoS) architecture that maps services with similar requirements to an evolved packet system (EPS) bearer. The EPS bearer mapping is based on packet filters, and the bearer is implemented as series of lower-level bearers in tandem, with each bearer applying the appropriate treatment to the traffic flowing through it. Of special interest, due to the bottleneck nature of the air interface, is the radio bearer, which uses eNodeB scheduling strategies to ensure the appropriate level of QoS for the traffic flow.

The overall protocol architecture, with the physical layer, MAC, RLC, PDCP, and RRC functions making up layers 1–3 of the protocol stack, is very similar to that of UMTS, with the major difference that all these entities are present at the eNodeB due to the flat LTE network architecture. The physical layer is based on OFDM downlink and SC-FDMA uplink. Both the uplink and the downlink use resource blocks made of a two-dimensional time-frequency grid of seven symbols for the normal cyclic prefix and 12 subcarriers. These resource blocks make up a 1 ms subframe in TDMA structure containing ten subframes per radio frame.

The subcarriers, or resource elements, within a resource block are divided into data and control resource elements and reference signals that help with the decoding of data. The resource elements are divided into different physical channels, some of which carry data, while others carry control signaling. In the downlink, the PDSCH carries the bulk of the unicast data traffic, with the PBCH and PMCH used for broadcast and multicast, respectively. The PDCCH, PCFICH, and PHICH are physical-layer channels to carry control information for the operation of the physical layer. In the uplink, the PUSCH carries data traffic, with the PUCCH and PRACH used for various control functions. There are several transmission modes defined in LTE that specify the MIMO configuration used by the system. The most commonly used are mode-3 and mode-4, which define open-loop and closed-loop spatial multiplexing, respectively, with dynamic switching to diversity transmissions via SFBC or rank-1 pre-coded transmissions.

Above the physical layer, the MAC sub-layer implements some key functions, including the transport channels that map to the physical layer channels, multiplexing SDUs from the RLC to different transport blocks on transport channels, HARQ operation, scheduling procedures, and transport block format selection. Vendor-specific cross-layer algorithms that include scheduling, link adaptation, and power control, are also implemented in the MAC layer.

Cross-layer algorithms play a major role in optimizing the performance of the LTE system by ensuring that the physical layer is operating as efficiently as possible. The MAC sub-layer services the RLC sub-layer, which is responsible for segmentation and concatenation of higher-layer SDUs, in-sequence delivery and duplicate detection, and retransmission of SDUs in error. The PDCP layer forms an interface between the RLC and the IP layer or the RRC and implements header compression, packet retransmissions for lossless handover, and ciphering. The RRC layer is responsible for control signaling for connection management and handover management. In addition, the RRC state machine is responsible for maintaining the state of the connection, so that a good trade-off between power consumption and responsiveness is achieved.

All of the basic architecture described in the preceding paragraphs was put in place as part of Release 8, the first release of LTE. The key features added in the next release, Release 9, were eMBMS and SON. With the use of eMBMS, the LTE system can be used to efficiently broadcast or multicast popular content to the users in a group of cells called an MBSFN area. The MBSFN area uses a single frequency with identical content sent from different eNodeBs and combined at the UE, using a larger than normal cyclic prefix. This makes eMBMS more efficient, and also provides better coverage. The SON features introduced in Release 9 include mobility load balancing, mobility robustness optimization, and coverage and capacity optimization. These features specify the signaling needed to distribute information required at the eNodeBs to tune the load distribution, handover parameters, and coverage parameters, so that network reliability and efficiency are optimized.

The next release of LTE, Release 10, is considered to be the release that introduces LTE-Advanced, a set of technology features built on top of LTE that meets the requirement of IMT-Advanced as specified by the ITU. The main features introduced as part of this release include carrier aggregation, support for heterogeneous networks with small cells, enhanced DL and UL MIMO, and relays. Of these, carrier aggregation and small cells are likely to be deployed first. Carrier aggregation presents an elegant option for service providers to stitch distinct frequency channels together into one composite channel that provides both an increase in peak throughput, and an improvement in spectral efficiency via trunking gains at low to medium traffic loads.

Small cells provide a cost-effective way for service providers to add focused capacity or coverage where needed within the nominal footprint of a macro-cell, via the use of low-power nodes. The LTE air interface is robust enough to allow the operation of such low-power small cells directly in the interference environment of a high-power macro-cell. However, in some environments, such as dense urban deployments, the footprint of the small cells may be too small for cost-effective offload of macro-cell users. As a result, the range of the small cell may need to be expanded to where the macro-cell signal may be several times stronger than that of the small cell. In order to operate the network in such a severe interference environment, a time-domain partitioning of subframes between the macro-cell and small cell, called eICIC, was introduced in Release 10.

Beyond Release 10, the next releases, Release 11 and beyond, continue to enhance the LTE-Advanced system by introducing more features that can be considered to form a toolkit of options for service providers to offer improved throughputs, latency, and capacity, thereby delivering better user experience. These features can be mixed and matched as appropriate, based on the existing network configuration and service requirements, and therefore provide a flexible set of design options to the LTE network architect. The key features introduced in Release 11 include further enhancements to eICIC, CoMP, and improved control signaling in the downlink via ePDCCH. CoMP permits multiple eNodeBs to transmit or receive collaboratively to UEs, thereby improving cell edge performance and capacity.

The evolution of LTE-Advance continues with Release 12, slated for completion in 2014, and promises to add features that improve HetNet, DL-MIMO, and carrier aggregation performance. The relatively quick deployment and penetration of LTE mainly for data services, before managed voice services are even introduced, is a testament to the explosion of user demand in the mobile marketplace. Through continuous evolution of the technology, the elegantly designed LTE and LTE-Advanced system finds itself well poised to serve the needs of the mobile computing world for several years to come.

Further Readings

1. 3GPPTS 36.300. Evolved Universal Terrestrial Radio Access (E-UTRA) and Evolved Universal Terrestrial Radio Access Network (E-UTRAN); Overall description; Stage 2.
2. 3GPPTS 36.211. Evolved Universal Terrestrial Radio Access (E-UTRA); Physical channels and modulation.
3. 3GPPTS 36.212. Evolved Universal Terrestrial Radio Access (E-UTRA); Multiplexing and channel coding.
4. 3GPPTS 36.213. Evolved Universal Terrestrial Radio Access (E-UTRA); Physical layer procedures.
5. 3GPPTS 36.321. Evolved Universal Terrestrial Radio Access (E-UTRA); MAC protocol specification.
6. 3GPPTS 36.322. Evolved Universal Terrestrial Radio Access (E-UTRA); Radio Link Control (RLC) protocol specification.
7. 3GPPTS 36.323. Evolved Universal Terrestrial Radio Access (E-UTRA); Packet Data Convergence Protocol (PDCP) Specification.
8. 3GPPTS 36.331. Evolved Universal Terrestrial Radio Access (E-UTRA); Radio Resource Control (RRC); Protocol specification.

18

Conclusions Regarding Broadband Access Networks and Technologies

Residential broadband access has rapidly become an important and integral part of the culture in much of the world. For many people, it has become virtually as essential as power and water services.

The key reasons for this broadband access explosion include the growing number of applications, the incredible volume of content available through the internet, and also the increasingly affordable means to access it. The availability of the applications and content has driven user desire for broadband access, and the increasing availability of broadband access has accelerated the internet's growth and usefulness. In turn, the customer's desire to use internet's content and applications has driven industry and research institutions to explore the potential means for providing access in an economical manner. Insatiable user demand and the rapidly expanding market for broadband access has fueled great innovation and creativity in the scientific and engineering communities. As the reader of this book can see, many different methods have received serious exploration, and a significant number of them have been widely deployed around the world.

The variety of different protocols and technologies for providing broadband access raises two obvious questions:

- Which protocol/technology is the "best?"
- Which protocol/technology will "win" in the market?

The short answer to these questions is that "it depends ...". In other words, there is no single technology or protocol that is best for all cases. The intention of this concluding chapter is to provide a brief review of some of the factors that determine which technology or combination of technologies is best suited to a given situation.

First of all, the right technology depends on the nature of the customer base. The important factors here include:

- Residential vs. enterprise customers.
- Urban vs. rural vs. suburban.
- How much mobility is required.
- Whether the existing telecom access infrastructure is sufficient to serve customers, or a new one must be deployed.

Broadband Access: Wireline and Wireless – Alternatives for Internet Services, First Edition.
Steven Gorshe, Arvind Raghavan, Thomas Starr, and Stefano Galli.
© 2014 John Wiley & Sons, Ltd. Published 2014 by John Wiley & Sons, Ltd.

- Availability of capital to invest in infrastructure (to satisfy current demand or build a future-proof network).

Ultimately, the cost to build and operate a network is largely driven by the access technology, and affordability is just as important as delivering enough bandwidth. For example, virtually no one would buy a 250 kbit/s service, nor would they buy a $250 per month service at any bit rate. The cost of building and operating the access network depends on the factors mentioned above, and also the extent and nature of the legacy access infrastructure.

Where a substantial amount of relatively new wireline telephone connections is in place, DSL technologies will typically have cost advantages. DSL also allows the carrier to avoid the especially costly changes to the last 100 m to the home, where new cable installation can involve digging up subscribers' landscaping. Similarly, if the MSO cable network is present and in good shape, DOCSIS broadband access may be more cost-effective. However, the capacity of cable access networks is shared among users, and it may not be able to satisfy user demand at all times, especially during busy periods in densely populated areas.

On the other hand, in brand new deployments, or scenarios where the existing cable plant needs rehabilitation, a fiber-based technology may be more cost-effective in some cases, depending on the strategy used to cover the last 100 m to the home. For example, terminating the fiber potion of the network at a cabinet and then delivering high-speed access though DSL has proven to be a very cost effective way of exploiting fiber access.

VDSL2 with vectoring and bonding, and G.fast are providing dedicated per-customer bit-rates comparable to the common per-customer rates of technologies like fiber PON and DOCSIS cable, which provide a shared capacity. When looking at "future-proofing" the access network investment, it is important to look at what each customer gets, and not the total capacity shared by a large number of customers. From this point of view, DSL technology is the only broadband access technology that provides dedicated (not shared) capacity to every user, making DSL resilient to its own success, because its per-user data rate is independent of the number of customers served.

Wireless technology is taking over an increasing share of the broadband access market. Customers are choosing mobile devices such as smartphones and tablets as their primary "screens" to consume internet content and services, and it is not at all uncommon for such customers to opt out of wireline service for voice. However, cellular broadband access, based on the use of licensed spectrum, is fundamentally limited in the capacity that it can provide using a single tower to cover large areas, because of the shared nature of the medium. As a result, cellular service providers will find it difficult to sustain unlimited or "all-you-can-eat" plans for cellular access, and consumers may notice a marked degradation in the quality of experience as the spectrum resources become scarce due to over-utilization.

One way around this bottleneck is to use a large number of small cells to provide focused capacity and coverage where needed. The widespread proliferation of WiFi access is a clear indication of this trend, and cellular providers are poised to adopt this model for cellular access on a large scale. However, providing broadband backhaul access becomes challenging in such a wholesale deployment of small cells. In effect, one has just transferred the problem of providing high-capacity mobile access back to that of providing economical and ubiquitous wired broadband access for the backhaul!

Therefore, both wired and wireless technologies are likely to be tightly interwoven and complementary in providing broadband access, although the end user or device may increasingly use a wireless link as the last "leg" for access. For example, multi-tenant residences with many large screens consuming HD video at more than a gigabyte per day on average will necessarily require a wired broadband service to serve the demand for content economically, even though the screens may be wirelessly connected to the network via WiFi.

Wireless technology is also attractive in developing nations or more rural areas where there is little, if any, wireline infrastructure. The wireless network can be built relatively quickly in these scenarios. Similar considerations apply to PLC as well. Although PLC has failed to prove itself as a competitive access technology when other means of broadband access are available, there are good reasons to believe that PLC can be a good candidate for providing broadband access in developing countries or rural areas characterized by a very low penetration of telecom infrastructure. Microwave backhaul links, or even in-band relays using cellular technology, can minimize the amount of cable that needs to be deployed to serve the cell towers. Satellite-based broadband technology has matured to where it is possible to deliver broadband access over entire continents, but latency will continue to be a challenge for quick-response applications such as gaming.

Fiber technology has a clear advantage in terms of bandwidth capacity and ongoing maintenance costs. The bandwidth capacity makes it the best technology in terms of "future-proofing" the access network investment. However, it will often have the highest initial installation costs. The lowest-cost scenarios for fiber access are typically where the subscribers are relatively close to each other, such as urban multi-tenant buildings or suburban neighborhoods. However, this may not always be true, because digging rights in urban areas can be very high. Unfortunately, while optical components have been decreasing in price, they cannot take advantage of Moore's law that decreases the cost of electronic components more quickly.

As discussed in this book, fiber is often used as the infrastructure for other access technologies, including connectivity to wireless cell sites, to DSL serving nodes, and DOCSIS fiber nodes. Thus, fiber is becoming an integral part of the most future access networks.

Sometimes, the choice of access technology depends on factors independent of technical considerations, and this is the case in countries that require legacy carriers to "unbundle" their access networks. This situation occurs where the legacy telephone carrier has enjoyed a regulated monopoly for wireline telephone access. As these countries have promoted competition to drive broadband access, they have required the legacy carrier to provide their access network infrastructure to the competing carrier on the same cost terms as they provide it to their own access service divisions. The rationale is that the legacy carrier was effectively subsidized by the government or subscribers to build their access networks, while new competing carriers cannot receive any subsidies to build their infrastructure. Consequently, the legacy telephone carriers are required to unbundle their services so that a competitor can provide individual services such as high-speed data access. If the legacy carriers are required also to provide unbundled service over new infrastructure, they have very little incentive to deploy new fiber networks. It makes more economic sense for them to use DSL technologies over their existing infrastructure, rather than to deploy new networks that they must both pay for completely, then share at the same cost with competitors.

In many cases, a subscriber will desire both wireline and wireless broadband access service. This combination allows optimization for capacity at home over the wireline network, with the availability of mobility using the wireless network. It also provides a backup capability. For example, as one of the authors was writing his last chapter for this book, he was prevented from reaching a website due to an unusual network congestion situation affecting his FTTH connection. Little time was lost, as the author pulled up the web page using his iPhone's LTE wireless connection instead.

Due to these different considerations, each of the technologies discussed in this book will play a role in future broadband access. Fiber, DSL, DOCSIS, and wireless technologies, and also PLC, each have important applications for which they are the most appropriate choice, either separately or in combination.

Index